TECHNICAL COMMUNICATION
A Reader-Centred Approach

First Canadian Edition

Paul V. Anderson
Miami University (Ohio)

Kerry Surman
Algonquin College

Australia Canada Mexico Singapore Spain United Kingdom United States

Technical Communication; A Reader-Centred Approach, First Canadian Edition

by Paul V. Anderson, Kerry Surman

Associate Vice President, Editorial Director:
Evelyn Veitch

Editor-in-Chief, Higher Education:
Anne Williams

Executive Editor:
Cara Yarzab

Acquisitions Editor:
Bram Sepers

Marketing Manager:
Sandra Green

Developmental Editor:
Natalie Barrington

Photo Researcher/Permissions Coordinator:
Indu Ghuman

Content Production Managers:
Anne Macdonald, Wendy Yano

Copy Editor:
Mathew Kudelka

Proofreader:
Laurel Sparrow

Indexer:
Elizabeth Bell

Senior Manufacturing Coordinator:
Ferial Suleman

Design Director:
Ken Phipps

Interior Design:
Katherine Strain

Cover Design:
Johanna Liburd

Cover Image:
Top: Stockbyte; Upper middle: GOODSHOOT/Alamy; Lower middle: Image Source/Alamy; Bottom left: Stockbyte; Bottom right: Blend Images/Alamy

Compositor:
Nelson Gonzalez

Printer:
Quebecor World

Printed and bound in the United States
1 2 3 4 09 08 07 06

For more information contact Nelson, 1120 Birchmount Road, Toronto, Ontario, M1K 5G4. Or you can visit our Internet site at http://www.nelson.com

Library and Archives Canada Cataloguing in Publication Data

Anderson, Paul V.
Technical communication: a reader-centred approach/Paul V. Anderson, Kerry Surman.—1st Canadian ed.

Includes bibliographical references and index.
ISBN 0-17-625184-7

1. Communication of technical information—Textbooks.
2. Technical writing—Textbooks.
3. Technical English—Textbooks.
I. Surman, Kerry, 1965– II. Title

PE.1475.A628 2007 808'.0666
C2006-903302-1

PREFACE

Welcome to the first Canadian edition of *Technical Communication: A Reader-Centred Approach.* In the past few years there have been substantial changes in the methods, tools, and products of technical communication and in the audiences for these products. This edition offers a Canadian perspective on these core topics, an updated emphasis on the application of technology to technical communication, a contemporary treatment of international technical communication, and enhanced pedagogical tools and relevant Canadian examples and situations that make the book even more effective for current teaching and learning.

The first Canadian edition takes a distinctive reader-centred approach, whose hallmarks are that it:

- **Teaches highly transferable strategies.** The simplified six-section organization helps students take a thoughtful, resourceful approach to all communications that they will be required to prepare in their careers.
- **Provides in-depth coverage in an easy-to-learn manner.** Because all of this book's advice grows from a common set of reader-centred principles and processes, the book enables students to integrate, remember, and apply discussions on a wide array of topics as they move through the six sections.
- **Benefits students in many disciplines.** The strategic emphasis, the wide applicability of the twin principles of usability and persuasiveness, and the broad coverage of work environments in the text's examples, exercises, and cases make this book well suited to students in many technical, business, engineering, and scientific fields.

FEATURES OF THE FIRST CANADIAN EDITION

Among the many features introduced in this edition, three are particularly notable because of the ways they increase the book's breadth and effectiveness for teaching and learning:

- **Application of Canadian societal context.** This edition gives examples and situations from the current Canadian social, cultural, and legislative environment. Its materials build on the increasing diversity of ethnicities and religions that comprise the Canadian social fabric, and the resulting effects on technical communication.

- **Provision of diverse employment situations.** This edition presents examples and situations from a broad array of Canadian workplaces: from the private to the public sector, from the small business to the large corporation, from the resource and technology sectors to the recreation and hospitality fields. A consistent linkage of the book's tools and strategies to technical communication in the workplace is enhanced by this diversity.
- **Application of relevant international contexts.** This edition provides a continuity of application, from chapter to chapter, of international contexts of particular interest to Canadian students as they recognize the importance of and engage in global technical communication.

ORGANIZATION AND COVERAGE OF THIS EDITION

This book's six sections combine attention to communication processes and products. Throughout these sections, the book instructs students on how to apply effective, reader-centred strategies in the situations that they will encounter in the workplace.

4 PART I DEFINING YOUR COMMUNICATION'S OBJECTIVES

CHAPTER 1

DEFINING OBJECTIVES
APPLYING
CHOOSING
PLANNING
DEVELOPING
EVALUATING

From the perspective of your career, communication is one of the most valuable subjects you will study in college or university.

Why? Imagine what your days at work will be like. If you are majoring in a technical field, you will spend much of your time using the special knowledge and skills you learned in school to answer questions asked by coworkers and to complete projects assigned by managers. Furthermore, you will generate many good ideas on your own. Looking around, you'll discover ways to make things work better or do them less expensively, to overcome problems that have stumped others, or to make improvements others haven't begun to dream about.

For additional chapter resources, visit Chapter 1 at www.techcomm.nelson.com.

Yet all your knowledge and ideas will be useless unless you communicate them to someone else.

Consider the examples of Sarah Berlou and David Thayer. A recent mechanical engineering technology graduate, Sarah has spent three weeks analyzing pistons that broke when her employer tested an experimental automobile engine. Although her analysis was skillful, what she learned about why the pistons failed will be useless to her employer unless she communicates her results clearly and usefully to the engineers who must redesign the pistons. Similarly, David, a newly hired information systems analyst, must communicate to accomplish his goal of increasing efficiency in the railway's intermodal freight office where he works. David's ideas will reduce costs and improve service to freight customers only if he presents his recommendations persuasively to the people who have the power to implement them.

COMMUNICATION EXPERTISE WILL BE CRITICAL TO YOUR SUCCESS

Like Sarah and David, you will make your work valuable to others only if you communicate it effectively to them. Consequently, your expertise in writing and speaking will be just as critical to your on-the-job success as your expertise in your specialized field. This fact is well known to employees and employers alike. In one survey, 94 percent of the graduates from seven departments reported that the ability to "write well" (not just write, but *write well*) is important to them in their jobs (P. V. Anderson, 1985). Another survey found that writing and speaking effectively are among the major factors that employers consider when evaluating job performance (Beer & McMurrey, 1997). Recognizing the importance of good writing, Fortune 100 companies spend an average of $300,000 per year to provide writing instruction for their employees (Kiggins, 1998). Communication expertise is one of the key qualifications employers look for when hiring.

Graduates typically spend one day a week—or more—writing.

Moreover, communication will consume a large portion of your time at work. Numerous studies indicate that the typical graduate spends about 20 percent of his or her on-the-job time writing (Barnum & Fischer, 1984; Beer & McMurrey, 1997; Northy, 1990; Pinelli, Glassman, Oliu, & Barclay, 1989; Roth, 1982). That's one day out of every five-day work week! And it doesn't include the additional time spent talking—whether on the phone or in person, whether in groups and meetings or one-on-one.

This book's goal is to help you develop the communication expertise you must have to realize the full potential of your expertise in your specialized field.

NEL

- **Defining Your Communication's Objectives.** Chapter 1 helps students understand the differences between the communication skills and strategies at work and at school. Chapter 2 shows students how to focus on usability and persuasiveness as they match their purpose with their communication.
- **Applying a Reader-Centred Approach.** Chapter 3 guides students through the critical step of beginning a communication, while Chapter 4 helps them develop an effective communication style. The students are then ready to act on the advice of these chapters as they craft email messages and Web communications in Chapter 5.
- **Choosing a Communication Format.** This section takes a reader-centred approach to three of the most common types of workplace communications: instructions (Chapter 6), proposals (Chapter 7), and reports (Chapter 8). Detailed advice helps students craft each element of these communications in ways that meet their readers' needs while achieving the writers' goals.

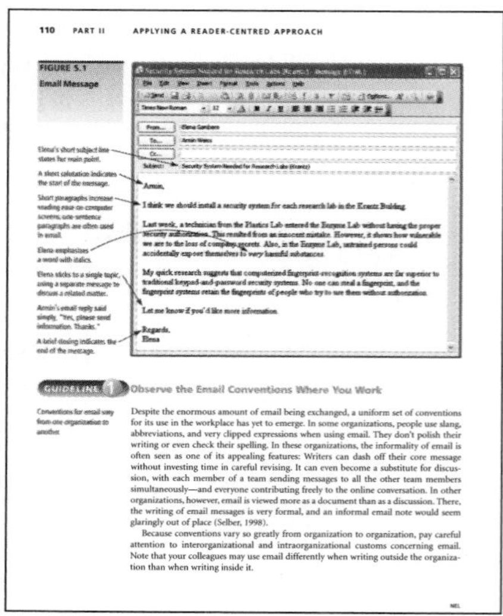

110 PART II APPLYING A READER-CENTRED APPROACH

FIGURE 5.1
Email Message

Observe the Email Conventions Where You Work

Conventions for email vary from one organization to another.

Despite the enormous amount of email being exchanged, a uniform set of conventions for its use in the workplace has yet to emerge. In some organizations, people use slang, abbreviations, and very clipped expressions when using email. They don't polish their writing or even check their spelling. In these organizations, the informality of email is often seen as one of its appealing features: Writers can dash off their core message without investing time in careful revising. It can even become a substitute for discussion, with each member of a team sending messages to all the other team members simultaneously—and everyone contributing freely to the online conversation. In other organizations, however, email is viewed more as a document than as a discussion. There, the writing of email messages is very formal, and an informal email note would seem glaringly out of place (Selber, 1998).

Because conventions vary so greatly from organization to organization, pay careful attention to interorganizational and intraorganizational customs concerning email. Note that your colleagues may use email differently when writing outside the organization than when writing inside it.

NEL

- **Planning a Communication Strategy.** This section refocuses students' attention on usability (Chapter 9) and planning a persuasive strategy (Chapter 10) before applying these strategies to the creation of highly effective résumés and job application letters (Chapter 11). Chapter 12 provides advice for creating a communication collaboratively using various team structures. Chapter 13 teaches students how to plan and conduct efficient, effective research while heeding the relevant aspects of intellectual property and copyright law.
- **Developing the Communication Elements.** This section leads students to become confident, resourceful communicators as they draft the common elements in a variety of communication media and situations, including paragraphs and sections (Chapter 14), graphics (Chapter 15), websites (Chapter 16), and oral presentations (Chapter 17).
- **Evaluating a Communication.** This section enables students to increase the usability and persuasiveness of their communications as it guides them through the activities of checking, reviewing, and testing their drafts (Chapter 18) before making their revisions.

244 PART IV PLANNING A COMMUNICATION STRATEGY

Figure 10.1 shows a marketing brochure that uses organizational objectives to persuade its intended readers (beekeepers). Notice the boldfaced statements that proclaim how the system will increase profit potential, reduce costs, and reduce labour.

FIGURE 10.1 Brochure That Stresses Organizational Objectives

NEL

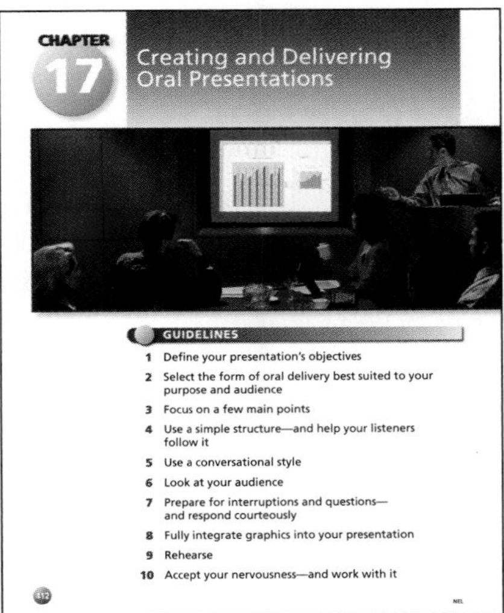
CHAPTER 17

Creating and Delivering Oral Presentations

GUIDELINES

1. Define your presentation's objectives
2. Select the form of oral delivery best suited to your purpose and audience
3. Focus on a few main points
4. Use a simple structure—and help your listeners follow it
5. Use a conversational style
6. Look at your audience
7. Prepare for interruptions and questions—and respond courteously
8. Fully integrate graphics into your presentation
9. Rehearse
10. Accept your nervousness—and work with it

412 NEL

PEDAGOGICAL FEATURES OF THIS EDITION

This edition provides many teaching and learning features:

- **Cases for each part.** These new and revised cases, which are based on such topics as employee Internet usage and outsourcing, allow the students to integrate and apply the principles in the chapters. They are suitable for in-class or online discussion, or for individual or collaborative assignments.
- **Reference guides for each part.** These Reference Guides highlight technical communication topics of particular relevance to students, such as patterns for organizing information, ways of ending communications, common types of workplace

reports, research methods used on the job, types of graphics, and steps for revising.

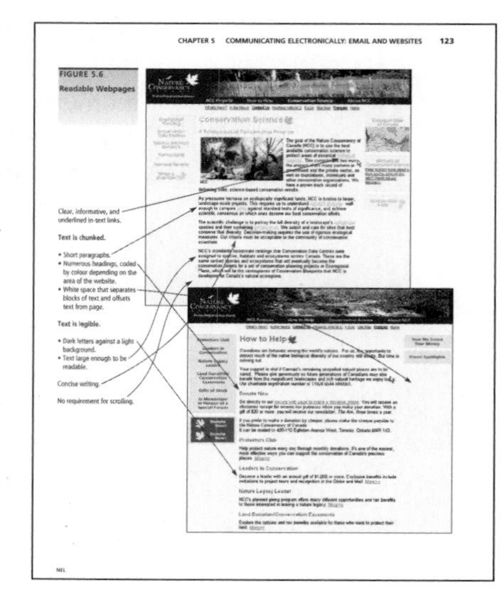
CHAPTER 5 COMMUNICATING ELECTRONICALLY: EMAIL AND WEBSITES 123

FIGURE 5.6
Readable Webpages

Clear, informative, and underlined in-text links.

Text is chunked.

- Short paragraphs.
- Numerous headings, coded by colour depending on the area of the website.
- White space that separates blocks of text and offsets text from page.

Text is legible.

- Dark letters against a light background.
- Text large enough to be readable.

Concise writing.

No requirement for scrolling.

- **Sample communications with margin notes.** An abundance of well-annotated examples of technical communications from a diverse range of Canadian workplaces, such as websites, email messages, memos, letters, and many types of reports, are enhanced by margin notes to demonstrate the chapters' advice for students' learning.
- **Integrated discussion of communication ethics.** This treatment ensures that students and professors include ethics as a consistent theme throughout their technical communication courses, rather than as the topic for one day's reading. This integrated treatment is reflected in "ethics guidelines" and in special exercises contained in most chapters.

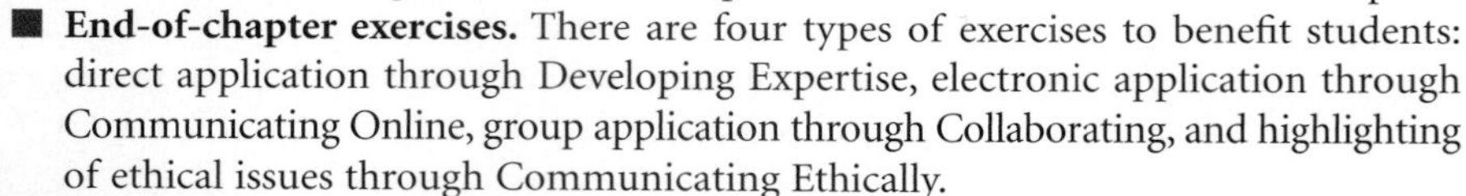

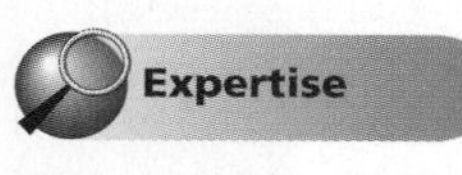

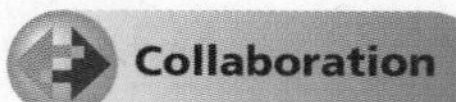

- **End-of-chapter exercises.** There are four types of exercises to benefit students: direct application through Developing Expertise, electronic application through Communicating Online, group application through Collaborating, and highlighting of ethical issues through Communicating Ethically.
- **Guidelines for each chapter.** Students will find the list of guidelines on the opening page of chapters helpful as a summary of what the chapter contains, and of what has been presented as they review course materials. Chapters 6, 7, and 8, which present the structures of instructions, proposals, and reports, include a summary list of the components of these structures.

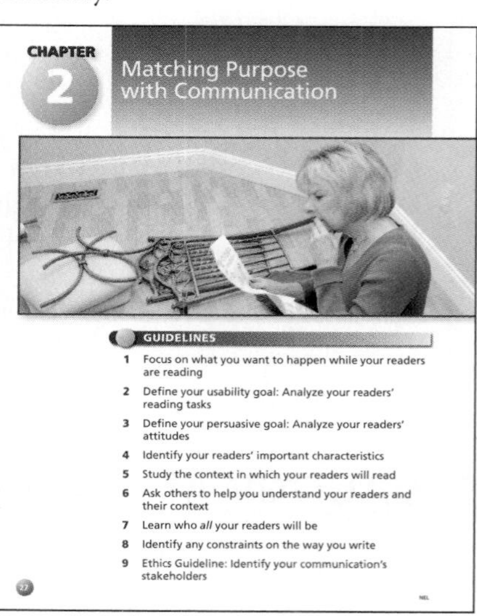
CHAPTER 2
Matching Purpose with Communication

GUIDELINES

1. Focus on what you want to happen while your readers are reading
2. Define your usability goal: Analyze your readers' reading tasks
3. Define your persuasive goal: Analyze your readers' attitudes
4. Identify your readers' important characteristics
5. Study the context in which your readers will read
6. Ask others to help you understand your readers and their context
7. Learn who *all* your readers will be
8. Identify any constraints on the way you write
9. Ethics Guideline: Identify your communication's stakeholders

- **Enhanced appendixes, including four documentation styles.** Appendix A contains contemporary Canadian examples of commonly used formats for email messages, letters, memos, and reports. Appendix B explains the four most common documentation styles used in technical communication: APA, MLA, CMS, and IEEE. Appendix C provides a wide selection of workplace-oriented projects appropriate for individual or collaborative work.

SUPPLEMENTS TO ACCOMPANY THIS EDITION

Several supplements accompany the first Canadian edition to promote effective teaching and learning.

- **Instructor's Manual with Test Bank (0-17-625185-5).** Accompanying this edition of *Technical Communication* is an instructor's manual prepared by Kerry Surman, the Canadian co-author. It includes a thorough introduction to the material; advice on teaching for the exercises, cases, and projects; strategies for integrating the content into course learning requirements and learning outcomes; and more. The manual also includes a bank of test questions for each chapter, with solutions. It is available for downloading on the instructor website: www.techcomm.nelson.com.

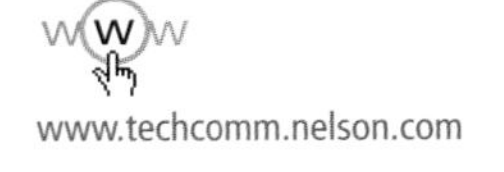

www.techcomm.nelson.com

- **PowerPoint® presentations.** Slides organized by chapter may be used by professors for discussion of chapter content in their lectures. The chapter guidelines, checklists, and planning guides included on these slides are a valuable resource for summarizing chapter content and for providing a foundation for assignments and projects. The PowerPoint slides are available for downloading on the instructor website.
- **Website. www.techcomm.nelson.com** A student website provides additional study resources as well as links from the text, quizzes, additional exercises, citation resources, and more.

ACKNOWLEDGMENTS

Many people contributed to the first Canadian edition of *Technical Communication: A Reader-Centred Approach*. It is my pleasure to thank those people, while hoping that their assistance will continue through subsequent editions.

After reading the American text, I was convinced that Paul Anderson (Miami University) had produced an appealing and comprehensive foundation on which to base a Canadian edition.

Once I began to write, I received excellent advice from my colleague, Devon Galway (Algonquin College). I also benefited from the expertise of Neeta Sharma (Algonquin College) regarding technical communication practices in India, Petra Watzlawik-Li (Carleton University) and Stephen Li regarding technical communication practices in China, and Margaret O'Brien (Algonquin College) and Ray Gotceitas regarding technical communication practices in the workplace. Also, I received many valuable suggestions from the following individuals, who helped shape this Canadian adaptation: Carol Acton, St. Jerome's University; Roger Legge, Mount Royal College; and Panteli Tritchew, Kwantlen University College.

While developing this edition, I received outstanding help from the team at Thomson Nelson. Natalie Barrington, developmental editor, calmly reassured me as I tentatively incorporated suggestions and reorganized material. Additional thanks to

Anne Macdonald, production editor; Matthew Kudelka, copy editor; Laurel Sparrow, proofreader; Indu Ghuman, photo research and rights; Bram Sepers, acquisitions; and Sandra Green, marketing manager.

Through it all, Peter, Charlie, Margaret, and Mum and Dad encouraged and supported me.

Thank you.

KERRY SURMAN
Ottawa, Ontario

BRIEF CONTENTS

CONTENTS

APPLYING A READER-CENTRED APPROACH 65

DEVELOPING THE COMMUNICATION ELEMENTS 339

EVALUATING A COMMUNICATION 463

Defining Your Communication's Objectives

CHAPTER 1

Communication, Your Career, and This Book

CHAPTER OVERVIEW

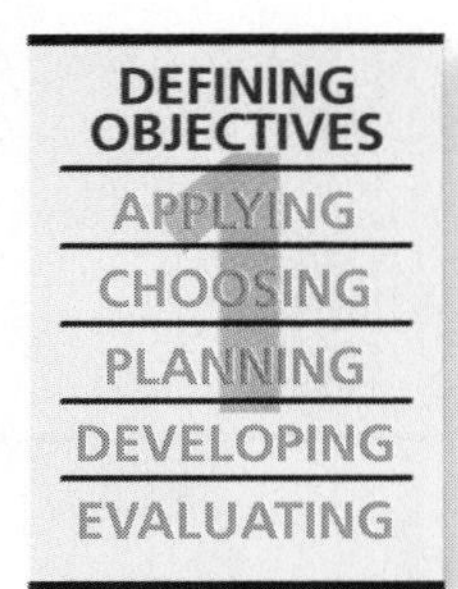

From the perspective of your career, communication is one of the most valuable subjects you will study in college or university.

Why? Imagine what your days at work will be like. If you are majoring in a technical field, you will spend much of your time using the special knowledge and skills you learned in school to answer questions asked by coworkers and to complete projects assigned by managers. Furthermore, you will generate many good ideas on your own. Looking around, you'll discover ways to make things work better or do them less expensively, to overcome problems that have stumped others, or to make improvements others haven't begun to dream about.

For additional chapter resources, visit Chapter 1 at www.techcomm.nelson.com.

Yet all your knowledge and ideas will be useless unless you communicate them to someone else.

Consider the examples of Sarah Berlou and David Thayer. A recent mechanical engineering technology graduate, Sarah has spent three weeks analyzing pistons that broke when her employer tested an experimental automobile engine. Although her analysis was skillful, what she learned about why the pistons failed will be useless to her employer unless she communicates her results clearly and usefully to the engineers who must redesign the pistons. Similarly, David, a newly hired information systems analyst, must communicate to accomplish his goal of increasing efficiency in the railway's intermodal freight office where he works. David's ideas will reduce costs and improve service to freight customers only if he presents his recommendations persuasively to the people who have the power to implement them.

COMMUNICATION EXPERTISE WILL BE CRITICAL TO YOUR SUCCESS

Like Sarah and David, you will make your work valuable to others only if you communicate it effectively to them. Consequently, your expertise in writing and speaking will be just as critical to your on-the-job success as your expertise in your specialized field. This fact is well known to employees and employers alike. In one survey, 94 percent of the graduates from seven departments reported that the ability to "write well" (not just write, but *write well*) is important to them in their jobs (P. V. Anderson, 1985). Another survey found that writing and speaking effectively are among the major factors that employers consider when evaluating job performance (Beer & McMurrey, 1997). Recognizing the importance of good writing, Fortune 100 companies spend an average of $300,000 per year to provide writing instruction for their employees (Kiggins, 1998). Communication expertise is one of the key qualifications employers look for when hiring.

Graduates typically spend one day a week—or more—writing.

Moreover, communication will consume a large portion of your time at work. Numerous studies indicate that the typical graduate spends about 20 percent of his or her on-the-job time writing (Barnum & Fischer, 1984; Beer & McMurrey, 1997; Northy, 1990; Pinelli, Glassman, Oliu, & Barclay, 1989; Roth, 1982). That's one day out of every five-day work week! And it doesn't include the additional time spent talking—whether on the phone or in person, whether in groups and meetings or one-on-one.

This book's goal is to help you develop the communication expertise you must have to realize the full potential of your expertise in your specialized field.

WWW For more information, visit Chapter 1 at www.techcomm.nelson.com.

Developing your communication expertise will be even more important if you have chosen a career as a technical, scientific, medical, or professional communicator. Private corporations, non-profit organizations, and government departments all employ such communication specialists. Typically, these specialists work in jobs where they communicate on topics such as computers, medicine, or environmental science that match their personal interests. Many prepare print communications, such as instruction manuals, proposals, and technical and scientific reports. Others create interactive media, such as websites, online documentation, and software-based training programs. Some work with both print and interactive media.

Besides being essential to your career, communication expertise will enable you to make valuable contributions to your school or community. Volunteer organizations, service clubs, and committees of municipal government will welcome your assistance in writing clear and compelling reports, proposals, and other documents. When confronting complex decisions about environmental standards, economic policy, and other issues, the citizens of your community will be grateful for your ability to explain technical, scientific, and other specialized subjects lucidly.

WRITING AT WORK DIFFERS FROM WRITING AT SCHOOL

This book assumes that you already know many things about effective communication that will be indispensable to you in your career. It also assumes that you must learn new skills—and even new ways of thinking about communication—to develop the expertise required on the job. That's because writing in the workplace differs from writing at school in many important ways. The following sections discuss the key differences with respect to writing. Similar differences exist for oral communication.

SERVES PRACTICAL PURPOSES

At work, people write for practical purposes.

As a student, you write for *educational* purposes. Professors ask you to compose term papers, prepare laboratory reports, and take written exams to help you learn the course material and enable you to demonstrate your mastery of a subject. They read your papers, projects, and tests primarily to assess your knowledge and assign a grade. They are unlikely to rely on what you say as a guide for their own beliefs or actions. On the job, in contrast, people write for *practical* purposes, such as helping their employer improve a product or increase efficiency. Readers are coworkers, customers, or other individuals who *need* the writer's information and ideas in order to pursue their own practical goals.

At work, writers include only what their readers will find useful or persuasive.

These different purposes profoundly affect the kinds of communications you need to produce. Consider just one example. In school, where your aim is to show how much you know, one of your writing strategies is probably to say as much as you can about your subject. At work, where you will write to support or influence other people's actions, your strategy should be to include only the information your readers need—no matter how much more you know. Extra information will only clog your readers' paths to what they need, thereby decreasing their efficiency and creating frustration. Developing skill at determining exactly what your readers need is one of the major steps in acquiring expertise in workplace communication.

ADDRESSES COMPLEX AUDIENCES

When you write a paper in school, you most often write to a single person: your professor. At work, however, you will often create a single communication that addresses a wide variety of people who differ from one another in many important ways, including their familiarity with your specialty, the way they will use your information, and their professional and personal concerns. The audience for David's report recommending changes to the railway's intermodal freight office is complex in this way. His readers include his supervisor, who will want to know how operations in her area would have to change if David's recommendations were adopted; the vice-president of finance, who will want to analyze David's cost estimates; the director of human resources, who will want to know how job descriptions will need to be rewritten; and members of the union, who will want assurances that the information system David is proposing will not eliminate jobs. When you are writing in situations like David's, you will need expertise at constructing one communication that simultaneously satisfies an array of individuals, each of whom will read it with a different set of concerns and goals in mind.

Even a complex audience is made up of specific individuals with particular reasons for reading.

GLOBAL AUDIENCES

Also, when writing at work, you may often address readers from other countries. Many organizations have international clients, customers, and suppliers. Corporate websites are accessed by people around the planet. Even when communicating to coworkers at your own location, you may address a multicultural audience—people of diverse national and ethnic origins.

USES DISTINCTIVE TYPES OF COMMUNICATION

In addition, on the job, people create a wide variety of communications that aren't usually prepared at school, including memos, business letters, instructions, project proposals, and progress reports. Each of these types of communication has its own conventions, which you must follow to write successfully.

EMPLOYS GRAPHICS AND VISUAL DESIGN TO INCREASE EFFECTIVENESS

When writing at school, you may be accustomed to writing assignments that involve only text, only words. At work, however, tables, charts, drawings, photographs, and other graphics are as important as written text in communicating facts and shaping attitudes. To write effectively, you will need expertise at creating graphics and at arranging your graphics and text on a page or screen in ways that make your communications visually appealing, easy to understand, and easy to navigate. Figure 1.1 shows a page from an instruction manual that illustrates the importance of graphics and visual design.

REQUIRES COLLABORATION

Collaborative writing is more common at work than in school.

You are much more likely to write collaboratively at work than at school. For long documents, the number of cowriters is sometimes astonishingly large. Martin Marietta Corporation's multivolume proposal to build the international space station contained text and drawings by more than 300 engineers (Mathes & Stevenson, 1991). Even when

FIGURE 1.1 A Communication That Illustrates the Importance of Graphics and Visual Design

Visual features of this page help readers perform their task.

The large heading explains what readers will learn.

The drawings show exactly what the readers need to do; they even show a hand performing these tasks.

They include arrows to indicate the direction of movement.

Each numbered drawing corresponds to the step with the same number.

To help readers match each drawing with its step, the numbers in the drawings are very large, and the numbers for the steps are bold and in a column of their own.

Dark lines highlight the cautions; the word *caution* is printed in bold.

The rectangle at the side of the page helps readers flip to the specific information they need.

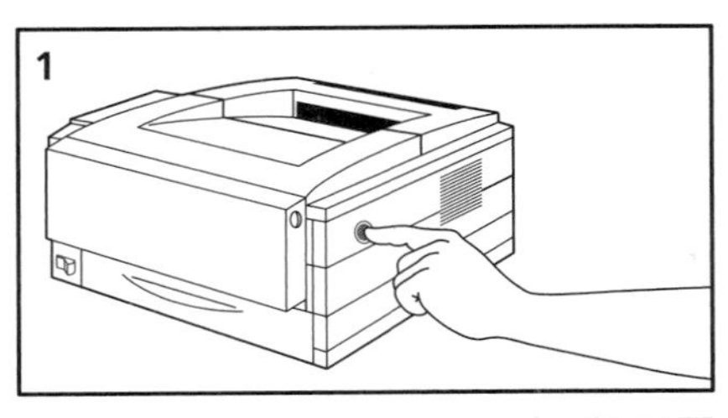

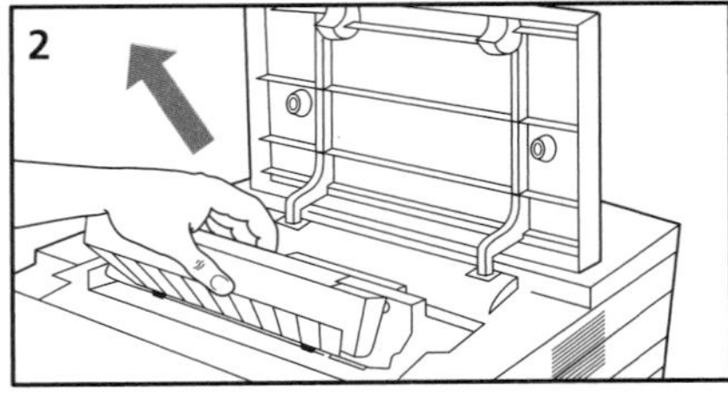

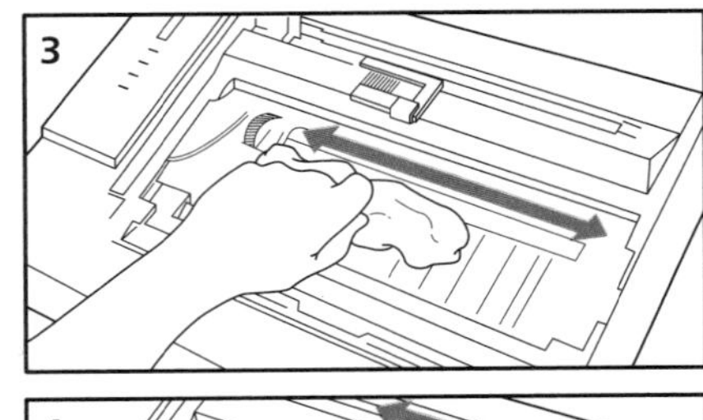

To Clean the Printer

1 Turn the printer off and unplug the power cable, and then open the printer's top cover by pressing the top cover release on the side of the printer.

2 Remove toner cartridge.

Caution
Because light damages the cartridge's photosensitive drum, do not expose the cartridge to light for more than a few minutes.

3 With a dry lint-free cloth, wipe any residue from the paper path area and the toner cartridge cavity as shown.

4 Remove the cleaning brush from the shoulder above the toner cartridge area. Place the flat part of the brush on the shoulder while allowing the brush to be inserted below the shoulder where the mirror is located. Move the brush from side to side several times to clean the mirror.

5 Replace the brush and toner cartridge, close the top cover, plug in the power cable, and then turn the printer on.

Caution
Do not touch the transfer roller (shown in the illustration) with your fingers. Skin oils on the roller can cause print quality problems.

Caution
If toner gets on your clothes, wipe it off with a dry cloth and wash your clothes in **cold** water. Hot water sets toner into fabric.

4 Troubleshooting and Maintenance

Troubleshooting and Maintenance 4-19

you prepare communications alone, you may consult your coworkers, your boss, and even members of your intended audience as part of your writing process.

At work, writing often must be reviewed and approved by others.

In one common form of collaboration, you will need to submit drafts of some of your communications for review by managers and others who have the power to demand changes. The number of reviewers may range from one to a dozen or more, and some drafts go through many cycles of review and revision before obtaining final approval. Communication expert Carolyn Boiarsky (1993) describes one memo that went through more than a hundred drafts!

SHAPED BY SOCIAL AND POLITICAL FACTORS

At work, writers must shape their communications according to their relationships with their readers.

Every communication situation has social dimensions. In the writing done at school, the key social relationship is that of a student to the teacher who assigned the paper or project. At work, you will have a much wider variety of relationships with your readers, such as manager and subordinate, customer and supplier, coworker and coworker. Sometimes these relationships will be characterized by cooperation and goodwill. At other times, they will be fraught with competitiveness as people strive for recognition, power, or money for themselves and their departments. You will need to attune the style, tone, and overall approach of each communication to these social and political considerations.

SHAPED BY ORGANIZATIONAL CONVENTIONS AND CULTURE

Different organizations have different writing customs.

In addition, each organization has a certain style that reflects the way it perceives itself and presents itself to outsiders. For example, an organization might be formal and conservative or informal and innovative. Individual departments within organizations may also have their own styles. On the job, you will be expected to understand and reflect the style of your organization in your writing.

MUST MEET DEADLINES

"It's better to be 80 percent complete than 100 percent late."

At work, your deadlines for completing communications are much more significant—and changeable—than the deadlines for the papers you write at school. For example, when a company prepares a proposal or sales document, it must reach the client on time. Otherwise, it may not be considered at all—no matter how good it is. Employers sometimes advise that "it's better to be 80 percent complete than 100 percent late."

PRODUCED WITH ADVANCED COMPUTER TECHNOLOGY

On the job, writers often use advanced software and employ advanced features of less specialized software. For example, Microsoft Word enables you to compare drafts to identify every change and to work on a communication at the same time that other writers are also working on it. Similarly, you can link an online report or a PowerPoint presentation to an Excel spreadsheet so that when the data in the spreadsheet are changed (perhaps through the automatic operation of a database), the report changes as well.

SENSITIVE TO LEGAL AND ETHICAL ISSUES

Under the law, most documents written by employees represent the position and commitments of the organization itself. Company documents can even be subpoenaed as evidence in disputes over contracts and in product liability lawsuits. These are among the reasons why certain documents are carefully reviewed before being sent to their intended readers.

Ethics are discussed further on page 18.

Even when the law does not come into play, many communications written at work have moral and ethical dimensions. The decisions and actions they advocate can affect many people for better or worse. Because of the importance of the ethical dimension of workplace writing, this book incorporates in most chapters a discussion of ethical issues that may arise in your on-the-job communications.

AT WORK, WRITING IS AN ACTION

As you can infer from the preceding section, there is tremendous variety among communications written on the job, depending on such variables as their purposes and readers; organizational conventions and cultures; and the political, social, legal, and ethical contexts in which they are prepared. Some people are hindered in their ability to write effectively in these multifaceted, shifting situations because they mistakenly think of writing as an afterthought, as merely recording or transporting information they developed while acting as specialists in their chosen fields.

Nothing could be further from the truth.

When you write at work, you act. You exert your power to achieve a specific result, to change things from the way they are now to the way you want them to be. Consider, again, the examples of Sarah and David. Sarah wants to help her team develop a successful engine. Acting as a mechanical engineering technologist, she has tested the faulty pistons to determine why they failed. Her ultimate purpose is to help her employer design and produce pistons without flaws. To contribute to the success of the engine, she must perform a writing act. She must compose sentences, construct tables of data, and perform other writing activities to present her results in a way the engineers will find useful. Similarly, David believes that the railway's intermodal freight office is run inefficiently. Acting as an information systems analyst, he has devised a plan for improving its operation. For his plan to be put into effect, however, David must perform an act of writing. He must write a proposal that will persuade the railway's decision makers to implement his plan.

The most important thing to remember about the "writing acts" you will be performing at work is that they are social actions. Every communication you write will be an interchange between particular, individual people: you and your readers. Perhaps you will be a supervisor telling a coworker what you want done, an adviser trying to persuade your boss to make a certain decision, or an expert helping another person operate a certain piece of equipment. Your reader may be an experienced employee who is uncertain of the purpose of your request, a manager who has been educated to ask certain questions when making a decision, or a machine operator who has a particular sense of personal dignity and a specific amount of knowledge about the equipment to be operated.

Even when writing to a large group of people, your communication will establish an individual relationship between you and each person in the group. Each person will read with his or her own eyes, and react with his or her own thoughts and feelings.

THE MAIN ADVICE OF THIS BOOK: THINK CONSTANTLY ABOUT YOUR READERS

The observation that writing is a social action leads to the main advice of this book: When writing, think constantly about your readers. Think about what they want from you—and why. Think about how you want to help or influence them and how they will react to what you have to say. Think about them as if they are standing right there in front of you and you are talking together.

You may be surprised that this book emphasizes the personal dimension of writing more than such important characteristics as clarity and correctness. Although clarity and correctness are important, they cannot, by themselves, ensure that something you write at work will be successful.

For example, if David's proposal for modifying the railway's information management system is to succeed, he will have to explain the problems created by the present operation in a way that his readers find compelling, he will have to address the kinds of objections his readers will raise to his recommendations, and he will have to deal sensitively with the possibility that his readers may feel threatened by having a new employee suggest improvements to a system they themselves set up. If his proposal fails to do these things, it will not succeed, no matter how "clear and correct" the writing is. A communication may be perfectly clear and perfectly correct, yet be utterly unpersuasive, utterly ineffective.

The same is true for all the writing you will do at work: What matters is how your readers respond. That's the reason for taking the reader-centred approach described in this book. This approach focuses your attention on the ways you want to help and influence your readers and teaches specific strategies that you can use to achieve those goals.

The importance of thinking constantly about your readers is highlighted by studies in which researchers compared the ways students approach their assignments with the ways successful workplace communicators approach their writing tasks (Beaufort, 1999; Bereiter & Scardamalia, 1993; Dias & Paré, 2000; Spilka, 1993). These studies reveal that the distinctive feature of the experts' approach is that they consider their readers while deciding about nearly every detail of their communications. When they aren't thinking specifically about the readers themselves, the experts are thinking about the conventions, expectations, and situations that influence their readers' responses to their writing. This book will assist you in developing the same reader-centred approach that is used by these expert communicators.

QUALITIES OF EFFECTIVE ON-THE-JOB COMMUNICATION: USABILITY AND PERSUASIVENESS

A first step in developing your expertise at on-the-job writing is to focus your attention on the two qualities that a workplace communication must have to be successful: usability and persuasiveness. Both qualities, of course, must be defined from the readers' perspective.

Usability defined

Usability refers to a communication's ability to help readers do what it is intended to help them do. As explained above, people read at work to gain information they need to *do something.* Their tasks may be physical, such as installing a new memory card in a computer. Or their tasks may be mental, as when Sarah's readers use her report to redesign the pistons and David's readers compare his recommended procedures with those currently used in the railway. No matter what the readers' task, a workplace communication is highly usable if it enables readers to:

- Locate quickly the information they need to accomplish their goal.
- Understand the needed information easily and accurately.
- Use the information to complete their task with minimum effort.

If readers have difficulty locating, understanding, or using the information they need, the communication is not as usable as it should be.

Persuasiveness defined

A communication's persuasiveness is its ability to influence its readers' attitudes and actions. It's easy to imagine why persuasiveness is important in proposals and recommendation reports. However, persuasiveness is also an indispensable quality in *all* on-the-job communications, including ones we often think of as purely informational. For example, many people are impatient with instructions. They read them carelessly, if at all. Thus, an instruction manual must persuade potential readers to consult it rather than try things out on their own, possibly damaging products, fouling the equipment, or harming themselves.

WWW To view other websites and print documents that combine usability and persuasiveness, visit Chapter 1 at www.techcomm.nelson.com.

As the preceding paragraphs suggest, every communication written at work must have *both* usability and persuasiveness to succeed. In most communications, one of these qualities dominates—usability in instructions, for example, and persuasiveness in proposals. However, the two are always inextricably intertwined. Instructions are effective only if the intended readers are persuaded to use them. A proposal can persuade only if its readers can easily find, understand, and analyze its content. To see how you can combine usability and persuasiveness in a single communication, look at the webpage and the memo shown in Figure 1.2 (page 12) and Figure 1.3 (page 13).

THE DYNAMIC INTERACTION BETWEEN YOUR COMMUNICATION AND YOUR READERS

Of course, the people who actually decide whether your on-the-job communications are usable and persuasive will be your readers. Therefore, the more you know about how your readers will read your communications, the better prepared you will be. Consequently, the detailed suggestions in this book are based, in large part, on what researchers have learned about how people read. The following paragraphs describe the three research findings that will be the most useful to you.

- Readers construct meaning.
- Readers' responses are shaped by the situation.
- Readers react moment by moment.

All three research findings stress that writing is not a passive activity for readers, but a dynamic interaction between readers and texts.

READERS CONSTRUCT MEANING

Readers construct the meaning of a communication by actively interacting with it.

When researchers say that readers *construct* meaning, they are emphasizing the fact that the meaning of a written message doesn't leap into our minds solely from the words we see. Instead, to derive meaning from the message, we actively interact with it. In this interaction, we employ a great deal of knowledge that's not on the page, but in our heads. Consider, for example, the knowledge we must possess and apply to understand this simple sentence: "It's a dog." To begin, we must know enough about letters and language to decipher the three printed words (*it's, a,* and *dog*) and to understand their grammatical relationships. Because they do not possess this knowledge, young children and people from some other cultures cannot read the sentence. Often, we must also bring other kinds of knowledge to a statement in order to understand it. For instance, to understand the meaning of the sentence "It's a dog," we must know whether it was made during a discussion of Jim's new pet or during an evaluation of ABC Corporation's new computer.

FIGURE 1.2

Webpage That Combines Usability and Persuasiveness

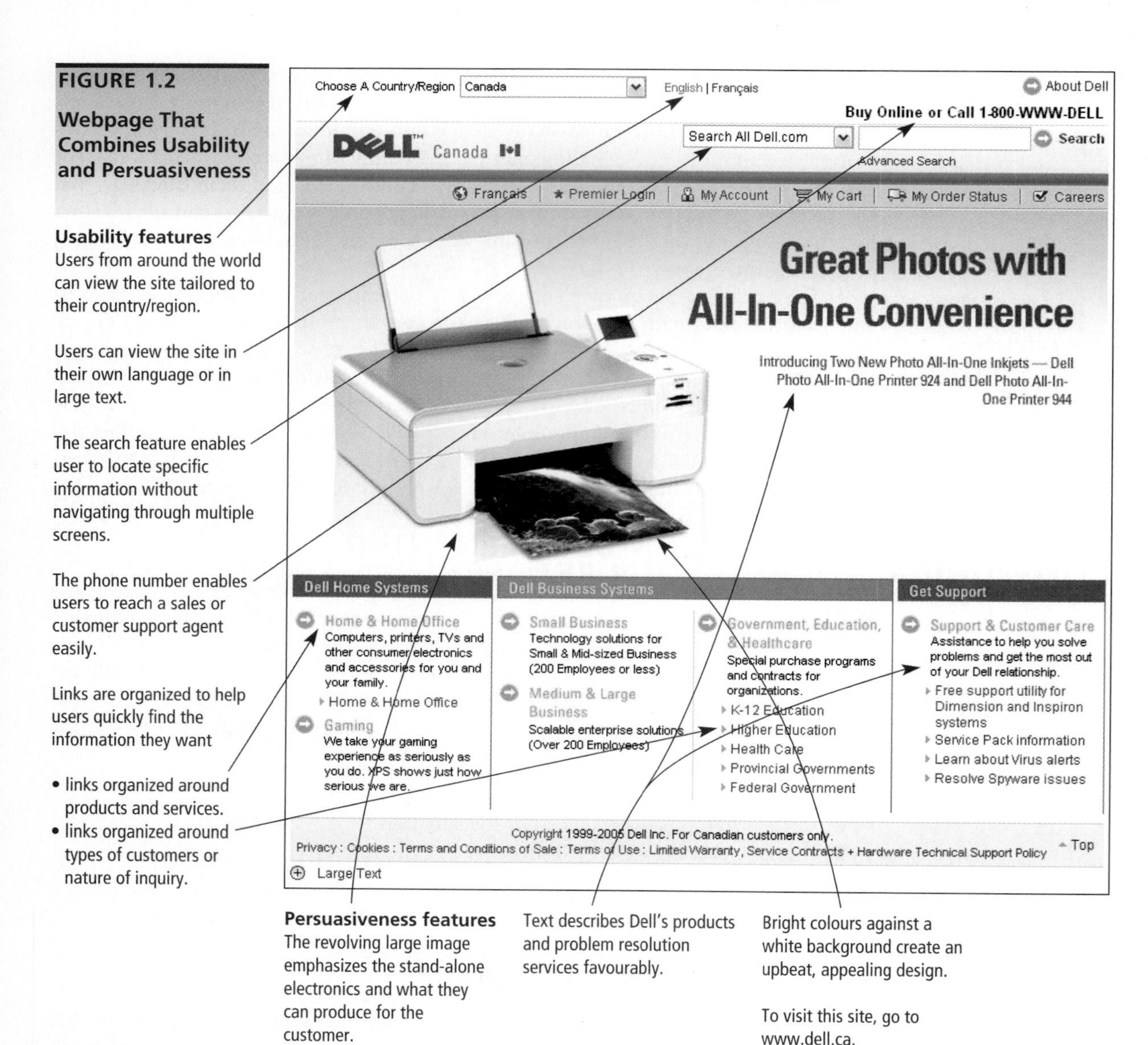

In addition to constructing the meanings from individual words and sentences, we build these smaller meanings into larger structures of knowledge. These structures are not merely memories of words we have read. They are our own creations. To demonstrate this point, write a sentence that explains the following heading: "Every communication written at work aims at change." Next, look for a sentence in the book that exactly matches yours. Most likely, you won't find one. The sentence you wrote is not one you remembered. Rather, it is the meaning you constructed through your interaction with the text.

FIGURE 1.3

Memo That Combines Usability and Persuasiveness

PIAGET HOUSEHOLD PRODUCTS
Intracompany Correspondence

October 12, 2007

To Alyssa Wyatt
From Stephen Li
Subject Test of Salett 321 Bottles for Use with StripIt

We have completed the tests you requested to find out whether we can package StripIt Oven Cleaner in bottles made of the new plastic, Salett 321. We conclude that we cannot, chiefly because StripIt attacks and begins to destroy the plastic at 40°C. We also found other significant problems.

Test Methods
To test Salett 321, we used two procedures that are standard in the container industry. First, we evaluated the storage performance of filled bottles by placing them in a chamber for 28 days at 20°C. We stored other sets of 24 bottles at 40°C and 50°C for the same period. Second, we tested the response of filled bottles to environmental stress by exposing 24 of them for 7 days to varying humidities and varying temperatures up to 60°C.

We also subjected glass bottles containing StripIt to the same test conditions.

Results and Discussion
In the 28 day storage tests, we discovered three major problems:

- StripIt attacked the bottles made from Salett 321 at 40°C and 50°C. At 50°C, the damage was particularly serious, causing localized but severe deformation. Most likely, StripIt's ketone solvents weakened the plastic. The deformed bottles leaned enough to fall off shelves in retail stores.
- The sidewalls sagged slightly at all temperatures, making the bottles unattractive.
- StripIt yellowed in plastic bottles stored at 50°C. No discoloration occured in glass bottles at this temperature. We speculate that StripIt interacted with the resin used in Salett 321, absorbing impurities from it.

In the environmental test, StripIt attacked the bottles at 60°C.

Conclusion
Salett 321 is not a suitable container for StripIt. Please call me if you want additional information about these tests.

Usability features

The subject line tells the reader the memo's exact topic.

The first sentence explains the memo's relevance to the reader: it reports on tests she requested.

The next two sentences present the information the reader needs most: the new plastic won't work.

Headings and topic sentences tell the reader what each section contains, thereby helping her locate specific information quickly.

The use of boldface for the headings makes them stand out.

The list promotes rapid reading.

The writer explicitly states the significance of the facts he presents.

Persuasive features

The writer describes the test method in detail to build credibility in the test results.

Specific data support all conclusions.

Lively, active verbs (highlighted in colour) create a vivid, action-oriented writing style.

Neatness and correctness bolster confidence in the writer.

The fact that readers construct the meaning they derive from a communication has many implications for writers that are explored later in this book. An especially important one is this: You should learn as much as possible about the knowledge your readers will bring to your communication so that you can determine how to help them construct the meanings you want them to build.

READERS' RESPONSES ARE SHAPED BY THE SITUATION

A second important fact about reading is that people's responses to a communication are shaped by the total situation in which they read—including such things as their purpose for reading, their perception of the writer's purpose, their personal stake in the subject discussed, and their past relations with the writer.

Kate's response to the message will be shaped by her situation.

For example, Kate's employer asked her to select the new notebook computer to be purchased for fifty field engineers. After completing her research, Kate read a message that said the ABC computer is "a dog." Her response to this statement would depend on many things. Did the other opinions and data she had gathered support this assessment? Was the statement made by a computer specialist, a salesperson for one of ABC's competitors, or the president of Kate's company? Has she already announced publicly her own assessment of the computer, or is she still undecided about it? Depending on the answers to these questions, Kate's response might range anywhere from pleasure because her own judgments have been supported by a well-respected person to embarrassment from having her publicly announced judgment called into question.

The range of situational factors that can affect a reader's response is obviously unlimited. The key point is that to predict how your readers might respond, you must understand thoroughly the situation in which they will read your message.

READERS REACT MOMENT BY MOMENT

The third important fact about reading is that readers react to communications moment by moment. When we read a comic novel, we chuckle as we read a funny sentence. We don't wait until we finish the entire book. Similarly, people react to each part of a memo, report, or proposal as soon as they come to it. The following demonstration illustrates this point.

Imagine that you manage a factory's human resources department. A few days ago, you discussed a problem with Brian Pryzblo, who manages the information processing department. Recently, the company's computer began issuing some payroll cheques for the wrong amount. Your department and Pryzblo's work together to prepare each week's payroll in a somewhat antiquated way. First, your clerks collect a time sheet for each employee, review the information, and transfer it to time tickets, which they forward to Pryzblo's department. His clerks enter the information into a software application that calculates each employee's pay and prints the cheques. The whole procedure is summarized in the following diagram:

Factory → time sheets → Your department (Human Resources) → time tickets → Brian Pryzblo's department (Information Processing) → database → Software program → payroll cheques

In your discussion with Pryzblo, you proposed a solution he did not like. Because you two are at the same level in the company, neither of you can tell the other what to do. When you turn on your computer this morning, you find an email message from Pryzblo.

Your task in this demonstration is to read the message *very slowly*—so slowly that you can focus on the way you react, moment by moment, to each statement. First, turn to the email message shown in Figure 1.4 on page 16. Cover it with a sheet of paper. Then slide the paper down the page, stopping after you read the first sentence. Immediately record your reactions (in your role as manager of the human resources department). Proceed in this way through the rest of the message.

Readers respond on a moment-by-moment basis.

Finished? Now look over your notes. Most people who participate in this demonstration find themselves responding strongly to almost every sentence. For example, they react to the quotation marks that surround the word *errors* in the first sentence. The word *insinuated* in the second sentence also draws an immediate response from most readers. (They laugh if they forget to play the role of human resources manager; they cringe if they remember to play the role.)

Readers' reactions in one moment shape their subsequent reactions.

The fact that readers respond to a communication moment by moment is important to you as a writer because their reaction to any one sentence will influence their reaction to everything they read from that point forward. For example, most people who read Brian Pryzblo's message while playing the role of human resources manager grow defensive the moment they see the quotation marks around the word *errors,* and they become even more so when they read the word *insinuated.* After they read the third paragraph, their defensiveness hardens into a grim determination to resist any recommendation Pryzblo may make.

When writing, keep your readers foremost in mind.

A few readers are more even tempered. Instead of becoming defensive, they become skeptical. As they read the first two sentences, they realize Pryzblo is behaving emotionally rather than intellectually, so they evaluate his statements very carefully. When they read his accusation that the human resources department clerks are miscopying the time sheets, they want to know what evidence supports that claim. When the next sentence fails to provide any evidence, they feel disinclined to go along with any recommendations from Pryzblo.

Thus, even though different readers react to the first few sentences of this memo differently, their early reactions shape their responses to the sentences that follow. Consequently, even though Pryzblo's recommendation seems sensible enough, no reader I've met feels inclined to accept it.

Of course, Pryzblo may have had some other purpose in writing. For example, he may have wanted to inflame the human resources manager into responding rashly so he or she would get in trouble with the boss. For that purpose, his message might have worked very well. But even in that case, the basic points of this demonstration would remain unchanged: People react to what they read on a moment-by-moment basis, and their reactions at each moment shape their reactions to what follows.

The preceding discussion underscores the importance of taking a reader-centred approach to writing, one in which you think constantly about your readers. Each reader creates his or her own response to your communications. To write effectively, you must predict these responses and design your messages accordingly. You will be best able to do this if you keep your readers—their needs and goals, feelings and situations, preferences and responsibilities—foremost in mind throughout your work on each communication.

FIGURE 1.4

Email Message for Demonstration

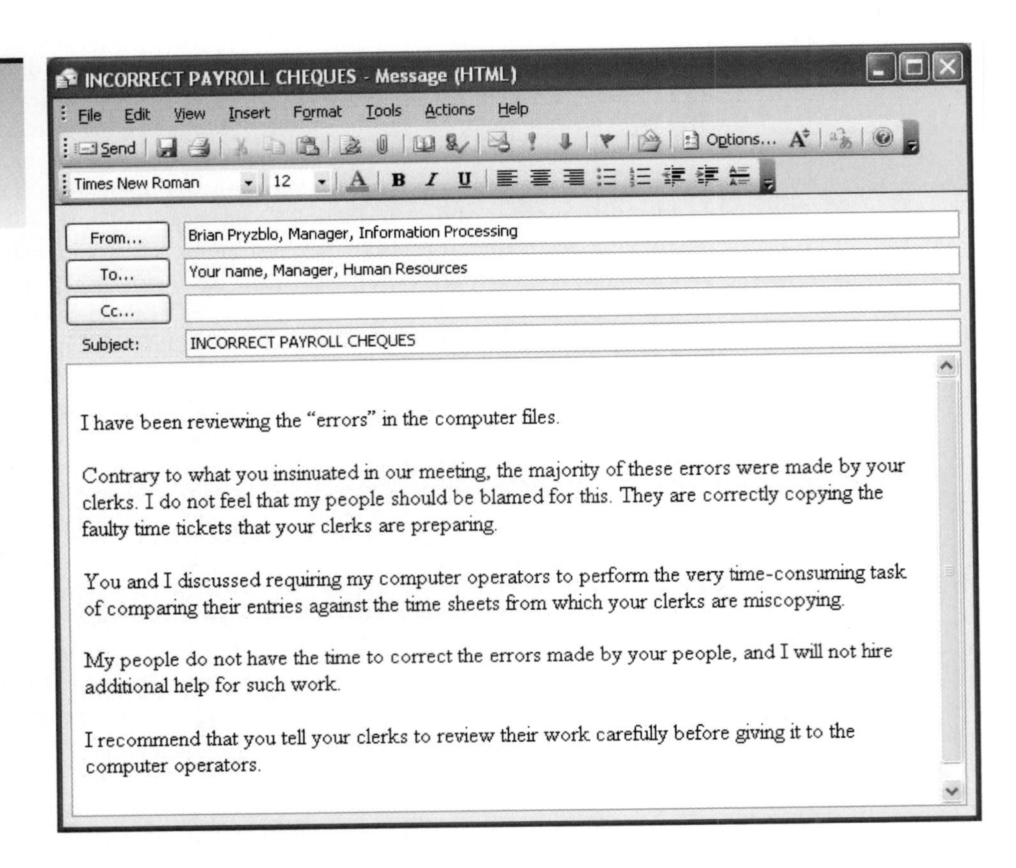

INCORRECT PAYROLL CHEQUES - Message (HTML)

From... Brian Pryzblo, Manager, Information Processing

To... Your name, Manager, Human Resources

Cc...

Subject: INCORRECT PAYROLL CHEQUES

I have been reviewing the "errors" in the computer files.

Contrary to what you insinuated in our meeting, the majority of these errors were made by your clerks. I do not feel that my people should be blamed for this. They are correctly copying the faulty time tickets that your clerks are preparing.

You and I discussed requiring my computer operators to perform the very time-consuming task of comparing their entries against the time sheets from which your clerks are miscopying.

My people do not have the time to correct the errors made by your people, and I will not hire additional help for such work.

I recommend that you tell your clerks to review their work carefully before giving it to the computer operators.

SOME READER-CENTRED STRATEGIES YOU CAN BEGIN USING NOW

Despite the many ways readers' goals, concerns, feelings, and likely responses can vary from situation to situation, readers approach almost all on-the-job communications with several widely shared aims and preferences. The following discussion briefly introduces several reader-centred strategies you can begin using immediately that address these common aims and preferences. All are discussed more fully later in this book. Many are illustrated in the webpage and memo shown in Figure 1.2 (page 12) and Figure 1.3 (page 13).

- **Help your readers find the key information quickly.** State your main points at the beginning rather than the middle or end of your communications. Use headings, topic sentences, and lists to guide your readers to the specific information they want to locate. Eliminate irrelevant information that can clog the path to what your readers want.
- **Use an easy-to-read writing style.** Trim away unnecessary words. Use the active voice rather than the passive. Put action in verbs rather than other parts of speech. To test the effectiveness of these three techniques, researchers James Suchan and Robert Colucci (1989) created two versions of the same report. The high-impact version used the techniques. The low-impact did not. The high-impact version reduced reading time by 22 percent, and tests showed that readers understood it better.

- **Highlight the points your readers will find to be persuasive.** Present the information your readers will find more persuasive before you present the information they will find to be less persuasive. Show how taking the actions you advocate will enable them to achieve their own goals. When selecting evidence to support your arguments, look specifically for items you know your readers will find to be credible and compelling.

Another reader-centred strategy is to talk with your readers. Before you begin work on a report or set of instructions, ask your readers, "What do you want in this communication? How will you use the information it presents?" When planning the communication, share your thoughts or outline, asking for their reaction. After you've completed a first draft, ask for their feedback.

Create an imaginary conversation if you can't actually talk with your readers.

If you cannot speak directly with your readers, talk with them imaginatively. Think of your communication as a conversation in which you make a statement and your reader responds. Write each sentence, each paragraph, each chapter, to create the interaction that will bring about the final outcome you desire. When following this strategy, it's crucial that you talk *with* your readers, not *to* them. When you talk *to* other people, you are like an actor reciting a speech: You stick to your script without regard to the way your audience is responding. When you talk *with* other people, you adjust your statements to fit their reactions. Does someone squeeze his brows in puzzlement? You explain the point more fully. Does someone twist her hands impatiently? You abbreviate your message. Figure 1.5 shows how one writer benefited from "talking" with her reader in this way while drafting one step in a set of instructions. This reader-centred strategy of talking imaginatively with your readers enables you to anticipate their moment-by-moment responses to your message and write accordingly.

FIGURE 1.5

How One Writer Used a Mental Portrait of Her Reader Reading

Marty's Use of Her Mental Portrait of Her Reader Reading

Marty, a civic engineer, was writing instructions for calibrating an instrument for testing the strength of building materials. One of the statements she drafted read like this:

15. Check the reading on Gauge E.

Marty then imagined how a typical user of her instructions would react to that statement. She saw her reader look up and ask, "What should the reading be?" So Marty knew she would have to tell her readers to look for the correct reading in the Table of Values.

Then, she imagined her reader asking: "Where is that table?" so she decided to give its page number.

Next, Marty imagined that the reader looked at the Table of Values and discovered that the reading on Gauge E was incorrect. The reader asked, "What do I do now?"

In the end, Marty revised her instructions to read as follows:

15. Determine whether the reading on Gauge E matches the appropriate value listed in the Table of Values (page 38).
    - If the value *does not* match, follow the procedures for correcting imbalances (page 27).
    - If the value does match, proceed to the next section.

COMMUNICATING ETHICALLY

To read more about ethics, see the Ethics Guidelines at the end of other chapters.

So far, this chapter has introduced concepts and strategies that will serve as the springboard from which you can understand and apply the rest of this book's advice for creating effective communications at work. Because effective communications create change—make things happen—they also have an important ethical dimension. When things happen, people are affected. Their happiness and even their health and well-being may be impacted. For example, you may write a proposal for a new product that can cause physical harm—at least if not handled properly. You may prepare a report that managers will use to make other people's jobs significantly more—or less—desirable, or even determine whether these people will continue to be employed. Ethical considerations can also apply to impacts on people outside your employer's organization, even future generations, and the environment.

Because technical communications have an important ethical dimension, in addition to providing advice for creating usable, persuasive communications, this book also provides advice for creating ethical ones.

WHAT IS ETHICAL?

To act ethically, you must first have a sense of what is ethical. At work, you have three major sources for guidance. First, professionals in your specialty have probably developed a code of ethics. Most professions have. Second, your employer may also have developed an ethics code. Many employers have. Some have even hired professional ethics specialists whom employees may consult. Third, you have your own sense of values, the ones you developed in your home, community, and studies.

Because they want to avoid conflicts over values in their careers, some employees have decided that their personal values have no place on the job. But that is a dangerous course. It can lead you into going along with actions at work that you would condemn at home. Furthermore, as companies decide what to do in certain situations, they sometimes discuss quite explicitly the ethical dimensions of the actions they might take. In these discussions, you can influence your employer's organization to act in accordance with your own ethical views, but only if you have brought your values with you to work—and only if you have the communication expertise required to present your view in ways that others find persuasive.

THIS BOOK'S APPROACH TO ETHICS

A basic challenge facing anyone addressing ethical issues in workplace settings is that different people have different values and, consequently, different views of the right actions to take in various situations. This fact should not surprise you. For thousands of years, philosophers have offered various incompatible ethical systems. They have yet to reach agreement. Moreover, as you know, people from different cultural backgrounds and different nations adhere to different values. In your own class, you may know other students with whom you disagree on ethical issues.

The same thing happens in the workplace. Consequently, this book won't tell you what your values ought to be. Instead, it seeks to help you act in accordance with your own values. Toward that end, it seeks to enhance your sensitivity to often subtle and difficult-to-detect ethical implications so that you don't inadvertently end up preparing a communication that affects people in ways you would wish it hadn't. This book also presents

Major Parts of This Book

- **The Writing Process.** Parts I to VI are oriented around the six major activities of writing:
  - Defining your communication's objectives
  - Applying a reader-centred approach
  - Choosing a communication format
  - Planning your communication strategy, including elements of usability and persuasiveness
  - Developing your communication elements, including text and graphical elements
  - Evaluating your communication, through checking, reviewing, testing, and revising

  All chapters emphasize a reader-centred approach to guide the many decisions you will make. They also describe the most important strategies for constructing communications that are used by workplace writers, and they discuss the communication conventions that apply on the job. Additionally, the chapters discuss ethical issues that will arise when you are carrying out the various activities of the writing process.
- **Special Applications.** Chapter 5 outlines supplementary guidelines for preparing electronic communications such as email messages. Chapter 12 provides advice for working on a team that is preparing a written or oral communication collaboratively. Chapter 16 explains how to create informational websites of the types used by organizations in the public and private sectors. Chapter 17 tells how to prepare and deliver oral presentations.
- **Structures.** Chapters 6, 7, and 8 introduce you to the general frameworks used for constructing workplace communications such as instructions, proposals, and reports.
- **Formats.** Appendix A describes the conventional formats for workplace email messages, memos, letters, and similar communications. Appendix B helps you use four common formats for citing sources.
- **Reference Guides.** Each section contains a reference guide that provides annotated examples demonstrating the section's advice in action. These reference guides cover the following topics:
  - Six Patterns of Organizing, as you define your communication's objectives
  - Seven Ways to End a Communication, as you apply a reader-centred approach
  - Three Types of Special Reports, as you choose your communication format
  - Five Research Methods, as you plan your communication strategy
  - Thirteen Types of Graphics, as you develop your communication elements
  - Four Steps for Revising, as you evaluate your communication

some ways of looking at the ethical aspects of various writing decisions—such as the way you use colours in graphs—that you may not have considered before. Ultimately, however, the book's goals are to help you communicate in ways that, after careful consideration, you believe to be ethical and to enable you to build the communication expertise needed to influence others when you want to raise ethical questions.

WHAT LIES AHEAD IN THIS BOOK

Throughout this book, you'll find detailed advice and information to help you create highly usable, highly persuasive, and thoroughly ethical communications on the job. The book is organized into six parts.

This book's guidelines are not rules.

All of the chapters offer guidelines, brief summary statements designed to make the book's suggestions easy to remember and use. As you read the guidelines, remember that they are just that: guidelines. They are not rules. Each one has exceptions, and some of them may even seem to conflict. An essential part of your communication expertise will be your ability to use good sense and creativity, guided always by thoughts of your readers and the specific ways you want them to respond to your communication.

To help you learn how to apply the guidelines, the chapters describe many sample situations and show many examples of memos, letters, reports, and other communications. These samples reflect typical workplace concerns and practices. However, what's *typical* is not what's universal. The readers and circumstances you encounter in your job will certainly differ to some extent from those described here. In fact, you may work for a boss or client whose regulations, values, or preferences are very untypical. To write successfully, then, you may need to ignore one or more of this book's guidelines.

No matter what situation you find yourself in, however, the book's overall advice will guide you to success: Size up your situation by focusing on your readers and the way you want to affect them, use your knowledge of your readers to shape your message in a way they will find highly usable and highly persuasive, and treat ethically all those who might be affected by your message. If you learn how to apply this advice to the typical situations described here, you will have mastered the strategies needed to write successfully in any atypical situation you encounter.

EXERCISES

For additional exercises, visit www.techcomm.nelson.com.

Expertise

1. Interview someone who holds a job you might like to have. Ask about the kinds of communications the person writes, the readers he or she addresses, the writing process and technology the person uses, and the amount of time the person spends writing. Supplement these questions with any others that will help you understand how writing fits into this person's work. According to your professor's directions, bring either notes or a one-page report to class.

2. Find a communication written by someone who has the kind of job you want, perhaps by asking a friend, a family member, or your own employer. Explain the communication's purposes from the point of view of both writer and readers. Describe some of the writing strategies the writer has used to achieve these purposes.

Online

Explore websites created by two organizations in the same business (airlines, computers, museums, etc.) or two employers for whom you might like to work. Compare the strategies used to make the sites usable and persuasive. Note ways their usability and persuasiveness might be increased.

Collaboration

Working with another student, rewrite the email message by Brian Pryzblo (Figure 1.4, page 16) so that it will be more likely to persuade the human resources manager to follow Pryzblo's recommendation. Assume that Pryzblo knows that the manager's clerks are miscopying because he has examined the time sheets, time tickets, and computer files associated with 37 incorrect payroll cheques; in 35 cases, the clerks made the errors. Take into account the way you expect the human resources manager to react upon finding an email from Pryzblo in his or her inbox. Make sure that the first sentence of your revision addresses a person in that frame of mind and that your other sentences lead effectively from there to the last sentence, which you should leave unchanged.

Ethics

As a first step in bringing your personal values to your on-the-job communication, list your values that you think will be especially important in your career. Explain situations in which you think it may be especially important for you to be guided by them.

CHAPTER 2

Matching Purpose with Communication

GUIDELINES

1. Focus on what you want to happen while your readers are reading
2. Define your usability goal: Analyze your readers' reading tasks
3. Define your persuasive goal: Analyze your readers' attitudes
4. Identify your readers' important characteristics
5. Study the context in which your readers will read
6. Ask others to help you understand your readers and their context
7. Learn who *all* your readers will be
8. Identify any constraints on the way you write
9. Ethics Guideline: Identify your communication's stakeholders

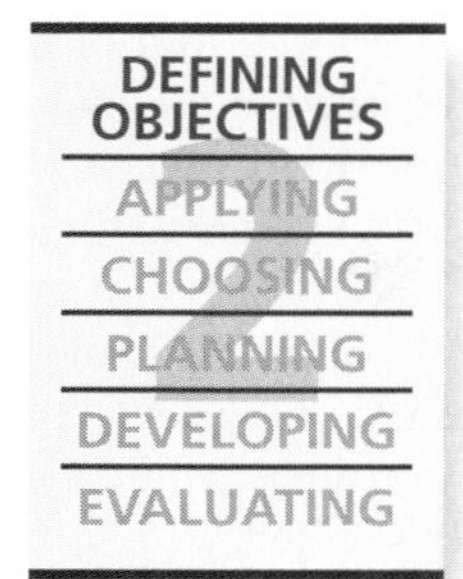

All writing activities are important, but defining objectives deserves your special attention. To see why, consider Haitham's situation.

A recent graduate, Haitham has been investigating ways of applying a more chip-resistant paint to his employer's products, which are all-terrain vehicles, dirt bikes, and other recreational vehicles. Haitham's boss told him that two vice-presidents have asked Haitham to write a report describing his progress. "How should I write this report?" Haitham asks himself. "What should I tell the vice-presidents and how should I say it?"

WWW For additional chapter resources, visit Chapter 2 at www.techcomm.nelson.com.

Haitham needs a basis for deciding how to write his report.

To answer these questions, Haitham must determine what his report must accomplish to be successful: He must define its objectives. Doing so will require some effort, but the result will guide him throughout the rest of his work on the report. When he plans and develops his draft, a carefully defined set of objectives will help him decide how long to make the report, what to say in it, what to present in text and what in graphics—how, in fact, to handle every aspect of his message. Similarly, when Haitham evaluates his draft, his objectives will help him determine what needs improvement. And when he revises, they will help him focus on the goals toward which all his revisions are directed.

This chapter presents nine guidelines that will enable Haitham—and you—to develop expertise at defining objectives to make sound decisions about the content and design of your communications.

GUIDELINE 1 Focus On What You Want to Happen While Your Readers Are Reading

Expertise involves a special way of seeing.

To develop expertise at defining objectives that will help you in these ways, you must gain the ability to see your communications in a certain way. In fact, the ability to see in a certain way is essential to expertise in every field. Research shows that physicians who are highly skilled at diagnosis consider the same number of a patient's symptoms as do less skilled diagnosticians; however, the expert physicians are better at focusing on the specific symptoms that point to the underlying disease (Patel & Groen, 1991).

Similarly, expertise in defining the objectives of a workplace communication requires the ability to see it from a reader-centred perspective. People who don't possess this expertise in defining objectives often describe the purpose of their communications by stating their topic: "I'm going to write about travel outside the solar system." They might name the type of communication they are going to prepare: "I'm going to write a proposal" or "I'm going to create a website." To develop expertise in defining the objectives of your communication, you must focus not on your topic or communication type, but on your readers in the act of reading your message.

The aim of all workplace writing is to bring about change.

Here's why. As explained in Chapter 1, when you write on the job, your goal will be to bring about change. You will be endeavouring to transform some aspect of the current situation—the way things are now—into a more desirable state.

Change requires the aid of other people.

The desire for this change may originate from your boss, other people in your employer's organization, clients, even you yourself. No matter where it originates, however, you will be able to achieve this change only through your communication's impact on your readers. After all, if you could make the change by yourself, you'd have no need to write.

Suppose you wish to obtain funding for a project. Your project will be funded only if your proposal persuades your boss to allocate the money you desire. Maybe you wish to help one of your employer's clients choose between two computers. The client will be able to select wisely only if your report enables readers to understand these alternatives fully and compare them meaningfully.

The outcome depends on what happens when your readers are reading.

Consequently, the true objectives of your communication are to affect your readers in the ways necessary to bring about the desired change. And there's only one time when your communication has a direct impact on readers: while the readers are reading it. If all goes well then, your communication will succeed.

However, if things don't go well during your readers' reading moments, your communication won't be successful.

Because the success or failure of your writing depends on your readers' responses during this critical time, focus your objectives on what must happen while your readers are reading your message.

Create a mental portrait of your readers in the act of reading.

An excellent way to establish this focus is to follow a suggestion offered in Chapter 1: Create a mental portrait or "movie" of your readers in the act of reading. This portrait will enable you to consolidate the important insights you develop while defining your objectives. It will also help you predict the way your readers will probably react to each writing strategy you think about using as you plan, develop, and evaluate your communication.

The next four guidelines describe techniques for constructing such a mental portrait of your readers.

GUIDELINE 2 Define Your Usability Goal: Analyze Your Readers' Reading Tasks

Begin constructing your mental portrait of your readers by determining what will make your communication *usable* in their view.

Usability is the ease with which readers can use your communication to perform their tasks.

As you learned in Chapter 1, usability is one of the two essential ingredients for success in a workplace communication. Usability is defined as the ease with which readers can use your communication to perform their tasks. The evolution of computer manuals provides an illustration. Years ago, most manuals were organized around the features of the software or hardware. For instance, they would describe the function each desktop icon, menu choice, or F-key enabled users to perform. This organization proved extremely frustrating to users, who didn't want to learn the functions of isolated icons or F-keys but rather the sequence of steps needed to complete their work. Computer manuals improved immensely once companies realized that they needed to study users' tasks and organize their manuals around them.

Similarly, to make *any* type of communication usable, you must first understand and support the tasks your readers wish to perform as they read what you've written for them. The following four-step procedure will enable you to do so.

Identifying Your Readers' Tasks

1. Identify your readers' purpose for reading.
2. Identify the information your readers want from your communication.
3. Determine how your readers will look for this information.
4. Determine how they will use the information *while reading.*

Identifying Your Readers' Purpose For Reading

At work, people write for practical purposes.

First, identify the specific purpose your readers will have in mind when they pick up and begin to read your communication. Almost certainly, they will be trying to gain information they can use in some practical way. A manager might read a report to gather information needed to make a decision. A programmer might consult a software reference manual to solve a specific coding problem. To create a usable communication, you must design it in ways that enable your readers to easily locate and use the information they need in order to achieve their purpose.

Identifying the Information Your Readers Want from Your Communication

Imagine the questions your readers will ask.

To ensure that your communication includes the information your readers need, you must first determine what that information is. An excellent way to identify this information is to imagine the questions your readers will have in mind as they pick up your communication. Their questions, of course, are determined by their purpose in reading. For example, Haitham has identified two reasons the vice-presidents might have for requesting that he report on his progress in his investigation of new methods for applying chip-resistant paint. First, they might want to decide which method they want him to concentrate on in his future work. In this case, their questions are likely to be "Which alternative appears to produce the best results?" "Which ones appear to be least costly?" and "How will our production line need to be altered if we employ the most promising alternative?" Alternatively, the vice-presidents may have asked Haitham for a progress report because they are worried that he won't complete his work in time for the company to use his results on a new line of products it will begin manufacturing next year. In this instance, their questions will probably be "What work do you still have to do?" "How long will it take you?" and "If you can't be done by April 30 without extra help, how much help will you need?"

To write effectively, Haitham must accurately identify his readers' questions.

To write a report that helps the vice-presidents achieve their purpose, Haitham must pinpoint the specific questions they want his report to answer. If Haitham's report answers one set of questions but the vice-presidents are asking the other set, he will have failed to prepare a communication his readers find usable.

Determining How Your Readers Will Look for Information

Once you've identified the information your readers will seek, imagine the ways they will look for it. Here are three search strategies readers often use on the job, together with some strategies you can use to assist readers:

Three ways readers search for information

For more information on graphics and page design, see Chapters 15 and 16.

- **Thorough, sequential reading.** When you write to readers who will read each sentence and paragraph in turn, you can systematically build ideas from one sentence, paragraph, and section to the next.
- **Selective reading.** When you write to readers who will scan for key points only, you can use lists, tables, boldface, and other strategies to make key points stand out.
- **Reference reading.** When you write to readers who will seek only specific pieces of information (as when troubleshooting a problem that has cropped up with a piece of equipment), you can use headings, contents, lists, indexes, and page design to guide your readers rapidly to the information they seek.

Determining How Your Readers Will Use the Information while Reading

Readers want to use the information you provide.

Once readers find the information they are seeking, they want to use it. By learning how they will use your information *while they are reading,* you can determine the most helpful way to organize and present it. Here are three common ways people use information while reading at work—together with example writing strategies appropriate to each:

- **To compare alternatives.** If the vice-presidents will read Haitham's report to identify the most promising painting process, he should organize in a way that will enable them to compare the alternatives easily on a point-by-point basis (cost of each, implementation time for each, etc.).
- **To determine how the information will affect them and their organization.** The engineer in charge of the production line in Haitham's company might want to know how the new process would affect her responsibilities. To meet this engineer's needs, Haitham would not write a progress report, but rather another kind of communication that focuses on the types of changes that might be involved, such as those in equipment and staff.
- **To perform a procedure.** A technician might need directions for performing some part of the new process. Haitham could most effectively assist this person by preparing step-by-step instructions in a numbered list.

Advice throughout the rest of this book describes ways to make your communications usable by employing your knowledge of the information your readers seek, the ways they will look for it, and the things they will do with it once they find it.

GUIDELINE 3 Define Your Persuasive Goal: Analyze Your Readers' Attitudes

When creating your mental portrait of your readers, you should also tell how you want your communication to alter their attitudes. In doing so, you identify the persuasive goal of your communication. As you learned in Chapter 1, persuasiveness is the second of the two essential qualities of all on-the-job writing. Every communication you write at work must be persuasive, even ones you don't normally think of as persuasive. For example, the vice-presidents will not rely on Haitham's progress report as a basis for their decisions unless they are persuaded that it provides accurately all of the relevant information. Instructions cannot achieve their purposes if people are not persuaded to read and follow them.

Persuasion is all about influencing readers' attitudes. When their attitudes change, their actions will follow. When writing at work, focus on your readers' attitudes toward the following four objects:

Things That Your Readers' Attitudes Focus On

- **Your subject matter.** Many of the topics you write about will be ones your readers have thought about already and so have already formed attitudes about.
- **You.** As one recent graduate observed, "As soon as I started my job, I realized that every time I communicate, people judge me based upon how well they think I write or speak." Even when you are merely answering factual questions, you will want your communication to inspire a favourable opinion of you as a person.
- **Your department or organization.** When you write to people outside your immediate working group, your communications will represent not only you but also your coworkers. Impressions your readers have of your department or company will shape their response to your message, so you will certainly want to influence your readers' attitude toward your organization.
- **The communication itself.** Readers can come to a communication with very high—or very low—motivation to read it. For example, many people are uninterested in reading instructions. When writing instructions, therefore, you will want to employ strategies that create the missing motivation.

You may wish to change, reinforce, or shape attitudes.

A change in attitudes always involves a shift from the present attitude to the desired one. This means that when analyzing your audience you should not only specify what you want the readers' final attitude to be, but also determine what their present attitudes are. Often, you will need to *reverse* negative attitudes and also *reinforce* positive ones. You may also need to *shape* your readers' attitudes on topics they don't have opinions about at present.

The distinctions among these three kinds of attitude change will be important to you because different kinds of change are best addressed by different persuasive strategies. To reinforce an attitude, you can build on the existing attitude and expect little resistance. Consequently, you can often present only positive points without trying to rebut arguments against your position. If you discuss negative points, it will be for the purpose of persuading readers that you have thoroughly investigated your topic. In contrast, if you want to reverse an attitude, you can expect your readers to be resistant. Consequently, you must not only cite positive points, but also address the counterarguments your readers will raise. Other writing techniques for overcoming resistance are discussed in Chapter 3.

Figure 2.1 (page 28) illustrates some ways you might want to alter your readers' attitudes in several typical communications.

FIGURE 2.1 **Ways Communications Prepared at Work Can Alter Readers' Attitudes**

The Way Things Are Now	You Write	The Way You Want Them to Be
Your reader is a manager who wants to decide whether to purchase a certain piece of equipment. →	You write a memo evaluating the equipment in terms of the benefits it will bring the company. →	The manager decides to buy the equipment and feels confident that she made a good decision based on information you provided.
Your reader is the director of a plant that is using an outdated process. →	You write a report on problems with the current process and the ways they can be overcome with various new processes. →	After reading your report, the plant director feels that the process being used now may be faulty and that one of the new ones is worth investigating further.
Your readers are customer service agents who will use a new computer system. They fear it is more complicated than the present system. →	You write a procedure manual that shows how easy the system is to use. →	The agents feel relaxed and self-confident after learning from your manual how to use the new system.

GUIDELINE 4 Identify Your Readers' Important Characteristics

Different readers respond differently to the same writing strategies. Consequently, when you construct your mental portrait of your readers, you should incorporate those characteristics of your particular readers that will influence the way they respond to your communication. The following paragraphs discuss several reader characteristics that are especially important to consider.

Professional Specialty

People with different professional specialties ask different questions.

People with different professional specialties ask different questions and use the answers differently. For instance, when reading a report on the industrial emissions from a factory, an environmental engineer working for the factory might ask, "How are these emissions produced, and what ideas does this report offer for reducing them?" whereas the corporate lawyer might ask, "Do these emissions exceed standards set by the Ministry of the Environment? And, if so, how can we limit our legal liability for these violations?"

The most obvious clues to your readers' specialties are their job titles—systems analyst, laboratory technician, director of public relations. However, don't settle for obtaining only a general sense of your readers' specialties. Determine as precisely as possible why each person will be reading your communication and exactly what information he or she will be looking for.

Organizational Role

Regardless of the particularities of their professional specialties, people at work usually play one of three organizational roles: decision maker, adviser, or implementer. Each role leads to a different set of questions.

Decision makers The decision makers' role is to say how the organization will act when it is confronted with a particular choice. Decision makers determine what the company should do in the future—next week, next month, next year. Consequently, decision makers usually ask questions shaped by their need to choose between alternative courses of action.

Typical Questions Asked by Decision Makers

Decision makers want conclusions, not details.

- **What are your conclusions?** Decision makers are much more interested in your conclusions than in the raw data you gathered or the particulars of your procedures. Conclusions can serve as the basis for decisions; details cannot.
- **What do you recommend?** Decision makers usually ask you about a topic because you have special knowledge of it. This knowledge makes your recommendation especially valuable to decision-making readers.

Decision makers look to the future.

- **What will happen?** Decision makers want to know what will occur if they follow your recommendations—and what will happen if they don't. How much money will be saved? How much will production increase? How will customers react?

Advisers Advisers provide information and advice for decision makers to consider when deciding what the organization should do. Unlike decision makers, advisers are interested in details. They need to analyze and evaluate the evidence supporting your general conclusions, recommendations, and projections.

Advisers want the details, not just generalizations.

Consequently, advisers ask questions that touch on the thoroughness, reliability, and impact of your work.

Typical Questions Asked by Advisers

- Did you use a reasonable method to obtain your results?
- Do your data really support your conclusions?
- Have you overlooked anything important?
- If your recommendation is followed, what will be the effect on other departments?
- What kinds of problems are likely to arise?

Implementers Decisions, once made, must be carried out by someone. Implementers are these individuals. The most important questions asked by implementers are these.

Typical Questions Asked by Implementers

Implementers want to know what to do—and why.

- **What do you want me to do?** Whether you are writing step-by-step instructions, requests for information, or policies that others must follow, implementers want you to provide clear, exact, easy-to-follow directions.
- **What is the purpose of the actions you are asking me to perform?** To produce satisfactory results, implementers often must know the reason for the policy or directive they are reading. Imagine, for instance, the situation of the managers of a factory who have been directed to cut by 15 percent the amount of energy used in production. They need to know whether they are to make long-term energy savings or compensate for a short-term shortage. If the latter, they might take temporary actions, such as altering work hours and

curtailing certain operations. However, if the reduction is to be long-term, they might purchase new equipment and modify the factory building.

- **How much freedom do I have in deciding how to do this?** People often devise shortcuts or alternative ways of doing things. They need to know whether they have this freedom or whether they must do things exactly as stated.

Name the deadline.

- **When must I complete this task?** To be able to adjust their schedules to include a new task along with their other responsibilities, implementers need to know when the new task must be completed.

Familiarity with Your Topic

Readers unfamiliar with your topic need background information.

Your readers' familiarity with your topic—company inventory levels, employee morale on the second shift, problems with new software—will determine how much background information you need to provide to make your communication understandable and useful to them. When you are writing about a topic they know well, you may not need to include any background at all. However, if you are treating a topic they don't know much about, you may have to explain the general situation before proceeding to the heart of your message. If your readers are unfamiliar with your topic, you may also need to explain how it relates to them so that they can judge whether to bother reading about it.

Knowledge of Your Specialty

To use the information you provide, readers need to understand the terms and concepts you employ. Consider, for example, an instruction manual for a computer-controlled machine that directs the reader to "zero the tool along the Z-axis." Readers unfamiliar with this type of equipment might ask, "What is zeroing? What is the Z-axis?" If the instructions do not answer these questions, the readers will have to ask someone else for help, which defeats the purpose of the instructions. On the other hand, the instructions writer would not want to provide these definitions if all of his or her readers are already familiar with these terms. Coming upon such explanations, these readers would ask, "Why is this writer making me read about things I already know?"

Relationship with You

When you are having a conversation, you adjust your speech to your relationship with the other person. You talk with a friend more informally than with a professor you don't know well—and both your friend and your professor might be startled if you didn't make such adjustments. Similarly, at work you should write in a way that reflects the relationships you have with your readers.

Personal Preferences

Accommodate your readers' preferences when you can.

A variety of personal characteristics can also influence readers' responses to your writing. For example, you may be writing to an individual who detests the use of certain words or insists on particular ways of phrasing certain statements. Or you might be writing to someone who is keenly interested in information you would not have to supply to most people. Some readers favour brief messages, and some insist on more detail. It only makes sense to accommodate such personal preferences where feasible.

People's preferences are shaped by the customary practices employed where they work. This is important to keep in mind as you begin to work for a new employer or a different department. When people say, "That's not the way we write here," they may be expressing preferences shared by many other people in the organization.

Cultural Background

Different cultures have radically different writing customs.

People's cultural backgrounds can also influence their response to your writing. For example, people in the dominant business culture in Canada prefer direct, brisk communication, whereas communication in China and India is much more indirect and polite. When you are addressing readers in other countries, it is vital that you learn as much as you can about their business customs and culture. Where possible, it can be helpful to ask someone from the culture to read over your draft to help you avoid sending an ineffective, difficult-to-understand, or even inadvertently offensive communication.

Special Considerations

This is a catchall category. It is a reminder that each reader is unique. You should always be on the lookout for reader characteristics you would not normally need to consider. For example, you may be addressing individuals with an especially high or low reading level, weak eyesight, or colour blindness—and it will be essential for you to take these characteristics into consideration when writing.

Sometimes you need to consider where your readers will read.

It is also wise to consider the setting in which your readers will be reading, in case this factor suggests ways you can adapt your communication to make it more useful for them. For example, if you are writing a software manual, you might want to design it in a small format to help conserve desk space for your readers. If you are writing a repair manual for hydraulic pumps, you might print your manual on paper coated to resist the moisture, oil, and dirt that are unavoidable in that reading environment.

GUIDELINE 5 Study the Context in Which Your Readers Will Read

Each message is part of an ongoing story.

At work, people interpret what they read as a chapter in an ongoing story. Consequently, they respond to each message in light of prior events as well as their understanding of the people and groups involved. Fill out your mental portrait of your readers by imagining how the following circumstances might influence their response to your communication:

- **Recent events related to your topic.** Maybe you are going to announce the reorganization of a department that has just adjusted to another major organizational change. You'll need to make a special effort to present the newest change in a positive light. Or maybe you are requesting money to attend an important professional meeting. If your department has just been reprimanded for excessive travel expenses, you will have to make an especially strong case.
- **Interpersonal, interdepartmental, and intraorganizational relationships.** If you are requesting cooperation from a department that has long competed with yours for company resources, you will need to employ special diplomacy. Political conflicts between individuals and groups can also create delicate writing situations in which certain ways of expressing your message can appear to support one faction and weaken another even if you have no intention of doing so.

GUIDELINE 6

Ask Others to Help You Understand Your Readers and Their Context

How can you learn all of the facts about your readers and their situation that are described in Guidelines 2 through 5? Experienced employees have often gained that knowledge through years of interacting with their readers. If you are a newcomer, you may have to ask others for help.

The best sources for most of the information are your actual readers. If possible, interview them as a first step in formulating your objectives. Better than anyone else, they can tell you about their needs, attitudes, preferences, and situations. Most readers will welcome the opportunity to tell you what they would find to be usable and persuasive in a communication you are writing for them.

If you cannot contact your readers, ask for help from your boss, coworkers, or anyone else who might have valuable insights about your readers and their circumstances. Like most activities of writing, defining objectives is one where you can benefit greatly from the assistance of other people.

Learn Who *All* Your Readers Will Be

So far, this chapter has assumed that you will know from the start just who your readers will be. That may not always be the case. Communications you prepare on the job may find their way to many people in many parts of your organization. Numerous memos and reports prepared at work are routed to one or two dozen people—and sometimes many more. Even a brief email message you write to one person may be forwarded, copied, or shown to others. To write effectively, you must learn who all your readers will be so that you can keep them all in mind when you write. The following discussion will help you identify readers you might otherwise overlook.

Phantom Readers

The most important readers of a communication may be hidden from you. That's because at work, written communications addressed to one person are often used by others. Those real but unnamed readers are called *phantom readers.*

Examples of times you might have phantom readers

Phantom readers are likely to be present behind the scenes when you write communications that require some sort of decision. One clue to their presence is that the person you are addressing is not high enough in the organizational hierarchy to make the decision your communication requires. Perhaps the decision will affect more parts of the organization than are managed by the person addressed, or perhaps it involves more money than the person addressed is likely to control.

Much of what you write to your own boss may actually be used by phantom readers. Many managers accomplish their work by assigning it to assistants. Thus, your boss may check over your communications, then pass them along to his or her superiors.

After working at a job for a while, employees usually learn which communications will be passed up the organizational hierarchy. However, a new employee may be chagrined to discover that a hastily written memo has been read by executives at very high levels. To avoid such embarrassment, identify your phantom readers, then write in a way that meets their needs as well as the needs of the less influential person you are addressing.

Future Readers

Your future readers may include lawyers and judges.

Your communications may be put to use weeks, months, or even years after you imagined their useful life was over. Lawyers say that the email messages, memos, reports, and other documents that employees write today are evidence for court cases tomorrow. Most company documents can be subpoenaed for lawsuits concerning product liability, patent violation, breach of contract, and other issues. If you are writing a communication that could have such use, remember that lawyers and judges may be your future readers.

Your future readers also may be employees of your company who may retrieve your old communications for information or ideas. By thinking of their needs, you may be able to save them considerable labour. Even if you are asked to write something "just for the record," remember that the only reason to have a record is to provide some future readers with information they will need to use in some practical way that you should understand and support.

Complex Audiences

Complex audiences are very common at work.

Writers sometimes overlook important members of their audience because they assume that all their readers have identical needs and concerns. Actually, audiences often consist of diverse groups with widely varying backgrounds and responsibilities.

That's partly because decisions and actions at work often affect many people and departments throughout the organization. For instance, a proposal to change a company's computer system will affect people throughout the organization, and people in different areas will have different concerns—some with recordkeeping, some with data communication, some with security, and so on. People in each area will examine the proposal.

Even when only a few people are affected by a decision, many employers expect widespread consultation and advice on it. Each person consulted will have his or her own professional role and area of expertise, and each will play that role and apply that expertise when studying your communication.

When you address a group of people who will be reading from many perspectives, you are addressing a *complex audience.* To do that effectively, you need to write in a way that will meet each person's needs without reducing the effectiveness of your communication for the others. Sometimes you may have to make a tradeoff by focusing on the needs and concerns of the most influential members of your audience. In any case, the first step in writing effectively to a complex audience is to identify each of its members or groups.

GUIDELINE 8 Identify Any Constraints on the Way You Write

So far, this chapter has focused your attention on developing a full understanding of your readers as you define your communication's objectives. As you gather the information that will form the basis for the way you craft your communication, you should also learn about any expectations, regulations, or other factors that may constrain what you can say and how you can say it. In the working world, expectations and regulations can affect any aspect of a communication—even tone, use of abbreviations, layout of tables, size of margins, and length (usually specifying a maximum length, not a minimum). It will be important for you to find out about these constraints and take them into account as you create your communication.

Reasons for organizational expectations about writing

Some of these constraints come directly from the employer, reflecting such motives as the company's desire to cultivate a particular corporate image, to protect its legal interests, or to preserve its competitive edge. In addition, most organizations develop writing customs—"the way we write things here." Writing constraints also can originate from outside the company—for instance, from government regulations that specify how patent applications, environmental impact assessments, and many other types of documents are to be prepared. Similarly, scientific, technical, and other professional journals have strict rules about many aspects of the articles they publish.

www.techcomm.nelson.com

Be sure to find out what these constraints are when you are defining your objectives so that you can take them into account as you plan and draft your communication. Many companies publish style guides that set forth their expectations for written materials. Ask if your employer has one. You can also learn about these constraints by asking coworkers and by reading communications similar to yours that your coworkers have written in the past.

GUIDELINE 9 Ethics Guideline: Identify Your Communication's Stakeholders

There are many strategies for ensuring that on-the-job writing is ethical. Some writers use an *alarm bell strategy.* They trust that if an ethical problem arises in their writing, an alarm bell will go off in their heads. Unless they hear that bell, however, they don't think about ethics. Other writers use a *checkpoint strategy.* At a single, predetermined point in the writing process, they review their work from an ethical perspective.

In contrast, this book teaches a more active and thorough *process strategy* for ethical writing. In it, you integrate an ethical perspective into every stage of your work on a communication. It's important to follow a process strategy because at every step of writing you make decisions that shape the way your communication will affect other people. Accordingly, at every step you should consider your decisions from the perspective of your personal ethical beliefs about the ways you should treat others.

Stakeholders

In a process approach to ethical writing, no step is more important than the first: defining objectives. When you are defining your communication's objectives, you are (in part) identifying the people you will keep in mind throughout the rest of your writing effort. When you follow the reader-centred approach to writing that is explained in this book, you begin by identifying your readers.

To write ethically, you must also identify another group of people: the individuals who will gain or lose because of your message. Collectively, these people are called stakeholders because they have a stake in what you are writing. Only by learning who these stakeholders are can you ensure that you are treating them in accordance with your own ethical values.

How to Identify Stakeholders

Because communications written at work often have far-reaching effects, it's easy to overlook some stakeholders. If that happens, a writer risks causing accidental harm that could have been avoided if only the writer had thought through all the implications of his or her communications.

To identify the stakeholders in your communications, begin by listing the people who will be directly affected by what you say and how you say it. In addition to your readers, these individuals may include many other people. For instance, when Dev was preparing a report for his managers on the development of a new fertilizer, he realized that his stakeholders included not only the managers but also the farmers who would purchase the fertilizer and the factory workers who would handle the chemicals used to manufacture it.

Next, list people who will be affected indirectly. For example, because fertilizers run off the land into lakes and rivers, Dev realized that the stakeholders of his report included people who use these lakes and rivers for drinking water or recreation. Indeed, as is the case with many other communications, the list of indirect stakeholders could be extended to include other species (in this case, the aquatic life in the rivers and lakes) and the environment itself.

Finally, think of the people who may be remotely affected. These people may include individuals not yet born. For example, if Dev's fertilizer does not break down into harmless elements, the residue in the soil and water may affect future generations.

Additional Steps

As just explained, the reason for identifying stakeholders at the very beginning of your work on a communication is to enable you to take them into account throughout the rest of the writing process. The ethics guidelines in later chapters will explain how to do that. By following their advice, you will be able to identify and address the sometimes subtle ethical dimensions of your on-the-job writing, and you will be able to keep your stakeholders constantly in mind, just as you keep your readers constantly in mind, while you are writing.

EXPERTISE IN ACTION: AN EXAMPLE

To see how developing your expertise at defining the objectives of your communications can help you write successfully on the job, consider the following example.

Amber works for the Winnipeg office of a non-profit organization whose volunteers provide Braille translations for textbooks and other reading materials requested by people who are visually impaired. The office assigned all translations to the volunteers on a rotating basis. As a result, some urgently needed translations weren't completed on time because they were assigned to slow-working volunteers. Amber decided to write a memo to Mrs. Lang, her boss, to recommend that urgent translations be assigned to the fastest and most reliable volunteers, rather than to the next people on the list.

WWW To download a copy of the Worksheet for Defining Objectives that you can fill out, visit Chapter 2 at www.techcomm.nelson.com.

Amber began work on her memo by filling out a worksheet based on this chapter's nine guidelines. Her completed worksheet is shown in Figure 2.2 (page 36).

By defining her memo's persuasive objectives, Amber developed an understanding of her reader that enabled her to choose effective writing strategies.

By completing each part of the worksheet, Amber gained important insights about the most effective way to draft her memo. For example, when responding to the questions about her reader's attitudes, she realized that Mrs. Lang would probably react defensively to any recommendation about the current system. Mrs. Lang had created the system and believed that it worked very well. Furthermore, she was the type of person who resists suggestions. By focusing on these characteristics of Mrs. Lang, Amber concluded that Mrs. Lang might even resist the suggestion that a problem existed. To

FIGURE 2.2 Amber's Completed Worksheet for Defining Objectives

DEFINING OBJECTIVES

Overall Purpose

What are you writing?
A proposal for a new method of assigning Braille translations in which the most urgent requests go to the quickest and most reliable volunteers.

What prompts you to write?
I believe the new method will reduce the number of urgent translations not completed on time.

What outcome do you desire?
I would like the new method to be put in effect, at least on a trial basis.

What outcome does your reader desire?
Initially, at least, she'll prefer that the current system remain in place.

Reader Profile

Who is your primary reader?
Mrs. Lang.

What is your reader's relationship to you?
She is my boss and likes to maintain a formal superior–subordinate relationship.

What are your reader's job title and responsibilities?
She is director of the Braille Division, responsible for recruiting and maintaining a large group of volunteer Braille translators, advertising translation services, and responding to requests for translations by assigning volunteers the work of making them.

Who else might read your communication?
Rich Juryn and Mina Williams, Mrs. Lang's chief assistants.

How familiar is your reader with your subject?
Mrs. Lang knows the present system of assigning Braille very well because she set it up and has run it for the past twelve years. She does not know that I am thinking of proposing an alternative but does know that some other offices of the Society for the Blind use systems similar to mine.

How familiar is your reader with your specialty?
Very familiar.

Does your reader have any communication preferences you should take into account?
She likes all communications to be "businesslike." She does not like informality.

Should you take into account any other things about your reader when writing?
Mrs. Lang gives the impression of being very sure of herself but feels threatened by suggestions for change.

Situational Analysis

What events and circumstances influence the way you should write?
Mrs. Lang was recently asked by three board members to retire. She successfully resisted.

FIGURE 2.2 *(continued)*

Usability Objectives (Reader's Tasks)

What are the key questions your reader will ask while reading?
What makes you think anything is wrong with the present system?
How, exactly, would your proposed system work?
What would I have to do differently?
How would the operations of the office be changed?
How would we determine which translations deserve highest priority?
How would we decide which translators are placed in our top group?
What would it cost?

How will your reader search for the answer? (The reader may use more than one strategy.)

__X__ Sequential reading from beginning to end
__X__ Selective reading, as when scanning for key points only
_____ Reference reading, as when using a manual to seek specific pieces of information
_____ Other (explain)

How will your reader use the information you provide?

__X__ Compare point by point (what will be the points of comparison?)
__X__ Attempt to determine how the information you provide will affect him or her
__X__ Attempt to determine how the information you provide will affect his or her organization
__X__ Follow instructions step by step
__X__ Other (explain)
She will compare her system with mine in terms of cost and speed of producing translations. Although she won't exactly look for instructions in my proposal, she will want to know in detail how it will work.

Persuasive Objectives (Reader's Attitudes)

What is your reader's attitude toward your subject? Why? What do you want it to be?
Mrs. Lang thinks the present system runs as well as possible. I want her to see that a better system is possible.

What is your reader's attitude toward you? Why? What do you want it to be?
Although I have worked for her for three years, Mrs. Lang still thinks of me as a newcomer who knows little and has impractical ideas. I want her to think that I am a helpful, knowledgeable, sensible person.

Stakeholders

Who, besides your readers, are stakeholders in your communication?
The translators—including those who would be given priority assignments and those who wouldn't.

How will they be affected by it?
I don't know. I'll have to investigate this.

Constraints

What expectations, regulations, or other factors limit the way you can write?
None.

write effectively, Amber would have to demonstrate that the current system could be improved—without seeming to criticize Mrs. Lang.

Amber realized that she needed to be especially careful to identify and answer all of Mrs. Lang's questions about the change's impact on the organization.

The worksheet also helped Amber focus on the fact that Mrs. Lang believed that Amber didn't understand all the issues involved in managing the agency. Therefore, Amber realized, Mrs. Lang would probably respond to her memo by looking for holes in Amber's reasoning. Consequently, as she filled out the worksheet's section about her memo's usability objectives, Amber concentrated on questions Mrs. Lang would ask as she searched for these possible holes. By identifying these questions, Amber pinpointed the information she would have to include in her memo in order to persuade Mrs. Lang that she had carefully considered all the organizational ramifications of her recommendation.

By thinking about the situation in which Mrs. Lang would read, Amber avoided an ineffective writing strategy.

Similarly, when she filled out the situational analysis section of the worksheet, Amber noted that two months earlier Mrs. Lang had successfully resisted pressure from several members of the agency's board of directors to force her to retire so that a younger person might take over. Amber realized that any reference to these board members or their wishes for streamlined operations would probably arouse a very hostile response from Mrs. Lang.

By considering the stakeholders in her situation, Amber found a way to avoid hurting them.

Finally, while trying to fill out the section concerning stakeholders, Amber discovered that she needed to talk with some of the translators to find out what they thought of her proposal. Through these interviews, she learned that those who might be judged less reliable and less speedy would be deeply offended. To avoid hurting their feelings, Amber modified her proposed plan in the following way: The agency would ask all the translators to tell how many pages they could commit to translating in a week. Urgent translations would go to those who made the largest commitments. Translators who failed to meet their original commitment would be invited to specify a lower commitment that would better suit their personal schedules.

Because of her expertise at defining objectives, Amber was able to write a detailed, diplomatic, four-page memo. After several months of deliberation, Mrs. Lang accepted Amber's proposal.

CONCLUSION

As Amber's example illustrates, expertise at defining objectives is an indispensable element in the ability to write successfully at work. Without the insights she gained by defining the objectives of her memo to Mrs. Lang, Amber would have had a much smaller chance of persuading Mrs. Lang to alter the organization's procedures. By following this chapter's guidelines, you will be able to develop a similar level of expertise in this critical part of the reader-centred writing process.

Note, by the way, that you should follow these guidelines throughout your work on a communication, not just as your first step. As you plan, develop, and evaluate, you will often refine your sense of what you want to accomplish and deepen your understanding of your readers. Make good use of what you learn. Let your new insights guide your future work, even when this means modifying your objectives and, therefore, changing a plan or draft you've already created.

EXERCISES

For additional exercises, visit www.techcomm.nelson.com.

Expertise

1. Find an example of a communication you might write in your career. Following the guidelines in this chapter, define its objective. Be sure to identify each of the following items:

   - The readers and their characteristics
   - The stakeholders and the ways they might be affected by the communication
   - The final result the writer desires
   - The communication's usability goals
   - The communication's persuasive goals

   Then, explain how the communication's features have been tailored to fit its objectives. If you can think of ways the communication might be improved, make recommendations.

2. Using the worksheet shown in Figure 2.2 (page 36), define the objectives of an assignment you are preparing for your technical communication class. To download a copy of the worksheet, visit Chapter 2 at www.techcomm.nelson.com.

Online

Find a website that could be used as a resource for a person in the profession for which you are preparing. Describe the target readers, usability objectives, and persuasive objectives that the website's creators might have had for it. Evaluate their success in achieving these objectives.

Collaboration

Working with another student, pick a technical or scientific topic that interests you both. Next, one of you should locate an article on the topic in a popular magazine such as *Wired,* and the other should locate an article on the topic in a professional or specialized journal. Working individually, you should each study the ways your article has been written so that its target audience will find the article to be usable and persuasive. Consider such things as the way your article opens, the language used, the types of details provided, and the kind of visuals included. Next, meet to compare the writing strategies used to meet the needs and interests of the two audiences. Present your results in the way your professor requests.

Ethics

A variety of websites present case studies that describe ethical issues that arise in business, engineering, science, and other fields. Locate, read, and respond to one such case. For links to ethics cases, visit Chapter 2 at www.techcomm.nelson.com.

Reference Guide 1: Six Patterns for Organizing

CONTENTS

This Reference Guide describes six patterns for organizing information and arguments. If you are writing a brief communication, you might use only one of these patterns to arrange your entire message. In longer communications, the patterns are typically woven together, with different ones employed in different paragraphs or sections. Your pattern of organization will depend on how you have defined your objectives.

WWW For additional resources, visit Reference Guide 1 at www.techcomm.nelson.com.

In this Reference Guide, each pattern is discussed separately so that you can quickly access advice about the one you need to use for the specific communication or part of a communication that you are preparing. A final section discusses the ways that several patterns can be integrated and coordinated in longer communications.

Patterns for Organizing

CLASSIFICATION (GROUPING FACTS)

When you are writing on the job, you will sometimes have to deal with what seems to be a miscellaneous set of facts. Your challenge will be to arrange them in a way that will be meaningful and useful to your readers.

One way to address this challenge is to use a strategy called *classification,* in which you arrange your material into parallel groups of related items. Your goal would be to create the groups so that they meet the following criteria:

Criteria for classification

- **Every item has a place.** In one group or another, every item fits.
- **Every item has only one place.** If there were two logical places for an item, either the communication would be redundant because the item would be mentioned twice or readers would have to guess which of the two locations really presented the item.
- **The groupings are useful to your readers.** Items that readers will use together are grouped together.

There are two types of classification patterns: formal and informal.

FORMAL CLASSIFICATION

Formal classification involves use of a *principle of classification.*

In formal classification, you group items according to a *principle of classification*—that is, according to some observable characteristic that every item in the group possesses. For example, Amanda was asked to write a brochure that tells consumers about sixty adhesives manufactured by her employer, a chemical company. As a principle of classification, she could use any characteristic possessed by all the adhesives, such as price, colour, and application (that is, whether it is used to bond wood, metal, or ceramic). To decide which of these possible principles would be most useful to her readers, she focused on the way consumers would use her brochure. Realizing that they would be looking for the adhesive that was best for the particular job they were doing, Amanda organized around the type of material each adhesive was designed to bond.

When you are organizing a large set of facts, you can divide your groups into subgroups. For instance, within her sections on adhesives for wood, metal, and ceramic, Amanda decided to subdivide the adhesives according to price. Where appropriate, you can subdivide again and again to create a multilevel, reader-centred hierarchy.

Guidelines for Formal Classification

1. **Choose a principle of classification that is suited to your readers and your purpose.** Follow Amanda's example in considering your readers' needs.
2. **Use only one principle of classification at a time.** In order to create a hierarchical organization that has only one place for each item, you must use only one principle of classification at a time. For example, you might classify the cars owned by a large corporation as follows:

Valid classification
- Cars built in Canada
- Cars built in other countries

In this grouping, you use only one principle of classification—the country in which the car was manufactured. Because each car was built in only one country, each would fit into only one group. Suppose you classified the cars this way:

Faulty classification
- Cars built in Canada
- Cars built in other countries
- Cars that are expensive

This classification is faulty because two principles are being used simultaneously—country of manufacture and cost. An expensive car built in Canada fit into two categories.

Of course, you can use different principles at different levels in a hierarchy:

Corrected classification
- Cars built in Canada
  - Expensive ones
  - Inexpensive ones
- Cars built in other countries
  - Expensive ones
  - Inexpensive ones

In this case, an expensive car built in Canada would have only one place at each level of the hierarchy.

In classification segments, writers often use tables to make their ideas clear to readers. See pages 432–37.

Figure PO.1 shows a passage organized according to formal classification. It describes various methods of detecting coronary heart disease. The writer chose to organize around the extent to which the methods require physicians to introduce something into the

www.techcomm.nelson.com

FIGURE PO.1

Passage Organized According to Formal Classification

The principle of classification is the extent to which the methods involve placing something in the person's body ("invading" it).

The first group of methods involves no invasiveness.

The second group does involve invasiveness.

The second group is subdivided according to the amount of invasiveness.

Detecting Coronary Artery Disease

Coronary artery disease (CAD) is the leading cause of death in industrialized countries. For this reason, the early detection of CAD has long been regarded as among the most vital areas of medical research, and several moderately invasive diagnostic methods have been developed.

Non-invasive methods gather diagnostic information without introducing anything into a patient's body. These methods include traditional physical examinations and history taking, electrocardiography, and echocardiography (ultrasonic imaging).

In contrast, invasive methods involve introducing a substance or object into the person's body. A moderately invasive method is the thallium test, in which a compound of radioactive thallium-201 is injected into the patient and then distributes itself throughout the myocardium (heart muscle) in proportion to the myocardial blood flow. The low-flow regions are detectable as cold spots on the image obtained from a radioisotope camera set over the chest. Although quite sensitive and accurate, the thallium test is also costly and time-consuming.

As far as invasive techniques are concerned, the most reliable way to diagnose CAD is by means of cardiac catheterization, in which a catheter is inserted into a large artery (usually in the upper arm or thigh) and advanced to the heart. Once the catheter is positioned in the heart, a radio-opaque dye is released through it, making it possible to observe the condition of the coronary arteries with X-rays. Although it produces excellent images and definitive diagnoses, cardiac catheterization is expensive, painful, and time-consuming—and it carries an element of risk. For these reasons, an equally accurate, non-invasive method for early detection of coronary artery disease is greatly to be preferred.

[The article continues with a description of a new technology that may provide a highly accurate, non-invasive early detection method.]

patient's body. The writer selected this principle of classification because he wants to focus on a new method whose advantage is that nothing needs to be placed in the patient's body.

INFORMAL CLASSIFICATION

Calvin must organize items where there is no objective principle of classification.

In some situations, it is impossible or undesirable to classify according to objective characteristics. Consider, for example, the organization of a report Calvin needed to write. Calvin's employer asked him to analyze advertisements appearing in several trade journals in the heavy equipment industry. Calvin might have classified the ads according to such objective characteristics as size of ad or number of words of copy. However, because he knew that his employer wanted to begin advertising in those journals and wanted some advice about ad design, Calvin decided to classify the ads according to the type of advertising appeal used. Obviously, "type of advertising appeal" is not an objective characteristic. It requires subjective interpretation and judgment. When writers classify facts without using objective characteristics as their principle of classification, they are said to be using *informal classification.*

Guidelines for Informal Classification

1. **Group your items in a way that is suited to your readers and your purpose.** Calvin organized his analysis around "type of advertising appeal" because he knew his employer was looking for advice about the design of ads.
2. **Use parallel groups at each level.** For instance, if you were classifying advertisements, you wouldn't use a list of categories like this:

   Categories not parallel with the others

   Focus on price
   Focus on established reputation
   Focus on advantages over a competitor's product
   Focus on one of the product's key features
   Focus on several of the product's key features

   The last two categories are at a lower hierarchical level than the other three. To make the categories parallel, you could combine them in the following way:

   Categories correctly subordinated to a lower level

   Focus on price
   Focus on established reputation
   Focus on advantages over a competitor's product
   Focus on the product's key features
     Focus on one key feature
     Focus on several key features

3. **Avoid overlap among groups.** Even when you cannot use strict logic in classifying items, strive to provide one and only one place for each item. To do this, you must avoid overlap among categories. For example, in the following list the last item overlaps the others because photographs can be used in any of the other types of advertisements listed.

   Final item overlaps with the others

   Focus on price
   Focus on established reputation
   Focus on advantages over a competitor's product
   Focus on the product's key features
   Use of photographs

WWW To see other examples of informal classification, visit Reference Guide 1 at www.techcomm.nelson.com.

Figure PO.2 shows the outline of a research report organized according to informal classification. The writers describe the ways investigators should go about identifying water plants that might be cultivated, harvested, and dried to serve as fuel for power generators. Because the actual selection of plants would require the expertise of specialists from many fields, the writers have organized their communication in a way that

FIGURE PO.2

Outline of a Report Organized According to Informal Classification

IDENTIFYING EMERGENT AQUATIC PLANTS THAT MIGHT BE USED AS FUEL FOR BIOMASS ENERGY SYSTEM

Introduction

Botanical Considerations
- Growth Habitat
- Morphology
- Genetics

Physiological Considerations
- Carbon Utilization
- Water Utilization
- Nutrient Absorption
- Environmental Factors Influencing Growth

Chemical Considerations
- Carbohydrate Composition
- Crude Protein Content
- Crude Lipid Content
- Inorganic Content

Agronomic Considerations
- Current Emergent Aquatic Systems
  - *Eleocharis dulcis*
  - *Ipomoea aquatica*
  - *Zizania palustris*
  - *Oryza sativa*
- Mechanized Harvesting, Collection, Densification, and Transportation of Biomass
- Crop Improvement
- Propagule Availability

Ecological Considerations
- Water Quality
- Habitat Disruption and Development
- Coastal Wetlands

Economic Considerations
- Prior Research Efforts
  - *Phragmites communis*
  - *Arundo donax*
  - Other Research
- Production Costs for Candidate Species
  - Planting and Crop Management
  - Harvesting
  - Drying and Densification
  - Total Costs
- End Products and Potential Competition

Selection of Candidate Species

helps people from each field find the information they need: There is a section on chemistry for chemists, one on economics for economists, and so on.

DESCRIPTION OF AN OBJECT (PARTITIONING)

At work, you will often need to describe a physical object for your readers. If you write about an experiment, for example, you may need to describe your equipment. If you write instructions, you may need to describe the machines your readers will be using. If you propose a new purchase, you may need to describe the object you want to buy.

To organize such descriptions, you can use a strategy called *partitioning.* In partitioning, you divide your subject into its major components and (if appropriate) divide those into their subcomponents.

HOW PARTITIONING WORKS

Partitioning is a form of classification.

Partitioning involves the same basic procedure as classification. You think of the object as a collection of parts and use some principle to identify groups of related parts. Most often, that principle is either function or location. Consider, for instance, how each of those two principles might be used to organize a discussion of the parts of a car.

Partitioning by location

If you organize your discussion by location, you might talk about the interior, the exterior, and the underside. The interior would be those areas under the hood, in the passenger compartment, and in the trunk. The exterior would be the front, back, sides, and top. The underside would include the parts under the body, such as the wheels and transmission.

Partitioning by function

However, if you were to organize your discussion by function, you might focus on parts that provide power and parts that guide the car. The parts that provide power are in several places (the gas pedal and gear shift are in the passenger compartment; the engine is under the hood; and the transmission, axle, and wheels are on the underside). Nevertheless, you could discuss them together because they are related by function.

Of course, other principles of classification are possible. You could organize your description of the parts of a car according to the materials they are made from—a classification that could be useful to someone looking for ways to decrease the cost of materials in a car. You could also partition the parts of a car according to the countries in which they were manufactured—a classification that could be useful to an economist studying the effects of import tariffs or quotas on Canadian manufacturers.

Guidelines for Partitioning an Object

1. **Choose a basis for partitioning suited to your readers and purpose.** For instance, if you are trying to describe a car for new owners who want to learn about what they have purchased, you might organize according to location so that your readers will learn about the comforts and conveniences available to them when they are sitting in the passenger compartment, using the trunk, and so on. In contrast, if you are describing a car for mechanics who will have to diagnose and correct problems, you would organize your description according to function.

Photographs and drawings can help you describe objects clearly to your readers. See pages 444–51.

2. **Use only one basis for partitioning at a time.** To ensure that you have one place and only one place for each part you describe, use only one basis for partitioning at a time, just as you use only one basis for classifying at a time (see page 41).
3. **Arrange the parts of your description in a way your readers will find useful.** For instance, if you are partitioning by location, you might move systematically from left to right, front to back, or outside to inside. If you are partitioning by function, you might treat the parts in the order in which they function in an activity of interest to your readers.

EXTENDED DESCRIPTIONS OF AN OBJECT

WWW To see other examples of partitioning, visit Reference Guide 1 at www.techcomm.nelson.com.

In some situations, you may need to describe an object in an extended fashion. For instance, in some instruction manuals, it's necessary to provide detailed information about the construction and function of the various parts of the equipment being used. Similarly, product descriptions sometimes involve detailed discussions that are organized by partitioning. In addition to information about the parts of the object, such descriptions may include a title, an introduction, a drawing or other graphic, and a conclusion.

Figure PO.3 (page 48) presents an example of an extended description of an object.

DESCRIPTION OF A PROCESS (SEGMENTATION)

A description of a process explains the relationship of events over time. You may have either of two purposes in describing a process:

Purposes of process descriptions

- **To enable your readers to *perform* the process.** For example, you may be writing instructions that will enable your readers to make a photovoltaic cell, apply for a loan, or run a software program.
- **To enable your readers to *understand* the process.** For example, you might want your readers to understand the following:
  - **How something is done.** For instance, how coal is transformed into synthetic diamonds.
  - **How something works.** For instance, how the lungs provide oxygen to the bloodstream.
  - **How something happened.** For instance, how Canada developed the IMAX motion picture projection system.

In either case, you want to explain the overall structure of the process to your readers. To do that, you *segment* the process. You begin with the list of steps or events involved in the process, and then you separate those steps or events into related groups. If the process is long enough, you may divide those groups of steps into subgroups, thereby creating an organizational hierarchy.

Use a principle of classification to group the steps.

As with partitioning, you need a principle of classification to guide you in segmenting a process. Commonly used principles include the *time* when the steps are performed (first day, second day; spring, summer, fall), the *purpose* of the steps (to prepare the equipment, to examine the results), and the *tools* used to perform the steps (for example, the function keys on a computer keyboard, the numeric pad, and so on).

FIGURE PO.3

Extended Description of an Object (Mechanism)

ADAPTIVE AIRCRAFT WINGS

Overall topic is introduced.

Until now, commercial airplanes have relied on rigid wings designed for a compromise between the requirements of take-off, landing, and cruising. In an effort to increase fuel efficiency, scientists at Daimler-Benz and the German Aerospace Research Center are experimenting with an "adaptive wing" that is capable of changing its geometry automatically during flight so that the best possible performance is always achieved. This new concept involves four major components: a sensor network, processor, intelligent flap, and adjustable contour region.

Forecasting statement indicates the organization of the rest of the description.

Figure helps to explain the components and their relationship.

Adaptive Wing

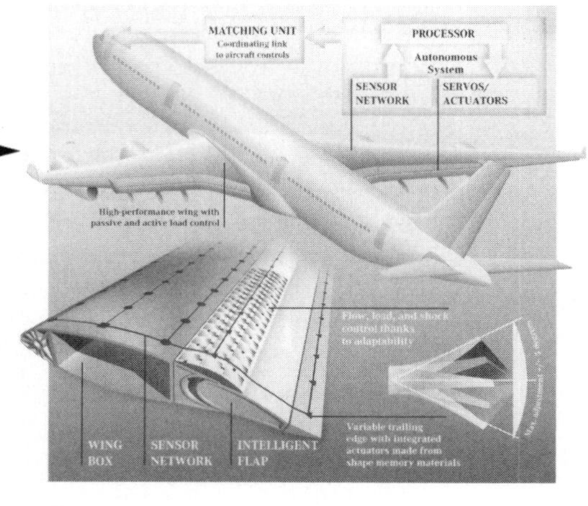

Sensor Network

First major component is described: construction and operation.

The sensor network consists of dozens of sensors located at widely differing points under the wing skin. Throughout a flight, they continuously collect information about such things as airflow, laminar flow length, possible turbulence, and pressure distribution.

Processor

Second major component is described: construction and operation.

Information from the sensor network is fed to a special computer processor, which is linked to—but separate from—the flight management computer through which the pilot flies the plane. The special processor calculates the ideal geometry for the wing, taking into account the plane's desired speed, altitude, and actual weight as provided by the pilot and the flight management computer. Once the processor has determined the ideal wing shape, it transmits the necessary instructions to the two remaining components of the system: the intelligent flap and the moveable contours.

Figure PO.3
(*continued*)

Third major component is described: construction, operation, benefits.

Intelligent Flap

Located at the rear of a wing, flaps are currently extended only during the take off and landing. The adaptive wing involves two important flap modifications. First, it puts the flaps into use throughout the flight, extending them in the manner calculated by the processor. Second, the flaps on the adaptive wing have a deformable trailing edge. This edge is constructed with small electrical actuators integrated into an elastic, shape-memory material. According to the amount of current that is applied, these materials change shape so that the trailing edge of the flap can be "tuned" to meet the demands of specific flight conditions.

A major benefit of the intelligent flap is that it could provide maximum lift with minimum drag for all flight situations, thereby saving fuel. Another benefit is an increased margin for safety during extreme in-flight maneuvers. For example, should a sudden loss of cabin pressure necessitate a rapid descent to an altitude where there is enough oxygen for breathing, the intelligent flap can reduce stressful loads on the wings. The intelligent flap can also alleviate the severity of the problems encountered when a plane flies into the vortex of turbulent air caused by another jet's engines.

Fourth major component is described: construction, operation, benefits.

Adjustable Contour Region

Midway back on its upper surface, the adaptive wing has an adjustable contour region that can be adjusted to lie smooth or bulge, thereby changing the wing's shape. Based on information provided by the sensor network, the processor would instruct actuators to thicken the upper surface of the wing at specific areas where doing so would increase the wing's aerodynamic efficiency. Scientists have identified two means of accomplishing this change in wing shape. First, tensioning elements placed under the wing's skin could act on a precalculated area of the wing like a bicep, causing the skin to bulge as they contract. Second, small electrical actuators integrated into an elastic wing surface could provide the same effect.

The advantage of the adjustable contour region is that it would weaken the effect of the "compression wave" that forms at the leading edge of a jet's wing. Even when an aircraft flies below the speed of sound, the convex profile of the upper wing surface forces air flowing over it to travel at supersonic speeds, creating this compression wave, which exerts draft on the wing.

Conclusion ties the entire segment together by highlighting benefits.

Conclusion

Although the adaptive wing is still under development, Project Leader Dr. Joseph Mertens estimates it will reduce wing resistence by fifteen percent. The benefits could snowball from there, he adds. For example, less drag means less fuel consumption, which in turn means less weight of fuel to be transported, which itself results in lower fuel consumption.

Guidelines for Segmenting

1. **Choose a principle for segmenting suited to your readers and your purpose.** If you are writing instructions, group them in ways that support an efficient or comfortable rhythm of work. If you are trying to help your readers understand a process, organize around concerns that are of interest or use to your readers.
2. **Make your smallest groupings manageable.** One of the most important things for you to do when you are segmenting a process is to make your groupings the smallest manageable. If they include too many steps or too few, your readers will not see the process as a structured hierarchy of activities or events but as a long, unstructured list of steps—first one, then the next, and so on. If you are writing instructions, the lack of structure will make it harder for your readers to learn the task. If you are describing a process, the lack of structure will make the process more difficult for them to comprehend.
3. **Make clear the relationships among the steps.** Finally, keep in mind that your segments should help your readers understand and remember the entire process. That means that readers will need to understand the relationships among the events and steps that make up the process.

Flowcharts and diagrams can help you describe processes clearly to your readers. See pages 451–54.

There are many ways of making those relationships clear. Where they are obvious, you can simply provide informative headings. At other times, you will need to explain them in an overview at the beginning of the segment, weave additional explanations into your discussions of the steps themselves, and explain the relationships again in a summary at the end.

SAMPLE PROCESS DESCRIPTIONS

WWW To see other examples of segmenting, visit Reference Guide 1 at www.techcomm.nelson.com.

Figure PO.4 shows a sample set of instructions that uses segmenting to identify the major parts of the process. Figure PO.5 (page 52) shows a group of paragraphs that describe one theory about the process by which planets are formed. Notice how the writer uses headings to help signal the major phases of the process.

COMPARISON

Purposes of comparisons

At work, you will often be writing comparisons between two or more things. Such occasions fall into two categories:

- **You want to help your readers make a decision.** The workplace is a world of choices. People are constantly choosing among courses of action, competing products, alternative strategies. To help them choose, you will often write comparisons.
- **You want to help your readers understand something by means of an analogy.** One of the most effective ways to help your readers understand something new is to explain the ways in which it resembles—or differs from—something they are familiar with.

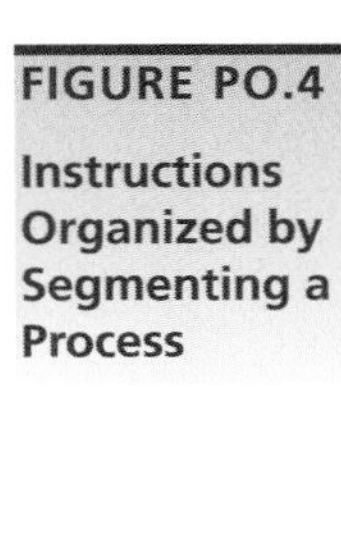

FIGURE PO.4

Instructions Organized by Segmenting a Process

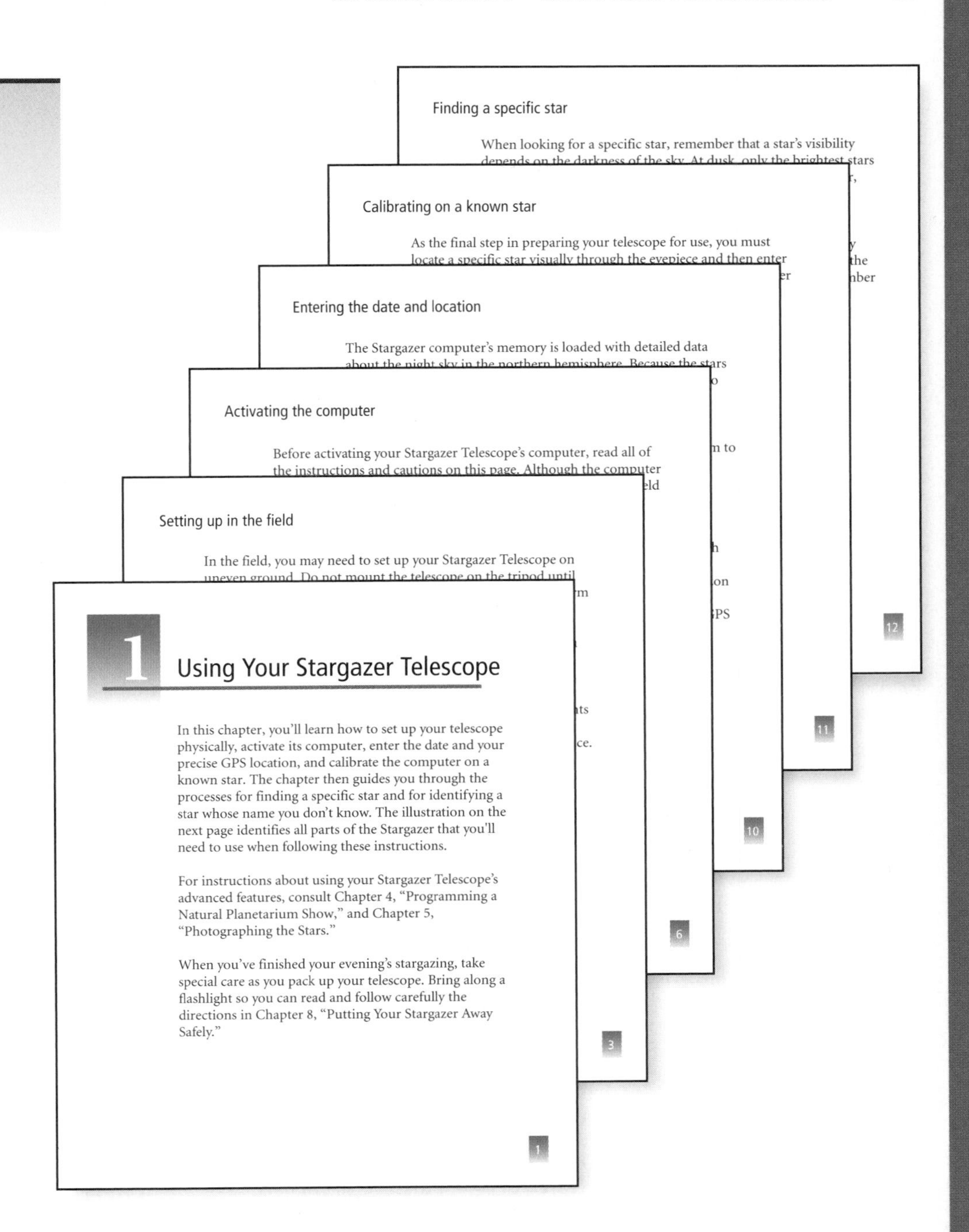

HOW COMPARISONS WORK

In some ways, a comparison is like a classification. You begin with a large set of facts about the things you are comparing and then group the facts around points that enable your readers to see how the things are like or unlike one another.

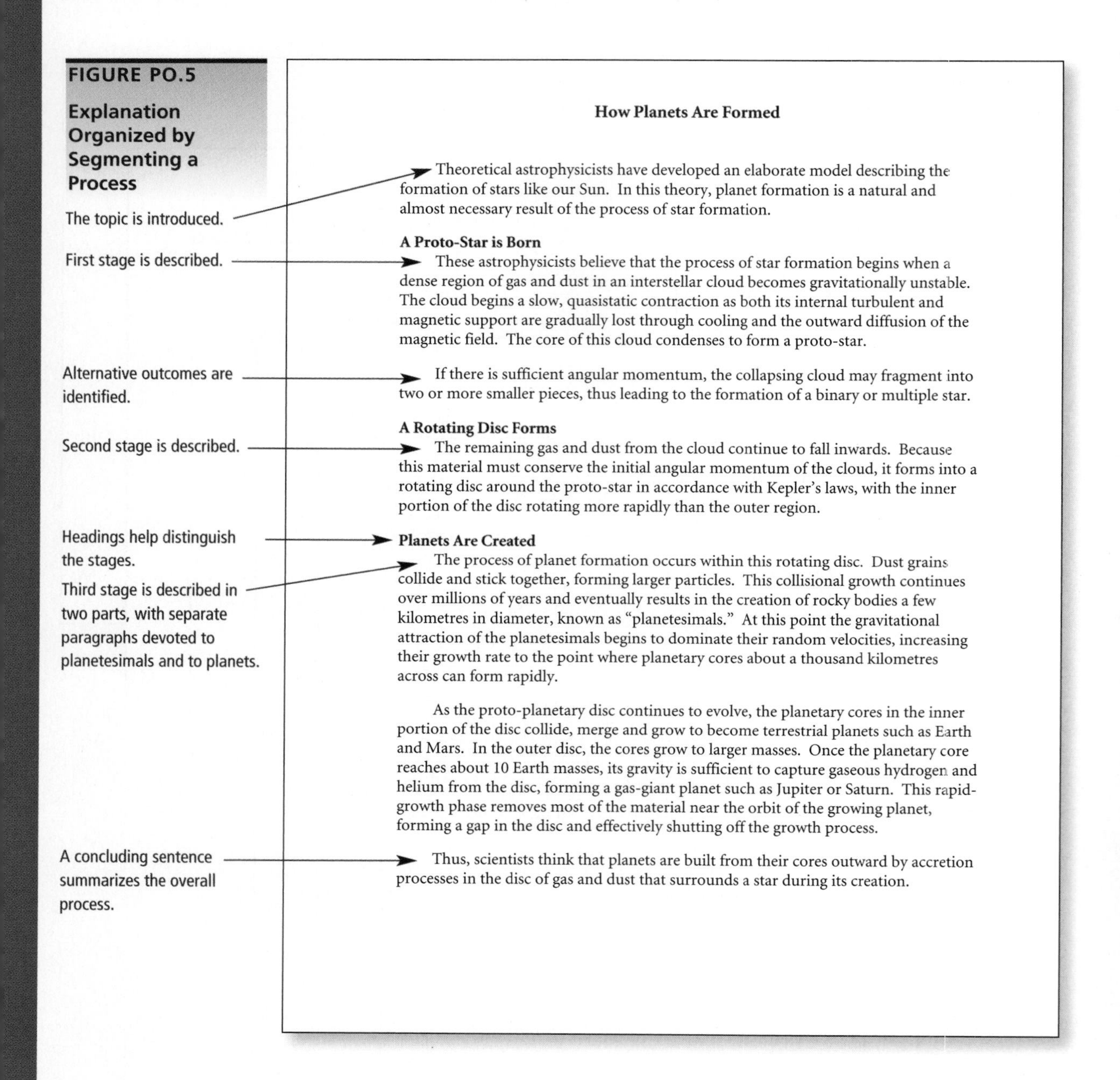
FIGURE PO.5

Explanation Organized by Segmenting a Process

The topic is introduced.

First stage is described.

Alternative outcomes are identified.

Second stage is described.

Headings help distinguish the stages.

Third stage is described in two parts, with separate paragraphs devoted to planetesimals and to planets.

A concluding sentence summarizes the overall process.

How Planets Are Formed

Theoretical astrophysicists have developed an elaborate model describing the formation of stars like our Sun. In this theory, planet formation is a natural and almost necessary result of the process of star formation.

A Proto-Star is Born

These astrophysicists believe that the process of star formation begins when a dense region of gas and dust in an interstellar cloud becomes gravitationally unstable. The cloud begins a slow, quasistatic contraction as both its internal turbulent and magnetic support are gradually lost through cooling and the outward diffusion of the magnetic field. The core of this cloud condenses to form a proto-star.

If there is sufficient angular momentum, the collapsing cloud may fragment into two or more smaller pieces, thus leading to the formation of a binary or multiple star.

A Rotating Disc Forms

The remaining gas and dust from the cloud continue to fall inwards. Because this material must conserve the initial angular momentum of the cloud, it forms into a rotating disc around the proto-star in accordance with Kepler's laws, with the inner portion of the disc rotating more rapidly than the outer region.

Planets Are Created

The process of planet formation occurs within this rotating disc. Dust grains collide and stick together, forming larger particles. This collisional growth continues over millions of years and eventually results in the creation of rocky bodies a few kilometres in diameter, known as "planetesimals." At this point the gravitational attraction of the planetesimals begins to dominate their random velocities, increasing their growth rate to the point where planetary cores about a thousand kilometres across can form rapidly.

As the proto-planetary disc continues to evolve, the planetary cores in the inner portion of the disc collide, merge and grow to become terrestrial planets such as Earth and Mars. In the outer disc, the cores grow to larger masses. Once the planetary core reaches about 10 Earth masses, its gravity is sufficient to capture gaseous hydrogen and helium from the disc, forming a gas-giant planet such as Jupiter or Saturn. This rapid-growth phase removes most of the material near the orbit of the growing planet, forming a gap in the disc and effectively shutting off the growth process.

Thus, scientists think that planets are built from their cores outward by accretion processes in the disc of gas and dust that surrounds a star during its creation.

Lauren must organize her data so that her readers can make their decision easily.

When writing a comparison, you can choose from two basic patterns: *alternating pattern* and *divided pattern.* Both include the criteria (or points of comparison) and information about each alternative in terms of each criterion. Consider, for example, Lauren's situation. Lauren works for a steel mill that has decided to build a new blast furnace. She has been asked to study the two types of furnaces the mill is considering and then present her results in a report that will help the mill's upper management decide which furnace to construct. Having amassed hundreds of pages of information, Lauren must now decide how to organize the body of her report.

For organizing her comparison, Lauren has the two basic choices: the divided pattern and the alternating pattern.

Divided Pattern	Alternating Pattern
Furnace A	**Cost**
Cost	Furnace A
Efficiency	Furnace B
Construction time	**Efficiency**
Air pollution	Furnace A
Et cetera	Furnace B
Furnace B	**Construction time**
Cost	Furnace A
Efficiency	Furnace B
Construction time	**Air pollution**
Air pollution	Furnace A
Et cetera	Furnace B
	Et cetera
	Furnace A
	Furnace B

The alternating pattern is best for point-by-point comparisons.

As you can see, the alternating pattern is organized around the criteria, making it ideal for readers who want to make point-by-point comparisons. After thinking about how her readers would use her long report on the blast furnaces, Lauren selected this pattern, which allows her readers to find the information about the costs or efficiency of both blast furnaces without the flipping back and forth that the divided pattern would require.

The divided pattern works best when readers want to size up all facts about an alternative at once.

The divided pattern is well suited to situations where readers want to see all the information about each alternative at one time. Typically, this occurs when both the general nature and the details of each alternative can be described in a short space—say, one page or so. The divided pattern was used, for instance, by an employee of a restaurant who was asked to investigate the feasibility of buying a new sound system for the restaurant. He described each of the systems in a single page.

Whether you use the alternating or divided pattern, you can assist your readers by incorporating two kinds of preliminary information:

- **Description of the criteria.** This information lets your readers know from the start what the relevant points of comparison are.
- **Overview of the alternatives.** This information provides your readers with a general sense of what each alternative entails before they focus on the details you provide.

In both patterns, the statement of criteria would precede the presentation of details.

To learn about graphics that can help you make comparisons clearly and persuasively, see the following discussions:

- tables, pages 432–37
- bar graphs, pages 437–40
- pictographs, pages 440–41
- pie charts, page 443

Guidelines for Writing Comparisons

1. **Choose points of comparison suited to your readers and your purpose.** When you are preparing comparisons for decision makers, be sure to include not only the criteria they consider important but also any additional criteria that you—with your expert knowledge—regard as significant.
   When you are writing comparisons to create analogies, be sure to compare and contrast only those features that will help your readers understand the points you are trying to make. Avoid comparing extraneous details.
2. **Arrange the parts of complex comparisons hierarchically.** For example, group information on all aspects of cost (purchase price, operating cost, maintenance cost, and so on) in one place, information on all aspects of performance in another place, and so on.
3. **Arrange the parts in an order your readers will find helpful.** When your comparison is intended to help readers make a decision, lead off with the criteria that reveal the most significant differences between the things you are comparing. When your comparison is designed to aid understanding, discuss points of similarity first. In this way, you begin with what your readers will find familiar and then lead them to the less familiar.

WWW To see other examples of comparison, visit Reference Guide 1 at www.techcomm.nelson.com.

SAMPLE COMPARISON

Figure PO.6 shows a segment that compares two methods of data collection.

CAUSE AND EFFECT

Purposes of cause-and-effect explanations

You may use the cause-and-effect strategy in two different ways:

- **To help your readers understand the cause or consequences of some action or event.** For example, you may need to explain the causes of solar flares, earthquakes, or the birth of identical twins.
- **To persuade.** For example, you might need to persuade your readers that the damage to a large turbine generator (effect) resulted from metal fatigue in a key part (cause) rather than from a failure to provide proper lubrication. Or you might try to persuade your readers that cutting the selling price of a product (cause) will increase sales and produce a greater profit (effect).

FIGURE PO.6

Passage Organized by Comparison

The need for knowledge of data collection methods is explained.

Comparison of two types of survey methods is announced.

First type is discussed; strength, then weakness implied (that of pattern reproduction).

Second type is discussed; strength, then weakness of one type of direct method.

DSS (DEMOGRAPHIC SURVEILLANCE SYSTEMS)
METHODS OF DATA COLLECTION

Knowledge of the methods for collecting or compiling data at the DSS sites is essential because these methods influence the ways that data are processed, analyzed, and interpreted. The most common demographic methods used in data collection are censuses, sample surveys, and vital-events registration systems. The last method, however, is nonexistent or only partially applied in many developing countries. Given the paucity of vital-events registration and knowledge on population or health status trends in such settings, demographic and health surveys have been introduced for health planning, practice, evaluation, and allocation of resources. Demographic estimates undertaken in developing countries have employed both indirect and direct methods, using retrospective single-round surveys and prospective multiround ones (Tablin 1984).

Indirect estimation methods rely on information obtained from subjects not directly at risk of a particular demographic phenomenon. The indirect methods can be used to estimate levels and trends of fertility, mortality, and migration where data sources are defective or incomplete. An example of an indirect method is the estimation of infant and child mortality from proportions of surviving children or the estimation of adult mortality from those orphaned. Indirect estimation methods are also used to assess data collected using conventional methods. Such data are compared with other information to infer a certain pattern, on the basis of certain assumptions. If this pattern is reproduced, then data can be further inferred. Indirect estimation may, in addition, involve the fitting of demographic models to fragmentary and incomplete data (Pressat 1985). The results obtained are used to estimate a particular parameter.

Direct methods use data on the people at risk to establish a demographic measure and pattern. These methods rely on data obtained from censuses, surveys, and recorded data on the components of change—that is, births, deaths, and migration. Data obtained from these methods are used directly to provide estimates of demographic phenomena such as fertility, mortality, and migration. An example of a direct method is the use of the number of children born to women of a particular age group to estimate age-specific fertility rates.

In single-round surveys, a population is enumerated once during a survey, and retrospective data are gathered on past events (Kpedekpo 1982; Tablin 1984; Newell 1994), such as a birth or death that occurred in the past year (or a life and maternity history). This method may result in overestimation or underestimation of events, as a result of memory lapse. Respondents may exclude events from the reference period. It has been argued that an underestimation of 30–40% is likely using this method (Tablin 1984). Some examples of single-round surveys are the World Fertility Survey and the Demographic and Health Surveys.

To learn about graphics that can help you describe cause and effect clearly and persuasively, see the discussion of line graphs, pages 441–43.

Guidelines for Describing Causes and Effects

1. **Begin by identifying the cause or effect that you are going to describe.** Your readers will want to know from the beginning of your segment exactly what you are trying to explain so that they will know what they should be trying to understand as they read it. Sometimes a single sentence will be enough. At other times you may need several sentences.
2. **Carefully explain the links in the chain of cause and effect that you are describing.** Remember that you are not simply listing the steps in a process. You want your readers to understand how each step leads to the next step or is caused by the preceding step.
3. **If you are dealing with several causes or effects, group them into categories.** Categories help readers to understand a complex chain of events.

Guidelines for Persuading Readers to Accept Your View of Cause and Effect

1. **State your claim at the beginning of your passage.** Your claim will be that some particular effect was created by some particular cause, or that some particular cause will lead to some particular effect.
2. **Present your evidence and lines of reasoning.** Where possible, focus on undisputed evidence because your readers' willingness to agree with you depends largely on their willingness to accept your evidence. Use lines of reasoning that your readers will accept as logically sound and appropriate to the situation.
3. **Anticipate and respond to objections.** In cause-and-effect segments, as in any persuasive segment, your readers may object to your evidence or to your line of reasoning. Be particularly careful to avoid the *post hoc, ergo propter hoc* fallacy. In this form of faulty reasoning, a writer argues that *because* an event occurred after another event, it was *caused* by that event. Consider this example. In an attempt to persuade his employer, a furniture company, to use computerized machinery for some of its manufacturing operations, Clayton argued that a competitor's profits had risen substantially after it made that move. Clayton's boss pointed out that the increase in sales might have been caused by other changes made over the same period, such as new designs or a reconfiguration of its sales districts. To persuade his boss that the computerization had caused the increase in sales, Clayton would have to do more than simply state that it had preceded the increase.

Avoid the *post hoc, ergo propter hoc* fallacy.

SAMPLE CAUSE-AND-EFFECT SEGMENT

Figure PO.7 shows a segment in which the writer explains one theory about the cause of the extinction of dinosaurs.

WWW To see other examples of cause-and-effect organization, visit Reference Guide 1 at www.techcomm.nelson.com.

FIGURE PO.7

Passage Organized around Cause and Effect

Effect to be explained is announced.

Possible cause is announced.

Link between effect and cause is explained: asteroid created dust cloud that killed dinosaurs.

Evidence of link is presented: rare molecules in sediment indicate an asteroid may have hit Earth when dinosaurs died.

Additional evidence of link.

WHAT CAUSED THE DEATH OF THE DINOSAURS?

One theory is that a comet, asteroid or other huge extraterrestrial body slammed into the Earth 65 million years ago and ended the 160-million year reign of the dinosaurs. According to this theory, the extraterrestrial body raised a huge dust cloud. Within days the black cloud spread over the Earth, darkening the sun. The air turned cold, and many dinosaurs died. Snow fell. Freezing darkness gripped the Earth for weeks. Plants, cut off from the sunlight that feeds them, couldn't survive. Without plants, the rest of the herbivorous dinosaurs followed, and the carnivores soon afterward. Along with a number of other species, the dinosaurs were gone forever.

Although many leading paleontologists and evolutionary biologists now accept the asteroid-impact theory, and despite popular accounts implying that the question is settled, it is not. A scattering of critics continue to challenge the whole notion.

Still, the theory is compelling. Every few months a new piece of evidence is added to the list, and most, to the critics' consternation, support the idea of an extraterrestrial impact.

Just recently, for example, scientists at the Scripps Institute of Oceanography, in La Jolla, California, found evidence of organic molecules in the layer of sediments laid down at the time the dinosaurs died; the molecules are exceedingly rare on Earth but relatively common in some meteorites and so, presumably, in some asteroids.

To put the discovery in perspective, and to appreciate the arguments on both sides of the impact debate, one must first understand the nature of the original finding.

In 1980, Luis Alvarez, a Nobel laureate in physics, his son Walter, a geologist, both at the University of California, Berkeley, and two associates published the theory that a massive impact took place at the end of the Cretaceous Period. The team had found a rare substance in the thin layer of sedimentary clay deposited just on top of the highest, and therefore the most recent, stratum of rock contemporary with those bearing dinosaur fossils. It was the element iridium, which is almost nonexistent in the Earth's crust but 10,000 times more abundant in extraterrestrial rocks such as meteorites and asteroids. Deposits above and below the clay, which is the boundary layer separating the Cretaceous layer from the succeeding Tertiary, have very little iridium.

(continues)

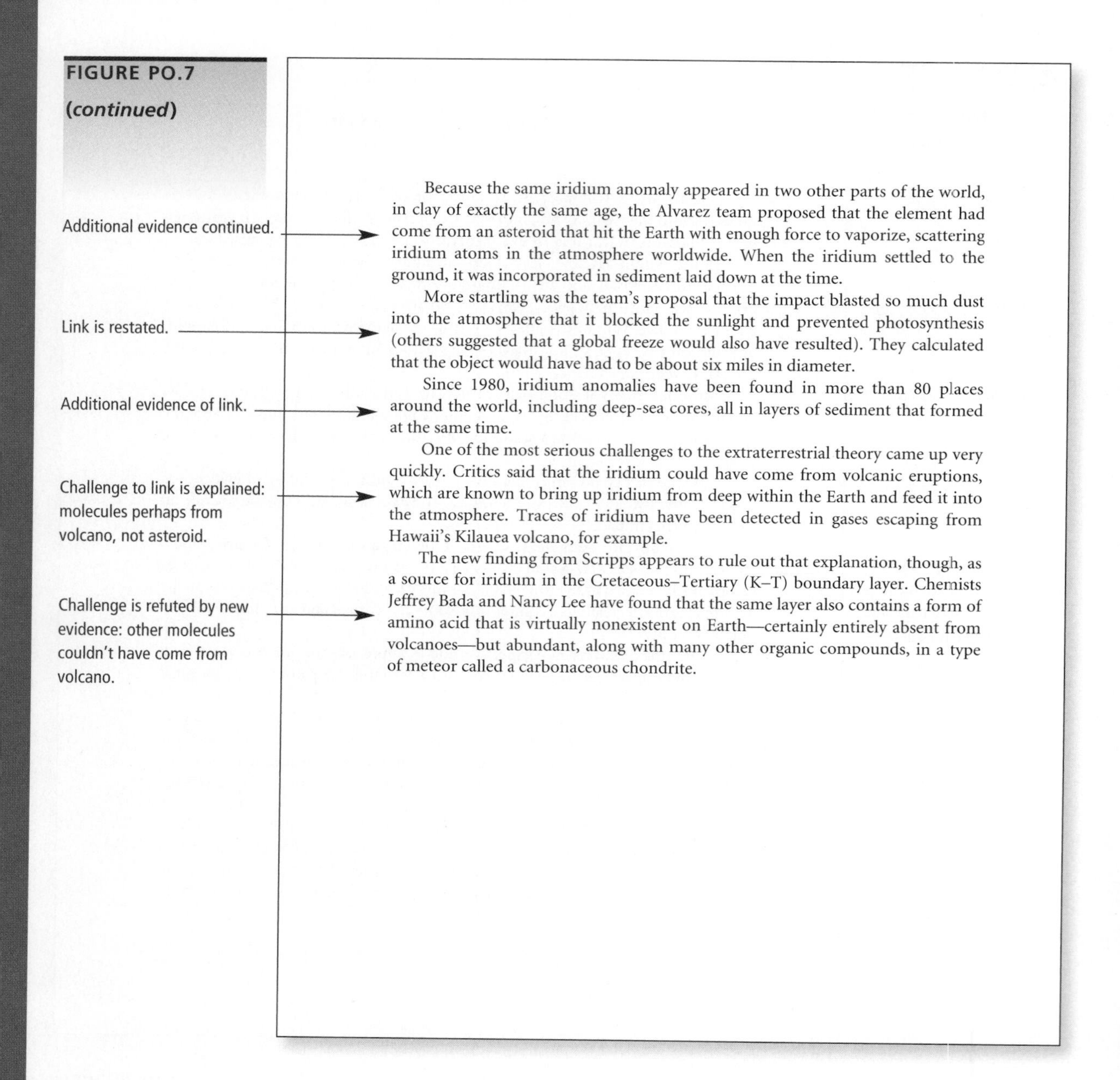

FIGURE PO.7 (*continued*)

Because the same iridium anomaly appeared in two other parts of the world, in clay of exactly the same age, the Alvarez team proposed that the element had come from an asteroid that hit the Earth with enough force to vaporize, scattering iridium atoms in the atmosphere worldwide. When the iridium settled to the ground, it was incorporated in sediment laid down at the time.

More startling was the team's proposal that the impact blasted so much dust into the atmosphere that it blocked the sunlight and prevented photosynthesis (others suggested that a global freeze would also have resulted). They calculated that the object would have had to be about six miles in diameter.

Since 1980, iridium anomalies have been found in more than 80 places around the world, including deep-sea cores, all in layers of sediment that formed at the same time.

One of the most serious challenges to the extraterrestrial theory came up very quickly. Critics said that the iridium could have come from volcanic eruptions, which are known to bring up iridium from deep within the Earth and feed it into the atmosphere. Traces of iridium have been detected in gases escaping from Hawaii's Kilauea volcano, for example.

The new finding from Scripps appears to rule out that explanation, though, as a source for iridium in the Cretaceous–Tertiary (K–T) boundary layer. Chemists Jeffrey Bada and Nancy Lee have found that the same layer also contains a form of amino acid that is virtually nonexistent on Earth—certainly entirely absent from volcanoes—but abundant, along with many other organic compounds, in a type of meteor called a carbonaceous chondrite.

PROBLEM AND SOLUTION

When you are talking about a problem and its solution, you may have either of two purposes:

- **To describe.** When talking about past events, you may want to describe the measures you, your coworkers, or your organization took to solve some problem that has now been eliminated.
- **To persuade.** When looking to the future, you may want to persuade your readers that the actions you are recommending will solve a problem they want to overcome.

Guidelines for Describing Problems and Their Solutions

1. **Begin by identifying the problem that was solved.** Make the problem seem significant to your readers and emphasize the aspects of the problem that were affected most directly by the solution.
2. **Explain the links between the problem and the solution.** You want your readers to understand how the problem was overcome by the solution.
3. **If the solution consisted of several actions, group them into categories.** The resulting hierarchy will help your readers understand the complex solution.

Guidelines for Persuading Readers to Accept Your Solution to a Problem

1. **Describe the problem in a way that makes it seem significant to your readers.** Remember that your aim is to persuade them to take the action you recommend. They will not be very interested in taking action to solve a problem they regard as insignificant.
2. **Present your evidence and indicate your line of reasoning.** Use evidence that your readers will find sufficient and reliable and a line of reasoning they will accept as logically sound and appropriate to the situation.
3. **Anticipate and respond to objections.** As with any segment designed to persuade, your readers may object to your evidence or your line of reasoning. Devote special attention to determining what those objections are so you can respond to them.

SAMPLE PROBLEM-AND-SOLUTION SEGMENT

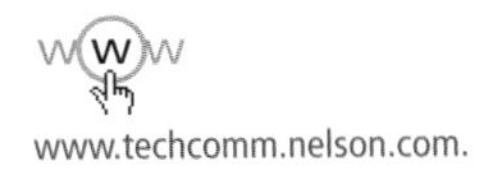

Figure PO.8 (page 60) shows a memo organized around a problem and a possible solution.

COMBINATIONS OF PATTERNS

Long communications usually mix several of the patterns described in this Reference Guide.

So far, this Reference Guide has treated the six organizational patterns in isolation from one another. In practice, they are often mixed with one another and with passages using other patterns.

This will happen, for instance, in a report that Justin is writing to tell decision makers in his organization about a new technique for applying coatings to bearings. Like many technical reports, his will employ a problem-and-solution pattern for its overall structure. First, he will describe the problem that makes the old technique undesirable for his

FIGURE PO.8

Memo Organized around a Problem and Its Solution

CANALLOY RESEARCH
Interoffice Memorandum

June 21, 2007

To Cory Gull

From Andy Mamalis

RE Suggestion to Investigate Kohle Reduktion Process for Steelmaking

As we have often discussed, it may be worthwhile to set up a project investigating steelmaking processes that could help the Canadian industry compete more effectively. I suggest we begin with an investigation of the Kohle Reduktion method, which I learned about in the April 2007 issue of *High Technology*.

Problem is identified. → A major problem for Canadian steelmakers is the process they use to make the molten iron ("hot metal") that is processed into steel. Problem is explained. → Relying on a technique developed on a commercial scale over 100 years ago by Sir Henry Bessemer, they make the hot metal by mixing iron ore, limestone, and coke in blast furnaces. To make the coke, they pyrolize coal in huge ovens in plants that cost over $100 million and create enormous amounts of air pollution.

Solution is announced. → In the Kohle Reduktion method, developed by Korf Engineering in Germany, the hot metal is made without coke. Solution is explained. → Coal, limestone, and oxygen are mixed in a gasification unit at 1400°C. The gas rises in a shaft furnace above the gasification unit, chemically reducing the iron ore to "sponge iron." The sponge iron then drops into the gasification unit, where it is melted and the contaminants are removed by reaction of the limestone. Finally, the hot metal drains out of the bottom of the gasifier.

Link between problem and solution is explained. → The Kohle method, if developed satisfactorily, will have several advantages. It will eliminate the air pollution problem of coke plants, it can be built (according to Korf estimates) for 25% less than conventional furnaces, and it may cut the cost of producing hot metal by 15%.

This technology appears to offer a dramatic solution to the problems with our country's steel industry: I recommend that we investigate it further. If the method proves feasible and if we develop an expertise in it, we will surely attract many clients for our consulting services.

employer, then he will describe the technique he has developed as a solution to that problem. Within this overall pattern, Justin will use many others, including the ones for describing an object (the equipment used in his technique), explaining a process (the way his technique works), and making a comparison (the performance of his new technique versus the performance of the current one). Similarly, when you write at work, you will

FIGURE PO.9

Outline of a Report That Interweaves Several Organizational Patterns

Like many technical communications, this report interweaves several organizational patterns.

To organize this part, the writers classified the companies according to their location.

They organized Chapter II around a problem and its possible solution.

When explaining why the solar cells collect heat, the writers organized by cause and effect.

The writers organized each of these subsections by partitioning the component into its major subparts.

They organized their discussion of the construction schedule by gathering the steps into related groups of steps (segmenting the process).

They organized the "Cost Comparison" section according to the alternating pattern for comparison (see page 53).

Solar Roofs for New Restaurants
A Feasibility Report for the Brendon's Restaurant Chain

I. Introduction
   A. The Brendon's Restaurant commissioned us to determine the feasibility of covering its restaurant roofs with photovoltaic cells (solar cells) that generate electricity
   B. Background
      1. Brendon's builds 30 new restaurants a year
      2. Brendon's is committed to environmental responsibility
      3. Federal government has goal of one million solar roofs in Canada by 2010
      4. Experiences of other companies
         a. Europe
         b. Asia
         c. United States
         d. Canada

II. Technical Assessment
   A. Problems
      1. Solar cells do not generate enough electricity to pay for themselves (though this technology is advancing rapidly)
      2. Solar cells collect heat, so additional power would be needed to cool the building
      3. Putting solar cells on top of the regular roof would add 15% to construction time
   B. Potential Solutions (Used by Applebee's Restaurant in Campbellford, Ontario)
      1. Use the collected heat to preheat the large amount of water used by the restaurant
         a. Collect the heat under the solar cells (instead of venting it)
         b. Use electricity from the solar cells to power fans that blow hot air to the preheating system
      2. Use large solar panels that can serve as the roof itself

III. Components of System
   A. Structure of the Solar Panels
   B. Structure of the Heat Ducts
   C. Structure of the Water Heating System

IV. Construction Schedule

V. Cost Comparison

VI. Conclusion

often weave together various patterns, each one suited to a special aim of a specific part of your message.

Figure PO.9 shows the outline for a report that mixes several of the organizational patterns.

EXERCISES

For additional exercises, visit www.techcomm.nelson.com.

Expertise

1. To choose the appropriate principle of classification for organizing a group of items, you need to consider your readers and your purpose. Here are three topics for classification, each with two possible readers. First, identify a purpose that each reader might have for consulting a communication on that topic. Then identify a principle of classification that would be appropriate for each reader and purpose.

   Types of instruments or equipment used in your field
   Student in your program
   Director of purchasing in your future employer's organization
   Intramural sports
   Director of intramural sports at your school
   Student
   Flowers
   Florist
   Owner of a greenhouse that sells garden plants

2. Use a principle of classification to create a hierarchy having at least two levels. Some topics are suggested below. After you have selected a topic, identify a reader and a purpose for your classification. Depending on your professor's request, show your hierarchy in an outline or use it to write a brief discussion of your topic. In either case, state your principle of classification. Have you created a hierarchy that, at each level, has one and only one place for every item?

   Boats
   Cameras
   Printers
   Physicians
   The skills you will need on the job
   Tools, instruments, or equipment you will use on the job
   Some groups of items used in your field (for example, rocks if you are a geologist, or power sources if you are an electrical engineer)

3. Partition an object in a way that will be helpful to someone who wants to use it. Some objects are suggested below. Whichever one you choose, describe a specific instance of it. For example, describe a particular brand and model of food processor rather than a generic food processor. Be sure that your hierarchy has at least two levels, and state the basis of partitioning you use at each level. Depending upon your professor's request, show your hierarchy in an outline or use it to write a brief discussion of your topic.

   Treadmill
   Graphing calculator
   Microwave oven
   Bicycle
   Some instrument or piece of equipment used in your field that has at least a dozen parts

4. Segment a procedure to create a hierarchy you could use in a set of instructions. Give it at least two levels. Some topics are listed below. Show the resulting hierarchy in an outline. Be sure to identify your readers and purpose. If your professor requests, use the outline to write a set of instructions.

   Changing an automobile tire
   Making homemade yogurt
   Starting an aquarium
   Rigging a sailboat
   Installing a software package
   Some procedure used in your field that involves at least a dozen steps
   Some other procedure of interest to you that includes at least a dozen steps

5. Segment a procedure to create a hierarchy you could use in a general description of a process. Give it at least two levels. Some suggested topics are listed below. Show the resulting hierarchy in an outline. Be sure to identify your readers and purpose. If your professor requests, use the outline to write a general description of the process addressed to someone unfamiliar with it.

   How the human body takes oxygen from the air and delivers it to the parts of the body where it is used
   How satellite signals from a broadcast in Toronto reach screens in other parts of the country
   How aluminum is made
   Some process used in your field that involves at least a dozen steps
   Some other process of interest to you that includes at least a dozen steps

6. One of your friends is thinking about making a major purchase. Some possible items are listed below. Create an outline with at least two levels that compares two or more good alternatives. If your professor requests, use that outline to write your friend a letter.

   Home gaming system
   Binoculars
   MP3 player
   Laptop computer
   Bicycle
   Car
   Some other type of product for which you can make a meaningful comparison on at least three important points

7. Think of some way in which things might be done better in a business, team, or some other organization. Imagine that you are going to write a letter to the person who can bring about the change you are recommending. Create an outline with at least two levels in which you compare the way you think things should be done and the way they are being done now.

8. A friend has asked you to explain the causes of a particular event. Some events are suggested below. Write your friend a brief letter explaining the causes.

   Contamination of groundwater
   Immunization from a disease
   Freezer burn in foods
   Yellowing of paper

9. Think of a problem you feel should be corrected. The problem might be noise in your school library, theft from a particular store, or the shortage of computer access labs on campus. Briefly describe the problem and list the actions you would take to solve it. Next, explain how each action will contribute to solving the problem. If your professor requests, use your outline to write a brief memo explaining the problem and your proposed solution to a person who could take the actions you suggest.

CASE

SELECTING THE RIGHT FORKLIFT

For additional cases, visit www.techcomm.nelson.com.

It has been two weeks since you received this assignment from your boss, Arit Philip, who is the production engineer at the manufacturing plant that employs you. "We've been having more trouble with one of our forklifts," she had explained.

"And Kamal's finally decided to replace one of them," she'd continued. Kamal is Arit's boss and top executive of the plant. His title is plant manager.

"What finally happened to make him decide that?" you had asked. "I thought he was going to keep trying to repair those wrecks forever."

"Actually, he wants to replace one of the newer ones we bought just two years ago," Arit had replied. "That particular forklift was manufactured by a company that has since gone bankrupt. Kamal's afraid we won't be able to get replacement parts. I think he's right."

"Hmm," you had commented.

"Anyway," Arit had said. "Kamal wants to be sure he spends the company's money more wisely this time. He's done a little investigation himself and has narrowed the choice to two machines. He's asked me to figure out which one is his best choice."

You could see what was coming. You'd had a hundred assignments like this before from Arit.

"I'd like you to pull together all the relevant information for me. Don't make any recommendation yourself. Just give me all the information I need to make my recommendation. Have it to me in two weeks."

"Okay," you had said, as you started to think about how you could squeeze this assignment into your already tight schedule.

YOUR ASSIGNMENT

It's now two weeks later. You've gathered the information given below in the "Notes on Forklifts." First, plan your final report by performing the following activities:

- List the specific questions Arit will want your report to answer. Note: Your boss does not want you to include a recommendation in your report.
- Underline the facts in your notes that you would include in your report; put an asterisk by those you would emphasize.
- Decide how you would organize the report.
- Explain which techniques from the list on pages 16–17 you would use when writing this report. What other things would you do to assist your readers?

Second, imagine that you have been promoted to Arit's job (production engineer). Tell how you would write the report that you would send to Kamal about the purchase of a new forklift. It must contain your recommendation.

NOTES ON FORKLIFTS

Present Forklift. The present forklift, which is red, moves raw material from the loading dock to the beginning of the production line and takes finished products from the packaging department back to the loading dock. When it moves raw materials, the forklift hoists pallets weighing 300 kg onto a platform 3 m high, so that the raw materials can be emptied into a hopper. When transporting finished products, the forklift picks up and delivers pallets weighing 100 kg at ground level. The forklift moves between stations at 6 km/h, although some improvements in the production line will increase that rate to 8 km/h in the next two months. The present forklift is easy to operate. No injuries and very little damage have been associated with its use.

Electric Forklift. The electric forklift carries loads of up to 500 kg at speeds up to 20 km/h. Although the electric forklift can hoist materials only 2 m high, a 1 m ramp could be built beneath the hopper platform in three days (perhaps over a long weekend, when the plant is closed). During construction of the ramp, production would have to stop. The ramp would cost $1,600. The electric forklift costs $37,250, and a special battery charger costs $1,500 more. The forklift would use about $2,000 worth of electricity each year. Preventive maintenance costs would be about $700 per year, and repair costs would be about $800 per year. While operating, the electric forklift emits no harmful fumes. Parts are available from a warehouse 800 km away. They are ordered by phone and delivered the next day. The electric forklift has a good operating record, with very little damage to goods and with no injuries at all. It comes in blue and red.

Gasoline Forklift. The gasoline forklift is green and carries loads of up to 0.5 tonnes as rapidly as 30 km/h. It can hoist materials 4 m high. Because this forklift is larger than the one presently being used, the company would have to widen a doorway in the cement wall separating the packaging department from the loading dock. This alteration would cost $800 and would stop production for two days. The gasoline forklift costs $49,000 but needs no auxiliary equipment. However, under regulations established by the *Occupational Health and Safety Act*, the company would have to install a ventilation fan to carry the exhaust fumes away from the hopper area. The fan costs $870. The gasoline forklift would require about $1,800 of fuel per year. Preventive maintenance would run an additional $400 per year, and repairs would cost about $600 per year. Repair parts are available from the factory, which is 10 km from our plant. Other owners of this forklift have incurred no damage or injuries during its operation.

Applying a Reader-Centred Approach

CHAPTER 3

Beginning a Communication

GUIDELINES

1. Give your readers a reason to pay attention
2. State your main point
3. Tell your readers what to expect
4. Encourage openness to your message
5. Provide necessary background information
6. Adjust the length of your beginning to your readers' needs
7. For longer communications, use a summary
8. Adapt your beginning to your readers' cultural background
9. Ethics Guideline: Begin to address unethical practices promptly—and strategically

CHAPTER 3

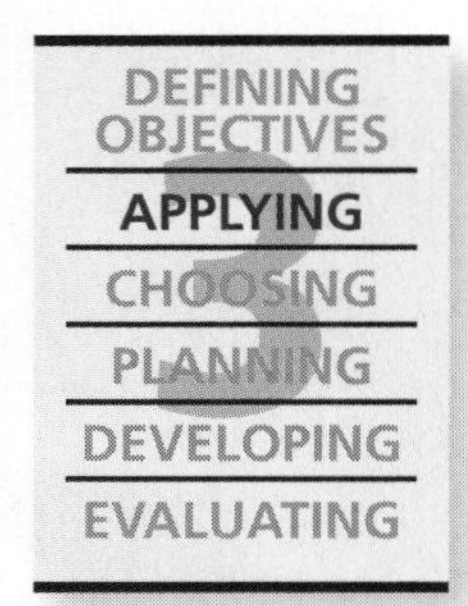

Nothing contributes more to the success of an on-the-job communication than the way it begins.

A well-written beginning can greatly increase a communication's persuasiveness and usability.

As explained in Chapter 1, reading is a dynamic interaction between readers and the words and graphics they read. In this interaction, a reader's response during one reading moment influences what happens during subsequent moments. Consequently, a communication's beginning has a huge impact on its persuasiveness and usability. A good beginning can go a long way toward persuading your readers to receive your communication's content favourably and to devote their full attention to your message rather than skimming it or—worse yet—setting it aside unfinished. Similarly, a thoughtfully crafted beginning can greatly increase your communication's usability by boosting the efficiency with which your readers comprehend and use your communication.

This chapter describes eight reader-centred strategies used by expert workplace writers to begin their communications in highly persuasive, highly usable ways. All eight require you to employ skillfully the knowledge about your readers that you developed when defining a communication's objectives. The strategies you choose will depend on the format and length of your communication, the relationship you have with your readers, and the receptiveness of your readers to your message.

WWW For additional chapter resources, visit Chapter 3 at www.techcomm.nelson.com.

The chapter's last guideline discusses ethical approaches to situations in which people at work sometimes wonder whether they should begin to try communicating at all.

GUIDELINE 1 Give Your Readers a Reason to Pay Attention

The most important function of a beginning is to persuade your readers to pay attention.

People sometimes feel they are too busy to read.

This may be a very difficult task. At work, people often complain that they receive too many memos, reports, and email messages. As they sift through the envelopes delivered to their desks or scan the lists of email messages they've received, the first question they ask about each item is, "Why should I read this?" If they don't find a persuasive answer quickly, they may file the communication unread or throw it away.

People are more likely to be persuaded by messages they think deeply about.

In the face of these difficult circumstances, you must strive to gain not merely *some* attention from your audience, but their *close* attention. Doing so will be especially important when your communication's overall purpose is primarily persuasive. Research has shown that the more deeply people think about a message while reading or listening to it, the more likely they are to hold the attitudes it advocates, the more likely they are to resist attempts to reverse those attitudes, and the more likely they are to act upon those attitudes (Petty & Cacioppo, 1986).

To persuade your readers to pay close attention, you must usually do two things at the very beginning of your communication:

- Announce your topic.
- Tell your readers how they will benefit from the information you are providing.

Be sure to do *both* things. Don't assume that your readers will automatically see the value of your information after you have stated your topic. The benefit that appears obvious to you may not be obvious to them. Compare the following sets of statements:

State reader benefits explicitly.

Statements of Topic Only (Avoid Them)	**Statements of Topic and Benefit (Use Them)**
This memo tells about the new technology for making computer memory chips.	This memo answers each of the five questions you asked me last week about the new technology for making computer memory chips.
This report discusses step-up pumps.	Step-up pumps can save us money and increase our productivity.
This manual concerns the Cadmore Industrial Robot 3000.	This manual tells how to prepare the Cadmore Industrial Robot 3000 for difficult welding tasks.

Two strategies are particularly effective in persuading people that they will benefit from reading your communication: referring to your readers' request and offering to help your readers solve a problem.

Refer to Your Readers' Request

At work, you will often write in response to a request. A simple reference to the request will establish the benefit of your communication:

References to the readers' request

- Here are the test results you asked for.
- As you requested, I am enclosing a list of the steps we have taken in the past year to tighten security in the Information Technology Department.
- Thank you for your inquiry about the capabilities of our Model 1770 colour laser printer.

Offer to Help Your Readers Solve a Problem

A second strategy for persuading readers to devote serious attention to your communication is to explain that your communication will help them solve a problem with which they are dealing. This strategy can be especially effective at work because most employees see themselves as problem solvers. Sometimes the problem will be technical, such as detecting flaws in airplane wings or preserving the freshness of the company's meat products without using harmful chemicals. Other problems may be organizational, such as improving morale or increasing the efficiency of the quality assurance department. Still others may be ethical problems, such as ensuring that temporary employees are treated fairly. Whatever the problem, your readers will welcome communications that help them find a solution.

Readers welcome problem-solving ideas.

Mei thought of her trip as a problem-solving effort.

For example, consider the way that Mei, a computer systems analyst, wrote the beginning of a report concerning her trip to Calgary, where she studied the billing system at a hotel that her employer recently purchased. As she prepared to draft the report's beginning, she identified the problem that her communication would help her readers solve: The Calgary hotel is making little money, perhaps because its billing system is faulty. Then she thought of what she did in Calgary and afterwards to help her readers solve this problem: She evaluated the billing system and formulated possible improvements. Finally, she determined what her report would offer her readers to help them perform their own problem-solving activities: It would help them choose the best course of action by providing data and recommendations on which they could base their decision.

Here is how Mei wrote the beginning of her report:

Mei names the problem her report will help solve.

Mei describes her work toward solving the problem.

Mei tells how her report will help her readers do their part toward solving the problem.

> Over the past two years, our Calgary hotel has shown a profit of only 4 percent, even though it is almost always 78 percent filled. A preliminary examination of the hotel's operations suggests that its billing system may be inadequate: It may be too slow in billing customers, and it may be inefficient and needlessly ineffective in collecting overdue payments. Therefore, I have thoroughly examined the hotel's billing cycle and its collection procedures, and I have considered ways to improve them.
>
> In this report, I present the results of my analysis, together with my recommendations. To aid in the evaluation of my recommendations, I have included a discussion of the costs and benefits of each.

Must You Always Provide All This Information? Beginnings that use this strategy may be shorter or longer than Mei's. The crucial point is to make sure your readers understand all three elements of the problem-solving situation. If one or more of the elements will be obvious to them, there's no benefit to discussing them in detail. For example, if the only people who will read Mei's report are thoroughly familiar with the details of the problem in Calgary and the work she was asked to do there, Mei could have written a very brief beginning, such as the following:

> In this report, I evaluate the billing system in our Calgary hotel and recommend ways of improving it.

She included more explanation because she knew her report would also be read by people who were hearing about her trip for the first time.

Similarly, before using an abbreviated beginning, be sure that all your readers will understand immediately your communication's relevance to them even if you don't state it explicitly. Here are some situations in which a full description of the problem-solving situation is usually desirable:

Situations in which you may need to describe the problem-solving situation in detail.

- **Your communication will be read or heard by people outside your immediate working group.** The larger your audience, the less likely it is that all of them will be familiar with the context of your message.
- **Your communication will have a binding and a cover.** Bound documents are usually intended for large groups of readers. Such documents may also be filed to make them available for future reference. Unless you define the problem-solving situation, many readers will have no idea of what it was.
- **Your communication will be used to make a decision involving a significant amount of money.** Such decisions are often made by high-level managers, who often need to be told of the organizational context of the reports they read.

Defining the Problem in Unsolicited Communications According to a survey, the majority of college and university graduates write on their own initiative (unsolicited) at least as often as they write on assignment (P. V. Anderson, 1985). When writing unsolicited requests or recommendations, you may sometimes encounter a special problem as you write the beginning: persuading your readers that a problem even exists.

Roberto had to persuade his boss that a problem existed that was important to him.

Consider the way Roberto accomplished this goal. He works for a company that markets computer programs used to control manufacturing processes. One program contained bugs that Roberto wanted to fix because he sympathized with the customers who called for help in overcoming problems caused by the bugs. However, Roberto knew that his boss did not want to assign computer engineers to fix the bugs. Instead, she wanted the engineers to spend all their time on her top priority, which was to develop new products rapidly. Consequently, Roberto wrote his boss an email that opened by discussing the difficulty the company had been having in releasing new products on time. He then showed how much time the computer engineers needed to spend helping customers overcome problems caused by the bugs rather than working on new products. By tying his request to a problem that his reader found significant (not to the problems that actually prompted him to write), Roberto succeeded in being assigned to fix the two most serious bugs.

In sum, to describe the benefit your readers will gain from reading your communication, adopt your readers' view of both the situation you are addressing and the communication itself. There is no task for which a reader-centred approach is more critical than explaining the significance of your message to your audience.

GUIDELINE 2 State Your Main Point

You can usually boost your communication's usability and persuasiveness by stating your main point in your beginning. There are three major reasons for doing so:

Reasons for stating your main point at the beginning

- You help your readers find what they most want or need.
- You increase the likelihood that your readers will actually read your main point instead of putting your communication aside before they get to it.
- You provide your readers with a context for viewing the details that follow.

Choose Your Main Point Thoughtfully

Choose the main point of your communication in the same way you choose the main point of each segment. If you are responding to a request, your main point will be the answer to the question your reader asked. If you are writing on your own initiative, your main point might be what you want your readers to think or do after reading your communication.

Here are some sample statements:

Sample beginnings that state the main point

From the beginning of a memo written in a manufacturing company: We should immediately suspend all purchases from Valley Manufacturing until it can guarantee us that the parts it supplies will meet our specifications.

From the beginning of an e-mail written to a department head in a food services company: I request $1,200 in travel funds to send one of our account executives to the client's Vancouver headquarters.

From a research report: The test results show that the walls of the submersible room will not be strong enough to withstand the high pressures of a deep dive.

GUIDELINE 3 Tell Your Readers What to Expect

In addition to stating your main point, the beginning of a communication should tell readers what to expect in the segments that follow. A forecasting statement that is positioned at the beginning of a communication should focus on the communication's organization and its scope.

Tell about Your Communication's Organization

By telling your readers about your communication's organization in your beginning, you provide them with a framework for understanding the connections among the various pieces of information you convey. This framework substantially increases your communication's usability by helping your readers see immediately how each new point you make relates to the points they have already read. It also helps skimming readers navigate quickly to the information they are seeking.

You can tell your readers about the organization of your communication in various ways:

Forecasting statement

> In this report, we state the objectives of the project, compare the three major technical alternatives, and present our recommendation. The final sections include a budget and a proposed project schedule.

Forecasting list

> This booklet covers the following topics:
> - Understanding the Principles of Sound Reproduction
> - Reviewing the Types of Speakers
> - Choosing the Speakers That Are Right for You
> - Installing the Speakers

Tell about Your Communication's Scope

Readers want to know from the beginning what a communication does and does not contain. Even if they are persuaded that you are addressing a subject relevant to them, they may still wonder whether you will discuss the specific aspects of the subject they want to know about.

Let your readers know what your communication contains—and, when appropriate, what it doesn't contain.

Often, you will tell your readers about the scope of your communication when you tell them about its organization: When you list the topics it addresses, you indicate its scope.

There will be times, however, when you will need to include additional information. That happens when you want your readers to understand that you are not addressing your subject comprehensively or that you are addressing it from a particular point of view. For instance, you may be writing a troubleshooting manual to help factory workers solve a certain set of problems that often arise with the manufacturing robots they monitor. Other problems—ones your manual doesn't address—might require the assistance of a computer programmer or an electrical engineer. In that case, you should tell your readers explicitly about the scope of your manual:

Statement of scope

> This manual treats problems you can correct by using tools and equipment normally available to you. It does not cover problems that require work by computer programmers or electrical engineers.

Use your judgment in deciding how much to say in your beginning about the organization and scope of your communication. For brief communications, readers don't need any information at all about such matters. But for longer ones, they will benefit from knowing at the start what lies ahead.

GUIDELINE 4 Encourage Openness to Your Message

Other chapters in this book have emphasized that readers can respond in a variety of ways as they read a communication. For example, when they read a set of recommendations you are making, they can try to understand your arguments or search for flaws. When they read a set of instructions you have prepared, they can follow your directions in every detail or attempt the procedure on their own, consulting your instructions only if they get stumped.

One of the most important goals of a beginning is to persuade your readers to read your message openly.

Because the way you begin a communication has a strong effect on your readers' response, you should always pay attention to the persuasive dimension of your beginnings. Always begin in a way that encourages your readers to be open and receptive to the rest of your communication.

Situations Vary

Special care is needed when readers might resist your message.

Ordinarily, you will have no trouble eliciting a receptive response because you will be communicating with fellow employees, customers, and others who want the information you are providing. In certain circumstances, however, your readers may have a more negative attitude toward your message. In such situations, you will need to take special care in drafting the beginning of your communication if you are to win a fair hearing for your message.

Your readers' initial attitude toward your message will be negative if the answer to any of the following questions is "yes." In this case, try to pinpoint the attitudes that are likely to shape your readers' reactions to your communication. Then devise your beginning accordingly.

Questions for determining whether readers might resist your message

- Does your message contain bad news for your readers?
- Does your message contain ideas or recommendations that will be unwelcome to your readers?
- Do your readers have any feelings of distrust, resentment, or competitiveness toward you, your department, or your company?
- Are your readers likely to be skeptical of your knowledge of your subject or of the situation?
- Are your readers likely to be suspicious of your motives?

The strategy that is most likely to promote a positive initial reaction or to counteract a negative one differs from situation to situation. However, here are four strategies that often work.

Strategies for Encouraging Openness

- **Present yourself as a partner, not as a critic or a competitor.** Suggest that you are working with your readers to help solve a problem they want to solve or to achieve a goal they want to achieve.

See also Chapter 10's discussion of the indirect pattern of organization (pages 253–54).

- **Delay the presentation of your main point.** An initial negative reaction may prompt your readers to aggressively devise counterarguments to each point that follows. Therefore, if you believe that your readers may react negatively to your main point, consider making an exception to Guideline 2, which tells you to state your main point in your beginning. If you delay the presentation of your main point, your readers may consider at least some of your other points objectively before discovering your main point and reacting against it.

See also Chapter 10's suggestions for building credibility (page 256).

- **Establish your credibility.** As Chapter 10 suggests, people are more likely to respond favourably to a message if they have confidence in the person who is delivering it. Consequently, you can promote openness to your message if you begin by convincing your readers that you are expert in your subject and knowledgeable about the situation. This does not mean, however, that you should announce your credentials in the beginning of *every* communication. If you needlessly present your credentials, you merely burden your readers with unnecessary information. Avoid discussing your qualifications when writing to people, such as your coworkers, who have already formed a favourable opinion of your expertise, and when taking a position with which your readers already agree.

- **Avoid embarrassment and resentment on the part of your reader by using the passive voice.** As Chapter 4 suggests, the passive voice can be used to avoid finger pointing and to maintain the confidentiality of an information source.

Tell Yourself a Story

Although the strategies suggested above will often encourage openness, don't employ them mechanically. Always keep in mind the particular attitudes, experiences, and expectations of your readers as you craft the beginning of a communication.

Telling a story helps you focus on your particular readers.

You might do this by telling yourself a story about your readers. The central figure in your story should be your reader if you are writing to one person, or a typical member of your audience if you are writing to a group. Begin your story a few minutes before this person picks up your communication and continue it to the moment he or she reads your first words. Although you would not actually include the story in your communication, creating it can help you decide how to begin.

Here is a sample story, written by Jolene, a manager in an insurance company. Jolene wrote this story to help herself understand the readers of an instruction manual she is preparing. The manual will teach new insurance agents how to use the company's computer system.

Jolene predicts her readers' attitudes by imagining a story about one of them.

> It's Monday afternoon. After half a day of orientation meetings and tours, Jon, the new trainee, sits down at the computer terminal to learn this system. He was a sociology major who has never used a data entry program. Now, in two hours, he is supposed to work his way through this manual and then enter some sample policy information. He feels rushed, confused, and quite nervous. He knows that the information is vital, and he does not want to make an error.
>
> Despite his insecurity, Jon will not ask questions of the experienced agent in the next office because (being new to the company) he doesn't want to make a bad impression by asking dumb questions.
>
> Jon picks up the instruction manual for the SPRR program that I am writing: He hopes it will tell him quickly what he needs to know. He wants it to help him learn the system in the time allotted without his making any mistakes and without his having to ask embarrassing questions.

This story helped Jolene focus on several important facts: The reader will be anxious, hurried, and uncertain. Those insights helped her write an effective opening for her manual:

Jolene adopts a helpful tone.

> This manual tells you how to enter policy information into our SPRR system. It covers the steps for opening a file for a new policy, entering the relevant information, revising the file, and printing a paper copy for your permanent records.

Jolene reassures her readers.

> By following these instructions carefully, you can avoid making time-consuming errors. In addition, the SPRR system is designed to detect and flag possible errors so that you can double-check them.

By identifying her readers' probable feelings, Jolene was able to reduce their anxiety and encourage them to be more open to her instructions.

GUIDELINE 5 Provide Necessary Background Information

As you draft the beginning of a communication, ask yourself whether your readers will need any background information to understand what you are going to tell them.

Here are some examples of situations that might require such information at the beginning:

Signs that your readers need background information

- **Your readers need to grasp certain general principles in order to understand your specific points.** For instance, your discussion of the feasibility of locating a new plant in a particular city may depend on a particular analytical technique that you will need to explain to your readers.
- **Your readers are unfamiliar with technical terms you will be using.** For example, as a specialist in international trade, you may need to explain certain technical terms to the board of directors before you present your strategies for opening up international markets.
- **Your readers are unfamiliar with the situation you are discussing.** For example, imagine that you are reporting to the executive directors of a large corporation about labour problems at one of the plants it recently acquired in a takeover. To understand and weigh the choices that face them, the directors will need an introduction to the plant and its labour history.

Not all background information belongs at the beginning of your communication. Information that pertains only to certain segments should appear at the beginning of those segments. In the beginning of your communication, include only background information that will help your readers understand your overall message.

GUIDELINE 6 Adjust the Length of Your Beginning to Your Readers' Needs

There is no rule of thumb that tells how long the beginning should be. A good, reader-centred beginning may require only a phrase or may take several pages. You need to give your readers only the information they don't already know. Just be sure they know the following:

What your readers need to know

- The reason they should read the communication (Guideline 1)
- The main point of the communication (Guideline 2)
- The organization and scope of the communication (Guideline 3)
- The background information they need in order to understand and use the communication (Guideline 5)

If you have given your readers all this information—and have encouraged them to receive your message openly (Guideline 4)—then you have written a good beginning, regardless of how long or short it is.

Here is an opening prepared by a writer who followed all the guidelines given in this chapter:

Brief beginning

> In response to your memo dated November 17, I have called Goodyear, Goodrich, and Firestone for information about the ways they forecast their needs for synthetic rubber. The following paragraphs summarize each of those phone calls.

The following opening, from a two-paragraph memo, is even briefer:

Briefer beginning

> We are instituting a new policy for calculating the amount that employees are paid for overtime work.

At first glance, this single sentence may seem to violate all the guidelines. It does not. It identifies the topic of the memo (overtime pay), and the people to whom the memo is addressed will immediately understand its relevance to them. It also declares the main point of the memo (a new policy is being instituted). Moreover, because the memo itself is only two paragraphs long, its scope is readily apparent. The brevity of the memo also suggests its organization—namely, a brief explanation of the new policy, and nothing else. The writer has correctly judged that his readers need no background information.

Examples of longer beginnings are shown in Figures 3.1 and 3.2.

Figure 3.1 shows a relatively long beginning from a report written by a consulting firm hired to recommend ways to improve the food service at a hospital. Like the brief beginnings given above, it is carefully adapted to its readers and to the situation.

Figure 3.2 (page 78) shows the long beginning of a 948-page diagnostic manual for the 2001 PRIUS hybrid car manufactured by Toyota.

FIGURE 3.1

Beginning of a Recommendation Report

Problem

Subparts of the overall problem

What the writers have done to help solve the problem

How this report will help the readers

How the report is organized

Scope

Main points

INTRODUCTION

Hotel Dieu Hospital has added 200 patient beds through construction of the new West Wing. Since the wing opened, the food-service department has had difficulty meeting this extra demand. The director of the hospital has also reported the following additional problems:

1. Difficulties operating at full capacity. The equipment, some of it thirty years old, breaks down frequently. Absenteeism has risen dramatically.
2. Costs of operation that are well above average for the hospital sector nationally and in this region.
3. Frequent complaints about the quality of the food from both the patients and the hospital staff who eat in the cafeteria.

To study these problems, we have monitored the operation of the food service department and interviewed patients, food service employees, and staff who eat in the cafeteria. In addition, we have compared all aspects of the department's facilities and operations with those at other hospitals of roughly the same size.

In this report, we discuss our findings concerning the food service department's kitchen facilities. We briefly describe the history and nature of these facilities, suggest two alternative ways of improving them, and provide a budget for each. In the final section of this report, we propose a renovation schedule and discuss ways of providing food service while the renovation work is being done. (Our recommendations about staffing and procedures will be presented in another report in thirty days.)

The first alternative costs about $730,000 and would take four months to accomplish. The second costs about $1,100,000 and would take five months. Both will meet the minimum needs of the hospital; the latter can also provide cooking for the proposed program of delivering hot meals to those unable to leave their homes.

FIGURE 3.2

Beginning of a Service Manual

Scope →

What this manual is about →

Organization →

Background information, including cautions →

Background information to help readers use the manual →

Additional background information that is implicit throughout the rest of the manual →

General Information **2001 PRIUS**

FOREWORD

This manual (Volume 1) contains maintenance, preparation, specifications and diagnostics procedures for the 2001 PRIUS.

Applicable models: NHW11 series

For repair procedures for the engine, chassis and body, and electrical service procedures, refer to VOLUME 2 (Pub. No. RM778U2).

The manual is divided into 6 sections with a thumb index for each section at the edge of the pages.

...

CAUTION

This manual does not include all the necessary items about repair and service. This manual is made for the purpose of the use for the persons who have special techniques and certifications.

...

HOW TO USE THIS MANUAL

GENERAL INFORMATION

1. INDEX
An INDEX is provided on the first page of each section to guide you to the item to be repaired. To assist you in finding your way through the manual, the section title and major heading are given at the top of every page.

2. PRECAUTION
At the beginning of each section, a PRECAUTION is given that pertains to all repair operations contained in that section. Read these precautions before starting any repair task.

3. TROUBLESHOOTING
TROUBLESHOOTING tables are included for each system to help you diagnose the problem and find the cause. The fundamentals of how to proceed with trouble

...

9. SI UNIT
The UNITS given in this manual are primarily expressed to the SI UNIT (International System of Unit), and alternately expressed in the metric system and in the English System. Example: Torque: 30 N m (310 kgf.cm, 22 ft.lbf)

Page 4

For Longer Communications, Use a Summary

At work, it's quite common for communications more than two pages long to start off with a brief summary of the entire message. These summaries serve two purposes:

Purposes of initial summaries

- They help busy managers learn the main points without reading the entire document.
- They help all readers build a mental framework for organizing and understanding the detailed information they will encounter as they read on.

In short communications, such as email messages, memos, and letters, opening summaries are often only a few sentences incorporated in the opening paragraph or paragraphs. In longer communications, they may be a page or longer and may be printed on separate sheets that precede the body of the document.

Features of a Good Summary

No single pattern is appropriate for all summaries. However, readers usually want summaries to tell them something from each of a communication's main sections, so your summary's organization should ordinarily parallel the organization of the communication itself.

Informative summaries state the main points.

What you report from each section depends on what your readers will find most useful. One type of summary, the *informative summary,* states the main point or points from each section and emphasizes results, conclusions, and recommendations. Informative summaries are ideally suited to the needs of readers who are seeking advice about a decision or course of action. These summaries are sometimes called *executive summaries* because they are favoured by decision makers (executives), who rely on their advisers to read and evaluate the rest of the communication. Figure 3.3 (page 80) shows an executive summary.

Descriptive summaries indicate the topics discussed.

Another type, the *descriptive summary,* identifies the topics covered in a communication without telling the main points made about each topic. Consequently, a descriptive summary resembles a contents list. The summaries included in printed or online bibliographic resources are often descriptive summaries. Because users of these resources are primarily interested in locating articles on the topic of their research, descriptive summaries—which focus on topics—are well suited to their needs. Figure 3.4 (page 81) shows a descriptive summary.

Summaries exhibit purposeful redundancy.

By convention, nothing appears in the summaries that isn't also in the body of the communication. This purposeful redundancy enables the summary to present a complete and understandable message to the reader who reads nothing else in the communication. It also means that the summary can't serve as the introduction, even though the introduction that follows it will seem somewhat repetitious.

Summaries that appear in printed and online bibliographic resources are usually called *abstracts.* In many scientific and engineering fields, the term *abstract* is also used for the summary that appears at the front of long reports and proposals. Regardless of the name, they follow the conventions just described.

How Long Should Your Summary Be? In many situations, someone (such as your boss) will specify the length of your summary. If you have no such guidance, you might follow this rule of thumb: Make your summary roughly 5 to 10 percent of the length of the entire communication.

FIGURE 3.3

Summary of a Report Directed Primarily to Decision Makers

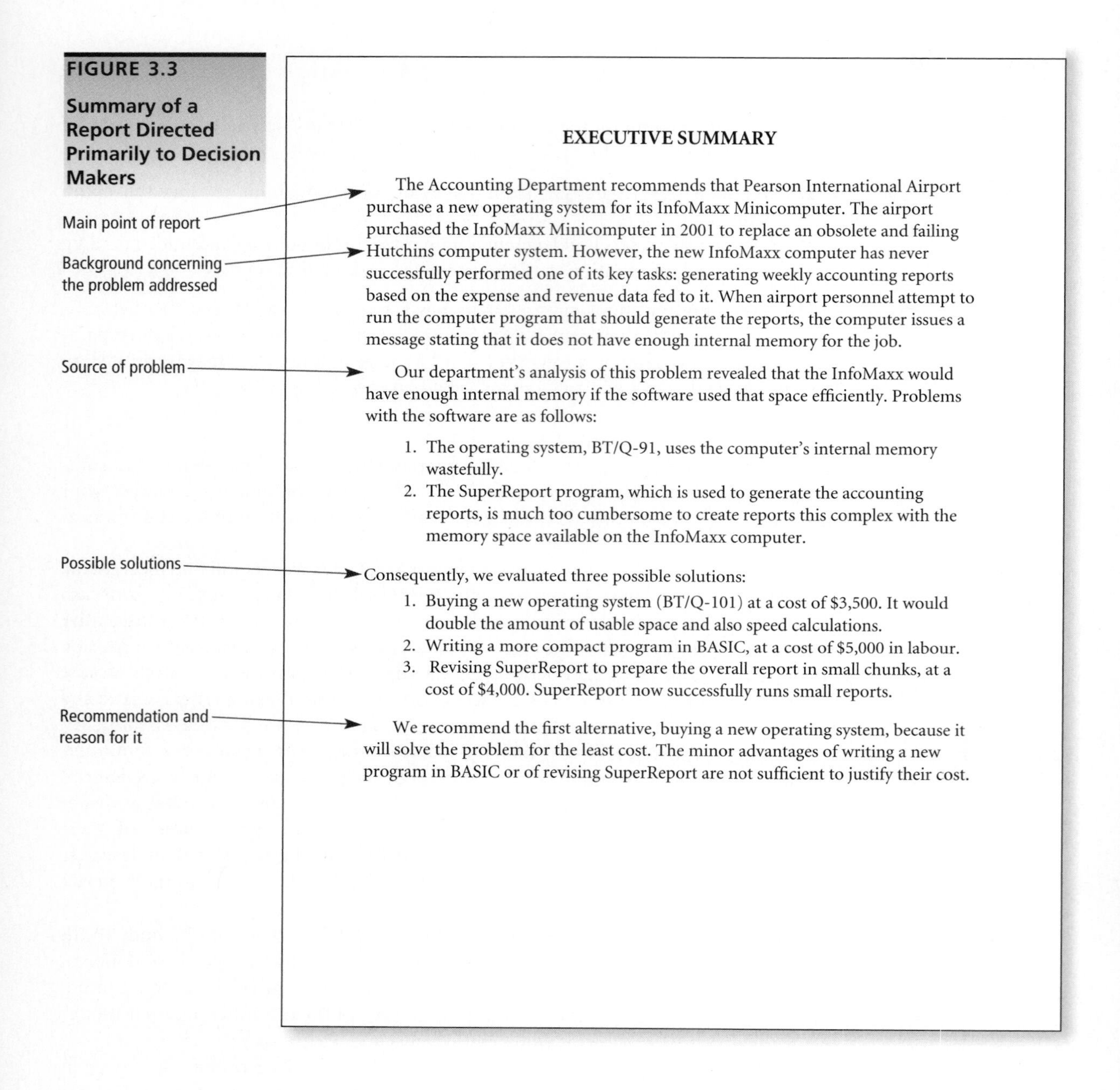

EXECUTIVE SUMMARY

The Accounting Department recommends that Pearson International Airport purchase a new operating system for its InfoMaxx Minicomputer. The airport purchased the InfoMaxx Minicomputer in 2001 to replace an obsolete and failing Hutchins computer system. However, the new InfoMaxx computer has never successfully performed one of its key tasks: generating weekly accounting reports based on the expense and revenue data fed to it. When airport personnel attempt to run the computer program that should generate the reports, the computer issues a message stating that it does not have enough internal memory for the job.

Our department's analysis of this problem revealed that the InfoMaxx would have enough internal memory if the software used that space efficiently. Problems with the software are as follows:

1. The operating system, BT/Q-91, uses the computer's internal memory wastefully.
2. The SuperReport program, which is used to generate the accounting reports, is much too cumbersome to create reports this complex with the memory space available on the InfoMaxx computer.

Consequently, we evaluated three possible solutions:

1. Buying a new operating system (BT/Q-101) at a cost of $3,500. It would double the amount of usable space and also speed calculations.
2. Writing a more compact program in BASIC, at a cost of $5,000 in labour.
3. Revising SuperReport to prepare the overall report in small chunks, at a cost of $4,000. SuperReport now successfully runs small reports.

We recommend the first alternative, buying a new operating system, because it will solve the problem for the least cost. The minor advantages of writing a new program in BASIC or of revising SuperReport are not sufficient to justify their cost.

FIGURE 3.4

Descriptive Summary

> This handbook is intended to assist staff involved with the design, construction, and installation of wells drilled for the purpose of monitoring groundwater for pollutants. It presents state-of-the-art technology that may be applied in diverse hydrogeologic situations and focuses on solutions to practical problems in well construction rather than on idealized practice. The information in the handbook is presented in both matrix and text form. The matrices use a numerical rating scheme to guide the reader toward appropriate drilling technologies for particular monitoring situations. The text provides an overview of the criteria that influence design and construction of groundwater monitoring in various hydrogeologic settings.

GUIDELINE 8 Adapt Your Beginning to Your Readers' Cultural Background

Readers' expectations and preferences about the beginning of a communication are shaped by their culture. The suggestions you have just read are suitable for readers in Canada. However, customs vary widely. For example, Chinese businesspeople expect a letter to open with a reference to the writer's "humble" company. The writer is then expected to make a brief mention of a happy event, such as a political happening, that adds harmony to the jointly shared business situation. Businesspeople in India, however, prefer correspondence to begin with a reference to the company's reputation. Any letter of introduction, such as that initiating a business relationship, should include proof of the company's success, such as profit margins or lists of clients. Clearly, you must have a good understanding of the communication customs of your readers' culture to create an effective opening.

GUIDELINE 9 Ethics Guideline: Begin to Address Unethical Practices Promptly—and Strategically

So far, the guidelines in this chapter have focused on the ways to begin a communication. In contrast, this last guideline concerns situations in which people at work sometimes hesitate to begin writing or speaking at all. Suppose you learn that your employer is engaged in an action you consider to be unethical. Or suppose you are asked to write something that violates your sense of what is ethical. Should you speak up or express your concerns in writing? New employees are sometimes advised to wait until they have achieved security and status before trying to bring about change. But that means you could spend years before addressing a practice you regard as unethical. Ignoring an unethical act could be seen as unethical in itself.

When determining how to draw attention to something you consider to be unethical, you face a challenge similar to that of figuring out how to begin a memo in which you will recommend a course of action with which you believe your readers will disagree. Here are three strategies that could enable you to open a discussion based on your values without jeopardizing your future with your employer's organization.

Strategies for changing unethical practices without risking your job.

- **Plant the seeds of change.** Instead of trying to alter the situation immediately, plant the seeds of change. For instance, if your employer is thinking of modifying working conditions in a factory, you might inquire, "How would this modification affect workers on the assembly line?" Or, you might ask, "Would this modification be fair to the assembly line workers?" By posing such questions, you extend the range of issues being considered, and you can subtly let others know the values that shape your understanding of the situation. In addition to asking questions, there are many other ways you can introduce your values into the discussion. Think creatively. Taking even a small step toward improving a situation is an ethical act. Moreover, in many circumstances, it's only through a long series of small steps that large improvements are achieved.
- **Use reason rather than accusation.** Your values are much more likely to prevail if you engage people with other views in a reasoned discussion than if you accuse and condemn them. It's not usually possible to persuade people by attacking them. They merely become defensive. Instead, ask them to share their sense of the values that apply to the situation. Appeal to their sense of fairness and of what is right and wrong.
- **Remain open to others' views.** One reason to avoid taking a rigid stand is that it impairs your ability to understand others' views of the situation. People regularly differ on ethical matters, and your own view is not necessarily shared by others. Strive for solutions that will satisfy both you and others. One possibility is to employ a strategy similar to that described in this chapter's discussion of Guideline 1: Identify the values the other people hold that would lead them to endorse the same course of action your values lead you to advocate.

In an extreme case, you may feel that you need to blow the whistle.

You may someday witness a practice that is so outrageous that you will be willing to risk future promotions and even your job in order to stop it. If you find yourself in that situation, seek the aid of influential people inside your company. If the practice you object to violates the law or a government regulation, alert the appropriate agency. This is called *whistleblowing*. A number of federal and provincial statutes contain whistleblowing provisions that apply to particular aspects of the private sector—that relate, for example, to the environment and to occupational health and safety. Also, disclosure protection legislation has been introduced to cover all federal public sector workers. Still, many whistleblowers do lose their jobs or continue to work under hostile conditions. If you are thinking of whistleblowing, consider the possibility of first attempting a non-confrontational approach to the problem.

CONCLUSION

The beginning is probably the most important segment of a communication. That's because it can influence the ideas and attitudes your readers derive from the rest of your communication. It can even determine whether or not they will read further.

This chapter has suggested that in writing a beginning you start by trying to identify your readers' attitudes toward your message and by determining what you can say to help them understand and use what follows. This reader-centred approach will enable you to create beginnings that prompt your readers to pay careful attention, encourage them to treat your information and ideas with an open mind, and help them read efficiently.

EXERCISES

For additional exercises, visit www.techcomm.nelson.com.

Expertise

1. Select a communication written to people in your field. This might be a letter, memo, manual, report, or article in a professional journal. (Do not choose a textbook.) Identify the guidelines from this chapter that the writer applied when drafting the communication's beginning. Is the beginning effective in achieving the writer's usability and persuasive goals? If so, why? If not, how could the beginning be improved?

2. The instructions for many consumer products contain no beginning section at all. For instance, the instructions for some lawnmowers, cake mixes, and detergents simply provide a heading that says "Instructions" and then start right in. Find such a set of instructions and—in terms of the guidelines given in this chapter—evaluate the writer's decision to omit a beginning.

Online

The home page of a website typically looks very different from the opening of a printed document. Nonetheless, the home page has many of the same reader-centred goals as the printed opening. Examine the ways that two websites follow this chapter's guidelines. Consider such things as the communication functions performed by the images, layout, text (if any), and words used for buttons and links. One of the home pages you study should be for a company that sells consumer products such as cars. The other should be for a non-profit organization or government agency. Present your analysis in the way your professor requests.

Collaboration

1. The following paragraphs are from the beginning of a report in which the manager of a production department asks for better quality products from the department in the company that provides abrasives. Is this an effective beginning for what is essentially a complaint? Explain. Working with another student, analyze this beginning in terms of this chapter's guidelines.

   I am sure you have heard that the new forging process is working well. Our customers have expressed pleasure with our castings. Thanks again for all your help in making this new process possible.

   We are having one problem, however, with which I have to ask once more for your assistance. During the seven weeks since we began using the new process, the production line has been idle 28 percent of the time. Also, many castings have had to be remade. Some of the evidence suggests that these problems are caused by the steel abrasive supplies we get from your department. If we can figure out how to improve the abrasive, we may be able to run the line at 100 percent of capacity.

   I would be most grateful for help from you and your people in improving the abrasive. To help you devise ways of improving the abrasive, I have compiled this report, which describes the difficulties we have encountered and some of our thinking about possible remedies.

2. Form a team of two or three students. Exchange the opening sections (paragraph or longer) of a project you are now preparing for your professor. First, suggest specific revisions that would make your partner's opening more effective at achieving his or her usability and persuasive goals. Next, suggest an alternative opening that might also be effective.

Ethics

It can be particularly challenging to write an effective opening for a communication in which you are advocating for change based on ethical grounds. Think of some organization's practice, procedure, or policy that your values lead you to believe should be changed. Identify the person or group within the organization that has the authority to make the change that you feel is needed. Then, following the advice given in this chapter, draft the opening paragraph of a letter, memo, email, report, or other communication on the topic that you could send to this person or group.

CHAPTER 4

Developing an Effective Style

GUIDELINES

Guidelines for Creating Your Voice

1. Find out what's expected
2. Consider the roles your voice creates for your readers and you
3. Consider how your attitude toward your subject will affect your readers
4. Say things in your own words
5. Ethics Guideline: Avoid stereotypes

Guidelines for Constructing Sentences

1. Simplify your sentences
2. Put the action in your verbs
3. Use the active voice unless you have a good reason to use the passive voice
4. Emphasize what's most important
5. Smooth the flow of thought from sentence to sentence
6. Vary your sentence length and structure

Guidelines for Selecting Words

1. Use concrete, specific words
2. Use specialized terms when—and only when—your readers will understand them
3. Use words accurately
4. Choose words with appropriate associations
5. Choose plain words over fancy ones
6. Ethics Guideline: Use inclusive language

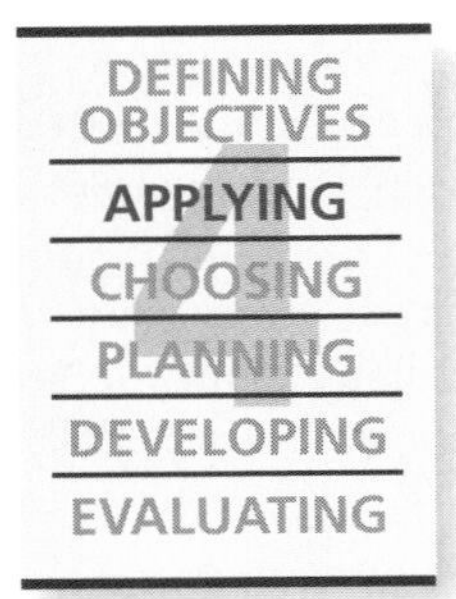

When we talk about "writing style," we can mean many things. For example, when we speak of Shakespeare's style or Stephen King's style, we mean the features that make that person's writing unique. In contrast, when we speak of legal style or scientific style, we are referring to writing characteristics shared by groups of people, such as lawyers or scientists. We also use the word *style* to talk about a communication's readability and impact, for example, we say that a communication is written in a clear or muddy style, an inspiring or boring style. We even use the word *style* to express judgments about the writer, not the writing, as when we say that a style is friendly or stuffy, relaxed or stiff, helpful or condescending.

WWW For additional chapter resources, visit Chapter 4 at www.techcomm.nelson.com.

At work, you must juggle all these dimensions of style at once, striving simultaneously to express your individuality, observe the stylistic conventions of your profession and your employer's organization, make reading easy for your readers, and create the impact you desire. The job isn't easy. These goals sometimes conflict. Furthermore, one style won't suit all occasions. You will need to use different styles in different situations.

As you balance these various considerations, take the same reader-centred approach that you apply to every other aspect of your writing: Consider your options in light of the way they will impact your communication's usability and persuasiveness in your readers' eyes. By following this chapter's three sets of guidelines, you can develop expertise at making reader-centred decisions about the three major building blocks of writing style: voice, sentence structure, and word choice.

CREATING YOUR VOICE

While reading something you've written, your readers "hear" your voice—and, based on what they hear, they draw conclusions about you and your attitudes. These conclusions can greatly affect the persuasiveness of your communications. Consequently, the ability to craft and control your voice is an area of expertise that is essential to your success at writing on the job.

GUIDELINE 1 Find Out What's Expected

To a large extent, an effective voice is one that matches your readers' sense of what's appropriate.

Here are three questions to ask yourself when determining how to match your voice to your readers' expectations:

Questions for determining what your readers expect

- **How formal do my readers think my writing should be?** An informal style sounds like conversation. You use contractions (*can't, won't*), short words, and colloquial words and phrasing. A formal style sounds more like a lecture or speech, with longer sentences, formal phrasing, and no contractions.
- **How subjective or objective do my readers believe my writing should be?** In a subjective style, you would introduce yourself into your writing by saying such things as "I believe . . ." and "I observed . . ." In an objective style, you would mask your presence by stating your beliefs as facts ("It is true that . . .") and by reporting about your own actions in the third person ("The researcher observed . . .") or the passive voice ("It was observed that . . .").
- **How much "distance" do my readers expect me to establish between them and me?** In a personal style, you appear very close to your readers because you do such things as use personal pronouns (*I, we*) and address your readers directly. In an impersonal style, you distance yourself from your readers—for instance, by avoiding personal pronouns and by talking about yourself and your readers in the third person ("The company agrees to deliver a fully operable model to the customer by October 1").

Here are some major factors that may influence your readers' expectations about style:

- Your professional relationship with your readers (customers? supervisors? subordinates?)
- Your purpose (requesting something? apologizing? advising? ordering?)
- Your subject (routine matter? urgent problem?)
- Type of communication (email? letter? formal report?)
- Your personality

- Your readers' personalities
- Customs in your employer's organization
- Customs in your field, profession, or discipline

To learn what style your readers expect, follow the advice in Chapter 2: Ask people who know (including even your readers), and look for communications similar to the one you are writing.

Expectations are Different in Other Cultures

You should learn about your readers' expectations when you address people in a different culture. The style of workplace writing varies considerably from country to country. For example, businesspeople in China write in a more personal yet formal style than businesspeople in Canada, whose impersonal, direct style might be found overly abrupt. Chinese readers prefer "softer" word choices, such as "our humble company" and "would you be so kind as to assist our firm," that make the receiver feel valued. Similarly, readers in India prefer a British formality, evidenced by the use of the titles Sir and Madam and by phrasing such as "your company would be of utmost help" in business correspondence. When writing to people in other countries, use the styles that are customary there to ensure that your communication remains reader-centred.

What If an Ineffective Style Is Expected?

Note that sometimes the expected style may be less effective than another style you could use. For example, in some organizations the customary and expected style is a widely (and justly) condemned style called *bureaucratese.* Bureaucratese is characterized by wordiness that buries significant ideas and information, weak verbs that disguise action, and abstract vocabulary that detaches meaning from the practical world of people, activities, and objects. Often, such writing features an inflated vocabulary and a general pomposity that slows or completely blocks comprehension. Here's an example:

Bureaucratese

> According to optimal quality-control practices in manufacturing any product, it is important that every component part that is constituent of the product be examined and checked individually after being received from its supplier or other source but before the final, finished product is assembled. (45 words)

The writer simply means this:

Plain English

> Effective quality-control requires that every component be checked individually before the final product is assembled. (16 words)

Another example:

Bureaucratese

> Over the most recent monthly period, there has been a large increase in the number of complaints that customers have made about service that has been slow. (27 words)

Plain English

> Last month, many more customers complained about slow service. (9 words)

Bureaucratese is such a serious barrier to understanding that the Canadian government's Communication Policy (1994) has declared that plain language and proper grammar must be used in all communication with the public to ensure clarity and consistency of information. This chapter's guidelines will help you avoid bureaucratese.

However, some managers and organizations want employees to use that puffed-up style, thinking it sounds impressive. If you are asked to write in bureaucratese, try to explain why a straightforward style is more effective, perhaps sharing this book. If you fail to persuade, be prudent. Use the style that is required. Even within the confines of a generally bureaucratic style, you can probably make improvements. For instance, if your employer expects a wordy, abstract style, you may still be able to use a less inflated vocabulary.

GUIDELINE 2 Consider the Roles Your Voice Creates for Your Readers and You

Through the voice you choose, you indicate the role you see yourself playing with respect to your readers. For example, when writing to workers in the department you manage, you might assume the voice of a stern authority or that of an open-minded leader. When instructing a new employee, you might assume the voice of a demanding instructor or a helpful guide.

The voice you choose indicates not only a role you assign yourself but also one you assign your readers. If you assume the voice of an equal writing to respected peers, your readers will probably accept their implied role as your equals. But if you assume the voice of a superior, unerring authority, they may resent their implied role as error-prone inferiors. Although they may not speak the words, they may think, "You have no right to talk to me like that." If readers respond to your voice negatively, they are not likely to receive your message favourably.

Your voice can come across in as short a space as a single sentence. Compare the following statements:

Supportive voice

> Let's meet tomorrow to see if we can figure out why people in your department had difficulty meeting last month's production targets.

Domineering voice

> I have scheduled an hour for us to meet tomorrow to discuss the failure of your department to meet production targets last month.

The first sentence portrays the writer as a supportive person who wants to work as an equal with the reader, who is in turn portrayed as someone who would like to solve a problem that stumps both individuals. The second sentence portrays the writer as a powerful person who considers the reader to be someone who can be bossed around and blamed—a role the reader probably does not find agreeable.

GUIDELINE 3 Consider How Your Attitude Toward Your Subject Will Affect Your Readers

In addition to communicating attitudes about yourself and your readers, your voice communicates an attitude toward your subject. Feelings are contagious. If you write about your subject enthusiastically, your readers may catch your enthusiasm. If you seem indifferent, they may adopt the same attitude.

Especially when your feelings about your subject are strong, carefully review the attitude you want to convey. Email presents a special temptation to be careless about voice because it encourages spontaneity. As Laura B. Smith (1993) says, "Staring at email can make users feel dangerously bold; they sometimes blast off with emotions that they probably would not use in a face-to-face meeting. It's sort of like being in the driver's seat of a car with the windows rolled up." The risk of writing something you'll regret is increased by the ease with which emails can be forwarded from one person to readers you never intended to see your message. Never include anything in an email that you wouldn't be prepared for a large audience to read. Check carefully for statements that you intend to have one tone of voice but that your readers might interpret as having another tone of voice.

GUIDELINE 4 Say Things in Your Own Words

No matter what style you choose, be sure to retain your own voice in your writing. You can do that even in your formal writing—for instance, in a scientific or engineering report. James Watson (1968), winner of the Nobel Prize for his role in discovering the structure of DNA, praised Linus Pauling, three-time winner of the same prize, for his distinctive writing style in highly technical papers. When you are using a formal style, the objective is not to silence your own voice; it's to let your style sound like *you,* writing in a formal situation.

Try reading your draft aloud.

To check whether you are using your own voice, try reading your drafts aloud. Where the phrasing seems awkward or the words are difficult for you to speak, you may have adopted someone else's voice—or slipped into bureaucratese, which reflects no one's voice. Reading your drafts aloud can also help you spot other problems with voice, such as sarcasm or condescension.

Sometimes it's appropriate to suppress your own voice.

Despite the advice given in this guideline, it will sometimes be appropriate for you to suppress your own voice. For example, when a report, proposal, or other document is written by several people, the contributors usually strive to achieve a uniform voice so that all the sections will fit together stylistically. Similarly, certain kinds of official documents, such as an organization's policy statements, are usually written in the employer's style, not the individual writer's style. Except in such situations, however, let your own voice speak in your writing.

GUIDELINE 5 Ethics Guideline: Avoid Stereotypes

The discussion of this guideline begins with a story. A man and a boy are riding together in a car. As they approach a railway crossing, the boy shouts, "Father, watch out!" But it is too late. The car is hit by a train. The man dies, and the boy is rushed to a hospital. When the boy is wheeled into the operating room, the surgeon looks down at the child and says, "I can't operate on him. He's my son."

When asked to explain why the boy would call the deceased driver "Father" and the living surgeon would say "He's my son," people offer many guesses. Perhaps the driver is a priest or the boy's stepfather or someone who kidnapped the boy as a baby. Few guess that the surgeon must be the boy's mother. Why? Our culture's stereotypes about the roles men and women play are so strong that when people think of a surgeon, many automatically imagine a man.

Stereotypes, Voice, and Ethics

What do stereotypes have to do with voice and ethics? Stereotypes are so deeply embedded in culture that most of us are prone to use them occasionally without realizing it. We may be particularly susceptible to using stereotypes when we draw on our less reflective oral ways of communicating as we seek to develop our distinctive voice for our workplace writing. Unfortunately, even these inadvertent uses of stereotypes have serious consequences for both individuals and groups. People who are viewed in terms of stereotypes lose their ability to be treated as individual human beings. Furthermore, if they belong to a group that is unfavourably stereotyped, they may find it nearly impossible to get others to take their talents, ideas, and feelings seriously.

The range of groups disadvantaged by stereotyping is quite extensive. People are stereotyped on the basis of their race, national or ethnic origin, colour, religion, age, gender, sexual orientation, and physical or mental disability. In some workplaces, manual labourers, union members, clerical workers, and others are the victims of stereotyping by people in white-collar positions.

For a discussion of stereotypes and word choice, see pages 103–4.

One way to treat people ethically is to treat them as individuals rather than as members of a stereotyped group. Another way is to avoid writing and speaking in ways that perpetuate stereotypes in the minds of your readers.

Avoiding Stereotypes

- **Avoid describing people in terms of stereotypes.** In your reports, sales presentations, policy statements, and other communications, avoid giving examples that rely upon or reinforce stereotypes. For example, don't make all the decision makers men and all the clerical workers women.
- **Mention a person's gender, race, or other characteristic only when it is relevant.** To determine whether it's relevant to describe someone as a member of a minority group, ask yourself if you would make a parallel statement about a member of the majority group. If you wouldn't say, "This improvement was suggested by Jane, a person without any physical disability," don't say, "This improvement was suggested by Margaret, a person with a disability." If you wouldn't say, "The Saskatoon office is managed by John, a hard-working white person," don't say, "The Saskatoon office is managed by Ben, a hard-working Aboriginal."
- **Avoid humour that relies on stereotypes.** Humour that relies on a stereotype reinforces the stereotype. Refrain from such humour not only when members of the stereotyped group are present, but at all times.

CONSTRUCTING SENTENCES

Many of the most valuable insights concerning effective sentence style have been produced by researchers who've studied the ways our minds process information when we read. Their studies have helped us understand how to construct sentences that significantly increase a communication's usability and persuasiveness—the two indispensable ingredients of successful on-the-job writing. Based primarily on this research, the following six guidelines explain ways to develop your expertise at constructing highly usable, highly persuasive sentences.

GUIDELINE 1 Simplify Your Sentences

There is no easier way to increase usability than to simplify your sentences. Reading is work. Psychologists say that much of the work of reading is done by the short-term memory, which must figure out how the words in each of your sentences fit together to create a specific meaning. By simplifying your sentences, you ease the work your readers' short-term memories must perform to understand your message, thereby making your communication easier to use. In addition, when the same message is conveyed more simply, it comes across more forcefully, more memorably, and more persuasively (F. Smith, 1993).

Simplifying Sentences

1. **Eliminate unnecessary words.** For short-term memory, every word represents work. Scour your writing for words that don't help convey your meaning. Consider this sentence:

   > The <u>physical size of the</u> workroom is too small <u>to accommodate</u> this equipment.

   With unnecessary words removed in two places, the sentence is just as clear and more emphatic:

   > The workroom is too small for this equipment.

2. **Substitute one word for several.** You can often replace an entire phrase with a single word. "Due to the fact that" can be shortened to "Because." Similarly, "They do not pay attention to our complaints" can be abbreviated to "They ignore our complaints." "At this point in time" is "Now."

3. **Keep modifiers next to the words they modify.** Short-term memory relies on word order to indicate meaning. If you don't keep related words together, your sentence may say something different from what you mean.

   > A large number of undeposited cheques were found in the file cabinets, which were worth over $41,000.

   According to the way the English language works, this sentence says that the file cabinets were worth over $41,000. Yet, the author meant that the cheques were worth that amount. Of course, readers would probably figure out what the writer meant because it is more likely that the cheques were worth that much money than that the file cabinets were. But readers arrive at the correct meaning only after performing work they would have been saved had the writer kept related words together—in this example, by putting *which were worth over $41,000* after *cheques,* rather than after *file cabinets.*

4. **Combine short sentences to clarify meaning.** Often, combining two or more short sentences makes reading easier because it reduces the total number of words and helps the reader see the relationships among the points presented.

   Separate: Water quality in the Ganaraska River declined in March. This decline occurred because of the heavy rainfall that month. All the extra water overloaded Northumberland County's water treatment plant.

   Combined: Water quality in the Ganaraska River declined in March because heavy rainfalls overloaded Northumberland County's water treatment plant.

GUIDELINE 2 Put the Action in Your Verbs

Most sentences are about action. Sales rise, equipment fails, engineers design, managers approve. Clients praise or complain, and technicians advise. Yet, many people bury the action in nouns, adjectives, and other parts of speech. Consider the following sentence:

Original | Our department accomplished the conversion to the new machinery in two months.

It could be energized by putting the action (*converting*) into the verb:

Revised | Our department converted to the new machinery in two months.

The revised version is briefer; it is also more emphatic and lively. According to researcher E. B. Coleman (1964), when you put the action in your verbs, you can make your prose up to 25 percent easier to read.

To create sentences that focus on action, do the following:

Focusing Sentences on Action

- **Avoid sentences that use some form of the verb *to be* (*is, was, will be,* etc.).** The verb *to be* often tells what something is, not what it does.

  Original | The sterilization procedure is a protection against reinfection.

  Revised | The sterilization procedure protects against reinfection.

- **Avoid sentences that begin with *It is* or *There are.***

  Original | It is because the cost of raw materials has soared that the price of finished goods is rising.

  Revised | Because the cost of raw materials has soared, the price of finished goods is rising.

  Original | There are several factors causing the engineers to question the dam's strength.

  Revised | Several factors cause the engineers to question the dam's strength.

- **Avoid sentences where the action is frozen in a word that ends with one of the following suffixes: *-tion, -ment, -ing, -ion, -ance.*** These words petrify the action that should be in verbs by converting them into nouns.

  Original | Consequently, I would like to make a recommendation that the department hire two additional programmers.

  Revised | Consequently, I recommend that the department hire two additional programmers.

Although most sentences are about action, some aren't. For example, topic and forecasting statements often introduce lists or describe the organization of the discussion that follows.

Topic sentence for which the verb *to be* is appropriate | There are three main reasons the company should invest money to improve communication between corporate headquarters and the out-of-province plants.

Learn to distinguish between the sentences for which the verb *to be* is appropriate and sentences in which the verb *to be* petrifies the action.

GUIDELINE 3 Use the Active Voice Unless You Have a Good Reason to Use the Passive Voice

Another way to focus your sentences on action and actors is to use the *active voice* rather than the *passive voice.* To write in the active voice, place the actor—the person or thing performing the action—in the subject position. Your verb will then describe the actor's action.

Active voice

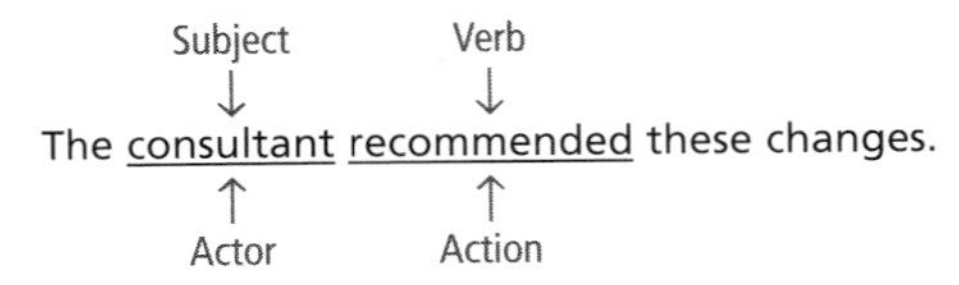

In the passive voice, the subject of the sentence and the actor are different. The subject is *acted upon* by the actor.

Passive voice

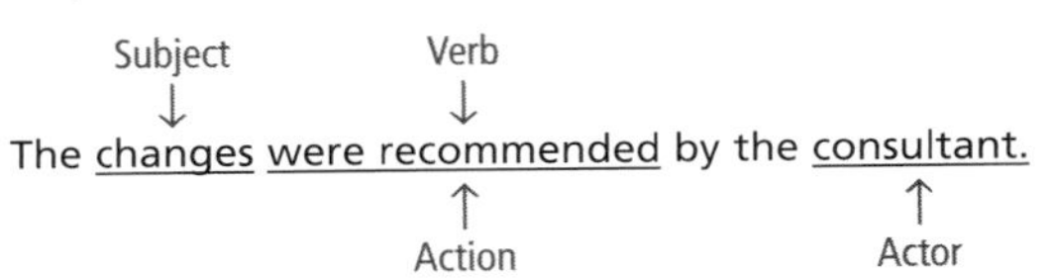

Here are some additional examples:

Passive voice | The Korean ore was purchased by us.

Active voice | We purchased the Korean ore.

Using the active voice increases the usability of your writing in two ways. First, the active voice speeds comprehension. Research shows that readers understand active sentences more rapidly than passive ones (Layton & Simpson, 1975). Second, the active voice avoids the vagueness and ambiguity that often characterize the passive voice. In the passive voice, a sentence can describe an action without telling who did it. For example, "The ball was hit" is a grammatically correct sentence even if it doesn't tell who or what hit the ball. With the active voice, the writer always identifies the actor: "Linda hit the ball."

To appreciate the importance of ensuring that readers of on-the-job writing understand who the actor is, consider the following sentence.

Passive voice

| The operating temperatures must be checked daily to ensure the motor is not damaged.

Will the supervisor of the third shift know that she is the person responsible for checking temperatures? In the passive voice, this sentence certainly allows her to imagine that someone else, perhaps a supervisor on another shift, is responsible.

There are some places where the passive voice is appropriate.

Although the passive voice generally reduces usability, it has some good uses. One occurs when you don't want to identify the actor. The following sentence is from a memorandum in which the writer urges all employees to work harder at saving energy but avoids causing embarrassment and resentment by naming the guilty parties.

Passive voice | The lights on the third floor have been left on all night for the past week, despite the efforts of most employees to help us reduce our energy bills.

Also, consider this sentence:

Passive voice | I have been told that you may be using the company phone for an excessive number of personal calls.

Perhaps the person who told the writer about the breach of corporate phone policy did so in confidence. If the writer decided that it would be ethically acceptable to communicate this news to the reader without naming the person who made the report, then she has used the passive voice effectively. (Be careful, however, to avoid using the passive voice to hide an actor's identity when it is unethical to do so—for instance, when trying to avoid accepting responsibility for your employer's actions.)

Another good reason for using the passive voice is discussed in Guideline 5.

Emphasize What's Most Important

Another way to write clear, forceful sentences is to direct your readers' attention to the most important information you are conveying.

Emphasizing What's Most Important

1. **Place the key information at the end of the sentence.** As linguist Joseph Williams (1994) points out, you can demonstrate to yourself that the end of the sentence is a place of emphasis by listening to yourself speak. Read the following sentences aloud:

   - Her powers of concentration are extraordinary.
   - Last month, he topped his sales quota even though he was sick for an entire week.

   As you read these sentences aloud, notice how you naturally stress the final words, *extraordinary* and *entire week.*

   To position the key information at the end of a sentence, you may sometimes need to rearrange your first draft.

   Original | The department's performance has been superb in all areas.

   Revised | In all areas, the department's performance has been superb.

   Original | The bright exterior design is one of the product's most appealing features to college-age customers.

   Revised | One of the product's most appealing features to college-age customers is its bright exterior design.

2. **Place the key information in the main clause.** If your sentence has more than one clause, use the main clause for the information you want to emphasize. Compare the following versions of the same statement.

   - Although our productivity was down, our profits were up.
   - Although our profits were up, our productivity was down.

   In the first version, the emphasis is on profits because *profits* is the subject of the main clause. The second version emphasizes productivity because *productivity* is the subject of the main clause. (Notice that in each of these sentences, the emphasized information is not only in the main clause but also at the end of the sentence.)

3. **Emphasize key information typographically.** Use boldface and italics. Be careful, however, to use typographical highlighting sparingly. When many things are emphasized, none stands out.

4. **Tell readers explicitly what the key information is.** You can also emphasize key information by announcing its importance to your readers.

> Economists pointed to three important causes of the stock market's decline: uncertainty about the outcome of last month's election, a rise in inventories of durable goods, and—*most important*—signs of rising inflation.

GUIDELINE 5 Smooth the Flow of Thought from Sentence to Sentence

As your readers begin reading each new sentence, they need to figure out how the new information it contains relates to the old information in the preceding sentence. An important way to make your communications more usable is to help your readers make these links by using the following strategies:

1. **Avoid needless shifts in topic.** The simplest relationship between two adjacent sentences is this: The first one says something about a particular topic, and the second says something more about the same topic.

> The links of the drive chain must fit together firmly. They are too loose if you can easily wiggle two links from side to side more than ten degrees.

Readers usually assume that the subject of a sentence is the topic. Consequently, you can help your readers rapidly detect the shared topic of adjacent sentences by putting that topic in the subject position of both sentences. For example, imagine that you have just written this sentence:

> **Topic (Subject)** Our company's new inventory system
> **Comment** reduces our costs considerably.

And suppose that in your next sentence you wanted to communicate information that could be expressed either in this way:

Version A

> **Topic (Subject)** Thousands of dollars
> **Comment** have been saved by the system this year alone.

or in this way:

Version B

> **Topic (Subject)** The system
> **Comment** has saved thousands of dollars this year alone.

Versions A and B contain the same information. However, Version B has the same topic—*system*—as the preceding sentence, and Version A does not. Therefore, your readers will be able to draw the link between Version B and the preceding sentence more easily than the link between Version A and the preceding sentence.

> Our company's new inventory system reduces costs considerably. The system has saved thousands of dollars this year alone.

You can often achieve the same easy bridge from one sentence to the next by keeping the same general topic in the subject position of the adjacent sentences, even if you don't keep the exact same word:

Focus is maintained on one topic

> The materials used to construct and furnish this experimental office are designed to store energy from the sunlight that pours through the office's large windows. The special floor covering stores energy more efficiently than wood. The heavy fabrics used to upholster the chairs and sofas also capture the sun's energy. Similarly, the darkly coloured panelling holds the sun's energy rather than reflecting it as lightly coloured walls would.

In this paragraph, the subject of the first sentence is *materials.* Although the same word is not the subject of the sentences that follow, the subjects of all those sentences are kinds of materials, namely the *special floor covering, heavy fabrics,* and *darkly coloured panelling.* Thus, although the specific word placed in the subject position of the various sentences changes, the general topic is the same.

The passive voice can help maintain focus

One important implication of the preceding discussion is that you sometimes will be able to follow Guideline 5 only by using the passive voice. In the discussion of Guideline 3, you learned that it is generally desirable to use the active voice, not the passive. However, you also learned that sometimes the passive is more appropriate than the active, even preferable. One such time occurs when the passive voice enables you to avoid a needless shift in the topic of two adjacent sentences. Consider the following paragraph:

Focus shifts in the third sentence

> Tom works in the Paint Department. On Tuesday, he finished lunch late, so he took a shortcut back to his workstation. Fifteen metres above the factory floor, a can of paint slipped off a scaffold and hit him on the left foot. Consequently, at the busiest part of the year, he missed seventeen days of work.

The topic of most of the sentences in this accident report is "Tom" or "he." However, the third sentence shifts the topic from Tom to the can of paint. Furthermore, because the third sentence shifts, the fourth must also shift to bring the focus back to Tom. The writer could avoid these two shifts by rewriting the third sentence so that it is about Tom, not about the can of paint. That means making Tom the grammatical subject of the sentence, and, as a result, making the verb passive:

Better third sentence

> He was hit on the left foot by a can of paint that slipped off a scaffold fifteen metres above the factory floor.

2. **Use transitional words.** The preceding discussion explains how you can help your readers follow your flow of thought when two adjacent sentences are about the same topic. In most communications, most sentences shift topics. One way to help your readers follow such shifts is to use transitional words. Here are some of the most commonly used transitional words:

Links in time	after, before, during, until, while
Links in space	above, below, inside
Links of cause and effect	as a result, because, since
Links of similarity	as, furthermore, likewise, similarly
Links of contrast	although, however, nevertheless, on the other hand

3. **Use echo words.** Another way to guide your readers from one sentence to the next is to use echo words. An echo word is a word or phrase that recalls to the readers' minds some information they've already encountered. For example:

> Inflation can be cured. The cure appears to require that consumers change their basic attitudes toward consumption.

In this example, the noun *cure* at the beginning of the second sentence echoes the verb in the first. It tells readers that what follows in the second sentence will discuss the curing they have just read about in the first.

There are many other kinds of echo words:

Pronouns

We had to return the copier. Its frequent breakdowns were disrupting work.

Another word from the same "word family" as the word being echoed

I went to my locker to get my lab equipment. My oscilloscope was missing.

In this example, *oscilloscope* in the second sentence echoes *lab equipment* in the first.

A word or phrase that recalls some idea or theme expressed but not explicitly stated in the preceding sentence

The company also purchased and retired 17,399 shares of its $2.90 convertible, preferred stock at $5.70 a share. These transactions reduce the number of outstanding convertible shares to 635,200.

In the second sentence, the words *these transactions* tell readers that what follows in that sentence concerns the purchasing and retiring that were discussed in the preceding sentence.

4. **Place transitional and echo words at the beginning of the sentence.** Transitional and echo words help readers most when they appear at the beginning of a sentence. In that position, they immediately signal the relationship between that sentence and the preceding one.

Note that if you use *this* or *that* as an echo word at the beginning of a sentence, you should follow it with a noun. If used alone at the beginning of a sentence, you can leave your readers uncertain about what *this* is.

Avoid the ambiguous use of *This* and *That* at the beginning of sentences.

Original

> Our client rejected the R37 compound because it softened at temperatures of about 500 degrees Celsius. This is what our engineers feared.

In this example, the reader would be unsure whether "This" refers to the client's dissatisfaction or to the softening of the R37. The addition of a noun after "This" clears up the ambiguity.

Revised

> Our client rejected the R37 compound because it softened at temperatures of about 500 degrees Celsius. This softening is what our engineers feared.

GUIDELINE 6 Vary Your Sentence Length and Structure

If all the sentences in a sentence group have the same structure, two problems arise: Monotony sets in, and (because all the sentences are basically alike) you lose the ability to emphasize major points and de-emphasize minor ones.

You can avoid such monotony and loss of emphasis in two ways:

- **Vary your sentence length.** Longer sentences can be used to show the relationships among ideas. Shorter sentences provide emphasis in a context of longer sentences.

Short sentences used for emphasis

> In April, many amateur investors jumped back into the stock market because they believed that another rally was about to begin. They noted that exports were increasing rapidly, which they felt would strengthen the dollar in overseas monetary markets and bring international investors back to Bay Street. Also, they observed that unemployment had dropped sharply, which they also predicted would be taken as an encouraging sign for the economy. They were wrong on both counts. Bay Street interpreted rising exports to mean that goods would cost more at home, and it predicted that falling unemployment would mean a shortage of workers, hence higher prices for labour. Where amateur investors saw growth, Bay Street saw inflation.

- **Vary your sentence structure.** For example, the grammatical subject of the sentence does not have to be the sentence's first word. In fact, if it did, the English language would lose much of its power to emphasize more important information and to de-emphasize less important information.

  One alternative to beginning a sentence with its grammatical subject is to begin with a clause that indicates a logical relationship.

Introductory clause

> After we complete our survey, we will know for sure whether the proposed site for our new factory was once an Aboriginal camping ground.

Introductory clause

> Because we have thoroughly investigated all the alternatives, we feel confident that a pneumatic drive will work best and provide the most reliable service.

SELECTING WORDS

When selecting words, your first goal should be to increase the usability of your writing by enabling your readers to grasp your meaning quickly and accurately. At the same time, you need to keep in mind that your word choices affect your readers' attitudes toward you and your subject matter, so you also need to choose words that will increase your communication's persuasiveness.

GUIDELINE 1 Use Concrete, Specific Words

Almost anything can be described either in relatively abstract, general words or in relatively concrete, specific ones. You may say that you are writing on a piece of *electronic equipment* or that you are writing on *a laptop computer connected to a colour laser printer*. You may say that your employer produces *consumer goods* or that it makes *cell phones*.

You can increase the clarity, and therefore the usability, of your writing by using concrete, specific words rather than abstract, general ones. Concrete, specific words help your readers understand precisely what you mean. If you say that your company produces television shows for a *younger demographic segment,* they won't know whether you mean *teenagers* or *toddlers.* If you say that you study *natural phenomena,* your readers won't know whether you mean *volcanic eruptions* or the *migration of monarch butterflies.*

Such vagueness can hinder readers from getting the information they need to make decisions and take action. Consider the following sentence from a memo addressed to an upper-level manager who wanted to know why production costs were up:

Original | The cost of one material has risen recently.

This sentence doesn't give the manager the information she needs to take remedial action. In contrast, the following sentence, using specific words, tells precisely what the material is, how much the price has risen, and the period in which the increase took place.

Revised | The cost of the bonding agent has tripled in the past six months.

Of course, abstract and general terms do have important uses. For example, in scientific, technical, and other specialized fields, writers often are concerned with making general points, describing the general features of a situation, and providing general guidance for action. Your objective when choosing words is not to avoid abstract, general words altogether, but rather to avoid using them when more specific words are required by your readers.

GUIDELINE 2 Use Specialized Terms When—and Only When—Your Readers Will Understand Them

You can also increase the usability and persuasiveness of your writing by using wisely the specialized terms of your own profession.

In some situations, specialized terms help you communicate effectively:

- **They convey precise, technical meanings economically.** Many terms have no exact equivalent in everyday speech and would take many sentences or even paragraphs to explain to someone who isn't familiar with them.
- **They help you establish credibility.** When you use the special terms of your field accurately, you show your fellow specialists that you are adept in it.

On the other hand, if you use technical terms when communicating to people who are unfamiliar with them, you will make your message very difficult to understand and use. Consider the following sentence:

> The major benefits of this method are smaller in-gate connections, reduced breakage, and minimum knock-out—all leading to great savings.

Although this sentence would be perfectly clear to any manager who works in a foundry that manufactures parts for automobile engines, it would be unintelligible to most other people because of the use of the specialized terms *in-gate connections* and *knock-out.*

How to Identify Words Your Readers Won't Know

To identify words you need to avoid, consider all specialized terms from your *readers'* point of view. Use the portrait of your readers that you created by following Chapter 2's guidelines to determine how familiar they are with your field.

The task of identifying words readers may not know can be complicated by the fact that in many fields—perhaps including yours—some specialized terms are widely known but others are not. For instance, most people are familiar enough with chemistry to know what an acid is, and many have some sense of what a base is. But far fewer know what a polymer is. When addressing people who are not familiar with your specialty, you must distinguish between those technical terms your readers know and those they don't.

When addressing readers from another culture, ask someone familiar with that culture to review the words you've chosen. English words that sound like words in another language can have a completely different meaning. Only after Chevrolet introduced its Nova car to Latin America did it realize that in Spanish "No va" means "It doesn't go." Sales were slight until the name was changed (Grosse & Kujawa, 1988). In the 1920s, when Coca-Cola introduced its beverage in China, the company selected for its logo a series of Chinese characters that, when pronounced, sounded like the name of the beverage: *Ke Kou Ke La.* Later, they learned that the characters mean "Bite the wax tadpole." The characters used on Chinese Coke bottles today mean "Happiness in the mouth" (Ricks, 1983).

How to Explain Unfamiliar Terms If You Must Use Them

So far, this discussion has advised you to avoid using specialized terms when addressing readers who do not understand them. However, sometimes you may need to use such terms—for instance, when addressing a large audience that includes some people in your field and some outside of it or when explaining an entirely new subject not familiar to any of your readers.

In such cases, there are several ways to define the terms for readers who are not familiar with them.

Defining Terms Your Readers Don't Know

1. **Give a synonym.** Example: On a boat, a rope or cord is called a *line.*
2. **Give a description.** Example: The *exit gate* consists of two arms that hold a jug while it is being painted and then allow it to proceed down the production line.
3. **Make an analogy.** Example: An atom is like a miniature solar system in which the nucleus is the sun and the electrons are the planets that revolve around it.
4. **Give a classical definition.** In a classical definition, you define the term by naming some familiar group of things to which it belongs and then identifying the key distinction between the object being defined and the other members of the group. Examples:

Word		Group	Distinguishing Characteristic
A crystal	is a	solid	in which the atoms or molecules are arranged in a regularly repeated pattern.
A burrow	is a	hole in the ground	dug by an animal for shelter or habitation.

GUIDELINE 3 Use Words Accurately

Whether you use specialized terms or everyday ones and whether you use abstract, general terms or concrete, specific ones, you must be careful to use all your words accurately. This point may seem obvious, but inaccurate word choice is all too common in on-the-job writing. For example, people often confuse *imply* (meaning to *suggest* or *hint,* as in "He implied that the operator had been careless") with *infer* (meaning to draw a *conclusion based upon evidence,* as in "We infer from your report that you do not expect to meet the deadline"). It's critical that you avoid such errors. They distract your readers from your message by drawing their attention to your problems with word choice, and they may lead your readers to believe that you are not skillful or precise in other areas—such as laboratory techniques or analytical skills.

How can you ensure that you use words accurately? There's no easy way. Consult a dictionary whenever you are uncertain. Be especially careful when using words that are not yet part of your usual vocabulary. Pay careful attention as well to the way words are used by other people.

GUIDELINE 4 Choose Words with Appropriate Associations

The three guidelines for choosing words that you have just read relate to the literal or dictionary meaning of words. At work, you must also consider the associations your words have for your readers. Two kinds of associations you should be especially sensitive to are *connotation* and *register.*

Connotation

Connotation is the extended or suggested meaning that a word has beyond its literal meaning. For example, according to the dictionary, *flatfoot* and *police detective* are synonyms, but they connote very different things: *flatfoot* suggests a plodding, perhaps not very bright cop, while *police detective* suggests a highly trained professional.

Verbs, too, have connotations. For instance, to *suggest* that someone has overlooked a key fact is not the same as to *insinuate* that she has. To *devote* your time to working on a client's project is not the same as to *spend* your time on it.

Research on the impact of connotation

The connotations of your words can shape your audience's perceptions of your subject matter. To demonstrate this effect, researchers Raymond W. Kulhavy and Neil H. Schwartz (1981) wrote two versions of a description of a company that differed from one another in only seven words scattered throughout the 246 words in the entire description. In one version, the seven words suggested flexibility, such as *asked* and *should.* In the second version, those seven words were replaced by ones that suggested stiffness, such as *required* and *must.* Consider the following sentence from the first version:

First version | Our sales team is constantly trying to locate new markets for our various product lines.

In the second version of this sentence, the researchers replaced the flexible word *trying* with the stiff word *driving.*

Second version | Our sales team is constantly driving to locate new markets for our various product lines.

None of the substitutions changed the facts of the overall passage.

The researchers found that people who read the flexible version believed that the company would actively commit itself to the welfare and concerns of its employees, actively promote employment equity programs for visible minorities, receive relatively few labour grievances, and pay its employees well. People who read that version also said they would recommend the company to a friend as a place to work. People who read the stiff version reported opposite impressions of the company. That people's impressions of the company could be affected so dramatically by just seven nonsubstantive words demonstrates the great importance of paying attention to the connotations of the words you use.

Register

Linguists use the term *register* to identify a second type of association exhibited by words. A word's register is the type of communication in which people expect the word to appear. At work, you need to use words whose register matches the type of communication you are preparing. In an advertisement, you might say that your restaurant gives *amazingly* good service, but you would not say the same thing about your engineering consulting firm in a letter to a prospective client. The word *amazingly* has the register of consumer advertising but not of correspondence with business clients.

If you inadvertently choose words with the wrong register, you may give the impression that you don't fully grasp how business is conducted in your field, and your credibility can be lost. As you choose words, be sensitive to the kinds of communications in which you usually see them used.

GUIDELINE 5 Choose Plain Words over Fancy Ones

Another way to make your writing easy to understand is to avoid using fancy words where plain ones will do. At work, people often do just the opposite, perhaps because they think fancy words sound more official or make the writer seem more knowledgeable. The following list identifies some commonly used fancy words; it includes only verbs but might have included nouns and adjectives as well.

Fancy Verbs	Common Verbs
ascertain	find out
commence	begin
compensate	pay
constitute	make up
endeavour	try
expend	spend
fabricate	build
facilitate	make easier
initiate	begin
prioritize	rank
proceed	go
terminate	end
transmit	send
utilize	use

There are two important reasons for preferring plain words over fancy ones:

- **Plain words promote efficient reading.** Research has shown that even if your readers know both the plain word and its fancy synonym, they will still comprehend the plain word more rapidly (Klare, 1977).
- **Plain words reduce your risk of creating a bad impression.** If you use words that make for slow, inefficient reading, you may annoy your readers or cause them to conclude that you are behaving pompously, showing off, or trying to hide a lack of ideas and information behind a fog of fancy terms. Consider, for instance, the effect of the following sentence, which one writer included in a job application letter:

Pompous word choices

> I am transmitting the enclosed résumé to facilitate your efforts to determine the pertinence of my work experience to your opening.

Don't misunderstand this guideline, however. It doesn't suggest that you should use only simple language at work. When addressing people with vocabularies comparable to your own, use all the words at your command, provided that you use them accurately and appropriately. This guideline merely cautions you against using needlessly inflated words that bloat your writing and may open you to criticism from your readers.

GUIDELINE 6 Ethics Guideline: Use Inclusive Language

For another discussion of stereotypes and ethics, see pages 89–90.

When constructing your voice, use inclusive language rather than sexist language. Sexist language supports negative stereotypes. Usually, these stereotypes are about women, but they can also adversely affect men in certain professions, such as nursing. By supporting negative stereotypes, sexist language can blind readers to the abilities, accomplishments, and potential of very capable people. The same is true of language that insensitively describes people with disabilities, illnesses, or other limitations.

What about Miss, Mrs., and Ms.?

People are sometimes unsure whether to use the traditional terms *Miss* or *Mrs.* or the newer term *Ms.* On one hand, people charge that using the older terms suggests that a woman's marital status is somehow relevant to her ability to perform her job. After all, they point out, all men, whether married or single, are addressed as *Mr.* On the other hand, some women prefer to be addressed as either *Mrs.* or *Miss*. If you know an individual's preference, follow it. If you don't know the individual's preference, use *Ms.*, which has been accepted as the non-sexist term in the workplace.

Using Inclusive Language

1. Use nouns and pronouns that are gender-neutral rather than ones containing the word *man.*

   Instead of: businessman, workman, mailman, salesman
   Use: businessperson, manager, *or* executive; worker; mail carrier; salesperson

   Instead of: manmade, man hours, man-sized job
   Use: synthetic, working hours, large job

2. Use plural pronouns or *he or she* instead of sex-linked pronouns when referring to people in general.

   Instead of: "Our home electronics cater to the affluent shopper. She looks for premium products and appreciates a stylish design."
   Use the plural: "Our home electronics cater to affluent shoppers. They look for premium products and appreciate a stylish design."

   Instead of: "Before the owner of a new business files the first year's tax returns, he might be wise to seek advice from a certified public accountant."
   Use *he or she*: "Before the owner of a new business files the first year's tax returns, he or she might be wise to seek advice from a certified public accountant."

3. Refer to individual men and women in a parallel manner.

   Instead of: "Mr. Sundquist and Anna represented us at the trade fair."
   Use: "Mr. Sundquist and Ms. Tokagawa represented us at the trade fair" or "Christopher and Anna represented us at the trade fair."

4. Revise salutations that imply the reader of a letter is a man.

   Instead of: Dear Sir, Gentlemen
   Use: The title of the department or company or the job title of the person you are addressing: Dear Human Resources Department, Dear Switzer Plastics Corporation, Dear Director of Research

5. When writing about people with disabilities, refer to the person first, then the disability.

   Instead of: the disabled, mentally retarded people
   Use: people with disabilities, people with developmental challenges

WWW For additional chapter resources, visit Chapter 4 at www.techcomm.nelson.com.

CONCLUSION

Your writing style can make a great deal of difference to the success of your writing. The voice you use, the sentence structures you employ, and the words you select affect both your readers' attitudes toward you and your subject matter and also the readability and impact of your writing. This chapter has suggested many things you can do to develop a highly usable, highly persuasive style. Underlying all these suggestions is the advice that you take the reader-centred approach of considering all your stylistic choices from your readers' point of view.

EXERCISES

For additional exercises, visit www.techcomm.nelson.com.

Expertise

1. Imagine that you are the head of the Safety and Security department at your school. Faculty and staff have been parking illegally, sometimes where there aren't parking spots. Sometimes individuals without disabilities are parking in spots reserved for those with disabilities. Write two memos to all employees announcing that beginning next week, the Safety and Security department will strictly enforce parking rules, something it hasn't been doing. Write the first memo in a friendly voice and the second in a stern voice. Then compare the specific differences in organization, sentence structure, word choice, and other features of writing to create each voice. (Thanks to Don Cunningham for the idea for this exercise.)

2. Without altering the meaning of the following sentences, reduce the number of words in them.
   a. After having completed work on the data-entry problem, we turned our thinking toward our next task, which was the processing problem.
   b. Those who plan federal and provincial programs for the elderly should take into account the changing demographic characteristics in terms of size and average income of the composition of the elderly population.
   c. Would you please figure out what we should do and advise us?
   d. The result of this study will be to make total white-water recycling an economical strategy for meeting federal regulations.

3. Rewrite the following sentences in a way that will keep the related words together.
   a. This stamping machine, if you fail to clean it twice per shift and add oil of the proper weight, will cease to operate efficiently.
   b. The plant manager said that she hopes all employees will seek ways to cut waste at the supervisory meeting yesterday.
   c. About 80 percent of our clients, which include over fifteen hundred companies throughout North and South America and a few from Africa, where we've built alliances with local distributors, find the help provided at our website to be equivalent in most cases to the assistance supplied by phone calls to our service centres.
   d. Once they wilt, most garden sprays are unable to save vegetable plants from complete collapse.

4. Rewrite the following sentences to put the action in the verb.
   a. The experience itself will be an inspirational factor leading the participants to a greater dedication to productivity.
   b. The system realizes important savings in time for the administrative staff.
   c. The implementation of the work plan will be the responsibility of a team of three engineers experienced in these procedures.
   d. Both pulp and lumber were in strong demand, even though rising interest rates caused the drying up of funds for housing.

5. Rewrite the following sentences in the active voice.
   a. Periodically, the shipping log should be reconciled with the daily billings by the Accounting Department.
   b. Fast, accurate data from each operating area in the foundry should be given to us by the new computerized system.
   c. Since his own accident, safety regulations have been enforced much more conscientiously by the shop foreman.
   d. No one has been designated by the manager to make emergency decisions when she is gone.

6. In three of the following pairs of sentences, the topic shifts from the first to the second sentence. Rewrite one or the other sentence so that the topics are the same.
   a. "Grab" samplers collect material from the floor of the ocean. Rock, sediment, and benthic animals can be gathered by these samplers at rates as high as 8,000 tonnes per hour.
   b. To fluoridate the drinking water, a dilute form of hydrofluorisilic acid is added directly to the municipal water supply at the main pump. An automatic control continuously meters exactly the right amount of the acid into the water.
   c. Fourteen variables were used in these calculations. The first seven concern the volume of business generated by each sales division each week.
   d. The city's low-income citizens suffer most from the high prices and limited selection of food products offered by commercial grocers. Furthermore, information concerning nutrition is difficult for many low-income citizens to find.

7. Create a one-sentence, classical definition for a word used in your field that is not familiar to people in other fields. The word might be one that people in other fields have heard of but cannot define precisely in the way specialists in your field do. Underline the word you are defining. Then circle and label the part of your definition that describes the familiar group of items that the defined word belongs to. Finally, circle and label the part of your definition that identifies the key distinction between the defined word and the other items in the group. (Note that not every word is best defined by means of a classical definition, so it may take you a few minutes to think of an appropriate word for this exercise.)

8. Create an analogy to explain a word used in your field that is unfamiliar to most readers. (Note that not every word is best defined by means of an analogy, so it may take you a few minutes to think of an appropriate word for this exercise.)

Online

Using your word processing software, examine the readability statistics for two communications. These might be two projects you're preparing for courses, or they might be a course project and a letter or email to a friend or family member. What differences, if any, do you notice in the statistics? What accounts for the differences? Are the statistics helpful to you in understanding and constructing an effective writing style in either case? Read the explanations of the scores that are provided with your word processing software; these may be provided in the program's Help feature. Do the interpretations of your scores agree with your own assessment of your communications? If not, which do you think is more valid? (To learn how to obtain the readability statistics with your program, use its Help feature.)

Collaboration

Working with another student, examine the memo shown in Figure 4.1. Identify places where the writer has ignored the guidelines given in this chapter. You may find it helpful to use a dictionary. Then write an improved version of the memo by following the guidelines in this chapter.

Ethics

1. Find a communication that fails to use inclusive language and revise several of the passages to make them inclusive.

2. The images in advertising often rely on stereotypes. Find one advertisement that perpetuates one or more stereotypes and one that calls attention to itself by using an image that defies a stereotype. Evaluate the ethical impact of each image. Present your results in the way your professor requests.

FIGURE 4.1

Memo for Collaboration Exercise

MEMO

July 8, 2007

TO: Gina MacIntyre, Vice-President, Eastern Region

FROM: Brent Willard, Branch Manager, Ottawa Area Offices

The ensuing memo is in reference to provisions for the cleaning of the six offices and two workrooms in the Elgin Street building in Ottawa. This morning, I absolved Thomas's Janitor Company of its responsibility for cleaning the subject premises when I discovered that two of Thomas's employees had surreptitiously been making unauthorized long-distance calls on our telephones.

Because of your concern with the costs of running the Ottawa area offices, I want your imprimatur before proceeding further in making a determination about procuring cleaning services for this building. One possibility is to assign the janitor from the St. Laurent Boulevard building to clean the Elgin Street building also. However, this alternative is judged impractical because it cannot be implemented without circumventing the reality of time constraints. While the St. Laurent janitor could perform routine cleaning operations at the Elgin Street establishment in one hour, it would take him another ninety minutes to drive to and fro between the two sites. This is more time than he could spare and still be able to fulfill his responsibilities at the Elgin Street building.

Another alternative would be to hire a full-time or part-time employee precisely for the Elgin Street building. However, that building can be cleaned so expeditiously, it would be irrational to do so.

The third alternative is to search for another janitorial service. I have now released two of these enterprises from our employ in Ottawa. However, our experiences with such services should be viewed as bad luck and not affect our decision, except to make us more aware that making the optimal selection among companies will require great care. Furthermore, there seems to be no reasonable alternative to hiring another janitorial service.

Accordingly, I recommend that we hire another janitorial service. If you agree, I can commence searching for this service as soon as I receive a missive from you. In the meantime I have asked the employees who work in the Elgin Street building to do some tidying up themselves and to be patient.

CHAPTER 5 Communicating Electronically: Email and Websites

GUIDELINES

Guidelines for Using Email

1. Observe the email conventions where you work
2. Keep your messages brief
3. Make your messages easy to read on screen
4. Provide an informative, specific subject line
5. Take time to revise
6. Remember that email isn't private

Guidelines for Creating Websites

1. Begin by defining your site's objectives
2. Provide quick and easy access to the information your readers want
3. Design pages that are attractive and easy to read

4 Design your site for global readers

5 Enable readers with disabilities to use your site

6 Help readers find your site on the Internet

7 Test your site on multiple platforms and browsers before launching it

8 Keep your site up to date

9 Ethics Guideline: Respect intellectual property and provide valid information

DEFINING OBJECTIVES
APPLYING
CHOOSING
PLANNING
DEVELOPING
EVALUATING

In this chapter, you will learn how to work effectively in a workplace that is becoming more reliant on computer-based communication. First, you'll learn how to use email effectively on the job, where conventions are distinctly different from those of the personal email you exchange with friends and family members. You'll also find advice for creating informational websites of the kind that organizations use to provide information for their employees, customers, clients, vendors, and investors.

This chapter's guidelines supplement rather than supersede the other guidelines in this book.

You should view this chapter's guidelines as supplementary to those for preparing paper-based communications. When writing for on-screen reading, you should follow the same basic process as described in other chapters: Learn about your audience and define your purpose, choose a format, plan a strategy, develop the elements, evaluate your draft, and revise your communication. The guidelines in this chapter are designed to provide additional advice for addressing the special opportunities and challenges of preparing electronic communications.

USING EMAIL

WWW For additional chapter resources, visit Chapter 5 at www.techcomm.nelson.com

Advantages of email over paper mail

Used effectively, email has some notable advantages over paper mail. Most importantly, you can get your message to your readers almost instantly and your readers can respond without having to print something out, put it into an envelope, and take it to the maildrop. Also, many people use email in situations where previously they made a phone call, because email allows them to exchange information quickly without needing to be free at the same time as the other person. In addition, email lets you send as attachments fully formatted documents, spreadsheets, and graphics over networks.

On the other hand, email has some disadvantages. For example, readers can't scribble notes in the margin, and it's easy to send off messages that, on reflection, you would prefer to have proofread again—or not sent at all.

The following guidelines will help you work effectively with the possibilities and limitations associated with on-screen correspondence. The email message shown in Figure 5.1 (page 110) illustrates their application.

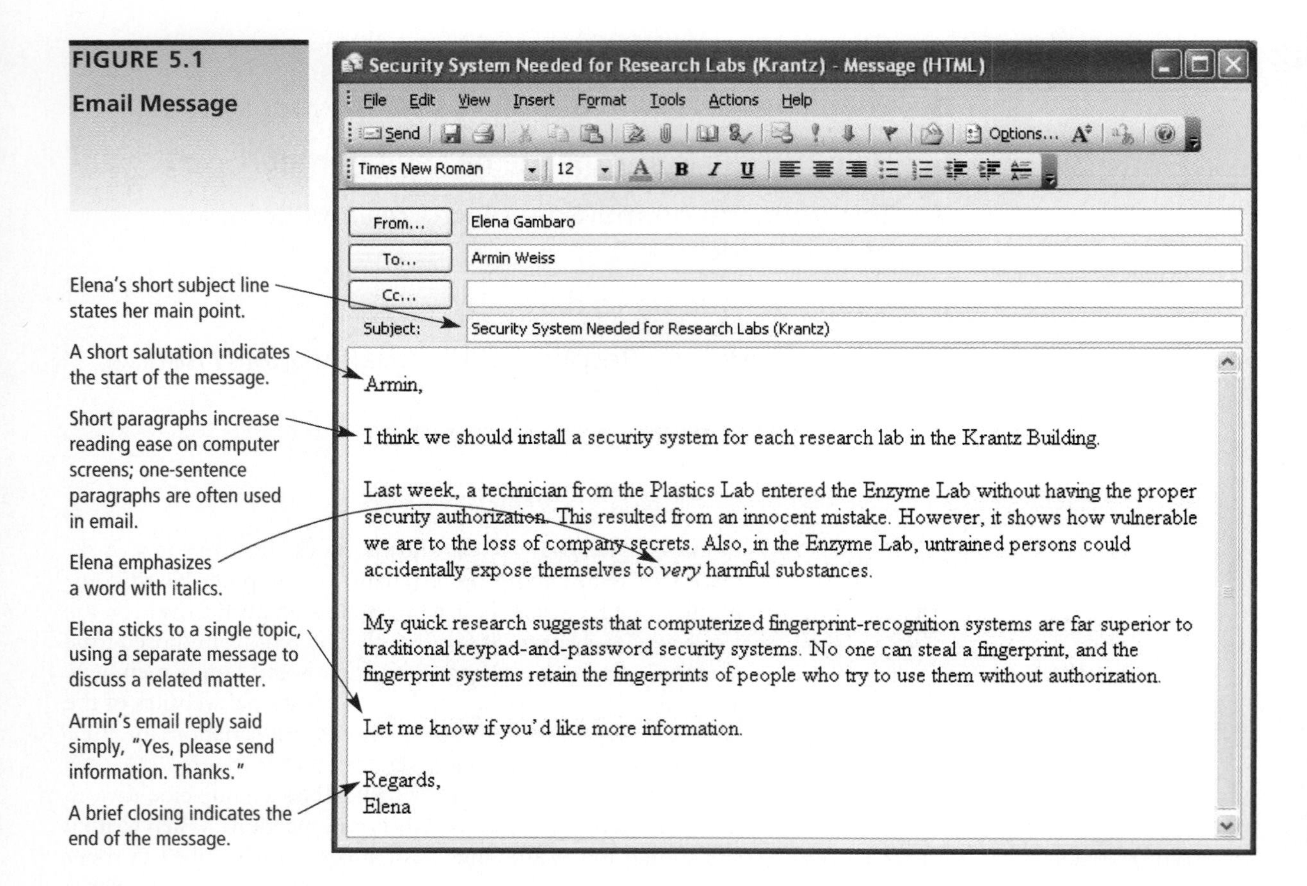

FIGURE 5.1
Email Message

GUIDELINE 1 Observe the Email Conventions Where You Work

Conventions for email vary from one organization to another.

Despite the enormous amount of email being exchanged, a uniform set of conventions for its use in the workplace has yet to emerge. In some organizations, people use slang, abbreviations, and very clipped expressions when using email. They don't polish their writing or even check their spelling. In these organizations, the informality of email is often seen as one of its appealing features: Writers can dash off their core message without investing time in careful revising. It can even become a substitute for discussion, with each member of a team sending messages to all the other team members simultaneously—and everyone contributing freely to the online conversation. In other organizations, however, email is viewed more as a document than as a discussion. There, the writing of email messages is very formal, and an informal email note would seem glaringly out of place (Selber, 1998).

Because conventions vary so greatly from organization to organization, pay careful attention to interorganizational and intraorganizational customs concerning email. Note that your colleagues may use email differently when writing outside the organization than when writing inside it.

Observing E-communication Conventions

- Be careful in your use of instant and text messaging, especially during meetings. Do not allow messaging to distract you from the meeting or conversation you are engaged in.
- Limit your use of emoticons, both graphical and typographical, to less formal communication, since the meaning intended by the emoticon's sender is often unclear to the reader.
- Print and mail any documents that you want your audience to read. Documents sent as email attachments are frequently unopened.

GUIDELINE 2 Keep Your Messages Brief

Brevity is almost always valued at work, but especially with email messages. Many people find it more difficult to read computer screens than printed documents. The letters of the words on screen are less distinct, the scrolling required for long communications makes it difficult to flip back and forth between sections, and people dislike sitting in one position for an extended period, as they must do when using a desktop computer. Here are some tips for keeping your messages brief (Berghel, 1997; Weisband & Reinig, 1995).

Keeping Email Messages Brief

- Use a simple, three-part structure. Succinctly state your topic, quickly summarize background information, and then state your main point and supporting material.
- Exclude any information that is not directly related to your topic.
- Stick to one topic. If you want to write about two things, send two messages.
- When you quote from a previous email message, include only the relevant lines, not the entire earlier message. However, forwarding a message will eliminate the need to repeat any text.

GUIDELINE 3 Make Your Messages Easy to Read On Screen

As with paper-based communications, there are a few things you can do to enhance the readability of your email messages.

Writing for On-Screen Reading

- Write short paragraphs.
- Put blank lines between paragraphs.
- Make headings for longer communications, putting each heading on a line of its own.
- Use lists.
- Use bolding or italics to emphasize words or phrases.

GUIDELINE 4 Provide an Informative, Specific Subject Line

Readers use subject lines to decide whether to read email messages.

Some people receive literally hundreds of email messages a day. When reviewing their inboxes, their first task is to decide which messages to open and which to file or delete unopened. To make this decision, they see who sent each message and what the subject line says. You can ensure that your important messages will be opened if you use your subject line to tell as exactly as possible what's in your message. When doing so, focus on making clear to your readers that your message's contents are valuable to them.

People often save some email messages so that they can refer back to them at a later time. By writing informative, specific subject lines you help your readers quickly locate your message when they want to read it a second time.

GUIDELINE 5 Take Time to Revise

Because email can be written so quickly and because in many workplaces it's acceptable to send fairly rough prose, many people simply key in their message and zip it off. If you do this, however, you risk sending unclear messages that will require additional correspondence to straighten out. Moreover, if you don't take time to reflect on your message, you may send a note while feeling anger or some other intense emotion. Messages sent under these circumstances can create hard feelings. Many email programs allow you to write a message but to postpone sending it. Use this feature when you need to review your message carefully or cool down before mailing it.

GUIDELINE 6 Remember That Email Isn't Private

Many people believe that their email is private. Perhaps they harbour this belief because the word *mail* suggests privacy since we know that the postal service is not permitted to read the letters we send. Or maybe, as Weisband and Reinig (1995) suggest, people assume that because they use a password to enter their email account, no one else has access to the mail in that account. Nonetheless, courts have held that employers have the right to look at all employee email that is on or sent through a company computer (Berghel, 1997).

WWW To download a checklist for writing email messages, visit www.techcomm.nelson.com.

Even if your employer doesn't read the email in your mailbox, your messages may be read by many people you didn't intend to see them. With a mere click of the mouse, the people to whom you send a message can forward it to anyone else who has email capability.

To protect your own privacy and avoid embarrassment or misunderstanding, do the following.

Protecting the Privacy of Your Email

- Never include anything in an email message that you aren't prepared for a large audience to read.
- Be careful about mixing personal and work-related topics in the same email message.
- Never write something about a person that you wouldn't say directly to that person.
- Do not send email when feeling a strong emotion.
- If an email exchange becomes emotionally charged, talk to the person directly.
- Never include confidential information.

CREATING WEBSITES

In your career, there's a good chance your employer will ask you to create or help create a website. Employers use websites for many purposes, such as advertising products and services, conducting business transactions, providing support to clients and customers, and enabling their employees to access data they must have to perform their jobs. In fact, many organizations have several sites, including one or more that are open to the world through the World Wide Web and others that only authorized employees can access because the sites contain proprietary research results, engineering plans, sales analyses, or other information the employer wants to keep confidential.

Websites have the same objectives as any other workplace communication: usability and persuasiveness.

To be effective, a website must possess the same essential qualities as all other workplace communications: usability and persuasiveness. Consequently, many of the strategies you use when creating reader-centred print documents will also help you construct effective websites. However, to construct successful, reader-centred websites, you must also develop new communication skills. That's because reading screens differs significantly from reading paper.

Reading a website is very different from reading a print communication.

- **Websites present readers with a less accommodating reading environment than the printed page.** The act of scrolling, the variable resolution of computer screens, and the fact that screens aren't as movable or flexible as paper makes reading on screen more difficult than reading print. People using your site rely on you to make it as easy as possible for them to read your webpages.
- **Websites require readers to search for information in different ways.** In print communications, every page is on a different surface (the front or back of a different sheet of paper). Consequently, readers can flip quickly through a communication to find the page they want. At a website, however, all pages are on the same

surface: the computer screen. As a result, readers can't flip through the pages of a website. Also, the pages of a print communication are arranged in a linear order, and pages remain in that order so that readers know what preceded and what follows each page. At a website, however, the pages do not have a fixed order, and moving from one specific page to another specific page can be difficult. Readers depend on you to construct a site that makes it as easy as possible to find what they want in this non-linear space.

- **A website's appearance is partly controlled by the reader.** When you design a printed document, you determine exactly what your readers will see. In contrast, the size of your readers' computer screens, the browsers they use, and the browser settings they choose all affect the way your website will look on their computers. Readers depend on you to create a website that will be easy to use and look good regardless of their hardware, software, and settings.

The rest of this chapter briefly explains how websites work, and presents nine guidelines that will help you build the communication expertise needed to create websites that readers find highly usable and highly persuasive.

HOW THE WORLD WIDE WEB WORKS

Only two simple components orchestrate the eye-catching and mind-catching array of features and capabilities presented at many websites on the World Wide Web.

- **Webpage files.** Webpage files consist of commands that tell the browser how to create the images that you see on your screen. These commands are written in a special computer language such as HTML (hypertext markup language) or XML (extensible markup language).
- **Web browser.** A Web browser performs three functions: It locates the webpage file for the webpage you want to visit, it copies the file to your computer's memory, and it uses the instructions in the file to construct the display you see on your screen.

The interaction of these two elements—webpage files and Web browsers—allows you to view millions of pages created by thousands of individuals and organizations in hundreds of countries around the world. Figure 5.2 shows a webpage that uses many common webpage features. Figure 5.3 (page 116) shows the file that created this page. Many online sources provide free guidance in writing the computer code illustrated in Figure 5.3. Alternatively, you can use a software program called a Web editor that lets you design the appearance of the page, and then translates your designs into a file with the necessary commands. Some of these programs, such as Dreamweaver and FrontPage, can be purchased separately; others are included in word processing programs, such as Microsoft Word, and browsers, such as Netscape Composer. Still other Web editors are available for free through the Internet.

FIGURE 5.2

Webpage Created by the File in Figure 5.3

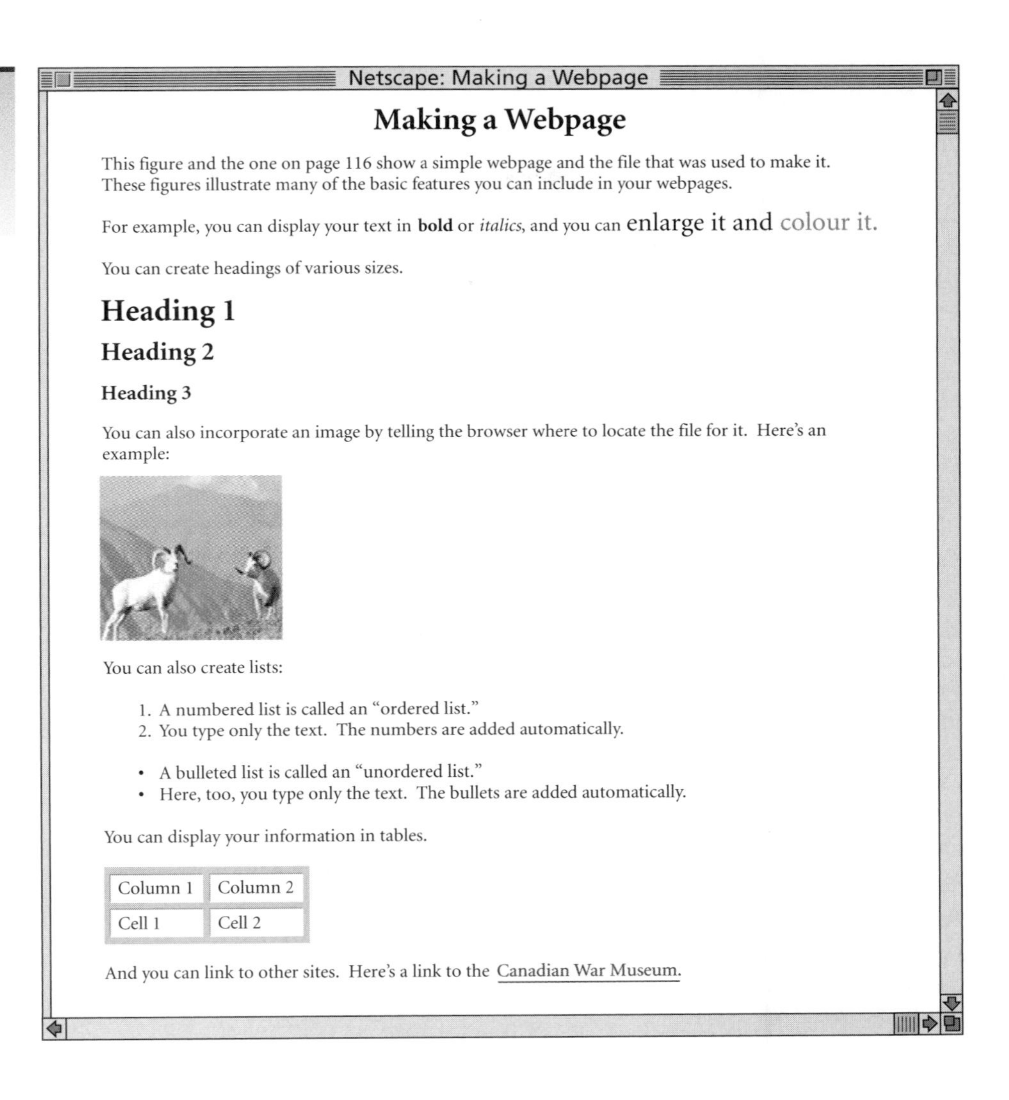

GUIDELINE 1 Begin by Defining Your Site's Objectives

Start by following Chapter 2's guidelines for defining your communication's objectives.

Begin creating a website in the same way you would begin fashioning any other work-related communication: by thinking about your audience and purpose. Don't let the novelty of the Web distract you from carefully defining your site's usability and persuasive objectives. Whom do you want to visit your site? What tasks will they want to accomplish while there? What attitudes will they bring to your site concerning your subject matter, your employer's organization, or websites in general? How do you want to influence their attitudes? Answer these questions as specifically as you can. Chapter 2's guidelines for defining objectives will be as helpful to you when you are creating a site as when you are preparing any written or oral communication.

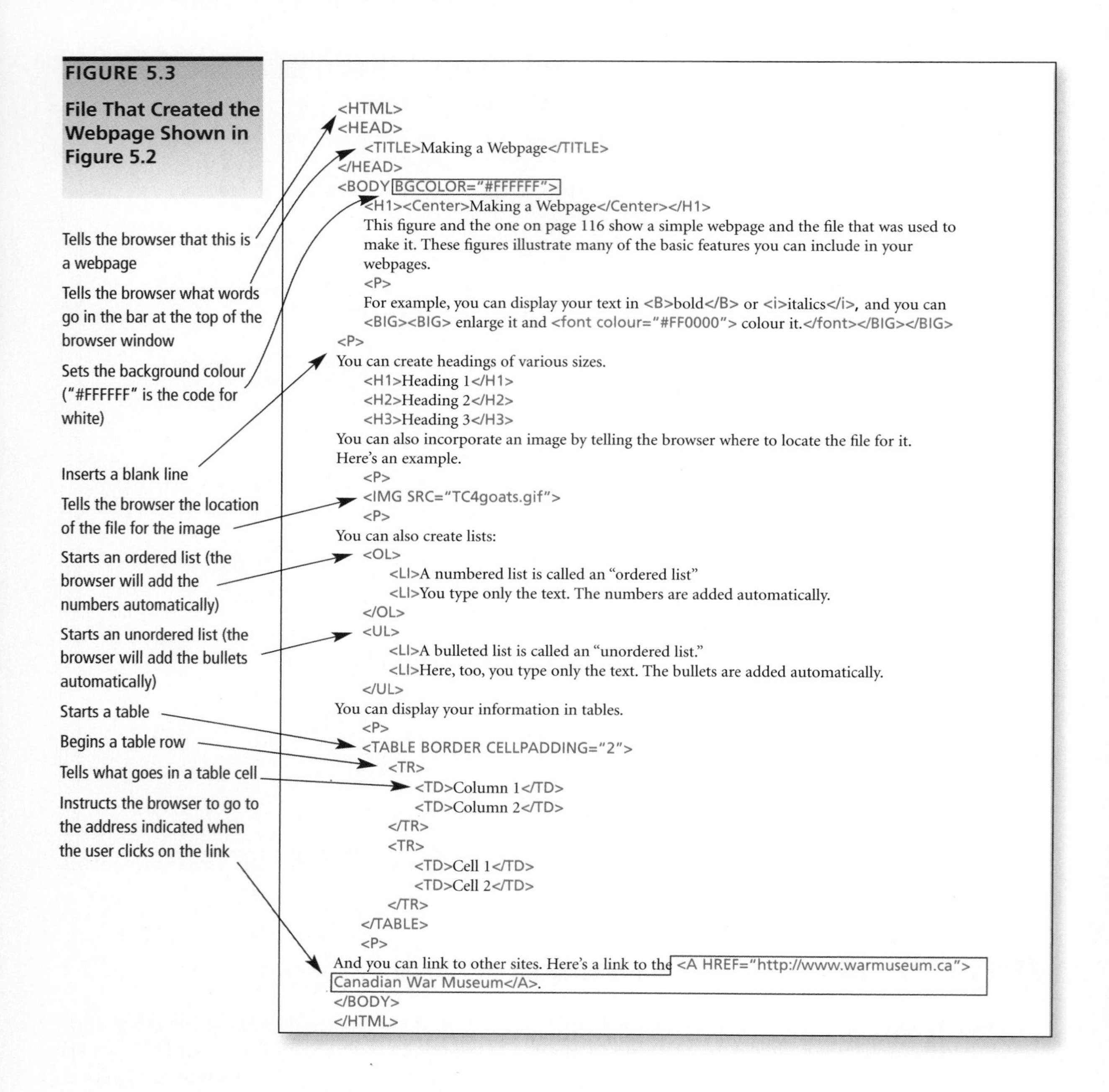

```
<HTML>
<HEAD>
   <TITLE>Making a Webpage</TITLE>
</HEAD>
<BODY BGCOLOR="#FFFFFF">
   <H1><Center>Making a Webpage</Center></H1>
   This figure and the one on page 116 show a simple webpage and the file that was used to
   make it. These figures illustrate many of the basic features you can include in your
   webpages.
   <P>
   For example, you can display your text in <B>bold</B> or <i>italics</i>, and you can
   <BIG><BIG> enlarge it and <font colour="#FF0000"> colour it.</font></BIG></BIG>
<P>
You can create headings of various sizes.
   <H1>Heading 1</H1>
   <H2>Heading 2</H2>
   <H3>Heading 3</H3>
You can also incorporate an image by telling the browser where to locate the file for it.
Here's an example.
   <P>
   <IMG SRC="TC4goats.gif">
   <P>
You can also create lists:
   <OL>
      <LI>A numbered list is called an "ordered list"
      <LI>You type only the text. The numbers are added automatically.
   </OL>
   <UL>
      <LI>A bulleted list is called an "unordered list."
      <LI>Here, too, you type only the text. The bullets are added automatically.
   </UL>
You can display your information in tables.
   <P>
   <TABLE BORDER CELLPADDING="2">
      <TR>
         <TD>Column 1</TD>
         <TD>Column 2</TD>
      </TR>
      <TR>
         <TD>Cell 1</TD>
         <TD>Cell 2</TD>
      </TR>
   </TABLE>
   <P>
And you can link to other sites. Here's a link to the <A HREF="http://www.warmuseum.ca">
Canadian War Museum</A>.
</BODY>
</HTML>
```


FIGURE 5.3
File That Created the Webpage Shown in Figure 5.2

When thinking about the audience and purpose of a website, however, you should consider a few issues not mentioned in Chapter 2.

- **What type of Internet connection will your readers use?** If your readers will access your site through a high-speed connection, you will be able to use large graphics, tables, movies, and other features that you would want to avoid if your readers will be using a slow connection.

- **What kind of monitor will your readers use?** You should adjust the height and width of the pages on your site to the resolution of the display your readers will be using.
- **What browsers will your readers use?** Enhancements to browsers are being developed constantly. Even though you can create features that use these enhancements, limit yourself to those supported by the browsers that all or almost all of your readers will be using.
- **Will your readers want to print information from your site?** If so, you can asssist them by including an alternative version of some or all of your information in a PDF (Acrobat) or other format suitable for printing and for reading in printed form.

GUIDELINE 2 Provide Quick and Easy Access to the Information Your Readers Want

Research shows that the usability of websites is readers' major concern. The overwhelming majority go to a website to find very specific information or a particular service. Unless you design your site so that they can locate what they want very, very rapidly, they will abandon your site and look elsewhere (Nielsen, 2000). You can speed your readers to their destination by building a reader-centred, three-click hierarchy, by providing plenty of reader-centred navigational aids, and by limiting the loading time of your pages.

Build a Reader-Centred, Three-Click Hierarchy

To review advice for creating a hierarchical organization, go to page 230.

If your site has only a few pages, readers will be able to locate rapidly what they want, provided that your home page includes a menu with links to every other page. If your site is larger, however, such a menu will be impractical. In this case, create a reader-centred hierarchy. In particular, construct your hierarchy so that readers need to make no more than three clicks from the home page to "drill down" to what they want. Creating a three-click hierarchy is a goal commonly adopted for all but the largest sites built in the workplace.

Example of a three-click hierarchy

For instance, imagine that a high school student visits the website for a community college to learn about its programs in fabricating and welding. On the home page, she can choose among Full-Time Programs, Part-Time Studies, Student Services, Community Services, and Alumni. By clicking on Full-Time Programs (click 1), she comes to a page that lists such programs as Applied Arts, Business, Health and Science, Hospitality, and Skilled Trades and Apprenticeships. By clicking on Skilled Trades and Apprenticeships (click 2), she comes to a page that lists all programs alphabetically. A click on Fabricating and Welding (click 3) brings her to an overview of that program. Additional clicks will lead her to additional details, but with her first three, she has drilled down to the specific area of the site that contains the information she is seeking.

Using intuitive categories in your hierarchy

Of course, readers can drill down in three clicks only if they don't make mistakes. Every wrong choice requires one or more clicks to get back to the starting point. To help your readers avoid errors, create categories they will intuitively think of when searching for a topic. For example, if you are creating a site about your college or university, place bylaws for the Students' Association in the area for Students (and in the subcategory for Student Organizations) rather than in the area for General College Policies. Intuitive categories increase readers' accuracy when predicting which link will lead them to the information they want (Schriver, 1997).

Provide Plenty of Reader-Centred Navigational Aids

After you have carefully organized your site in a reader-centred way, you have done half the work needed to help your readers locate the information they want. The other half is providing them with clear, efficient navigational aids that will enable them to move directly to their destinations within that structure. The following strategies, which are illustrated in Figure 5.4, will help you guide your readers quickly to any part of your site.

Helping Readers Navigate Your Site

- **Include the main menu on every page.** Enable your readers to start a new "drill down" from whatever page they are on.
- **On every page, provide a link to the home page.** If a link to your home page isn't in your main menu, be sure to provide it elsewhere on every page.
- **On each page, provide a menu for the category in which that page is located.** Once readers are in a category, they may want to explore the various types of information you provide in it. By providing the category menu, you enable them to do that without backtracking to a higher-level page.
- **Place navigational aids in the same location on every page.** Once readers have learned where your navigational aids are, let them use that knowledge throughout the time they spend at your site. Placing your navigational aids in the same place on every page also helps you create a visually unified site, as explained on page 398.
- **Use clear, intuitive labels for all menu items.** Links are helpful only if readers can accurately predict what they will find if they click on them.

Advice for writing in-text links

- **Make clear, informative in-text links.** When creating links in your text, signal clearly what your readers will find there.
  - *Highlight only the major words.* If you were to highlight all the words in the following sentence, your readers wouldn't know whether the link will lead to information on whales, the Bering Sea, or krill. Highlighting only major words makes clear what the topic of the link is.

    Coloured type indicates the link

    | When whales reach the northern Bering Sea, they feed on the plentiful krill.
  - *Supply explanations of the links.* These are especially helpful for links included in a list.

    | International projects—Read about our current projects in Europe and Asia.
  - *Use precise, specific wording.* Choose words that tell exactly what the linked page will show.
- **Use underlining only for links.** Readers are accustomed to the convention that links are signalled by underlining. To avoid confusion, don't use

underlining for anything but links. Similarly, do *not* use the feature of some webpage editors that would allow you to remove the underlining from your links unless you are considering replacing the underlining with the use of colour.

- **Include a site map.** Many readers use site maps as a navigation shortcut or as an aid to finding information they haven't otherwise been able to locate. See Figure 5.5 (page 120) for an example.

FIGURE 5.4

Navigational Aids for Website Readers

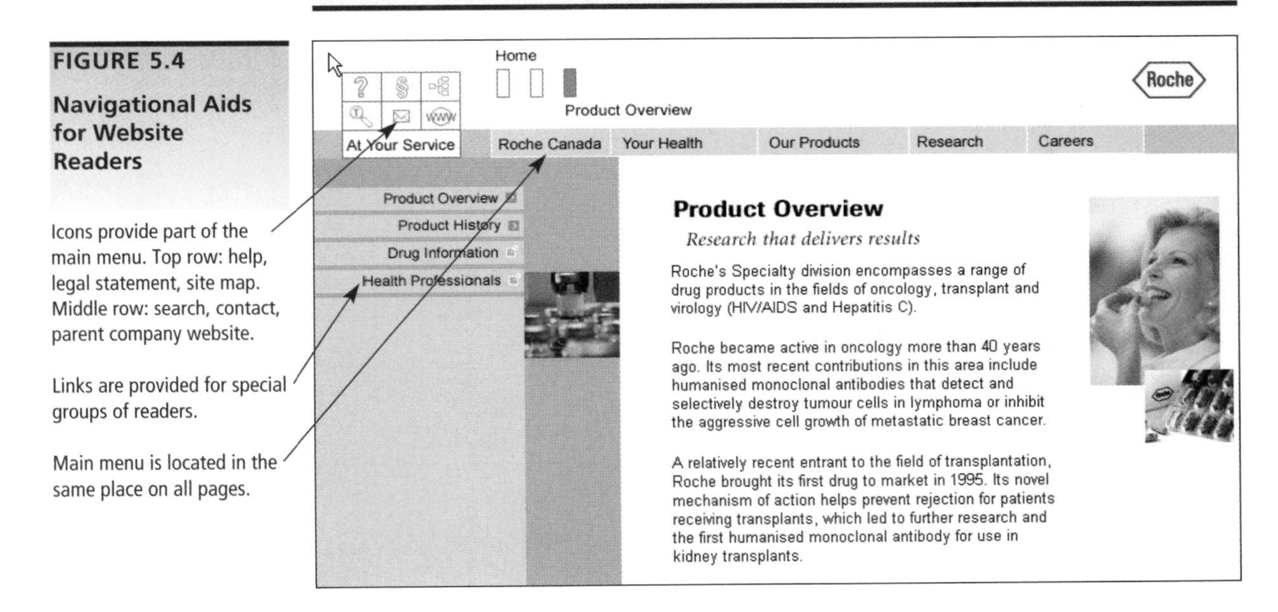

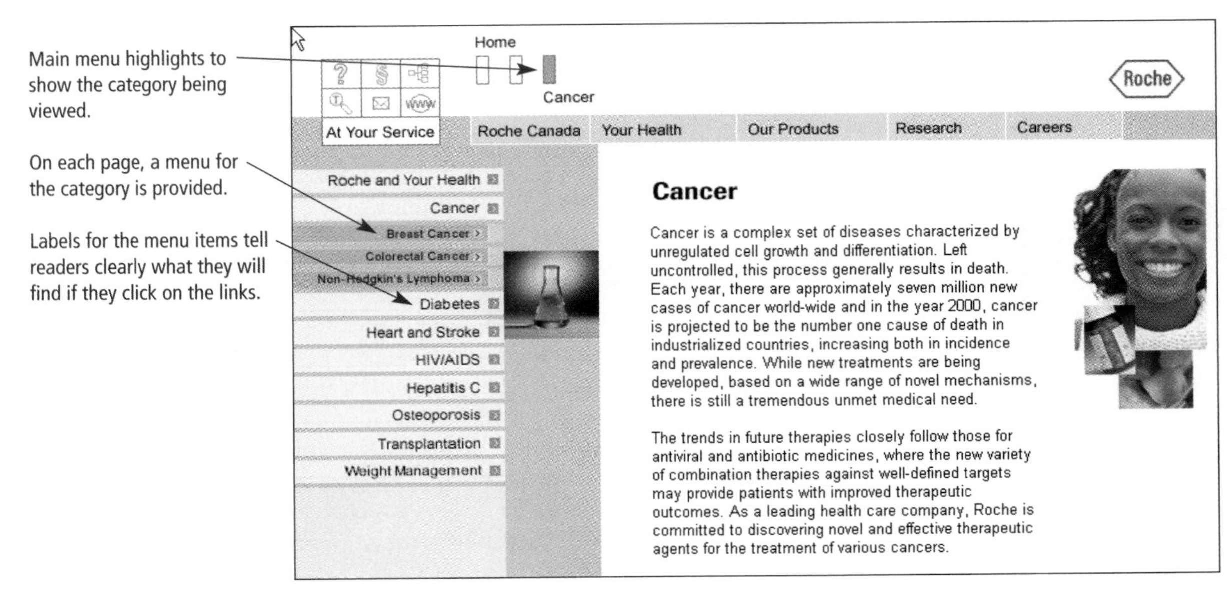

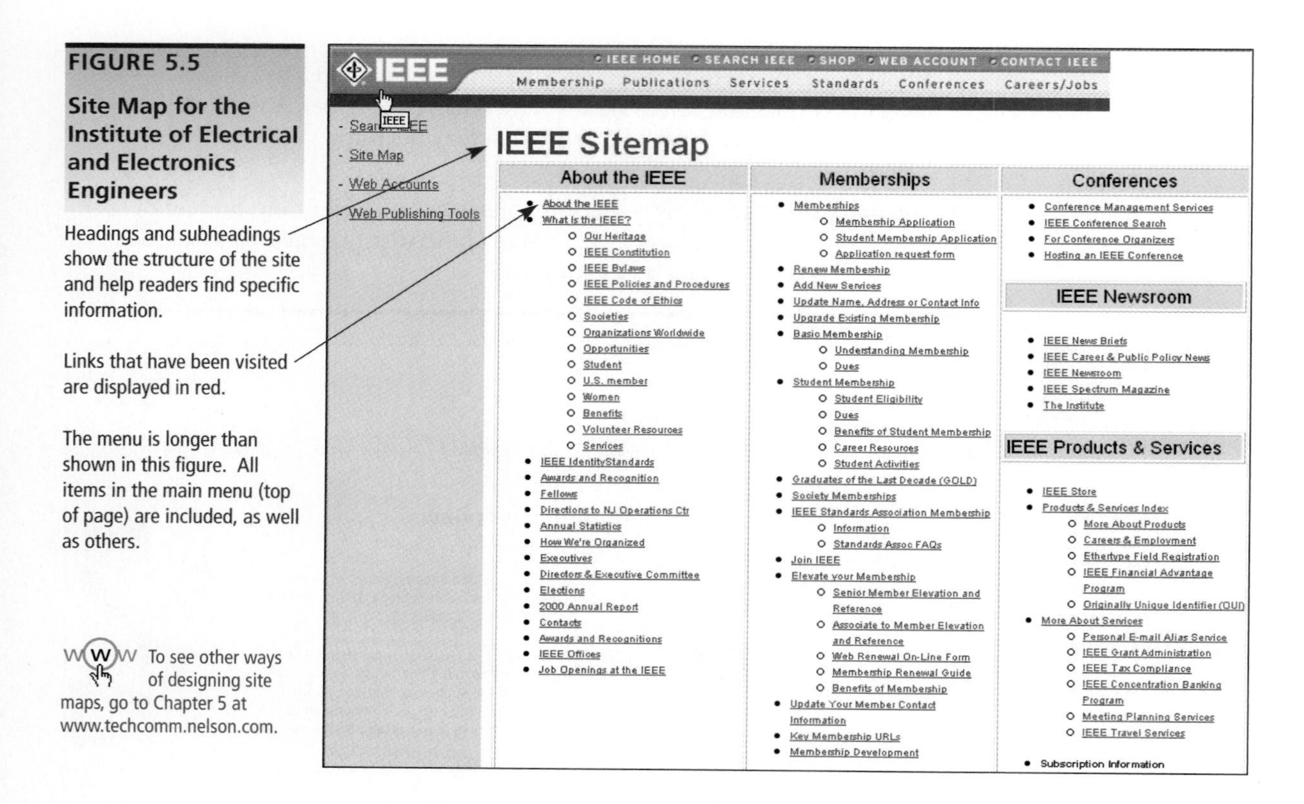

FIGURE 5.5

Site Map for the Institute of Electrical and Electronics Engineers

Headings and subheadings show the structure of the site and help readers find specific information.

Links that have been visited are displayed in red.

The menu is longer than shown in this figure. All items in the main menu (top of page) are included, as well as others.

WWW To see other ways of designing site maps, go to Chapter 5 at www.techcomm.nelson.com.

Limit the Loading Time of Your Pages

You can also increase your readers' ability to locate information quickly by creating pages that load rapidly. The slower the Internet connections your readers will use, the more critical rapid loading becomes. Loading time is determined by two factors:

Two factors affect loading time

- The time it takes to transmit a site's webpage files across the Internet from the server that hosts the site to the reader's computer.
- The time it takes the reader's computer to interpret the files into what the reader sees in his or her browser window.

The following strategies will help you limit the loading time for your webpages.

Limiting the Loading Time of Webpages

- **Where possible, use text rather than images.** You can present some content as an image or as text. For example, you can create a vertical set of links by creating the image of a button for each link or you can present the links in a list created as text, perhaps setting it against a colour background. Because text files are much smaller than image files, they transmit more rapidly over the Internet and they display much more quickly on a computer. Consequently, wherever you have images, ask whether you can achieve your communication objectives with text rather than an image.

- **Limit the number of images you use.** Images are an important element of most webpages, but using too many of them reduces a site's usability by increasing the loading time of each page.
- **Keep images as small as practical.** Smaller images use smaller files. Similarly, display animations and movies in a reasonably small size—and keep them as short as you can while still achieving your purpose.
- **Compress files.** If you are creating images in Photoshop, Illustrator, or a similar program, compress your files using the efficient .jpg and .gif formats.
- **Reuse the same images on different pages.** As explained on page 114, a computer downloads the files for the webpages a reader visits. Usually, this image is stored in the computer's memory. Consequently, when you use the same image on a second page, the reader's computer doesn't need to download that image file again. Where feasible, use the same image rather than alternative images throughout your site. Similarly, refrain from creating a new file in order to alter the colour or other detail of an image. Each new file you create is another file for your readers' computers to download.
- **Break large tables into smaller ones.** Browsers usually won't display a table until it is completely loaded. Break long tables into separate parts so your readers can view the first part while the rest is being processed.

GUIDELINE 3 Design Pages That Are Attractive and Easy to Read

When designing webpages, your goals are the same as when designing printed pages: to create a usable and attractive communication. The same design principles that apply to print communications will help you achieve these goals for your website:

To see how a consistent design unifies a website, turn to page 382.

- Create a grid that aligns related items within your page.
- Use white space to group related items visually.
- Adjust colour and size to establish hierarchy and focus.
- Use a consistent design to unify your site.

However, as explained on page 113, readers find it much more difficult to read communications displayed on screen rather than on paper. Among other things, the resolution of the text on a display may be poorer and the mechanics of scrolling may make it more difficult for a reader's eyes to keep their place in a text. The following strategies will help you address the special challenges of creating legible and attractive webpages. Figure 5.6 (page 123) shows how these strategies were used to create effective webpages.

Making Webpages Attractive and Easy to Read

- Make sure your text is legible.
  - *Use dark letters against a light background (or vice versa) to produce a strong contrast between the two.* White is always a good choice for the background.
  - *Use large enough text for legibility.* Test your type sizes to be sure they will be easy to read at higher resolutions, which make text smaller.
  - *Avoid intricate backgrounds that make it difficult to pick out the letters of the text.*
  - *Avoid italics.* These ornate letters can be very difficult to read on screen.
- Present your information in small, visually distinct chunks.
  - *Keep paragraphs short.* Break longer paragraphs into smaller ones.
  - *Use white space (blank lines) to separate paragraphs and other blocks of text,* keeping in mind that the "white space" on a webpage may actually be any background colour. Some sources recommend that only about 30 percent of the page be text.
  - *Edit your text for conciseness.* The fewer words, the better.
  - *Use plenty of headings.* They not only help readers locate information but also break the screen into meaningful chunks.
- Limit scrolling (it greatly increases the reading difficulty).
  - *Avoid vertical scrolling* except on pages where users will be looking for in-depth information.
  - *Avoid horizontal scrolling altogether.*

GUIDELINE 4 Design Your Site for Global Readers

Although some websites are designed for specific groups of users located in only one nation or culture, others aim to serve global readers. The following strategies will help you create a site that will be easily read by people for whom English is a second or third language and by the computer programs that international readers can use to automatically translate your text into their language. If your employer decides to make a second version of your site in another language, these strategies will also ease that work.

FIGURE 5.6

Readable Webpages

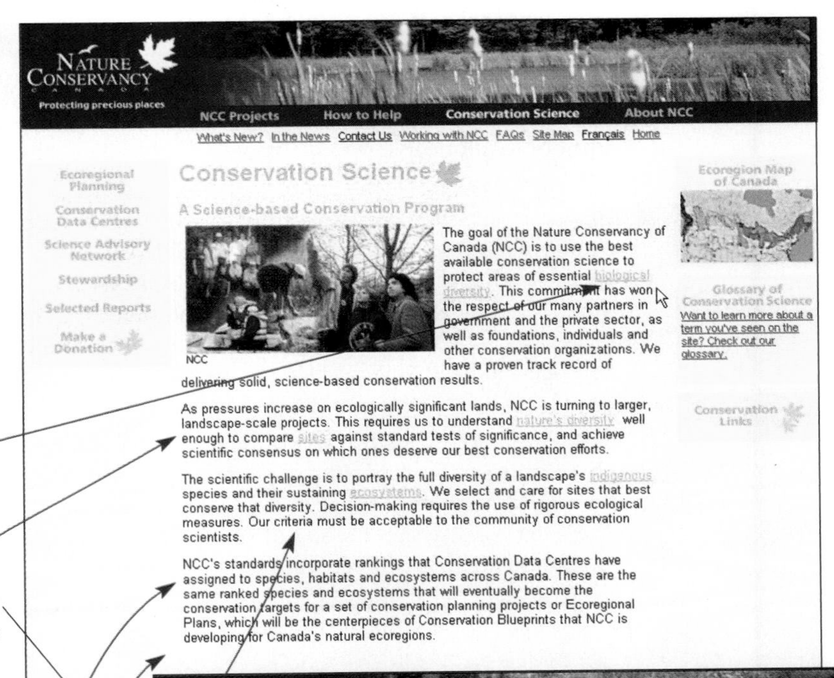

Clear, informative, and underlined in-text links.

Text is chunked.

- Short paragraphs.
- Numerous headings, coded by colour depending on the area of the website.
- White space that separates blocks of text and offsets text from page.

Text is legible.

- Dark letters against a light background.
- Text large enough to be readable.

Concise writing.

No requirement for scrolling.

Designing Pages for Global Readers

- **Text.** Write simply. Avoid slang, colloquial expressions, and metaphors that refer to sports, occupations, or other things specific to your culture. Check your draft with people who are knowledgeable about the cultures of your likely readers.
- **Images.** Images that are meaningful to others who share your culture may be incomprehensible—even offensive—to readers in other cultures. Take care when selecting your photographs, drawings, icons, and other images. Also, avoid including words in your images. Computerized translators cannot read images. And don't forget that people creating a version of your site in another language will need to recreate all images, including buttons in menus, that have words.
- **Page design.** Many English words are translated into much longer words in other languages. Similarly, the same statement may take as many as 30 percent more words in other languages. If you expect your site to be translated into other languages, design the size of your buttons, menus, and text areas to accommodate these differences.

GUIDELINE 5 Enable Readers with Disabilities to Use Your Site

A truly reader-centred website can be used by every reader who would like to use it. The international World Wide Web Consortium (W3C), as part of its Web Accessibility Initiative, has developed guidelines for making websites accessible to people with disabilities. Also, the Government of Canada has Common Look and Feel (CLF) standards for Web design. While the CLF standards apply primarily to government sites, it constitutes good business sense for organizations outside the federal domain to make their sites accessible to all readers. Some of the most important accessibility strategies are highlighted below.

Helping Readers with Disabilities Use Your Site

- **Visually impaired readers.** Use relative sizes, not point sizes, to designate how big your text will be. If you use relative sizes, readers with poor vision can use their browser settings to enlarge the text on your pages.

  Many people who can't see well use screen readers, which convert text to speech. To create a reader-centred site for these persons, include an <alt> tag with all images. This tag enables you to provide a text description of the image that can be synthesized into speech by screen readers. Also, when you create in-text links, highlight significant, descriptive words, not such phrases as "click here." For example, instead of writing, "To obtain more information about our newest products, click here," write, "You can obtain more information about our newest products." Screen readers aid visually impaired readers by pulling out all the links on a page and placing them in a list. A list in which every link says "Click here" will be useless. In addition, use each HTML tag only for its intended purpose. For instance, don't use the heading tags <H1>, <H2>, and so on to change type size. Some screen readers use a louder volume to signal that text is a heading.

For readers with colour blindness, supplement all information conveyed in colour with that same information in text. For example, when you present a warning in bold, red text, also include the word "Warning." In addition, because most colour blindness is red–green, avoid using this combination of colours to distinguish content.

When you include movies or animations, use a program such as QuickTime Pro to add captions for visually impaired readers.

If these and similar strategies don't succeed in making your site fully accessible to visually impaired readers, create a separate text-only version. Update it whenever you update the visual version of your site.

- **Mobility-impaired readers.** Some readers have difficulty controlling a mouse precisely. Make all clickable areas as large as possible.
- **Hearing-impaired readers.** If you include videos with sound, provide a volume control and captions. Also, provide concurrent captions with movies and other items that have sound. Supplement sound that gives feedback by providing visual information of the same sort. For instance, if your site provides a verbal "Thank you," also include a pop-up window with that message.

Several websites will test the accessibility of your webpages for free.

GUIDELINE 6 Help Readers Find Your Site on the Internet

Your website will achieve its usability and persuasive objectives only if your target readers find their way to it on the Internet. There are four ways you can help them do this.

Helping Readers Find Your Site

- **Register your site with Internet directories.** Internet directories, such as the one provided by Yahoo!, index selected sites that have been examined by the directories' professional staff. The websites for these directories include a link to pages that let you submit your site for inclusion in them.
- **Include keywords in your site's headings.** Every webpage file begins with a section called the "heading." In the heading, you can include information that won't be displayed by browsers but is used by search engines. The most important information of this type for you to include is a list of keywords that search engines will use to classify your site. Some search engines even give highest priority to these words (rather than words in the body of a webpage) when telling their users which sites are the "closest" matches to the search they ordered. Therefore, by choosing keywords wisely, you can boost the chances that your site will be one of the first listed when your target readers conduct an Internet search.
- **Include a description of your site in your home page's heading.** In the heading, you can also include a description of your site. In their search results, search engines normally display some arbitrarily selected number of words from your webpage. However, some will instead display the more informative and appealing description you include in your heading.
- **Include your site's URL on all publications, emails, and other communications from your organization.**

Test Your Site on Multiple Platforms and Browsers before Launching It

The same webpage may display differently on different computers and with different browsers on the same computer.

The way your website appears in your browser is not necessarily the way it will appear to all of your readers even if they are using only the default settings. For example, a PC and a Mac can process the same webpage file so differently that a page that looks fine in one can be seriously flawed in the other. Likewise, different browsers, such as Netscape and Internet Explorer, can handle webpage files in very different ways. For these reasons, you should test your site on computers with different operating systems and browsers before launching it. While testing, be sure to set the screen resolutions at the various settings that members of your target audience are likely to have. A webpage that looks good at a resolution of 1024 × 768 may be too wide to fit on a screen whose highest resolution is 800 × 600.

Posting your website to a server

Once you've tested your site, you are ready to post it on a server where it can be accessed by the readers you want to reach. As a student, you would probably post it on a server at your school. As an employee, you would probably post it on a server owned by your employer, although some employers pay companies called Internet service providers to host their sites. Often posting is done by means of FTP (file transfer protocol) programs, but posting can be accomplished in other ways as well. Once your site is posted, be sure to test it once again using several kinds of computers and more than one browser.

GUIDELINE 8 Keep Your Site Up to Date

Once you hand a print communication to your readers, its contents are fixed. With a website, however, you can continuously update the information you provide. Replace old information with new. Expand your content as you identify new information your readers will find useful or persuasive. And periodically check all external links to be sure the sites to which you are directing your readers are still active. You can even enlist your readers' help in updating your site by providing a link that lets them send you an email. So that your readers can know how current your information is, you can include a line at the bottom of your site that tells when it was last updated.

GUIDELINE 9 Ethics Guideline: Respect Intellectual Property and Provide Valid Information

When you are creating a website, you encounter two sorts of ethical issues: those involved with what you borrow from other people when making your site, and those involved with what you offer visitors to your site.

Ethics of Using Images and Text You've Downloaded

It's very easy to download images and text from someone else's website for incorporation into your own site. Some items available on the Web fall into the public domain, meaning that no one has a copyright on them and they can be used freely without permission.

However, much of what's on the Web is copyrighted. This includes not only text but also technical drawings and photographs or frames from your favourite characters from the comic pages of the newspaper. Copyrighted items also include many sound files, animations, and videos (Mohler, 1997). When you are thinking of using images from another site, check to see whether the site includes a copyright notice or request concerning use of its materials. Note, however, that what you may find at some sites has itself been copied from some other site, which may have restrictions that weren't observed by the creator of the site you are visiting.

Whenever you are unsure about whether permission is needed to use something, check with your professor or other knowledgeable person—or simply request the permission.

Even when permission isn't needed, acknowledge your source when you use something that clearly represents someone else's intellectual or creative effort. Appendix B tells how to cite Web sources.

Ethics of Posting

When you post something on the Web, keep in mind that your readers rely on your expertise and integrity. Be sure that you have checked your facts and that your information is up to date. Remember also that all the other ethical considerations discussed in this book apply to Web communications just as much as to print communications. Don't post anything that will harm other people.

To download a copy of a planning guide, visit Chapter 5 at www.techcomm.nelson.com.

PLANNING GUIDE AND OTHER RESOURCES

A planning guide you can download is available at the website for this book.

EXERCISES

For additional exercises, visit www.techcomm.nelson.com.

Expertise

Visit a website, perhaps one related to your program or to a project you are preparing in this course. Begin by identifying the site's usability and persuasive objectives. Next, evaluate one of the site's pages in terms of this chapter's guidelines for designing effective webpages. Finally, evaluate the site's overall design in terms of this chapter's guidelines for building effective websites. Report the results of your analysis in a memo to your professor.

Online

1. Using a search engine, locate four websites that provide advice for designing websites. Compare their advice to the advice given in this chapter. Report your results in a memo to your professor.

2. Create your own website, following the guidelines given in the website assignment in Appendix C.

Collaboration

Working with two or three other students, evaluate the usability and persuasiveness of your school's website from the perspective of a high school student who wants to learn about your school. Together, brainstorm the top six questions you think the student would have, and then look for the answers. Report your results in a memo to your professor.

Ethics

Create a webpage for use by students in your program at your school that provides links to useful information on ethical issues. Include links to your school's policies concerning plagiarism and related issues as well as information from professional organizations. Make the page visually appealing, informative, and easy to use.

Reference Guide 2: Seven Ways to End a Communication

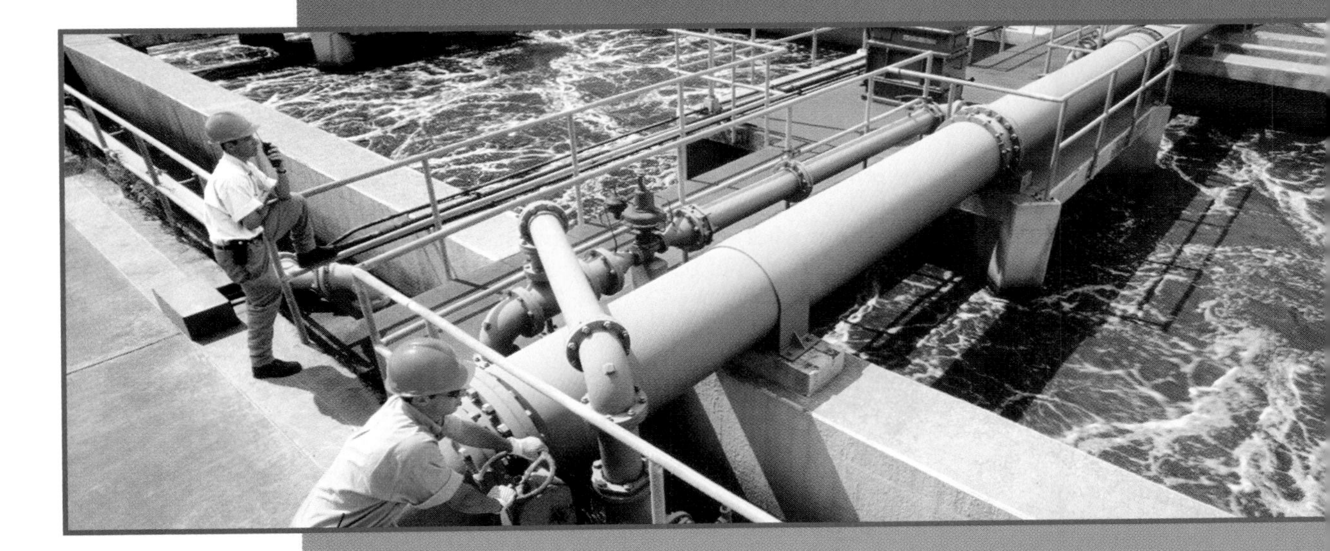

CONTENTS

Reasons why endings are important

This Reference Guide describes seven ways to end an oral or written communication. The ending of a communication can substantially increase its effectiveness. Researchers have found that readers are better able to remember things said at the end of a communication than things presented in the middle.

This guide's seven reader-centred strategies can enhance your communication's usability and persuasiveness. Despite the important role that endings play, however, your best strategy sometimes will be to say what you have to say and then stop, without providing a separate ending. Other times, you can use one or a combination of strategies.

WWW For additional resources, visit Reference Guide 2 at www.techcomm.nelson.com.

To decide which strategies to use, focus on your readers:

How to choose among possible strategies for endings

1. **Decide what you want your readers to think, feel, and do as they finish your communication.** Then select the strategy or strategies for ending that are most likely to create this response.
2. **Determine what kind of ending your readers might be expecting.** Look at what other people in your organization and your field have done in similar situations. This investigation can tell you what your readers are accustomed to finding at the end of the type of communication you are preparing.

NATURAL CONCLUSION

Sometimes you should end your communications without doing anything special at all. After you've made your last point, simply stop. This conclusion is used when you use a pattern of organization that brings you to a natural stopping place. Here are some examples:

Communications that you may want to end after the last point

- **Proposals.** You will usually end a proposal with a detailed description of what you will do and how you will do it. Because that's where your readers expect proposals to end, they will enjoy a sense of completion if you simply stop after presenting your last recommendation. Furthermore, by ending after your recommendations, you will have given them the emphasis they require.
- **Formal reports.** When you prepare a formal report (a report with a cover, title page, and binding), the convention is to end with either your conclusions or your recommendations—both appropriate subjects for emphasis.
- **Instructions.** You will usually end instructions by describing the last step.

These are just some examples of communications that you may decide to end directly after making your last point. However, if your analysis of your purpose, readers, and situation convinces you that you should add something after your last point, use the other strategies in this guide that will best help you meet your objectives.

SUMMARY OF KEY POINTS

An ending is a place of emphasis and a communication's final chance to influence readers' impressions of its subject matter. You can use the persuasive possibilities of the ending to focus your readers' attention on the points you want to be foremost in their minds as they finish reading.

Perhaps there is one main point that you consider to be of paramount importance and thus wish to repeat. Consider, for instance, the final paragraph of an article on "Preventing Wound Infections," written for family physicians (Mancusi-Ugaro & Rappaport, 1986). The point made in this final paragraph was stated in the abstract at the beginning of the article and again in the fourth paragraph, where it was supported by a table. It was also referred to several other times in the article. Nevertheless, the writers considered it to be so important that in the final paragraph they stated it again:

Ending that repeats the communication's main point

> Perhaps the most important concept to be gleaned from a review of the principles of wound management is that good surgical technique strives to maintain the balance between the host and the bacteria in favour of the host. The importance of understanding that infection is an absolute quantitative number of bacteria within the tissues cannot be overemphasized. Limiting, rather than eliminating, bacteria allows for normal wound healing.

You can use the same strategy in communications intended to help your readers make a decision or to persuade them to take a certain action. Here, for instance, is the final paragraph of a memo urging new safety measures:

Another example

> I cannot stress too much the need for immediate action. The exposed wires present a significant hazard to any employee who opens the control box to make routine adjustments.

Alternatively, instead of repeating only one main point, you may wish to summarize a number of key points. In summarizing, you are concerned that your audience has understood the general thrust of your *entire* communication. A skillfully written summary can increase a communication's usability by stating or reviewing points that readers will want to recall later.

Here, for example, is the ending of a 115-page book titled *Understanding Radioactive Waste,* which is intended to help the general public understand the impact of the nuclear power industry's plans to open new plants (Murray, 1982):

Key points are summarized

> It may be useful to the reader for us to now select some highlights, key ideas, and important conclusions for this discussion of nuclear wastes. The following list is not complete—the reader is encouraged to add items.
>
> 1. Radioactivity is both natural and manmade. The decay process gives radiations such as alpha particles, beta particles, and gamma rays. Natural background radiation comes mainly from cosmic rays and minerals in the ground.
> 2. Radiation can be harmful to the body and to genes, but the low-level radiation effect cannot be proved. Many methods of protection are available.
> 3. The fission process gives useful energy in the form of electricity from nuclear plants, but it also produces wastes in the form of highly radioactive fission products . . .

This list continues for thirteen more items, but this sample should give you an idea of how this author ended with a summary of key points.

Notice that a summary at the end of a communication differs significantly from a summary at the beginning. Because a summary at the beginning is meant for readers who have not yet read the communication, it must include some information that will be of little concern at the end. For example, the beginning summary of a report on a quality-control study will describe the background of the study. In contrast, the ending summary would focus sharply on conclusions and recommendations.

GOAL RESTATEMENT

Many communications begin by stating a goal and then describe or propose ways to achieve it. If you end a communication by referring to that goal, you remind your readers of the goal and sharpen the focus of your communication. In the following examples, note how the ending refers to the beginning.

This first example comes from a 17-page proposal prepared by operations analysts in a company that builds customized, computer-controlled equipment used in print shops and printing plants:

Beginning states a goal

> To maintain our competitive edge, we must develop a way of supplying replacement parts more rapidly to our service technicians without increasing our shipping costs or tying up more money in inventory.

Ending refers to the goal

> The proposed reform of our distribution network will help us meet the needs of our service technicians for rapidly delivered spare parts. Furthermore, it does so without raising either our shipping expenses or our investment in inventory.

The second example is from *Biotechnology,* a journal concerned with the synthesis of new organisms that are commercially useful (Filho et al., 1986). Note that the highly technical language is appropriate because the writers are addressing people knowledgeable in this field.

Beginning states a goal

> Given the necessity of producing alcohol as an alternative fuel to gasoline, especially in countries like Brazil, where petroleum is scarce, it is important to have a yeast strain able to produce ethanol directly from starchy materials.

Ending refers to the goal

> We are convinced that the stable pESA transformants can be of technological value in assisting ethanol fermentation directly from starchy materials, and we have described the first step towards this end. Continuing this work, genetic crosses with different *Saccharomyces distaticus* strains are presently being carried out to introduce maltase and glucomylase genes into the stable transformants that secrete functional a-amylase.

KEY FEELING FOCUS

Sometimes you may want to focus your readers' attention on a feeling rather than on a fact. For instance, if you are writing instructions for a product manufactured by your employer, you may want your ending to encourage your readers' goodwill toward the product. Consider this ending of an owner's manual for a clothes dryer. Though the last sentence provides no additional information, it seeks to shape the readers' attitude toward the company.

Ending designed to build goodwill

> The GE Answer Center™ consumer information service is open 24 hours a day, seven days a week. Our staff of experts stands ready to assist you anytime.

The following passage is the ending of a booklet published by the National Cancer Institute for people who have apparently been successfully treated for cancer but do not know how long the disease will remain in remission. It, too, seeks to shape the readers' feelings.

Ending designed to shape complex attitudes

> Cancer is not something anyone forgets. Anxieties remain as active treatment ceases and the waiting stage begins. A cold or cramp may be cause for panic. As 6-month or annual check-ups approach, you swing between hope and anxiety. As you wait for the mystical 5-year or 10-year point, you might feel more anxious rather than more secure.
>
> These are feelings that we all share. No one expects you to forget you have had cancer or that it might recur. Each must seek individual ways of coping with the underlying insecurity of not knowing the true state of his or her health. The best prescription seems to lie in a combination of one part challenging responsibilities that require a full range of skills, a dose of activities that seek to fill the needs of others, and a generous dash of frivolity and laughter.
>
> You still might have moments when you feel as if you live perched on the edge of a cliff. They will sneak up unbidden. But they will be fewer and farther between if you have filled your mind with other thoughts than cancer.
>
> Cancer might rob you of that blissful ignorance that once led you to believe that tomorrow stretched on forever. In exchange, you are granted the vision to see each today as precious, a gift to be used wisely and richly. No one can take that away.

REFERRAL

At work, a common strategy for ending a communication is to tell your readers how to get assistance or more information. These two examples are from a letter and a memo:

Endings that offer help

> If you have questions about this matter, call me at 523–5221.

> If you want any additional information about the proposed project, let me know. I'll answer your questions as best I can.

By ending in this way, you not only provide your readers with useful information, you also encourage them to see you as a helpful, concerned individual.

CALL TO ACTION

An ending is a transition. It leads readers out of the communication and into the broader stream of their activities. Therefore, it provides an excellent opportunity to help readers by answering the question, "What should I do now?"

One way to enact this strategy for effective endings is to tell your readers what you think should be done next. If more than one course of action is available, tell your readers how to follow up on each of them.

Ending that tells readers exactly what to do

> To buy this equipment at the reduced price, we must mail the purchase orders by Friday 11 December. If you have any qualms about this purchase, let's discuss them. If not, please forward the attached materials, together with your approval, to the Controller's Office as soon as possible.

Another way to enact this strategy for effective endings is to tell readers what needs to be found out next. This strategy is particularly useful for work done on the job that is completed in stages. For example, one study might answer preliminary questions. If the answers look promising, an additional study might then be undertaken.

Ending that identifies next question needing study

> This experiment indicates that we can use compound deposition to create microcircuits in the laboratory. We are now ready to explore the feasibility of using this technique to produce microcircuits in commercial quantities.

Such endings are often combined with summaries, as in the following example:

Another example

> In summary, over the past several months our Kingston plant has ordered several hundred electric motors from a supplier whose products are inferior to those we require in the heating and air conditioning systems we build. Not only must this practice stop immediately, but also we should investigate the situation to determine why this flagrant violation of our quality-control policies has occurred.

SOCIAL CONVENTION

All the strategies mentioned so far focus on the subject matter of your communications. It is also important for you to observe the social conventions that apply in a given situation.

Some of those conventions involve customary ways of closing particular kinds of communication. For example, letters usually end with an expression of thanks, a statement that it has been enjoyable working with the reader, or an offer to be of further help if needed. In contrast, formal reports and proposals rarely end with such gestures.

Other conventions about endings are peculiar to the organization in which they are found. For example, in some organizations writers rarely end their memos with the kind of social gesture commonly provided at the end of a letter. In other organizations, memos often end with such a gesture, and people who ignore that convention risk seeming abrupt and cold.

Social conventions also apply to personal relationships between you and your readers. Have they done you a favour? Thank them. Are you going to see them soon? Let them know that you look forward to the meeting.

EXERCISES

For additional exercises, visit www.techcomm.nelson.com.

Expertise

Select a communication written to people in your field. This might be a letter, memo, manual, technical report, or article in a professional journal. (Do not choose a textbook.) Identify the strategies from this Reference Guide that the writer applied when drafting the communication's ending. Is the ending effective in achieving the writer's usability and persuasive goals? If so, why? If not, how could it be improved?

Online

For many websites there is no single ending point. Visitors may leave the site at any point. In fact, links in many sites lead visitors directly to other sites. Visit three different kinds of sites, such as one that provides explanations or instructions, one that enables people to purchase products, and one created by an advocacy group. Within each site, where are places that one or more of this chapter's guidelines are employed effectively?

Collaboration

1. The following paragraphs constitute the ending of a report to the Ministry of the Environment concerning the economic and technical feasibility of using compost as a container growth medium. In this ending, the authors have used several of the strategies described in this chapter. Working with another student, identify these strategies.

Section 4. Conclusions

Replacement of peat-rich (70 to 80%) growth media with composts to the extent of 40 and 75% on a volume basis generated mixtures that had water retention and supplying capacities similar to those of a peat-rich medium alone, on a unit weight basis. On a unit volume basis, the compost mixtures carried and supplied more water to plants grown than either of the two peat media selected for the study as the best from the seven tested. The bulk densities of some effective compost mixtures differed only marginally from those of a peat medium.

Some of the mixtures with composts supported greater growth of Clematis vines than the peat-rich media themselves. The growth of Mums also showed that a substantial portion of the peat-rich media can be replaced by a proper compost to save costs without decreasing growth in some cases and with improvement in plant growth in others.

The plants grown on compost-containing media were generally richer in mineral nutrients, as were the media remaining in the pots, compared to the plants' growth on peat-rich media alone, and the peat media themselves after growth. The advantages of including composts in the growth media should therefore extend to marketing and transplanting of the potting plants.

It was therefore concluded that a good quality compost can substantially replace peat in contained growth studies. Such use of compost will:

- Increase plant growth;
- Improve survival of plants during marketing and upon transplanting;
- Widen profit margin for the container plant industry; and
- Be environmentally beneficial.

2. Working with another student, describe the strategies for ending used in the following figures in this text:

Figure	Page	Figure	Page
1.3	12	8.3	188
6.8	162	10.6	257
7.5	174	11.3	285

Ethics

Examine the student conduct code or values statement of your school. Which of the guidelines in this chapter does its ending follow? Why? Also, identify some of the specific values that underline the code's or value statement's provisions.

CASE

ANNOUNCING THE INTERNET USE POLICY

As you sit in your office in the large, one-storey building leased by your employer, Anodyne Computing, you look up from the draft of an email message that company president T.J. Paul sent you a few minutes ago. The email is announcing a new Internet acceptable-use policy. T.J. has asked for your opinion of it.

Anodyne is a 130-person software query tool company that employs a lot of young programmers, engineers, analysts, technical support staff, and salespeople. The employees sometimes spend long hours at work, testing and debugging software and solving technical support issues. Before being hired at Anodyne, many of these staff recently graduated from college or the university in St. John's, where Anodyne is located. Anodyne's founders, including T.J. Paul, started the company while they were still graduate students at the university.

The company has grown quickly—so quickly, in fact, that T.J. has had difficulty keeping track of all the new hires. Sometimes he longs for the old days, when all the employees could gather over a beer after work to make major decisions. As the company has increased in size, however, policies and procedures have had to be developed.

One of the policies that T.J. and the other founders have recently drafted is an acceptable-use policy for the Internet. This policy is needed because some of the employees have been using the company's Internet for online gaming and to download pornographic material. This usage was discovered when an IT staff member grew concerned about the slow speed of the company's Web server.

"T.J.'s email announcing this policy had better be good," you think as you prepare to review his draft. Although the employees know that T.J. has been meeting with you and others to work out a policy, the discussions have been kept secret. Nonetheless, feelings about the new policy have been running high. Some employees have talked of quitting if they can't surf the Net at work, and others have talked of quitting if an acceptable-use policy is not put in place. The controversies are harming morale—an impact that Anodyne wants to avoid since the company is behind schedule on a software release.

YOUR ASSIGNMENT

First, fashion a reasonable set of objectives for T.J.'s email message. Then complete a Worksheet for Defining Objectives (a downloadable version is available at www.techcomm.nelson.com). Second, evaluate T.J.'s draft in light of the objectives you have established. Third, evaluate T.J.'s draft in light of Chapter 5's guidelines for writing email messages. Finally, revise the draft to make it more effective. Be prepared to explain your revisions to T.J.

To: All employees
From: T.J. Paul
Subject: No more cyberslacking

I hereby notify you that beginning next month, Anodyne Computing will institute an acceptable-use policy for Internet use in the workplace. Anodyne's company Web server is to be used only for relevant and appropriate work-related tasks. Our IT staff will be monitoring the Web server to ensure that this policy is followed. I am really disappointed to be putting this policy in place, but it is unacceptable that some employees have been using our Web server for gaming and downloading pornography. Non-work-related surfing has no place during work hours. Any employee found contravening the acceptable-use Internet policy will be fired.

Choosing a Communication Format

CHAPTER 6

Instructions

CHAPTER 7

Proposals

CHAPTER 8

Reports

REFERENCE GUIDE 3

Three Types of Special Reports

CHAPTER 6

Instructions

COMPONENTS

Introduction

Description of the Equipment

List of Materials and Equipment Needed

Directions

Troubleshooting

For additional chapter resources, visit Chapter 6 at www.techcomm.nelson.com.

Instructions come in many lengths, shapes, and levels of complexity. They range from the terse directions on a shampoo bottle ("Lather. Rinse. Repeat.") to the huge manuals that are hundreds or thousands of pages long for servicing airplane engines, managing large computer systems, and performing biomedical procedures.

Most employees write instructions sometimes.

Although some instructions are prepared by professional writers and editors, most other employees also need to prepare instructions sometimes. Whether you are developing a new procedure, training a new coworker, or preparing to leave for vacation, you may need to provide written directions to someone else. You may also be called on to create instructions for people who will read and use them on a computer screen rather than on paper.

This chapter's advice applies to both printed and online instructions.

This chapter provides advice that you will find valuable regardless of the subject or size of the instructions you write. The advice given in the first part of this chapter applies equally to instructions written for paper and for computer screens. A special section at the end of the chapter provides additional suggestions for online instructions.

FOUR IMPORTANT POINTS

Usability is the primary goal of all instructions.

To succeed, instructions must possess the same qualities as are essential for all workplace communications: usability and persuasiveness. Writing usable instructions can be especially challenging because people read and use instructions differently than they read and use most other communications. When reading memos and reports, for instance, people usually read in a continuous stream, one sentence and paragraph after another. With instructions, however, they repeatedly shift their attention between reading and doing: They read a step, then do a step, read the next step, then do the next step. Consequently, you must help them find their place quickly each time they look back at your directions after completing another step. Also, people often want to use only one part of a set of instructions, so you must help them locate this information very rapidly. In addition, you must ensure that they understand your directions even if they are so new to the procedure you are describing that they don't know the vocabulary used to describe the actions they are to perform or the parts of the equipment they must use.

Instructions must also persuade in order to succeed.

Instruction readers also present special challenges to your ability to write persuasively. Many people dislike reading instructions, preferring to start right in on the task by relying on their own common sense. Consequently, your first challenge will often be to persuade your readers to read the instructions at all. When people do read instructions, they can be extremely impatient and easily frustrated. You must write in a way that entices them to look back at your instructions after each time they've looked away to perform a step. If you are preparing instructions for one of your employer's products, you'll also want to write in a way that persuades your readers to feel so good about your product that they will buy from your employer again and recommend that others do likewise.

This chapter describes the components of instructions, which will help you address these challenges. You will greatly increase your chances of meeting these challenges if you keep the following four points in mind.

For advice about writing clearly and succinctly, see Chapter 4.

1. **Write clearly and succinctly.** People can't use what they can't understand. Choose words that convey your meaning easily. Construct sentences your readers will

comprehend effortlessly. Use as few words as possible. More words make more work for your readers and increase the chances your readers will stop reading what you've written.

For advice about creating graphics, see Chapter 15.

2. **Use plenty of graphics.** For many purposes, well-designed graphics are much more effective than words. For example, words cannot show readers where the parts of a machine are located, how to grasp a tool, or what the result of a procedure should look like. It's especially helpful to use graphics when preparing instructions for readers who speak languages other than that in which the instructions were written. Sometimes graphics can even carry all the information your readers need (see Figure 15.1 on page 364).

For advice about designing pages and screens, see Chapter 16. Chapter 5 provides additional advice about screen design.

3. **Design usable and attractive pages and screens.** A carefully planned page or screen design can help your readers find their places as they bounce back and forth between reading about steps and performing those steps. Good page and screen design also helps readers see the connections between related blocks of information such as a written direction and the drawing that accompanies it. Attractive designs entice readers' eyes back to the instructions you want them to read.

For advice about testing the usability and persuasiveness of your drafts, see Chapter 18.

4. **Test your instructions.** Instructions can be surprisingly difficult to write. Words and illustrations that we think are easy to understand can baffle our readers. And we can too easily omit a step that seems obvious to us, only to find out later that our readers needed to have it explained. The page design we think will help our readers and appeal to them may, from our readers' perspective, be confusing and unappealing. The best way to prevent such outcomes is to ask members of your target audience to try out a draft of your instructions so you can determine whether your intended readers will really find them to be usable and persuasive.

COMPONENTS OF INSTRUCTIONS

The conventional structure for instructions contains the following elements:

Components of Instructions

Topic	Readers' Questions
Introduction	Will these instructions help me? How can I use them effectively?
Description of the equipment	Where are the parts that I'll be working with while I repair or use this equipment?
List of materials and equipment needed	What material and equipment should I collect before I begin?
Directions	Once I'm ready to start, what—exactly—do I do?
Troubleshooting	Something isn't working correctly: How can I fix it?

Many instructions contain only the directions.

The simplest instructions contain only the directions. More complex instructions contain some or all of the other four elements listed above. And some instructions also include such additional elements as covers, title pages, contents lists, appendixes, lists of references, glossaries, lists of symbols, and indexes.

To determine which elements to include in any instructions you write, follow this familiar advice: Consider your readers' aims and needs as well as their characteristics that will shape the way they read and respond to your communication.

INTRODUCTION

The effectiveness of your instructions can depend substantially on your decision about what—if anything—to provide as introductory material. An unneeded or overly long introduction may prompt readers to toss your instructions aside before reading even a single direction. Conversely, if you fail to provide needed information, you may make it impossible for your readers to perform their tasks.

The following sections describe the elements most commonly included in introductions, together with suggestions for deciding whether your readers need each of them.

Possible introduction topics

- Subject
- Aim (purpose or outcome of the procedure described)
- Intended readers
- Scope
- Organization
- Usage (advice about how to use the instructions most effectively)
- Conventions (explanations of short-cut expressions and other conventions used in the instructions)
- Motivation (reasons why readers should use the instructions rather than ignore them)
- Safety

Subject

Often the title of your instructions will fully convey their subject. Especially in longer instructions, however, you may need to announce the subject in an introduction. Here is the first sentence from the fifty-page operator's manual for a 10-tonne machine used in the manufacture of automobile and truck tires:

Opening sentence that announces the subject

> This manual tells you how to operate the Tire Uniformity Optimizer (TUO).

The first page of the manual for a word processing application reads:

> The Microsoft Word User's Guide contains detailed information about using Microsoft® Word for Windows™ and Microsoft Word for Apple® Macintosh®.

Aim

From the beginning, your readers will want to know the purpose or outcome of the procedure described in your instructions. When the purpose or outcome isn't obvious from the title, announce it in the introduction. You may be able to convey your instructions' aim by listing the major steps in the procedure or the capabilities of the equipment whose operation you are describing. Here is the second sentence of the manual for the Tire Uniformity Optimizer:

A list of the purposes for which readers can use the equipment

Depending upon the options on your machine, it may do any or all of the following jobs:

- Test tires
- Find irregularities in tires
- Grind to correct the irregularities, if possible
- Grade tires
- Mark tires according to grade
- Sort tires by grade

Intended Readers

Readers sometimes ask, "Is this manual for me?"

When they pick up instructions, people often want to know whether the instructions are directed to them or to people who differ from them in interests, responsibilities, level of knowledge, or some other variable.

Sometimes, they can tell merely by reading the instructions' title. For instance, the operator's manual for the Tire Uniformity Optimizer is obviously addressed to people hired to operate that machine.

Identify special knowledge that is required.

In contrast, people who consult instructions for a computer program may wonder whether the instructions assume that they know more (or less) about computers than they actually do. In such situations, answer their question in your introduction. Readers who don't already possess the required knowledge can then seek help or acquire the necessary background. To illustrate, here is part of the introduction to assistance in the form of online training for FrontPage 2003, an application for making webpages:

Description of the target audience from online assistance.

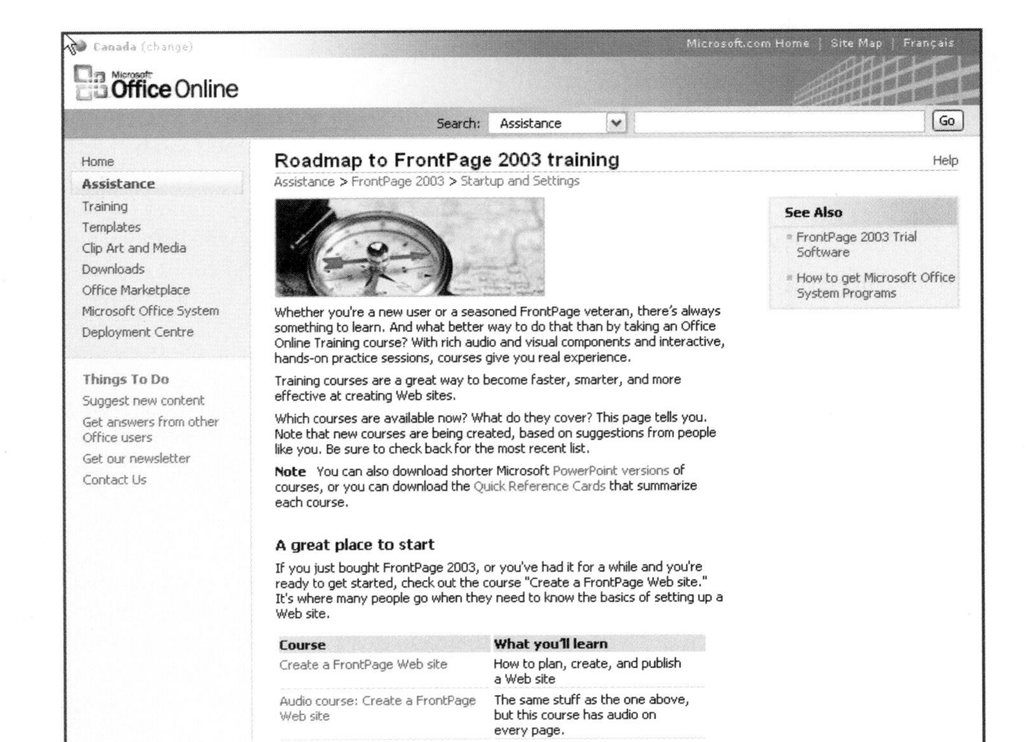

Scope

By stating the scope of your instructions, you help readers know whether the instructions contain directions for the specific tasks they want to perform. The manual for the Tire Uniformity Optimizer describes the scope of its instructions in the third and fourth sentences:

Statement of scope

> This manual explains all the tasks you are likely to perform in a normal shift. It covers all of the options your machine might have.

An online page for Mozilla, an open-source Internet application suite, describes the source of its Getting Started document this way:

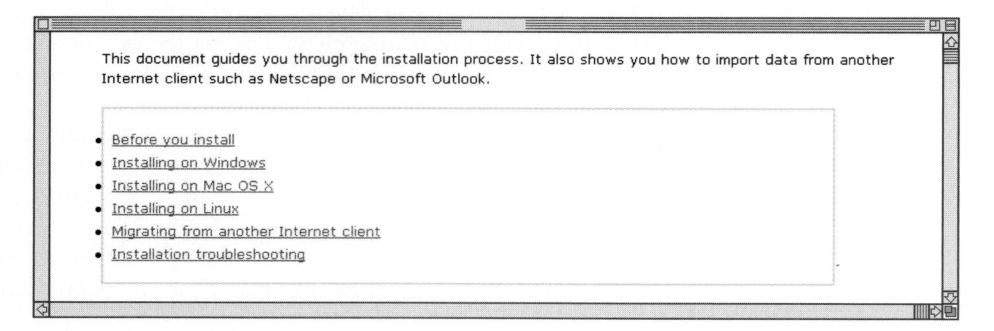

Organization

By explaining how the instructions are organized, an introduction can help readers understand the overall structure of the tasks they will be performing and locate specific pieces of information without having to read the entire set of instructions.

Often introductions explain scope and organization together. For example, the introduction to the Tire Uniformity Optimizer devotes several sentences to explaining that manual's organization, and this information also fills out the readers' understanding of the manual's scope.

Paragraph describing a manual's organization

> The rest of this chapter introduces you to the major parts of the TUO and its basic operation. Chapter 2 tells you step-by-step how to prepare the TUO when you change the type or size of tire you are testing. Chapter 3 tells you how to perform routine servicing, and Chapter 4 tells you how to troubleshoot problems you can probably handle without needing to ask for help from someone else. Chapter 5 contains a convenient checklist of the tasks described in Chapters 3 and 4.

Usage

Your readers will want to know how to use your instructions as efficiently and effectively as possible. If the best method is simply to follow the directions from beginning to end, you don't need to tell them anything at all. However, if you know of a better way to use your instructions, share it with your readers. For example, one of the end user documentation pages for Mozilla clarifies its purpose (and thus, its audience) in this way:

Advice about how to use the instructions

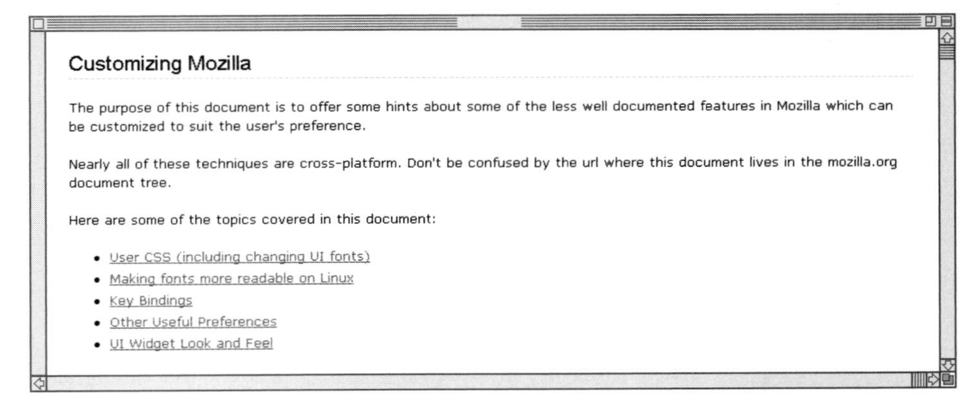

Customizing Mozilla

The purpose of this document is to offer some hints about some of the less well documented features in Mozilla which can be customized to suit the user's preference.

Nearly all of these techniques are cross-platform. Don't be confused by the url where this document lives in the mozilla.org document tree.

Here are some of the topics covered in this document:

- User CSS (including changing UI fonts)
- Making fonts more readable on Linux
- Key Bindings
- Other Useful Preferences
- UI Widget Look and Feel

Conventions

Some instructions use special conventions or terms that enable them to communicate succinctly. If you employ such shortcuts that your readers don't already know, be sure to include an explanation in your introduction.

The online manual for FrontPage 2003 provides an overview of form fields and their conventions, some of which are shown here:

Explanation of conventions used in an online manual

- **Text box**
  Use text boxes to collect a small amount of information in one line, such as first and last names or e-mail addresses. To collect larger amounts of information from site visitors, use text areas.

  someone@example.com

- **Option buttons**
  Use option buttons (also called radio buttons) when you want site visitors to select only one option from a group.

  A B C

- **Text area**
  Use text areas to collect one or more lines of text, such as a comment. This field scrolls to accommodate varying amounts of text. To collect a small amount of information from site visitors, use text boxes.

  Give feedback about this Web site:

Motivation

As pointed out above (and as you may know from your own experience), some people are tempted to toss instructions aside and rely on their common sense. A major purpose of many introductions is to persuade readers to read the instructions. You can accomplish this goal by employing an inviting and supportive tone and by creating attractive design. You can also include statements that tell readers directly why it is important to pay attention to the instructions. The following example is from instructions for a ceiling fan that purchasers install themselves.

Statement of scope

Motivation to read the instructions

We're certain that your Hampton Bay fan will provide you with many years of comfort, energy savings, and satisfaction. To ensure your personal safety and to maximize the performance of your fan, please read this manual.

Safety

Your readers depend on you to prevent them from taking actions that could spoil their results, damage their equipment, or cause them injury. Moreover, product liability laws require companies to pay for damages or injuries that result from inadequate warnings in their instructions.

To satisfy your ethical and legal obligations, you must provide prominent, easy-to-understand, and persuasive warnings. If a warning concerns a general issue that covers the entire set of instructions (e.g., "Don't use this electrical tool while standing on wet ground"), place it in your introduction. If it pertains to a certain step, place it before that step. The following principles apply to warnings in either location:

- **Make your warnings stand out visually.** Try printing them in large, bold type and surrounding them with a box (Velotta, 1987). Sometimes, writers use the following international hazard alert symbol to draw attention to the warning.

  You may also include an icon to convey the nature of the danger. Here are some icons developed by Westinghouse.

  Electrical Shock

  Fire

  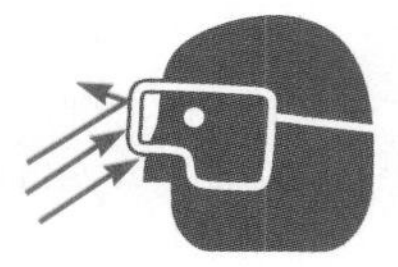

  Eye Protection

- **Place your warnings so that your readers will read them before performing the action the warnings refer to.** It won't help your readers to discover the warning after the step has been performed and the damage has been done.
- **State the nature of the hazard and the consequences of ignoring the warning.** If readers don't know what could happen, they may think that it's not important to take the necessary precautions.
- **Tell your readers what steps to take to protect themselves or avoid damage.** If possible, also tell your readers what action to take if they have ignored your warning.

The box and international hazard icon draw attention to the warning.
Initial statement tells what readers can do to avoid the hazard.
Readers are told the possible consequences of ignoring the warning.

Wear safety goggles when performing the next step. Flying chips could damage eyesight or cause blindness.

In many professions, writers distinguish among various types of warnings to indicate the severity of the hazard. Although these classification systems usually use the same words (*caution*, *warning*, and *danger*), there are no standard definitions for these words. However, here is one set of definitions that is widely used (Sides, 1984):

- **Caution.** Alerts readers that they may spoil the results if they do not follow the directions exactly.

Be sure to clean the flask with alcohol before adding the reagent. Otherwise, you may get false readings.

- **Warning.** Alerts readers that they may damage equipment if they do not follow the directions exactly.

Note that the word in the rectangle changes to reflect the severity of the hazard.

When performing the next step, **do not use detergents** to clean the reservoir. They contain chemicals that may damage the seals.

- **Danger.** Alerts readers to the danger of injury or death.

These canisters contain ammonia, which is highly explosive. Handle them with extreme caution. If they are damaged, they could explode, causing injury or death.

Sample Introductions

Figure 6.1 (page 148) shows the introduction to the instruction manual for the Tire Uniformity Optimizer. The introduction to another manual appears in Figure 3.2 (page 78).

FIGURE 6.1

Introduction to the Instruction Manual for the Tire Uniformity Optimizer

Chapter 1— Introduction

This manual tells you how to operate the Tire Uniformity Optimizer (TUO) and its controller, the Tire Quality Computer (TQC). The TUO has many options. Depending upon the options on your machine, it may do any or all of the following jobs:

- Test tires
- Find irregularities in tires
- Grind to correct the irregularities, if possible
- Grade tires
- Mark tires according to grade
- Sort tires by grade

This manual explains all the tasks you are likely to perform in a normal shift. It covers all of the options your machine might have.

The rest of this chapter introduces you to the major parts of the TUO and its basic operation. Chapter 2 tells you step-by-step how to prepare the TUO when you change the type or size of tire you are testing. Chapter 3 tells you how to perform routine servicing, and Chapter 4 tells you how to troubleshoot problems with the TUO. Chapter 5 contains a convenient checklist of the tasks described in Chapter 3.

Major Parts of the TUO

You can find the major parts of the TUO by looking at Figure 1-1. To operate the TUO, you will use the Operator's Control Panel and the Computer Panel.

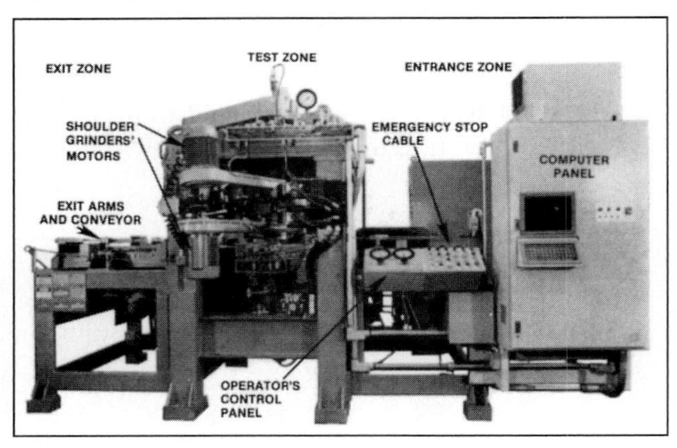

Figure 1-1. *Overview of the TUO and TQC.*

Page 1

DESCRIPTION OF THE EQUIPMENT

To describe equipment, use the pattern for describing an object (pages 45–47) or use a photograph or drawing (pages 444–47).

To be able to operate or repair a piece of equipment, readers need to know the location of its parts. Sometimes they need to know the function as well. For this reason, instructions often include a description of the equipment to be used. For example, the first page of the manual for the Tire Uniformity Optimizer displays a photograph of the machine with its major parts labelled. In some instructions, such illustrations are accompanied by written explanations of the equipment and its parts.

LIST OF MATERIALS AND EQUIPMENT NEEDED

Tell your readers what they need before they need it.

Some procedures require materials or equipment that readers wouldn't normally have at hand. If yours do, include a list of these items. Be sure to present the list *before* giving your step-by-step instructions. This will save your readers from the unpleasant surprise of discovering that they cannot go on to the next step until they have gone to the shop, supply room, or store to obtain an item that they didn't realize they would need.

DIRECTIONS

At the heart of a set of instructions are the step-by-step directions that tell readers what to do. Here are several detailed strategies for writing directions that your readers will find easy to understand and use. Figure 6.2 illustrates much of this advice.

FIGURE 6.2

Well-Designed Presentation of Procedures

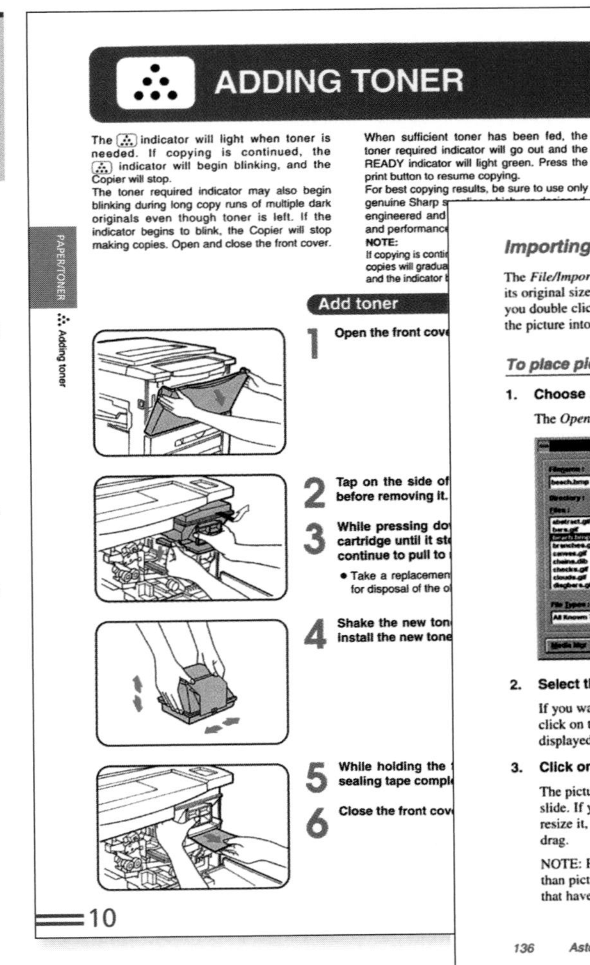

ADDING TONER

The indicator will light when toner is needed. If copying is continued, the indicator will begin blinking, and the Copier will stop.
The toner required indicator may also begin blinking during long copy runs of multiple dark originals even though toner is left. If the indicator begins to blink, the Copier will stop making copies. Open and close the front cover.

When sufficient toner has been fed, the toner required indicator will go out and the READY indicator will light green. Press the print button to resume copying.
For best copying results, be sure to use only genuine Sharp s
engineered and
and performanc
NOTE:
If copying is conti
copies will gradu
and the indicator

PAPER/TONER

Adding toner

Add toner

1 Open the front cov

2 Tap on the side of before removing it.

3 While pressing do cartridge until it st continue to pull to

- Take a replacemen for disposal of the o

4 Shake the new ton install the new tone

5 While holding the sealing tape compl

6 Close the front cov

10

Though very different in design, both are easy to read and use because they apply the principles discussed on pages 150–54.

Step numbers are prominent, easy to see.

Each step describes only one action.

Each direction is short and easy to comprehend.

Directions are on a line of their own, separate from explanations, to make them easy to read.

Figures provide additional guidance.

Importing Pictures

The *File/Import/Picture...* command places the image on the slide at its original size. If you have already loaded the picture into memory, you double click on the *Picture* tool, rather than load another copy of the picture into memory.

To place pictures using File/Import/Picture...

1. **Choose *File/Import/Picture...***

   The *Open Picture* dialog box appears.

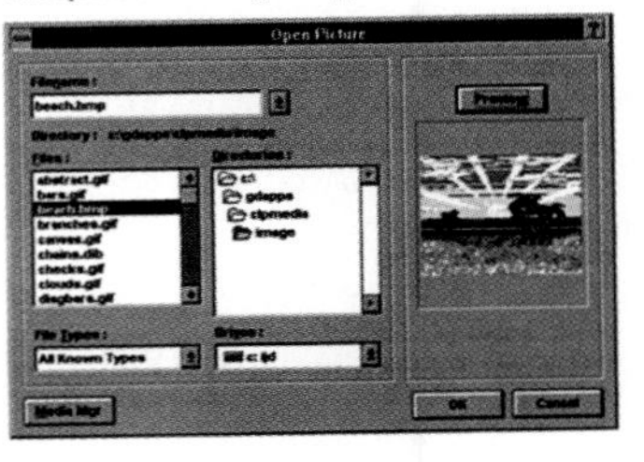

2. **Select the file you want to import.**

   If you want to preview the picture file that you have selected, click on the *Preview* button. The picture contained in the file is displayed in the *Preview* box.

3. **Click on the *OK* button.**

   The picture is placed in its original size in the center of the slide. If you press the [Shift] key while dragging the picture to resize it, the picture's aspect ratio will be maintained as you drag.

   NOTE: Pictures placed at their original size play more quickly than pictures placed at a size other than their original or pictures that have been resized.

136 *Astound*

1. **Write each direction for rapid comprehension and immediate use.** When using your instructions, readers will be impatient to complete their tasks. During the time they are reading (not doing), they will want to learn very quickly what they need to do next.
   - **In each direction, give your readers just enough information to perform the next step.** If you give more, they may forget some or become confused. If you give less, they may be unable to perform the step.
   - **Present the steps in a list.** A list format helps readers see exactly what they must read in order to perform the next step.
   - **Use the active voice and the imperative mood.** Active, imperative verbs give commands: "*Stop* the engine." (This statement is much simpler than "The operator should then stop the engine.")
   - **Highlight key words.** In some instructions, a direction may contain a single word that conveys the critical information. You can speed the readers' task by using boldface, all-capital letters or a different typeface to make this word pop off the page. Example: Press the **RETURN** key.
2. **Help your readers locate the next step quickly.** There are many things you can do to help your readers as they turn their eyes away from the task and back to your text:
   - **Number the steps.** With the aid of numbers, readers will not have to reread earlier directions to figure out which one they last read.
   - **Put blank lines between steps.** This white space makes it easy for readers to pick out a particular step from among its neighbours.
   - **Give one action per step.** It's easy for readers to overlook a direction that is tucked in with another direction and that does not have its own number.
   - **Put step numbers in their own column.** Instead of aligning the second line of a direction under the step number, align it with the text of the first line. Not this:

Step number is obscured

> 2. To quit the program, click the CLOSE button in the upper right-hand corner of
> the window.

But this:

Step number is in its own column

> 2. To quit the program, click the CLOSE button in the upper right-hand corner of
>   the window.

3. **Within steps, distinguish actions from supporting information.**
   - **Present actions before responses.** As the following example shows, you make reading unnecessarily difficult if you put the response to one step at the beginning of the next step.

The computer response obscures the action to be performed →

> 4. Press the RETURN key.
> 5. The Customer Order Screen will appear. Click on the TABS button.

Instead, place the response after the step that causes it.

Improved placement of the computer reaction lets the actions stand out →

> 4. Press the RETURN key.
>   The Customer Order Screen will appear.
> 5. Click on the TABS button.

- **Make actions stand out visually from other material.**
  In the following example, boldface is used to signal to the readers that the first part of Step 4 is an action and the second part is the response.

Use boldface and layout to make actions stand out

> 4. **Press the RETURN key.** The Customer Order Screen will appear.

You can also use layout to make such distinctions.

> 4. **Press the RETURN key.**
>   - The Customer Order Screen will appear.

And you can use similar techniques when explaining steps.

> 7. **Enter ANALYZE.** This command prompts the computer to perform seven analytical computations.

4. **Group related steps under action-oriented headings.** By arranging the steps into groups, you divide your procedure into chunks that readers are likely to find manageable. You also help them *learn* the procedure so that they will be able to perform it without instructions in the future. Moreover, if you use action-oriented headings and subheadings for the groups of steps, you aid readers who need directions for only one part of the procedure. The headings enable them to locate quickly the information they require.

   To create action-oriented headings, use verbs, not nouns, to describe the task. For example, use *collect* or *collecting* rather than *collection.*

   Here are some of the action-oriented topics from the online FrontPage 2003 manual:

The first word in each heading and subheading is a verb

> Collect and store form data entered by site visitors.
> Set up a form to save form results to an existing database.
> Set up a form to save form results to a file.
> Set up a form to save form results in an e-mail message.
> Set up a form to save form results by using a custom form handler.

Chapter 15 tells how to design effective graphics for instructions.

5. **Use many graphics.** Drawings, photographs, and similar illustrations often provide the clearest and simplest means of telling your readers such important things as:
   - **Where things are.** For instance, Figure 6.3 shows the readers of an instruction manual where to find four control switches.

FIGURE 6.3

Drawing That Shows Readers Where to Locate Parts of a Camcorder

This figure helps readers locate controls used to focus their camcorders.

The labels are placed far enough from the drawing to stand out distinctly.

To avoid ambiguity, arrows lead directly to each labelled part.

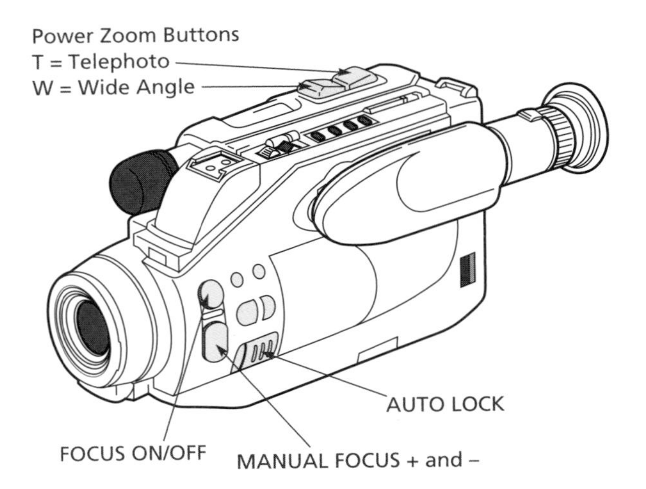

- **How to perform steps.** For instance, by showing someone's hands performing a step, you provide your readers with a model to follow as they attempt to follow your directions (see Figure 6.4).

FIGURE 6.4

Drawings That Show How to Do Something

These instructions tell people with diabetes how to obtain the drop of blood they need in order to test their insulin levels.

The Lancet is a sharp needle used to prick the skin.

Each drawing shows exactly how to hold the PENLET.

The drawing for Step 4 highlights the placement of the PENLET against the side of a finger.

The drawing for Step 5 emphasizes that the drop of blood must hang from the finger so that it may be applied to a test strip (in the next part of the procedure).

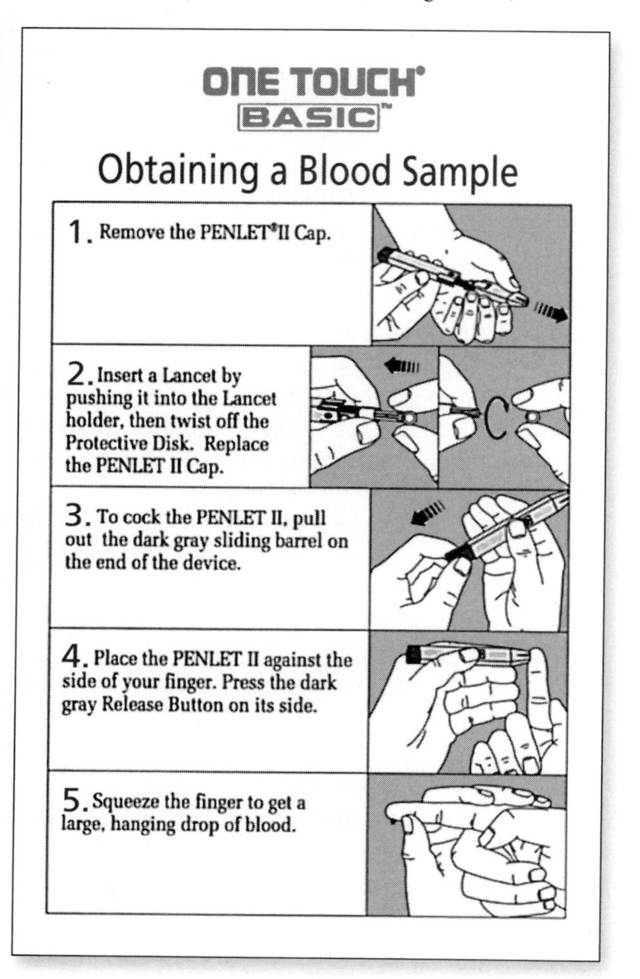

- **What should result.** By showing readers what should result from performing a step, you help them understand what they are trying to accomplish and help them determine whether they have performed the step correctly (see Figure 6.5).

Chapter 16 provides detailed advice for using page design to help readers see which figure goes with which text.

FIGURE 6.5

Drawing That Shows a Successful Outcome

This figure is part of the set-up instructions for a computer; it assists new owners in determining whether they have correctly connected their cables.

The monitor cable and port are highlighted because new owners can be confused about which of two ports receives that cable.

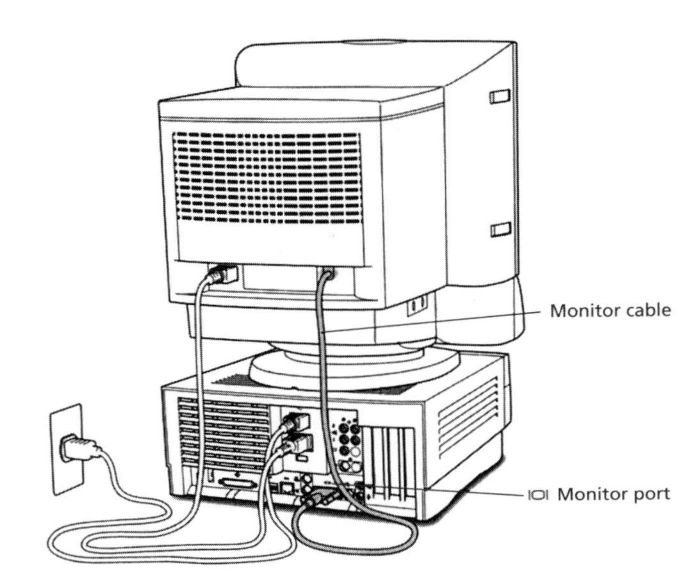

6. **Present branching steps clearly.** Sometimes instructions include alternative courses of action. For example, a chemical analysis might require one procedure if the acidity of a solution is at a normal level and another if the acidity is high. In such a situation, avoid listing only one of the alternatives.

Possibly confusing direction

> 6. If the acidity is high, follow the procedure described on page 20.

Instead, describe the step that enables readers to determine which alternative to choose (in the example, checking the acidity is that step) and then format the alternatives clearly:

Revised direction

> 6. Check the acidity.
> If it is high, follow the procedure described on page 20.
> If it is normal, proceed to Step 7.

Follow the same logic with other places where your instructions branch into two or more directions. The following example is from instructions for a computer program.

> 9. Determine which method you will use to connect to the Internet:
> If you will use PPP (Point to Point Protocol), see Chapter 3.
> If you will use SLIP (Serial Line Internet Protocol), see Chapter 4.

7. **Tell what to do in the case of a mistake or unexpected result.** Try to anticipate the places where readers might make mistakes in following your instructions. If it is not obvious how to correct or compensate for a mistake, tell them how to do so. Similarly, tell them what to do when a correct action does not produce the expected result:

> 5. Depress and release the RUN switches on the operator's panel.
>
> **NOTE:** If the machine stops immediately and the FAULT light illuminates, reposition the second reel and repeat Step 5.

TROUBLESHOOTING

The troubleshooting section of a set of instructions tells readers what to do if things don't work out as expected—if the equipment fails to work properly or if the results are unsatisfactory. Often, a table format works best. Figure 6.6 shows the chapter of the manual for the Tire Uniformity Optimizer that tells how to troubleshoot the TUO's Tire Quality Computer (TQC).

PHYSICAL CONSTRUCTION OF INSTRUCTIONS

The physical construction of instructions is an important element of their design. Computer manuals are often printed in a small format because readers use them on crowded desktops. Cookbooks are sometimes printed on glossy paper to withstand kitchen spills. Be sure to adapt your instructions to the environment in which they will be used.

ONLINE INSTRUCTIONS

WWW To view other examples of online instructions, visit Chapter 6 at www.techcomm.nelson.com.

As explained at the beginning of this chapter, all the advice you have read so far applies not only to printed instructions but also to online ones.

Online instructions are becoming increasingly common. Often they are included within sofware. Many sites on the World Wide Web also provide instructions meant for on-screen use. Some organizations provide online instructions to guide their employees through various technical processes.

To create online instructions, you can use any of several computer programs designed for that purpose. For a lot less money, you can design a website that presents the instructions your readers need.

FIGURE 6.6

Troubleshooting Section from the Manual for the Tire Uniformity Optimizer

Chapter 4—Troubleshooting

This chapter tells you what to check when troubleshooting the TQC. It lists the problems that may occur, the probable causes, and the remedies.

The first list in this chapter consists of the error messages that appear on the CRT when a problem occurs. Next to the error messages are the causes of the problem and the possible remedies. A list of all the error messages can be found in Appendix B. The second list consists of observable phenomena that are listed in order of normal TQC operation.

One easily solved problem is caused by entering entries too quickly to the TQC through the keyboard. If the operator does not wait for the TQC to respond to one request before entering another, errors and inaccurate data will result. Make sure you allow sufficient time for the TQC to respond to your input before you press another key.

Warning

EXTERNAL TEST EQUIPMENT CAN DAMAGE THE TQC. If you use external equipment to troubleshoot the TQC, make sure that it does not introduce undesired ground currents or AC leakage currents.

Troubleshooting with Error Messages

Power-up Error Messages

Error Message	*Probable Cause*	*Remedy*
BACKUP BATTERY IS LOW	1. Battery on Processor Support PCB.	1. Replace the battery on the Processor Support PCB.
CONTROLLER ERROR	1. PC interface PCB. 2. Processor Support PCB.	1. Swap the PC Interface PCB. 2. Swap the Processor Support PCB.
EPROM CHECKSUM ERROR	1. Configuration tables. 2. Analog Processor PCB.	1. Check the configuration tables. 2. Swap the Analog Processor PCB 88/40.
KEYBOARD MALFUNCTION: PORT	1. Keyboard or keyboard cable. 2. Processor Support PCB.	1. Check the keyboard and cable. 2. Swap the Processor Support PCB.
RAM FAILURE AT 0000:	1. Main Processor 86/30.	1. Swap the 86/30.
RAM FAILURE AT 1000:	1. Main Processor 86/30.	1. Swap the 86/30.
TIGRE PROGRAM CHECKSUM ERROR	1. TIGRE program.	1. Re-enter the TIGRE program or debug the program.

Table 4-1. *Power-up error messages.*

Page 59

The following strategies for creating online instructions supplement the guidelines given earlier in this chapter:

For detailed advice on creating online communications, see Chapter 5's discussion of website design.

Writing Online Instructions

1. **Keep each unit of instruction to screen size or smaller.** Eliminate scrolling by using a compact page design and by dividing larger tasks into smaller ones.
2. **Use a consistent design on all instructional screens.** A consistent design helps readers quickly locate the information they need.
3. **Let readers see their work areas simultaneously with your instructions.** Don't make the instructions window so large that it hides the window that readers are using to perform their task.
4. **Provide a map or contents list on the home page.** This will enable readers to locate the specific part of the instructions that interests them.
5. **Provide navigational aids within the instructions.** On every page include links to the home page and to other pages of interest.
6. **Conduct a usability test.** As with any instructions, the online variety should be tested by members of the target audience working under conditions identical to those of the anticipated actual use.

SAMPLE INSTRUCTIONS

Figure 6.7 shows a set of instructions written by a student. Figure 6.8 (page 162) shows part of the online instructions for Microsoft PowerPoint. For additional examples, see this book's website.

REVISION CHECKLIST, PLANNING GUIDE, AND OTHER RESOURCES

WWW To obtain a revision checklist and planning guide you can download, visit Chapter 6 at www.techcomm.nelson.com.

A "Revision Checklist for Instructions" that you can use in your course and on the job is available on the website for this book. The website also contains a "Planning Guide for Instructions."

FIGURE 6.7

Instructions Written by a Student

Determining the Percentages of Hardwood and Softwood Fibre in a Paper Sample

Importance of the procedure is explained.

These instruction tell you how to analyze a paper sample to determine what percentage of its fibres is from hardwood and what from softwood. This information is important because the ratio of hardwood to softwood affects the paper's physical properties. The long softwood fibres provide strength but bunch up into flocks that give the paper an uneven formation. The short hardwood fibres provide an even formation but little strength. Consequently, two kinds of fibres are needed in most papers, the exact ratio depending on the type of paper being made.

Overview of the procedure and important background information

To determine the percentages of hardwood and softwood fibre, you perform the following major steps: preparing the slide, preparing the sample slurry, placing the slurry on the slide, staining the fibres, placing the slide cover, counting the fibres, and calculating the percentages. The procedure described in these instructions is an alternative to the test approved by the Technical Association of the Pulp and Paper Industry (TAPPI). The TAPPI test involves counting fibres in only one area of the sample slide. Because the fibres can be distributed unevenly on the slide, that procedure can give inaccurate results. The procedure given here produces more accurate results because it involves counting all the fibres on the slide.

All equipment is listed before the directions.

EQUIPMENT

Microscope
Microscope slide
Microscope slide cover
Microscope slide marking pen
Acetone solvent
Clean cloth
Hot plate
Paper sample
Blender
Beaker
Eyedropper
Graff "C" stain
Pointing needle

(continues)

FIGURE 6.7
(continued)

Explanation of reason for the caution

Figure is placed immediately after its mention in text.

2

PREPARING THE SLIDE

1. **Clean slide.** Using acetone solvent and a clean cloth, remove all dirt and fingerprints. NOTE: Do not use paper towel because it will deposit fibres on the slide.

2. **Mark slide.** With a marking pen, draw two lines approximately 4 centimetres apart across the width of the slide.

3. **Label slide.** At one end, label the slide with an identifying number. Your slide should now look like the one shown in Figure A.

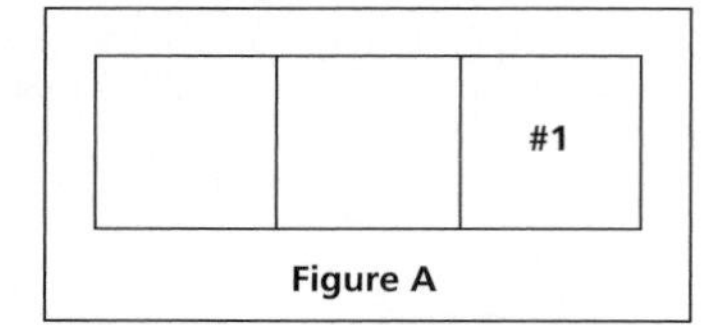

Figure A

4. **Turn on hot plate.** Set the temperature at warm. NOTE: Higher temperatures will "boil" off the softwood fibres that you will later place on the slide.

5. **Place the slide on hot plate.** Leave the slide there until it dries completely, which will take approximately 5 minutes.

6. **Remove slide from hot plate.** Leave the hot plate on. You will use it again shortly.

PREPARING THE SAMPLE SLURRY

1. **Pour 500 millilitres of water in blender.** This measurement can be approximate.

2. **Obtain paper sample.** The sample should be about the size of a dime.

3. **Tear sample into fine pieces.**

4. **Place sample into blender.**

5. **Turn blender on.** Set blender on high and run it for about 1 minute.

6. **Check slurry.** After turning the blender off, see if any paper clumps remain. If so, turn the blender on for another 30 seconds. Repeat until no clumps remain.

7. **Pour slurry into beaker.**

FIGURE 6.7
(*continued*)

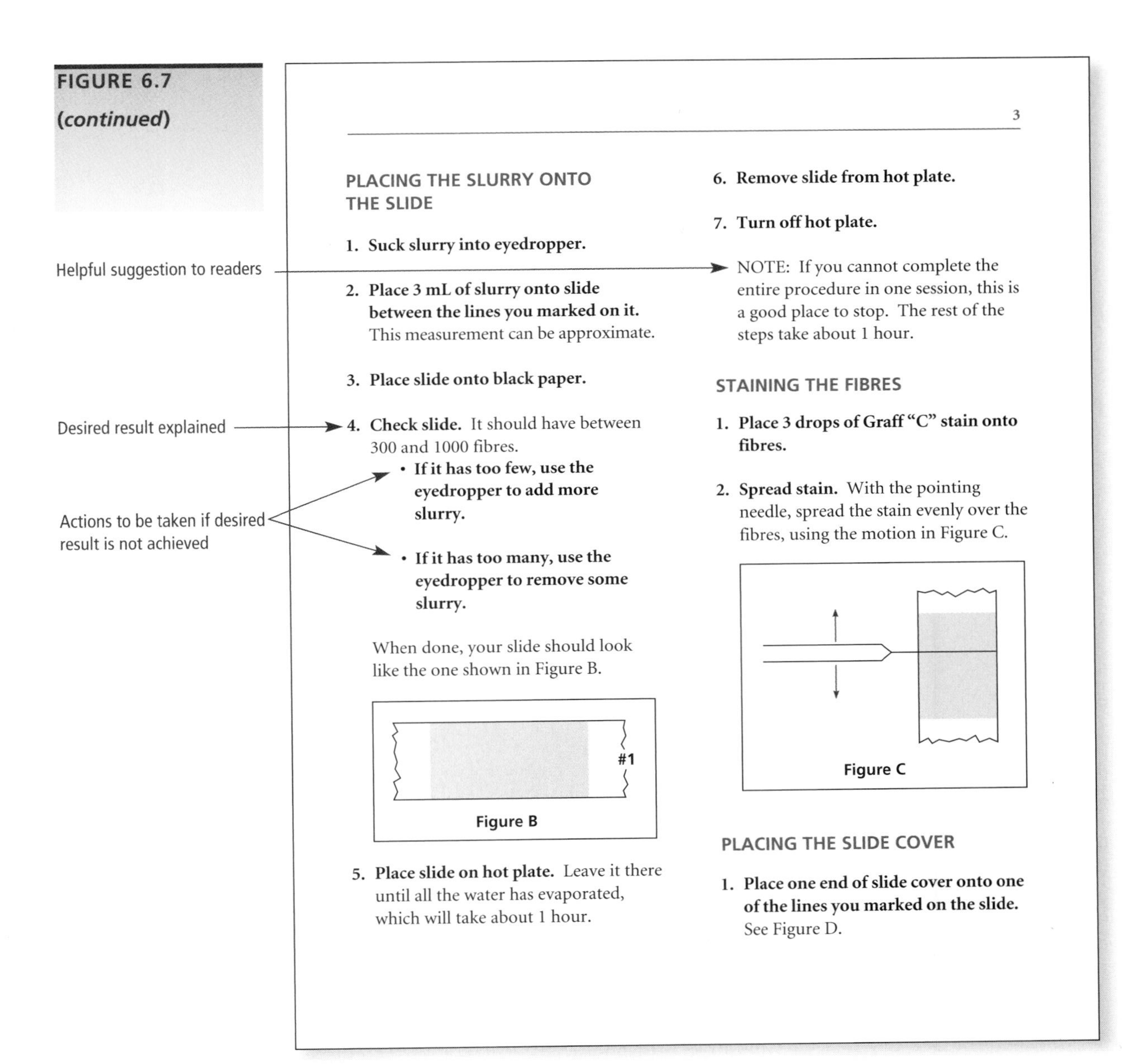

(*continues*)

FIGURE 6.7
(continued)

Figure is used to show a step not easily described in words.

Cover slide
Stain
Slide

Figure D

2. **Slowly lower the other side of the slide cover.** Be sure that no air gets trapped under the slide cover.

3. **Drain excess stain.** With a cloth underneath, turn the slide onto one of its longest edges so that the excess stain will run off.

4. **Clean slide.** Use acetone solvent to remove residue and fingerprints.

COUNTING THE FIBRES

1. **Place slide onto microscope.**

2. **Adjust magnification.** You should be able to distinguish black fibres from dark purple ones.

3. **Move slide to show upper left-hand corner of area with fibres.**

4. **Count whole fibres.** Move the slide so that your view of it changes in the manner shown in Figure E, counting

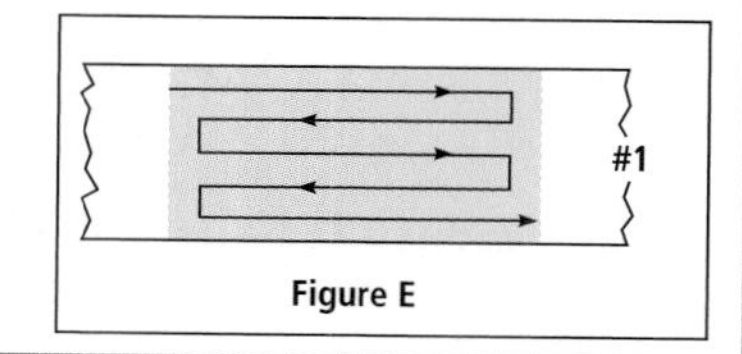

Figure E

the whole softwood and hardwood fibres you see. Ignore fragments of fibres, which you will count later.

* **Recognizing softwood fibres.** Softwood fibres are long and flat. They have blunt ends. The stain dyes the fibres colours that range from slightly purple (almost translucent) to a dark purple. See Figure F.

FIGURE 6.7
(continued)

Figure F

* **Recognizing hardwood fibres.** Hardwood fibres are much smaller than softwood fibres. Their ends come to a point, and the stain dyes them deep black. See Figure G.

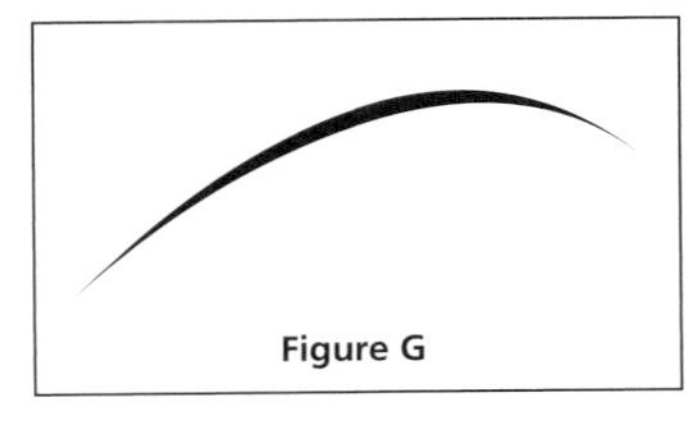
Figure G

5. **Count fragments.** Fragments result from refining. Count them while moving the slide in the way shown in Figure E. Count softwood fragments until they equal a typical whole softwood fibre on the slide. Do the same for the hardwood fragments.

CALCULATING THE PERCENTAGES

1. **Multiply the total number of hardwood and softwood fibres by the appropriate factors.**
   - Number of softwood fibres $\times$ 1.5 = X
   - Number of hardwood fibres $\times$ 0.2 = Y
2. **Determine the percentages.**
   - Percentage of softwood = X/(X + Y) $\times$ 100%
   - Percentage of hardwood = Y/(X + Y) $\times$ 100%

FIGURE 6.8
Online Instructions

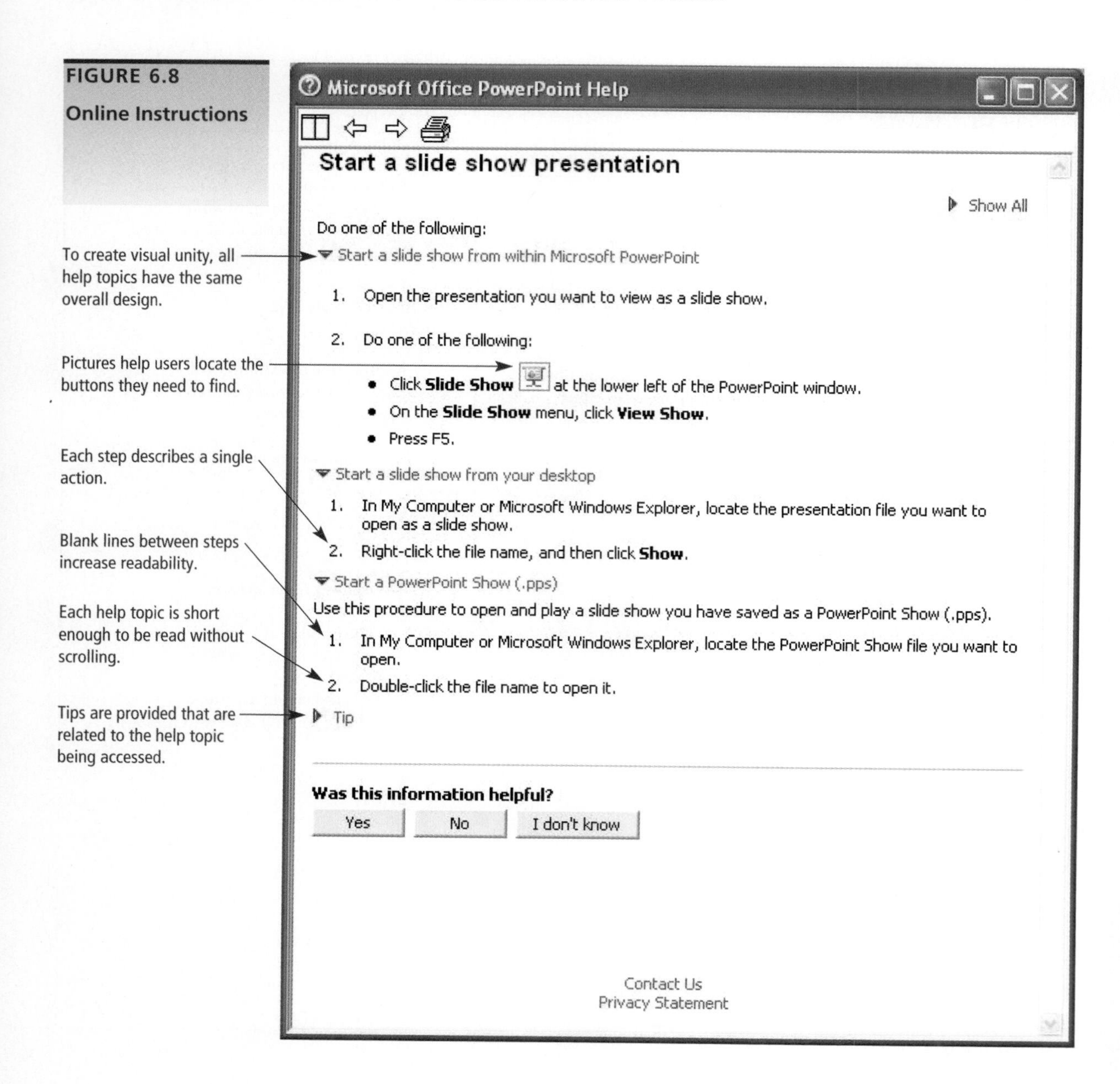

EXERCISES

For additional exercises, visit www.techcomm.nelson.com.

Expertise

1. Find and photocopy a short set of instructions (five pages or less). Analyze the instructions, noting how the writers have handled each component of the instructions. If they have omitted certain elements, explain why you think they did so. Comment on the page design and graphics (if any). Then evaluate the instructions. Tell what you think works best about them, and identify ways you think they can be improved.

2. Complete the "Instructions" project in Appendix C.

Online

Choose one:

a. Find a set of instructions designed to be read on the World Wide Web and discuss ways they might be changed to be effective in print.

b. Find a set of print instructions and discuss ways they might be changed to make them as effective as possible if viewed on the World Wide Web.

Collaboration

Working with another student, conduct an informal user test of a set of online instructions for a program on your computer. First, evaluate the screen design together. Next, while one of you uses part of the instructions, the other should record observations about parts of the text and graphics that work well and parts that could be improved. Present your results in a memo to your professor. Consider the guidelines given on pages 150–54 and your own experience as you prepare your memo.

CHAPTER 7

Proposals

COMPONENTS

Introduction

Problem

Objectives

Solution

Method

Resources

Schedule

Qualifications

Management

Costs

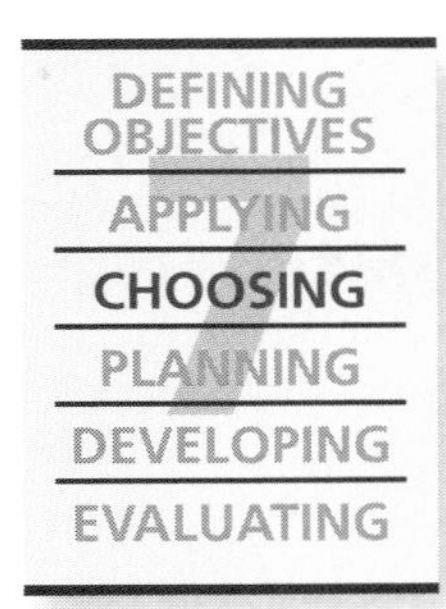

In a proposal, you make an offer and try to persuade your readers to accept it. You say that in exchange for money or time or some other consideration, you will give your readers something they want, create something they desire, or do something they wish to have done.

Throughout your career, you will have many occasions to make such offers. You may think up a new product that you could develop—if your employer will give you the time and funds to do so. Or you may devise a plan for increasing your employer's profits—if your employer will authorize you to put the plan into effect. You may even join one of the many companies that sell their products and services by means of proposals (rather than through advertising) so that your proposals are the means by which your employer generates income.

When one of your proposals has been accepted by your readers, it serves metaphorically—and, often, legally—as a contract. You and your readers are both obligated to provide what your proposal says you will each provide. Consequently, when writing a proposal, you have two goals:

The twin goals of a proposal: persuading and protecting

- **To persuade.** If you don't persuade your readers, you won't get the money, time, or other things you want.
- **To protect.** You must avoid writing in a way that offers, or seems to offer, more than you can provide or wish to provide. If you write vaguely or make overly ambitious promises, your proposal can cost your organization huge sums of money because your employer may be willing to perform large amounts of work rather than displease a client. In the extreme, your employer can be sued for failing to fulfill the contractual obligations made in your proposal.

WWW For additional chapter resources, visit Chapter 7 at www.techcomm.nelson.com.

Because proposals must both protect and persuade, expertise in writing them requires a dual vision. On one hand, you must determine how to present your offer in as appealing a way as possible. On the other hand, you must carefully define the limits of your offer so that no one thinks you are promising more than you can provide for the money, time, or other compensation you request from your readers. This chapter's discussion of the components of proposals will help you develop this expertise.

THE VARIETY OF PROPOSAL-WRITING SITUATIONS

You may be called on to write proposals in a wide variety of situations. Here are some ways in which these situations can differ:

- Your readers may be employed in your own organization or in other organizations.
- Your readers may have asked you to submit a proposal, or you may submit it to them on your own initiative.
- Your proposal may be in competition against others, or it may stand or fall on its own merits.
- Your proposal may have to be approved by people in your organization before you submit it to your readers, or you may be authorized to submit it directly yourself.
- You may have to follow regulations governing the content, structure, and format of your proposal, or you may be free to write your proposal as you think best.

- After you have delivered your proposal to your readers, they may follow any of a wide variety of methods for evaluating it.

To illustrate these variables in actual proposal-writing situations, the following paragraphs describe the circumstances in which two successful proposals were written. The information provided here will be useful to you later in this chapter, where several pieces of advice are explained through the example of these proposals.

EXAMPLE SITUATION 1

Brandon wanted permission to develop a customized course outline program that administration, faculty, and students could use to develop, store, and access course outlines at Badlands College. The present system of giving the schools within the college responsibility for their own course outlines had worked adequately until about six months ago. At that point, a lawsuit and the addition of more part-time faculty necessitated a change. Because Brandon is employed to write computer programs, he is well qualified to write this one. However, his work is assigned to him by his boss, and he cannot write the course outline program without permission from him and from his boss. Consequently, he wrote a proposal to them.

Brandon is proposing a project to his own employer. The only "cost" will be the time he spends on it.

As he wrote, Brandon had to think about only two readers because his boss and his boss's boss would decide about his proposed project without consulting other people. Because his employer had no specific guidelines for the way such internal proposals should be written, he could have used whatever content, structure, and format he thought would be most effective. Furthermore, he did not need anyone's approval to submit his proposal to people within his own department, although he would need approval before sending a proposal to another department.

Finally, Brandon did not have to worry about competition from other proposals because his would be considered on its own merits. However, his readers would approve his project only if they were persuaded that the time he would spend writing the program would not be better spent on his regular duties. (Brandon's entire proposal is presented at the end of this chapter, page 176.)

EXAMPLE SITUATION 2

Three people are proposing the provision of services to a school board, which will consider these proposals competitively.

www.techcomm.nelson.com

The second proposal was written under very different circumstances. To begin with, it was written by three people, not just one. The writers were a purchasing manager, a marketing manager, and a director of operations at a food services company that supplies various sites with cafeteria services. The writers had learned that the Thames Valley District School Board was seeking proposals for food service management of the board's cafeterias. To learn more about what the board wanted, the writers obtained a copy of the Request for Proposal (RFP). After studying the RFP, they decided to bid on the contract by submitting a proposal.

The formal review process used by the school board is used by many government agencies, departments, and organizations.

In their proposal, the writers addressed an audience very different from Brandon's. The board receives numerous proposals for every RFP it publishes. The evaluation is done by a committee of representatives from the board. All bids are first evaluated on a given list of criteria such as quality assurance, environmental policy, and legislative coverage. Only then are supplemental materials to the proposal considered. To win the bid, the writers of the food services proposal needed to generate a usable document that matched all the necessary and required conditions of the RFP (even to the extent of using the same numbering format in the proposal as that used in the RFP). Only then could the writers hope to add a persuasive element, particularly through the use of staff qualifications, references, and financial return offered.

Before the writers could even mail their proposal to the department, they had to obtain approval for it from the president of the food services company. That's because the proposal, if accepted, would become a contract between the company and the school board.

PROPOSAL READERS ARE INVESTORS

The descriptions of the proposals written by Brandon and by the employees of the food services company illustrate some of the many differences that exist in proposal-writing situations. Despite these differences, however, almost all proposal-writing situations have two important features in common—features that profoundly affect the way you should write your proposals:

Dynamics of proposal-writing situations

- In proposals, you ask decision makers to invest some resource, such as time or money, so that the thing you propose can be done.
- Your readers will make their investment decisions *cautiously*. Their resources are limited, so if they decide to invest in your project, resources will not be available for others. For example, to let Brandon spend six weeks creating the new course outline program, his bosses had to decide that he would not spend that time on the department's other projects. To award the food services contract to one company, the school board had to turn down other proposals competing for those same dollars.

THE QUESTIONS READERS ASK MOST OFTEN

The questions decision makers ask when they consider a proposal generally concern the following three topics:

- **Problem.** Your readers will want to know why you are making your proposal and why they should be interested in it. What problem, need, or goal does your proposal address—and why is it important to them?
- **Solution.** Your readers will want to know exactly what you propose to make or do and how it relates to the problem you describe. They will ask, "What kinds of things will a successful solution to this problem have to do?" and "How do you propose to do those things?" They will examine your responses carefully, trying to determine whether your overall strategy and your specific plans are likely to work.
- **Costs.** Your readers will want to know what it will cost to implement your proposal and whether the cost will be worth it to them.

In addition, if you are proposing to perform some work (rather than to supply a product), your readers will want an answer to this question:

- **Capability.** If your readers pay or authorize you to perform the work, how will they know whether they can depend on you to deliver what you promise?

STRATEGY OF THE STRUCTURE FOR PROPOSALS

The structure for proposals provides a framework for answering those questions—one that has been found successful in repeated use in the kinds of situations you will encounter on the job. This structure includes the ten topics described next. In some proposals you may need to include information on all ten, but in others you will need to cover only some of them. Even in the briefest proposals, however, you will probably need to treat the following four: introduction, problem, solution, and costs.

Structure for Proposals

Topic	Readers' Question	Your Persuasive Point
*Introduction	What is this communication about?	Briefly, I propose to do the following.
*Problem	Why is the proposed project needed?	The proposed project addresses a problem, need, or goal that is important to you.
Objectives	What features will a solution to the problem need in order to be successful?	A successful solution can be achieved if it has these features.
*Solution	What will your proposed solution look like?	Here's what I plan to produce, and it has the features necessary for success.
Method Resources Schedule Qualifications Management	Are you going to be able to deliver what you describe here?	Yes, because I have a good plan of action (method); the necessary facilities, equipment, and other resources; a workable schedule; appropriate qualifications; and a sound management plan.
*Costs	What will it cost?	The cost is reasonable.

*Topics marked with an asterisk are important in almost every proposal, whereas the others are needed only in certain ones.

To write a successful proposal, you must not only use the components for proposals but also follow this book's other reader-centred strategies.

Whichever question you are answering, remember the twin qualities of all workplace writing: usability and persuasiveness. To be usable, your proposal must enable your readers to find quickly and understand readily the answer to each of their questions. To be persuasive, your proposal must present each answer in a way that leads your readers to conclude that you will produce an outstanding result if your proposal is accepted. Consequently, to write a successful proposal, you must not only employ the proposal structure when determining your proposal's overall content and organization but also follow this book's other reader-centred strategies when developing and evaluating each section's textual and graphical elements. Your goal is to lead your readers through the following sequence of thought:

Ideal sequence of readers' thoughts while reading a proposal

1. The readers learn generally what you want to do. (Introduction)
2. The readers are persuaded that there is a problem, need, or goal that is important to them. (Problem)

3. The readers are persuaded that the proposed action will be effective in solving the problem, meeting the need, or achieving the goal that they now agree is important. (Objectives, Solution)
4. The readers are persuaded that you are capable of planning and managing the proposed solution. (Method, Resources, Schedule, Qualifications, Management)
5. The readers are persuaded that the cost of the proposed action is reasonable in light of the benefits the action will bring. (Costs)

There is no guarantee, of course, that your readers will read your proposal from front to back or concentrate on every word. Long proposals usually include a summary or abstract at the beginning. Instead of reading the proposal straight through, many readers will read the summary, and perhaps the first few pages of the body, and then skim through the other sections.

In fact, in some competitive situations, readers are prohibited from reading the entire proposal. For instance, companies competing for huge contracts to build parts of space shuttles for the National Aeronautics and Space Administration (NASA) submit their proposals in three volumes: one explaining the problem and their proposed solution, one detailing their management plan, and one analyzing their costs. Each volume (sometimes thousands of pages long) is evaluated by a separate set of experts: technical experts for the first volume, management experts for the second, and budget experts for the third.

Even when readers do not read your proposal straight through—and even when your proposal is only a page or two long—the account just given of the relationships among the parts will help you write a tightly focused proposal in which all the parts support one another.

Finally, note that in certain parts of a proposal you are describing what you will provide. These are places to write with protection in mind. Don't promise too much. Don't write in a way that permits your readers to think that you are offering more than you intend to provide.

COMPONENTS OF PROPOSALS

A structure is not an outline. For general advice about using structures, see pages 235–36.

The rest of this chapter describes in detail each of the ten components that comprise the conventional structure for proposals.

As you plan and write your proposal, remember that the ten topics identify kinds of information you need to provide, not necessarily the titles of the sections you include. In brief proposals, some parts may take only a sentence or a paragraph, or several sections may be grouped together. For instance, writers often combine their introduction, their discussion of the problem, and their explanation of their objectives under a single heading, which might be "Introduction," "Problem," or "Need."

INTRODUCTION

At the beginning of a proposal, you want to do the same thing that you do at the beginning of anything else you write on the job: tell your readers what you are writing about. In a proposal, this means announcing what you are proposing.

How long and detailed should the introduction be? In proposals, introductions are almost always rather brief. By custom, writers postpone the full description of what they are proposing until later, after they have discussed the problem their proposal will help solve.

You may be able to introduce your proposal in a single sentence, as Brandon did:

Brandon's introduction

> I request permission to spend six weeks writing, testing, and implementing a program to store and manage course outlines in the college.

When you are proposing something more complex, your introduction will probably be longer. You may want to provide background information to help your readers understand what you have in mind. Here, for example, is the introduction from the proposal written by the employees of the food services company to the Thames Valley Disctrict School Board:

Introduction to the food services company's proposal

For more ideas on writing introductions, see Chapter 3.

> Fulfords' Fine Foods is a local, family-owned food services corporation that has operated in southwestern Ontario for over forty years. We have provided corporate and institutional clients, such as the Grand River School Board, with reasonably priced catering and cafeteria operations and kitchen services, while ensuring high quality and service. Fulfords' Fine Foods proposes to provide cafeteria food services to the Thames Valley District School Board, as outlined in Request for Proposal #03-39S.

PROBLEM

Once you've announced what you're proposing, you must persuade your readers that it will address some problem, need, or goal that is significant to them. Your description of the problem is crucial to the success of your proposal. Although you might persuade your readers that your proposed project will achieve its objectives and that its costs are reasonable, you cannot hope to win approval unless you show that it is worth doing from your readers' point of view.

The following paragraphs offer advice that applies to each of three situations you are likely to encounter on the job: when your readers define the problem for you, when your readers provide you with a general statement of the problem, and when you must define the problem yourself.

When Your Readers Define the Problem for You

The food services company's writers read the RFP to determine the school board's goals.

You need to do the least research about the problem when your readers define it for you. For instance, your readers might issue an RFP that explains in complete detail some technical problem they would like your firm to solve. In such situations, your primary objective in describing the problem will be to show your readers that you thoroughly understand what they want. This was the situation for the food services company's writers as they constructed their proposal.

When Your Readers Provide a General Statement of the Problem

The writers used the RFP when deciding how to describe the problem they would solve.

At other times, your readers may describe the problem only vaguely. For example, an RFP originating from a government ministry might provide only the general statement that it perceives "a need for educational resources that could be used to teach about the relation of the natural and man-made environment to the total human environment."

Each group that wishes to submit a proposal then needs to identify a more specific problem that the ministry would find important.

Finding a problem that someone else thinks is important can be difficult. Often, it will require you to hunt for clues to your readers' values, attitudes, concerns, and opinions. You may have to interview people (including your target readers, where possible), visit the library, or search the Web.

When You Must Define the Problem Yourself

The reasons you offer your readers for supporting your proposal may be different from your reasons for writing it.

In other situations, you may not have the aid of explicit statements from your readers to help you formulate the problem. This is most likely to happen when you are preparing a proposal on your own initiative, without being asked by someone else to submit it. Describing the problem in such situations can be particularly challenging because the arguments that will be persuasive to your readers might be entirely distinct from your own reasons for writing the proposal.

Think about Brandon's situation, for instance. He originally came up with the idea of writing the program for course outlines because he felt frustrated on the many occasions when he was asked to retrieve an archived course outline. His bosses, however, are not likely to approve the scheduling project simply to help Brandon avoid frustration. For the project to appeal to them, it must be couched in terms that are allied to their responsibilities and professional interests.

First strategy: Identify goals that you will help your readers achieve.

In such situations, you can pursue two strategies to define the problem. The first is to think about how you can make your proposed project important to your readers. What goals or responsibilities do your readers have that your proposal will help them achieve? What concerns do they typically express that your proposal could help them address? A good place to begin is to think about some of the standard concerns of organizations: efficiency and profit.

When Brandon did this, he realized that from his employer's point of view, the problems involved with inconsistent course outlines were detrimental to the college's reputation. The quality of its educational offerings depended on course outlines.

Second strategy: Talk with your readers before you begin writing.

A second strategy for defining the problem is to speak with the people to whom you will send your proposal. This conversation can have two advantages. First, it will let you know whether or not your proposal has at least some chance of succeeding. If it doesn't, it's best to find out before you invest time writing it. Second, by talking with people who will be readers of your proposal, you can find out how the problem appears to them.

When Brandon spoke to his boss, his boss mentioned something he hadn't thought of before. Sometimes the course outlines are used for proposals to corporations. When prospective corporate clients see inconsistent course outlines, they may wonder whether the college has quality issues that would affect its course content.

OBJECTIVES

Here, as examples, are three of the objectives that the writers at Fulfords' Fine Foods devised for their cafeteria food services proposal. To help you see how these objectives grew out of the writers' statement of the problem, each objective is followed by the point from their problem statement that serves as the basis for it. (The writers did not include the bracketed sentences in the objectives section of their proposal.)

Objectives of the food services proposal

1. To provide food service management of the cafeterias of the board, including the preparation and service of morning food services, lunches, and other items of food and beverage as shall be required by the board for such persons as the board authorizes to be served. [This primary objective is based on the purpose section of the RFP.]
2. To operate a kitchen, servery, and eating area in each of these cafeterias that will provide employment and education opportunities for employment-aged students. [This objective is based on the evidence presented in the problem section that students at many of these schools require paid co-op job opportunities.]
3. To co-operate with all recycling and environmental procedures and initiatives established by government, the board, and the school. [This objective is based on the writers' argument in the problem section that students must be able to see their cafeterias as working models of recycling and environmental practices if they are to act in socially responsible ways themselves.]

The writers created similar lists of objectives for the other parts of their proposed project, each based on a specific point made in their discussion of the problem.

Like the writers at Fulfords' Fine Foods, Brandon built his objectives squarely on his description of the problem. He also took into account something he learned through his research: even with a standardized Web-based system, academic managers wanted a single person in each school to enter course outlines initially. They didn't want each faculty member to be able to enter changes into the system. Similarly, you will need to consider the concerns of your readers when framing the objectives for your proposals. Here are two of Brandon's objectives:

Brandon's objectives

- Provide a repository for all course outlines. [This objective relates to Brandon's discovery that part-time faculty and students are often unaware of the existence of course outlines.]
- Show the matching of program learning outcomes within a program to the courses in which those outcomes are delivered. [This objective corresponds to the college's concern regarding quality of its educational offerings.]

In proposals, writers usually describe the objectives of their proposed solution without describing the solution itself at all. The food services company, for example, wrote its objectives so that the readers could imagine achieving them through the operation of the cafeterias. Similarly, Brandon described objectives that might be achieved by many kinds of computer programs. He withheld his ideas about the design of his program until the next section of his proposal.

The purpose of separating the objectives from the solution is not to keep readers in suspense. Rather, this separation enables readers to evaluate the aims of the project separately from the writers' particular strategies for achieving those aims.

Proposal writers usually present their objectives in a list or state them very briefly. For instance, Brandon used only one paragraph to present his objectives. The members of the food services company presented all of their objectives in three pages of their 98-page proposal.

SOLUTION

When you describe your solution, you describe your plan for achieving the objectives you have listed. For example, Brandon described the various parts of the computer program he would write, explaining how they would be created and used. The food services company described each of the five components of its cafeteria food services.

When describing your solution, you must persuade your readers of two things:

- **That your solution will successfully address each of the objectives.** For instance, to be sure that the description of their proposed food services matched their objectives, the writers included detailed descriptions of the following: the selection and pricing of the food and beverage items (Objective 1), the job descriptions of the student positions (Objective 2), and the scope and methods of their recycling program (Objective 3).
- **That your solution offers a particularly desirable way of achieving the objectives.** For example, the food services company planned to use a local organic fruit and vegetable supplier, so its proposal offered an explanation of the advantages of this approach.

When describing your solution, you may find it helpful to use the strategies for describing an object or process (pages 45–50).

Of course, you should include such statements only where they won't be perfectly obvious to your readers. In his proposal, Brandon did not include any because he planned to use standard practices whose advantages would be perfectly evident to his readers, both of whom had spent several years doing exactly the kind of work Brandon is doing now.

The description of your proposed solution is one of the places where you must be very careful to protect yourself and your organization from seeming to promise more than you can deliver. The surest protection is to be very precise. For instance, Brandon included specific details about the capabilities of his program. It would support the entry of 1,000 course outlines linked by program (not more); it would provide storage capabilities for up to five years (not longer); and it would operate on an NT server (not necessarily other servers). If you want to identify things you might be able to do, explicitly identify them as possibilities, time and opportunity permitting, not as promises.

METHOD

Readers of proposals sometimes need to be assured that you can, in fact, produce the results that you promise. That happens especially in situations where you are proposing to do something that takes special expertise.

To assure themselves that you can deliver what you promise, your readers will look for information about several aspects of your project: your method or plan of action for producing the result; the facilities, equipment, and other resources you plan to use; your schedule; your qualifications; and your plan for managing the project. This section is about method; the other topics are discussed in the sections that follow.

To determine how to explain your proposed method, imagine that your readers have asked you, "How will you bring about the result you have described?"

In some cases, you will not need to answer that question. For example, Brandon did not talk at all about the programming techniques he planned to use because his readers were already familiar with them. On the other hand, they did not know how he planned to train people to use his program. Therefore, in his proposal he explained his plans for training.

When describing your method, you may find it helpful to use the strategies for describing a process (pages 47–50).

In contrast, Fulfords' Fine Foods needed to describe its method in detail to persuade readers that it would provide effective cafeteria food services. The writers outlined their quality assurance philosophy and program, including their annual survey to measure customers' opinions on pricing, food quality, service level, and catering staff. They also included a two-year priced menu with a four-week cycle.

RESOURCES

By describing the facilities, equipment, and other resources to be used for your proposed project, you assure your readers that you will use whatever special equipment is required to do the job properly. If part of your proposal is to request that equipment, tell your readers what you need to acquire and why.

If no special resources are needed, you do not need to include a section on resources. Brandon did not include one. In contrast, the food services writers needed many kinds of resources. In their proposal, they described the services and supplies provided by their company, such as debit bank services and supplies for cleaning all food preparation equipment, dishware, and cutlery. The writers also described their utilization of the services and supplies provided by the board, such as kitchen equipment (dishwashers, stoves) and custodial services.

SCHEDULE

Proposal readers have several reasons for wanting to know what your schedule will be. First, they want to know when they can enjoy the final result. Second, they want to know how you will structure the work so they can be sure that the schedule is reasonable and sound. In addition, they may want to plan other work around the project: When will your project have to coordinate with others? When will it take people's attention from other work? When will other work be disrupted and for how long? Finally, proposal readers want a schedule so they can determine if the project is proceeding according to plan.

For information on creating a schedule chart, see pages 455–56.

The most common way to provide a schedule is to use a schedule chart, which is sometimes accompanied by a text explanation of its important points.

In many projects, your ability to complete your work on time depends on your getting timely responses and cooperation from other people, including the client or other person or organization who approved your proposal. When discussing your schedule, be sure to include the dates when you will need something from your client, such as approvals of your drafts and information you must have in order to proceed with your work.

QUALIFICATIONS

When they are thinking about investing in a project, proposal readers want to be sure that the proposers have the experience and capabilities required to carry it out successfully. For that reason, a discussion of the qualifications of the staff involved with a project is a standard part of most proposals. For example, the food services writers discussed their qualifications in a section titled "Administrative Organization," in which they included an organization chart and the names and addresses of all company officers and managers. The writers also provided references from three current institutional clients and the names and phone numbers of the individuals whom the board could contact.

In other situations, much less information might be needed. For instance, Brandon's qualifications as a programmer were evident to his readers because they were employing him as one. If that experience alone were enough to persuade his readers that he could carry out the project successfully, Brandon would not have needed to include any section on qualifications. However, his readers might have wondered whether he was qualified to undertake the particular program he proposed because different kinds of programs require specialized knowledge and skills. Therefore, Brandon wrote the following:

Brandon's statement of his qualifications

> As you know, I am thoroughly familiar with the kind of database design that would be used in the course outline system. I recently completed a certificate in interactive media during which I learned how to program dynamic webpages that can be used to request a database search and display the results. In addition, as an undergraduate, I took several technical communication courses, which will help me with the documentation requirements of the project.

In some situations, your readers will want to know not only the qualifications of the people who will work on the proposed project, but also the qualifications of the organization for which they work.

To protect yourself and your organization, guard against implying that you and those working with you have more qualifications than you really have. Time spent learning new programs, gaining new knowledge, or otherwise gaining background needed to complete a project adds to the time (and money) required for completion.

MANAGEMENT

For information on creating an organizational chart, see pages 454–55.

When you propose a project that will involve more than about four people, you can make your proposal more persuasive by describing the management structure of your group. Proposal readers realize that on complex projects even the most highly qualified people can coordinate their work effectively only if a well-designed project management plan is in place. In projects with relatively few people, you can describe the management structure by first identifying the person or persons who will have management responsibilities and then telling what their duties will be. In larger projects, you might need to provide an organizational chart for the project and a detailed description of the management techniques and tools that will be used.

Because his project involved only one person, Brandon did not establish or describe any special management structure. However, the food services company writers did. Because they had a project involving many sites, they wrote a section titled "Staffing," in which they included a complete staffing proposal for each site within the board. For each site, the writers outlined the number of staff and the hours of each staff member. They also included a résumé for the executive chef, the site supervisor, and the supervisor of operations.

COSTS

For information on creating a budget statement, see pages 456–57.

As emphasized throughout this chapter, when you propose something, you are asking your readers to invest resources, usually money and time. Naturally, then, you need to tell them how much your proposed project will cost.

One way to discuss costs is to include a budget statement. Sometimes, a budget statement needs to be accompanied by a text explanation of any unusual expenses and the method used to calculate costs.

In proposals where dollars are not involved, information about the costs of required resources may be provided elsewhere. For instance, in his discussion of the schedule for his project, Brandon explained the number of hours he would spend, the time that others would spend, and so on.

In some proposals, you may demonstrate that the costs are reasonable by also calculating the savings that will result from your project.

SAMPLE PROPOSAL

Figure 7.1 shows the proposal written by Brandon. For additional examples, visit this book's website at www.techcomm.nelson.com.

FIGURE 7.1

Sample Proposal in the Memo Format

WWW To view additional proposals, visit Chapter 7 at www.techcomm.nelson.com.

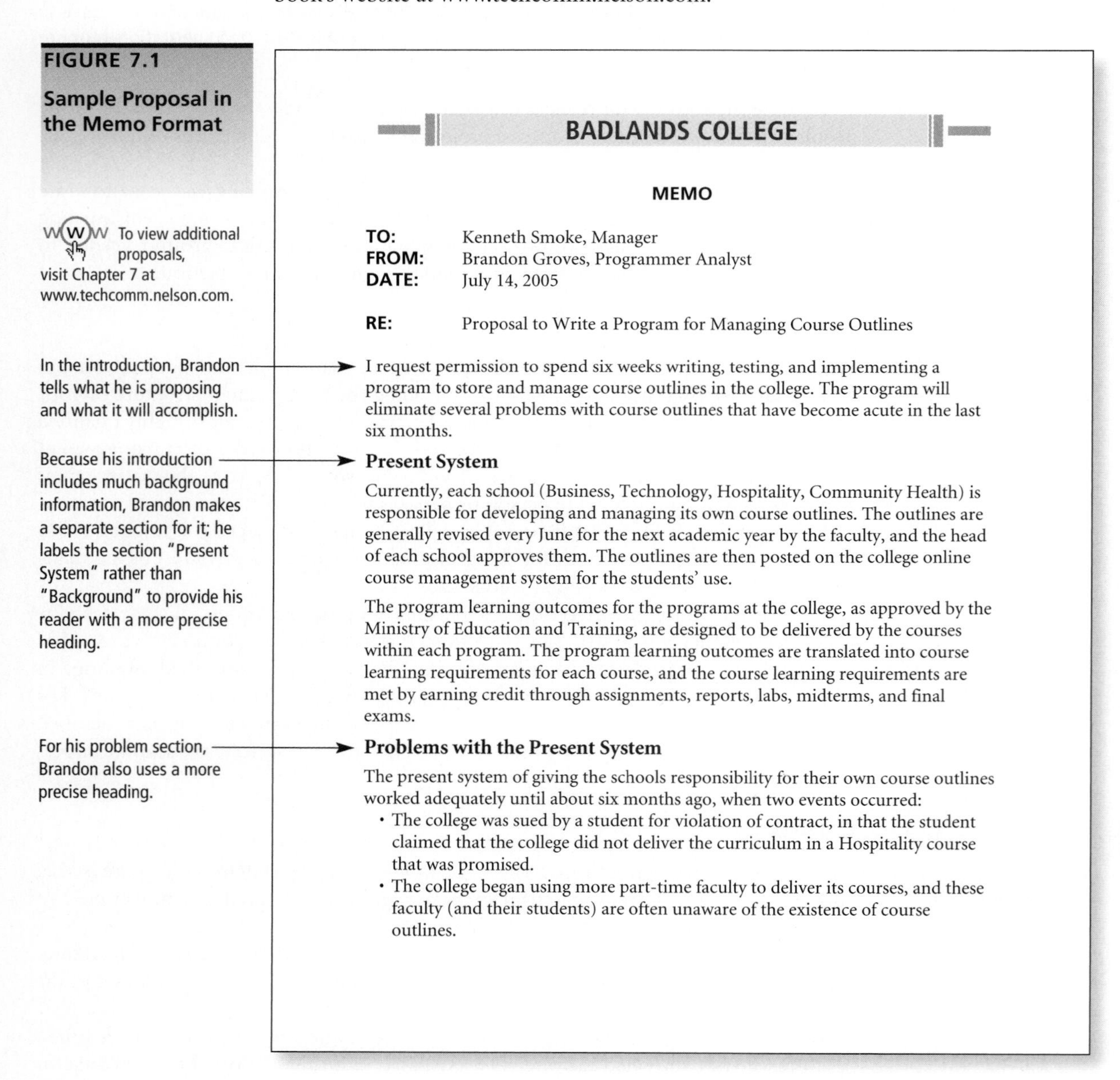

BADLANDS COLLEGE

MEMO

TO: Kenneth Smoke, Manager
FROM: Brandon Groves, Programmer Analyst
DATE: July 14, 2005

RE: Proposal to Write a Program for Managing Course Outlines

In the introduction, Brandon tells what he is proposing and what it will accomplish.

I request permission to spend six weeks writing, testing, and implementing a program to store and manage course outlines in the college. The program will eliminate several problems with course outlines that have become acute in the last six months.

Because his introduction includes much background information, Brandon makes a separate section for it; he labels the section "Present System" rather than "Background" to provide his reader with a more precise heading.

Present System

Currently, each school (Business, Technology, Hospitality, Community Health) is responsible for developing and managing its own course outlines. The outlines are generally revised every June for the next academic year by the faculty, and the head of each school approves them. The outlines are then posted on the college online course management system for the students' use.

The program learning outcomes for the programs at the college, as approved by the Ministry of Education and Training, are designed to be delivered by the courses within each program. The program learning outcomes are translated into course learning requirements for each course, and the course learning requirements are met by earning credit through assignments, reports, labs, midterms, and final exams.

For his problem section, Brandon also uses a more precise heading.

Problems with the Present System

The present system of giving the schools responsibility for their own course outlines worked adequately until about six months ago, when two events occurred:

- The college was sued by a student for violation of contract, in that the student claimed that the college did not deliver the curriculum in a Hospitality course that was promised.
- The college began using more part-time faculty to deliver its courses, and these faculty (and their students) are often unaware of the existence of course outlines.

FIGURE 7.1

(continued)

Problem

Brandon describes the consequences of the problem that are important to his readers.

Brandon ties the objectives of his proposed project directly to the points he raised when describing the problem.

After stating the general nature of his proposed project, Brandon links the features of his product to the objectives he identified earlier. Brandon protects himself from misinterpretations of what he will produce by precisely describing the capabilities of his proposed program.

Brandon provides an overview of his method, then explains each step in a way that shows how each will contribute to a successful outcome.

The second event has greatly complicated the offering of cohesive, integrated programs at Badlands College. The college's funding from the Ministry depends on its delivery of approved programs, containing all the program learning outcomes divided between the various courses within a program. The quality of a program, then, depends on the quality of its course outlines, and the distribution of those course outlines to administration, full- and part-time faculty, and students. If this quality is not assured, the college runs the risk of losing funding and of suffering additional lawsuits from students for breach of contract.

Objectives

To solve our course outline storage and management issues, we need a system that will do the following:

- Provide a course outline template for consistency across schools
- Provide a repository for all course outlines
- Show the matching of program learning outcomes within a program to the courses in which those outcomes are delivered

Proposed Solution

I propose to solve our course outline problems by creating a Web-based system for developing and storing our course outlines across the college. It will support the entry of 1,000 courses outlines linked by program and store them for up to five years. A designated person in each school will be able to enter and edit course outlines into a form through his or her Web browser. Located on an NT server, the program will automatically produce a PDF version of the course outline once it is approved by the appropriate manager. All approved course outlines will then be available to all administration, full- and part-time faculty, and students. A reporting function will enable managers to map program learning outcomes across courses to ensure diploma requirements are being met.

Method for Developing the Program

I propose to create and implement the program in three steps: writing it, testing it, and training people in its use.

Writing the Program

The program will have three features. The first will display the courses that exist within the college, and enable course outlines to be entered and edited. The second will allow the course outline to be approved, and will generate a PDF. The third will be a reporting function that will enable managers to map programs and courses.

(continues)

FIGURE 7.1
(continued)

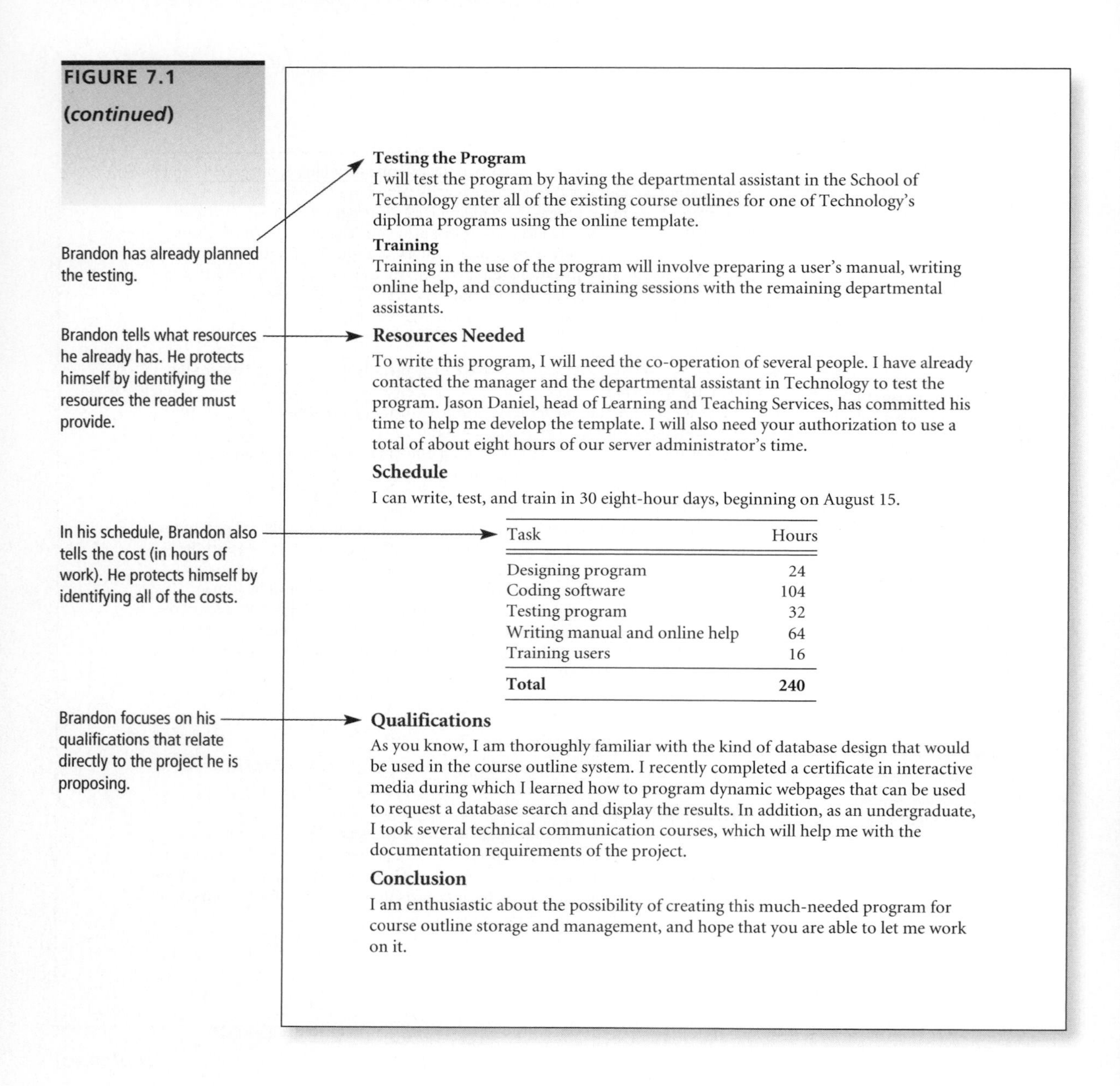

Testing the Program
I will test the program by having the departmental assistant in the School of Technology enter all of the existing course outlines for one of Technology's diploma programs using the online template.

Training
Training in the use of the program will involve preparing a user's manual, writing online help, and conducting training sessions with the remaining departmental assistants.

Resources Needed

To write this program, I will need the co-operation of several people. I have already contacted the manager and the departmental assistant in Technology to test the program. Jason Daniel, head of Learning and Teaching Services, has committed his time to help me develop the template. I will also need your authorization to use a total of about eight hours of our server administrator's time.

Schedule

I can write, test, and train in 30 eight-hour days, beginning on August 15.

Task	Hours
Designing program	24
Coding software	104
Testing program	32
Writing manual and online help	64
Training users	16
Total	**240**

Qualifications

As you know, I am thoroughly familiar with the kind of database design that would be used in the course outline system. I recently completed a certificate in interactive media during which I learned how to program dynamic webpages that can be used to request a database search and display the results. In addition, as an undergraduate, I took several technical communication courses, which will help me with the documentation requirements of the project.

Conclusion

I am enthusiastic about the possibility of creating this much-needed program for course outline storage and management, and hope that you are able to let me work on it.

REVISION CHECKLIST, PLANNING GUIDE, AND OTHER RESOURCES

WWW To obtain a revision checklist and planning guide you can download, visit Chapter 7 at www.techcomm.nelson.com.

A "Revision Checklist for Proposals" that you can use in your course and on the job is available on the website for this book. The website also provides a "Planning Guide for Proposals."

CHAPTER 8 Reports

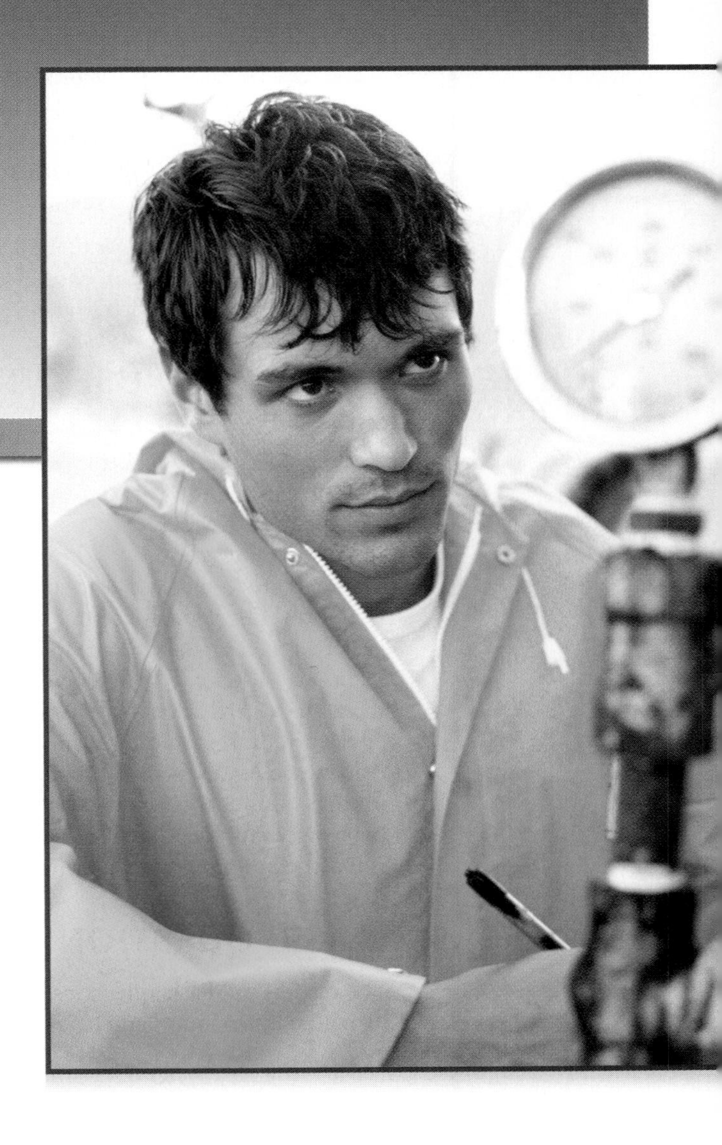

COMPONENTS

Introduction

Method of Obtaining Facts

Facts

Discussion

Conclusions

Recommendations

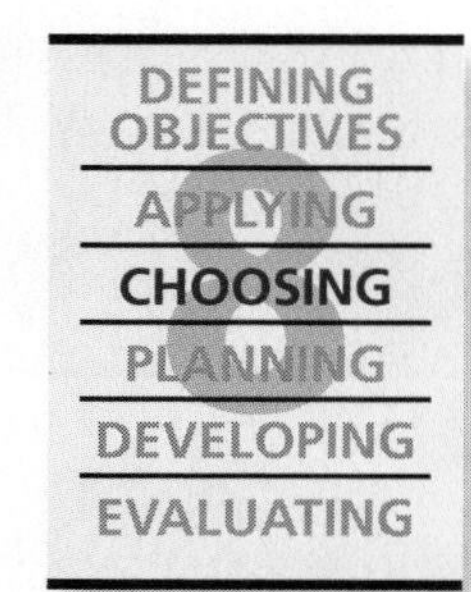

For additional chapter resources, visit Chapter 8 at www.techcomm.nelson.com.

Reports are among the most common types of communication prepared at work. They come in many varieties. Here are some examples:

- A 100-page report on a seven-month project testing a special method of venting high-thrust engines for use in space vehicles.
- A 12-page report based on library research concerning new techniques of genetic engineering used for food crops.
- A two-paragraph report based on a manufacturing engineer's visit to a new plant that is about to be put into service.

As these examples suggest, reports differ from one another in many ways, including subject matter, length, purpose, information sources, number and kinds of readers, and the circumstances that led to their preparation.

YOUR READERS WANT TO USE THE INFORMATION YOU PROVIDE

Despite their differences, almost all reports share one important feature: Their readers want to put the information the reports contain to some professional or practical use. The precise use varies, of course, from situation to situation. For example, readers sometimes want to use a report's information to solve an organizational problem (where typical goals are to increase efficiency and profit), a social problem (where a typical goal is to improve the general health and welfare of a group of people), or an ethical problem (where a typical goal is to see that people are treated justly and respectfully).

Despite the many differences in readers' goals, most successful reports prepared at work share certain characteristics. You can understand why if you follow Chapter 2's advice to think of your readers as people who ask questions. At a general level, report readers almost always ask the same basic set of six questions. The structure for reports found in this chapter is a conventional pattern that both readers and writers have found to be successful for answering these questions. In fact, each of the six report components is matched to one of the questions.

READERS' SIX BASIC QUESTIONS

The readers' six basic questions are as follows. All six reflect readers' overriding goal of using the information and ideas that you provide as a guide for future action:

Report readers' questions focus on the future

- **What will we gain from reading your report?** Most people at work want to read only those communications that are directly useful to them. Therefore, you need to explain your communication's relevance to the readers' responsibilities, interests, and goals.
- **Are your facts reliable?** Readers want to be certain that the facts you supply will provide a sound basis for their decisions or actions.
- **What do you know that is useful to us?** Readers don't want to read everything you know about your subject; they want you to tell them only the facts they can put directly to use. (Example: "The most important sales figures for this quarter are as follows: . . .")

- **How do you interpret those facts from our point of view?** Facts alone are meaningless. To give facts meaning, people must interpret them by pointing out relationships or patterns among them. (Example: "The sales figures show a rising demand for two products but not for two others.") Usually, your readers will want you to make those interpretations rather than leave that work to them.
- **How are those facts significant to us?** Readers generally want you to go beyond an interpretation of the facts to explain what the facts mean in terms of the readers' responsibilities, interests, or goals. (Example: "The sales of our handheld computer have fallen sharply as sales of our competitors' newest releases have skyrocketed. The falling sales may signal that our product is no longer competitive.")
- **What do you think we should do?** Because you will have studied the facts in detail, readers will usually assume that you are especially well qualified to make recommendations. (Example: "We should improve our handheld computer or else quit producing it.")

Of course, these six questions are very general. In large reports, writers need to take hundreds, even thousands, of pages to answer them. That's because readers often seek answers to these basic questions by asking a multitude of more specific, subsidiary questions. But these six questions are the general ones that should guide your overall work on any report.

STRUCTURE FOR REPORTS

The questions report readers ask are linked in the following way with the six elements of the report structure:

Report Structure

Report Element	Readers' Question
Introduction	What will we gain from reading your report?
Method of obtaining facts	Are your facts reliable?
Facts	What do you know that is useful to us?
Discussion	How do you interpret those facts from our point of view?
Conclusions	How are those facts significant to us?
Recommendations	What do you think we should do?

For a general discussion of how to use structures, see Guideline 5 in Chapter 9, pages 235–36.

As you think about drafting these elements, remember that your overall objective is the same as when preparing any other workplace communication: to create a document your readers will find highly usable and highly persuasive. Your report should enable your readers to find quickly and understand immediately your answers to their questions. It should also persuade them that your answers are credible and that your recommendations should be followed. Using the report structure will help you create a report that achieves these outcomes. However, it won't guarantee success. As you draft each element of your report, you must also follow the reader-centred advice given throughout this book.

Remember, too, that a structure is not the same as an outline. The six elements may be arranged in many ways, and one or more of them may be omitted if circumstances warrant. In some brief reports, for example, the writers begin with a recommendation, move to a paragraph in which the facts and conclusions are treated together, and state the sources of their facts in a concluding, single-sentence paragraph. Also, people sometimes present two or more of the six elements under a single heading. For instance, they may include in their introduction information about how they obtained their facts, and they frequently present and interpret their facts in a single section of their report.

The various ways of combining these elements into a particular report are discussed at the end of this chapter. First, however, each of the six elements is described individually, with particular attention to the ways each might be developed in a report in which all six appear separately.

INTRODUCTION

For additional advice on writing an introduction, see Chapter 3.

In the introduction of a report, you answer your readers' question, "What will we gain from reading your report?" In some reports, you can answer this question in a sentence or less. Consider, for instance, the first sentence of a report written by Efra, an employee of a university's fundraising office, who was asked to investigate the university's facilities and programs in outdoor and experimental education. Because her reader, Chander, had assigned her to prepare the report, she could tell him what he would gain from it simply by reminding him why he had requested it:

Efra's opening sentence

> In this report, I present the information that you wanted to have before deciding whether to place additional dedicated teaching space for the Outdoor and Experiential Education (OEE) program on next year's list of funding drives.

For a further explanation of this writing strategy, see pages 68–71.

In longer reports, your explanation of the relevance of your report to your readers may take many pages, in which you tell such things as (1) what problem your report will help solve, (2) what activities you performed toward solving that problem, and (3) how your readers can apply your information in their own efforts toward solving the problem.

Besides telling your readers what your communication offers them, your introduction may serve many other functions. The most important of these is to state your main points. In most reports, your main point will be your major conclusions and recommendations. Although you should postpone a full discussion of these topics to the end of your report, your readers will usually appreciate a brief summary of them in your introduction. Efra provided such a summary in the second, third, and fourth sentences of her OEE report:

Summary of conclusions

Summary of recommendations

> Overall, it seems that the space would make a good fundraising project because of the strength and reputation of the OEE program, the state of the current resources and facilities, and the existence of a loyal core of alumni who took OEE courses while undergraduates. The fundraising should focus on the purchase of a portable teaching space (a "yurt"), costing $12,000. An additional $10,000 could be sought for complementary resources, such as a trailer to store and transport the yurt, a small wind generator, and solar panels, but I recommend that these items be saved for a future fundraising drive.

In brief reports (for example, one-page memos), a statement of your main points may even replace the conclusions and recommendations that would otherwise appear at the end.

Four other important functions that an introduction may serve are to explain how the report is organized, outline its scope, encourage openness to your message, and provide background information the readers will need in order to understand the rest of the report. You will find a detailed discussion of these functions in Chapter 3.

METHOD OF OBTAINING FACTS

Your discussion of your method of obtaining the facts in your report can serve a wide variety of purposes. Report readers want to assess the reliability of the facts you present: Your discussion of your method tells them how and where you got your facts. It also suggests where your readers can find additional information. If you obtained your information from printed sources, for example, you can direct your readers to those sources. If you obtained your information from an experiment, survey, or other special technique, your account of your method may help others design similar projects.

Efra explained how she gathered her information this way:

Efra's methods

> I obtained the information given below from Dr. Nick Koutsouvanos, OEE Coordinator. Also, at last month's Alumni Weekend, I spoke with a half-dozen alumni interested in the OEE courses and facilities. Information about costs comes from Pacific Yurts of Cottage Grove, Oregon, the leading North American yurt manufacturer.

FACTS

When deciding how to present your facts, review Chapter 15's guidelines for presenting information in graphics rather than text.

Your facts are the individual pieces of information that you gathered. If your report, like Efra's, is based on interviews, your facts are what people told you. If your report is based on laboratory, field, or library research, your facts are the verifiable pieces of information you gathered: the laboratory data you obtained, the survey responses you recorded, or the knowledge you assembled from printed sources. If your report is based on your own efforts to design a new product, procedure, or system, your facts are the details of what you designed or created. In sum, your facts are the separate pieces of information you present as objectively verifiable.

You may present your facts in a section of their own, or you may combine your presentation of your facts with your discussion of them, as explained next.

DISCUSSION

Your discussion gives meaning to your facts.

Taken alone, facts mean nothing. They are a table of data, a series of isolated observations, or pieces of information without meaning. Therefore, an essential element of every report you prepare will be a discussion in which you interpret your facts in a way that is significant to your readers.

Relationship between facts and discussion.

Sometimes, writers have trouble distinguishing between the presentation and the discussion of their facts. The following example may help to make the distinction clear. Imagine that you observed that when the temperature on the floor of your factory is 17°C, workers produce 3 percent rejected parts; when it is 20°C, they produce 3 percent rejected parts; when it is 22°C, they produce 4.5 percent rejected parts; and when it is 25°C, they produce 7 percent rejected parts. Those would be your facts. If you were to say, "As the temperature rises above 20°C, so does the percentage of rejected parts," you would be interpreting those facts. Of course, in many reports you will be dealing with much larger and more complicated sets of facts that require much more sophisticated and extended interpretation. But the basic point remains the same: When you begin to make general statements based upon your facts, you are interpreting them for your readers. You are discussing them.

In many of the communications you write, you will weave your discussion of the facts together with your presentation of them. In such situations, the interpretations often serve as the topic sentences of paragraphs. Here, for example, is a passage in which Efra mixes facts and discussion:

Interpretation

Facts

> As one of only two OEE programs in Canada, Algoma's program is an international leader in OEE education, research, and development. For example, in 2004, program faculty and students worked with the Department of Education for the Government of Nunavut in the drafting of their Health and Personal Development module for Grades 7–12, using culturally appropriate experiential education tools and strategies.

Whether you integrate your presentation and discussion of the facts or treat them separately, it is important to remember that your readers count on you not only to select facts that are relevant to them, but also to discuss those facts in a way that is meaningful to them.

CONCLUSIONS

Like interpretations, conclusions are general statements based on your facts. However, conclusions focus on answering the readers' question, "How are those facts significant to us?" In her report, for instance, Efra provided many paragraphs of information about the OEE program, the current facilities, and the likely interest among alumni in contributing money for a dedicated teaching space. After reading her presentation and discussion of these facts, Efra's reader, Chander, might ask, "But how, exactly, does all this affect my decision about whether to start a fundraising project for the OEE program?" Efra answers that question in her conclusions:

Efra's conclusions

> In conclusion, my investigation indicates that the university's OEE courses and program could benefit substantially from a fundraising effort. However, the appeal of such an effort will be limited primarily to the very supportive alumni who took OEE courses or the OEE program while students.

Such brief, explicit statements of conclusions are almost always desired and welcomed by report readers.

RECOMMENDATIONS

Just as conclusions grow out of interpretations of the facts, recommendations grow out of conclusions. They answer the readers' question, "If your conclusions are valid, what do you think we should do?" Depending on many factors, including the number and complexity of the things you are recommending, you may state your recommendations in a single sentence or in many pages.

As mentioned previously, you can help your readers immensely by stating your major recommendations at the beginning of your report. In a short report, this may be the only place you need to present them. If your communication is long, or if a full discussion of your recommendations requires a great deal of space, you can summarize your recommendations at the beginning of your report and then treat them more extensively at the end. This is what Efra did when she summarized her recommendations in two sentences in the first paragraph of her introduction and then presented and explained them in three paragraphs at the end of her report. To be sure that her reader could readily find this fuller discussion, she placed it under the heading "Conclusions and Recommendations," and she began the paragraph with the words, "Therefore, I recommend."

Although readers usually want recommendations in reports, you may encounter some situations in which you will not want to include them. That might happen, for instance, in either of these situations:

When you should not include recommendations

- The decision being made is clearly beyond your competence, and you have been asked to provide only a small part of the information your readers need to make the decision.
- The responsibility for making recommendations rests with your boss or other people.

In the usual situation, however, your recommendations will be welcomed. If you are uncertain about whether to provide them, ask your boss or the person who asked you to prepare the report. Don't omit your recommendations out of shyness or because you don't know whether they are wanted.

SAMPLE OUTLINES AND REPORT

The sample outlines show some ways the structure is adapted for specific situations.

A structure is not the same thing as an outline. The six elements of the structure for reports can be incorporated into many organizational patterns. Figure 8.1 (page 186) illustrates one possibility. It shows the outline for a report written by Ryan, who works in the Human Resources Department of Grandmaison Corporation. Much to the credit of Ryan's department, the company's executive management recently decided that developing the full potential of all the company's employees would be a major corporate goal. As a first step, the executive management asked Ryan's department to study employee morale. Ryan has organized his report on the study into six sections that correspond directly to the six elements of the general structure.

Figure 8.2 (page 187) shows the much different outline for Efra's report to Chander on the university fundraising campaign. Like Ryan's report, Efra's has six sections. However, those sections do not bear a one-to-one relationship with the six elements of the general structure—although all six elements are present. In her opening section, Efra explains what the reader will gain from her communication and describes her methods for gathering facts. In each of the next four sections, she mingles her presentation of the facts with her discussion of them; each of these sections is built around one of the major focuses of her investigation. In her last section, Efra combines her conclusions and her recommendations.

As these two outlines suggest, when writing a report you should be certain that your communication contains all six elements of the structure and that you organize them in a way that is suited to your purpose and readers.

Figure 8.3 (page 188) shows Efra's full report, which she wrote in the memo format.

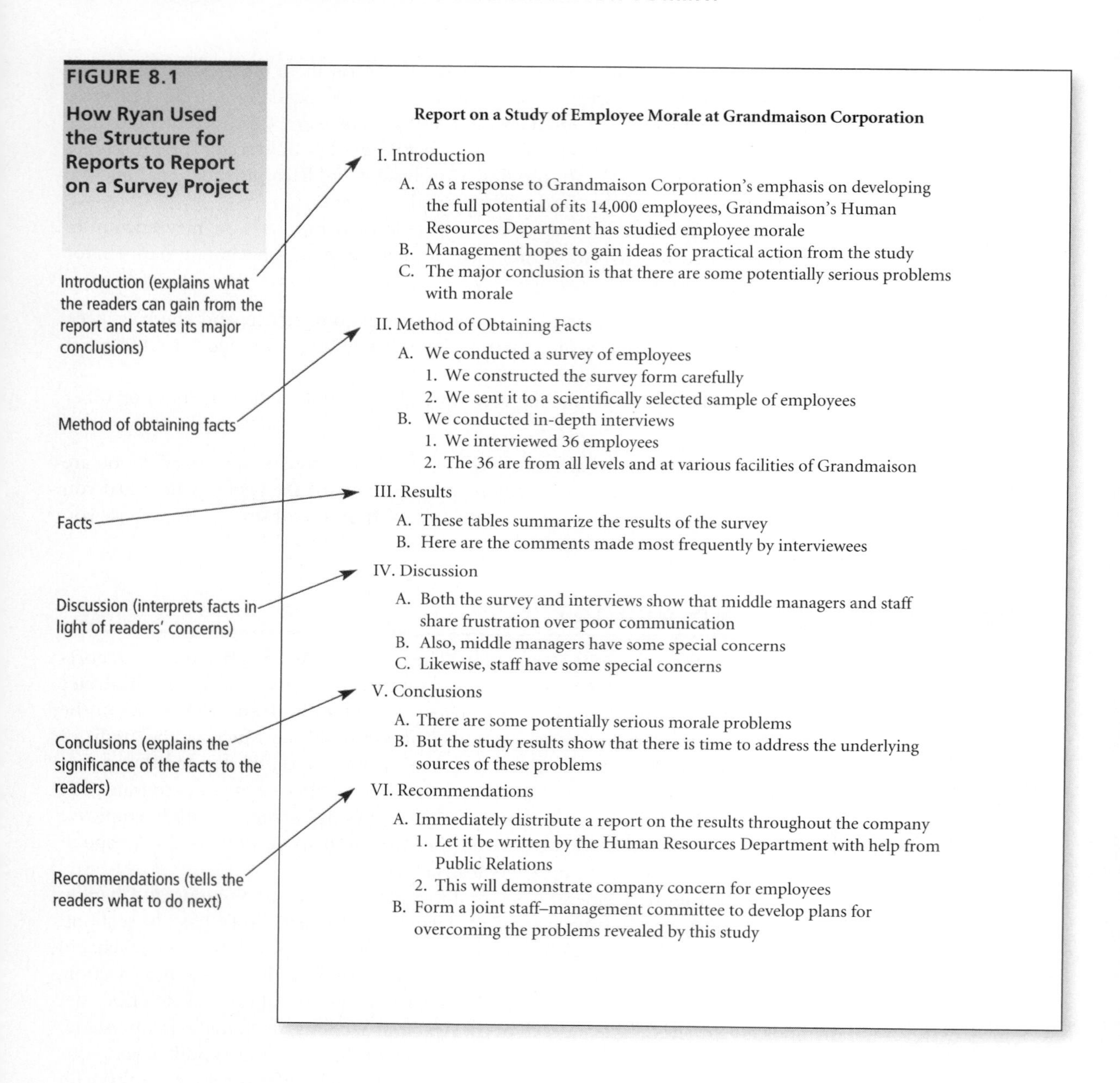

FIGURE 8.1

How Ryan Used the Structure for Reports to Report on a Survey Project

REVISION CHECKLIST, PLANNING GUIDE, AND OTHER RESOURCES

WWW To obtain a revision checklist and planning guide you can download, visit Chapter 8 at www.techcomm.nelson.com.

A "Revision Checklist for Reports" that you can use in your course and on the job is available on the website for this book. The website also provides a "Planning Guide for Reports."

FIGURE 8.2

How Efra Used the Structure for Reports to Report on a Fact-Finding Project

Introduction (tells what the reader will gain from reading the report; states the major conclusion and recommendation)

Method of obtaining facts (included in introduction)

Facts and discussion combined

Conclusions

Recommendations

Investigation of a Funding Drive for Additional Teaching Space

I. Introduction

A. As you requested, I've gathered information that will help you decide whether to launch a fund drive for additional teaching space for the Outdoor and Experiential (OEE) program.

B. I have concluded that a portable teaching space (a "yurt") would make a good fundraising project.

C. I recommend that you seek funds for a yurt, but not for a conventional structure.

D. I obtained my information from several sources.

II. Outdoor and Experiential Education program

A. University offerings in OEE have increased in breadth and depth.

B. The university's program is internationally ranked.

III. Needs

A. There is an increase in demand for the courses and the program track in OEE.

B. The Faculty of Education's teaching spaces do not reflect the OEE objectives and values.

C. The current portable teaching spaces are not suitable for use in winter.

IV. Possible Development Projects

A. A yurt.

B. A cabin at the field camp location.

C. A conventional portable classroom.

V. Prospects for Successful Fundraising

A. The OEE program does not appeal to general alumni.

B. Faculty of Education alumni who took OEE courses while students are very supportive.

VI. Conclusions and Recommendations

A. Conclusions

1. The OEE program would benefit from more dedicated teaching space.
2. The major contributors would be OEE alumni.

B. Recommendations

1. Conduct a fundraising drive for a yurt.
2. Do not seek funds for a conventional teaching space.

FIGURE 8.3

Report That Uses the General Structure

Efra's introduction explains the significance of her report to her reader and summarizes her conclusions and recommendations.

Efra explains her method of obtaining facts.

Efra begins the first of four sections that present and discuss her findings.

MEMORANDUM

Algoma University
Development Office

FROM: Efra Collins
TO: Chandar Modi, Director of Funding Drives
DATE: June 24, 2007

SUBJECT: POSSIBLE FUNDING DRIVE FOR ADDITIONAL TEACHING SPACE

In this report, I present the information that you wanted to have before deciding whether to place additional dedicated teaching space for the Outdoor and Experiential Education (OEE) program on next year's list of funding drives. Overall, it seems that the space would make a good fundraising project because of the strength and reputation of the OEE program, the state of the current resources and facilities, and the existence of a loyal core of alumni who took OEE courses while undergraduates. The fundraising should focus on the purchase of a portable teaching space (a "yurt"), costing $12,000. An additional $10,000 could be sought for complementary resources, such as a trailer to store and transport the yurt, a small wind generator, and solar panels, but I recommend that these items be saved for a future fundraising drive.

I obtained the information given below from Dr. Nick Koutsouvanos, OEE Coordinator. Also, at last month's Alumni Weekend, I spoke with a half-dozen alumni interested in the OEE courses and facilities. Information about costs comes from Pacific Yurts of Cottage Grove, Oregon, the leading North American yurt manufacturer.

OUTDOOR AND EXPERIENTIAL EDUCATION PROGRAM

In 1968, Algoma University's Outdoor and Experiential program started with some open country and experiential education courses on a university-owned property outside of the city. The innovative and challenging activities in the property's barn, such as a low-ropes course, served their purpose throughout the 1980s.

Because of funding and liability issues, the barn could no longer be used by 1992. Moreover, the initial co-op OEE option (an optional addition to a Bachelor

FIGURE 8.3 (continued)

Efra begins the second section that reports and discusses her findings.

June 24, 2007 Page 2

of Education program) was replaced in 1998 by an OEE track program. This program focuses on the experiential and outdoor activities that can be created in any schoolyard—not just at an outdoor centre or site. In addition to this program, there are also four OEE courses available to all B.Ed. students.

As one of only two OEE programs in Canada, Algoma's program is an international leader in OEE education, research, and development. For example, in 2004, program faculty and students worked with the Department of Education for the Government of Nunavut in the drafting of their Health and Personal Development module for Grades 7–12, using culturally appropriate experiential education tools and strategies.

NEEDS

Dr. Koutsouvanos and the alumni identify three challenges facing the OEE program: a significant increase in demand for OEE courses and specialization, a shortage of teaching spaces that reflect the OEE objectives and values, and the lack of a portable teaching space that can be used in the winter.

Increase in Demand

The most pressing challenge facing the OEE program is the number of students enrolling in its courses and specialization. While there were still vacancies in the two open OEE courses in the 2004–05 school year, by the next academic year, there was a waiting list of students for both of these courses. Even more telling, while there were five applicants for every one spot in the dedicated OEE program in the 2002–03 year, there were double that number one year later. Even though an additional section of students had been added, many excellent candidates had to be turned away.

Shortage of Teaching Space

Even when the number of students was lower, there has always been a shortage of teaching space in the Faculty of Education building that reflected the less traditional OEE objectives and values. For example, OEE teacher candidates are encouraged to experience the value of taking students outdoors, and outside of the classroom, to learn. Candidates also explore ways of building and using alternative and portable learning spaces.

Lack of a Winter-Friendly Portable Teaching Space

The OEE program has never owned a portable teaching space that is usable in the winter. While the program uses tents, gazebos, and lean-tos in the more temperate months, these spaces have been too cold to use in temperatures below freezing.

(continues)

Figure 8.3
(continued)

Efra begins her third section on her findings.

June 24, 2007 Page 3

POSSIBLE DEVELOPMENT PROJECTS

Dr. Koutsouvanos has identified three subjects for a possible fund drive: a yurt, a cabin, and a portable classroom.

Yurt

Although generally classified as a tent, the yurt is stronger and weathertight. It is a modern adaptation of an ancient shelter used by Central Asian nomads for centuries. The yurt is a circular structure that consists of a durable fabric cover, a tension band, and a wood frame that includes a lattice wall, radial rafters, central compression ring, and framed door.

According to both Dr. Koutsouvanos and the alumni I spoke with, an adequate yurt would be 10 m in diameter, 79 m^2 in area, and 4 m high at the centre. This structure would cost $10,000. Other options, such as additional windows, insulated window covers, and insulation liners, could be considered. A wooden platform deck, at $2,000, would also be required.

Cabin

The OEE program holds a field camp in early September before the start of the regular education year. This course allows OEE students to create a community in which they will experience the value derived from living with others in an outdoor setting. The camp takes place at the offsite university-owned property 60 km from the Faculty of Education.

A cabin could be constructed on the offsite property. This cabin could be used year-round for students in the OEE courses and program. Assuming that a building permit would be granted and a site inspection would warrant construction, a two-storey 7 m × 7 m cedar cottage resting on a concrete slab could be constructed. This structure would cost $20,000. Other options, such as a porch, insulation, and drywalling, could be considered.

Portable Classroom

The Faculty of Education has room and permission for up to four portable classrooms on its site. A 'tilt-up' construction technique using reinforced concrete panels, poured onsite, could be used to build an expandable portable classroom. This classroom could be relocated and built within one day. This structure would cost $15,000, and includes the platform, windows and doors, and roof.

FIGURE 8.3
(continued)

This is the last section in which Efra presents and discusses her findings.

Efra explicitly states her conclusions.

Efra presents her detailed recommendations.

June 24, 2007 Page 4

PROSPECTS FOR SUCCESSFUL FUNDRAISING

When I spoke with alumni at last month's Alumni Weekend, I found that a fundraising drive for the additional teaching space dedicated to the OEE program would not be widely appealing to Algoma's general alumni. Many alumni perceive that contributions to such a drive would benefit only a relatively small number of students, primarily those in the OEE program.

However, I also found that alumni of the Faculty of Education who had taken OEE courses are very supportive, especially alumni from the OEE program. Dr. Koutsouvanos has a long list of alumni who have expressed support for the program.

CONCLUSIONS AND RECOMMENDATIONS

In conclusion, my investigation indicates that the university's OEE courses and program could benefit substantially from a fundraising effort. However, the appeal of such an effort will be limited primarily to the very supportive alumni who took OEE courses or the OEE program while students.

Therefore, I recommend that we conduct a fundraising drive focused on the alternative and portable yurt option, leaving the trailer, solar panels, and wind generator for a future program. The drive should be announced to all alumni in one of the brief sketches in the brochure we send to all alumni each year to describe our development plans. The description should emphasize the classes taught and the need to provide additional dedicated space for the OEE offerings within the Faculty of Education.

A more extensive description of the yurt should be prepared for Faculty of Education alumni known to be interested in the OEE courses and program. This description can be posted on the OEE website and provided in a separate mailing. In addition to the information provided to all alumni, the description should emphasize the benefits of the yurt, such as

- its utility for classroom lessons on or off campus
- its value in imparting lessons embedded in creating a circular learning space
- its ability to be transferred to other school sites for teacher candidates to share and promote the lessons
- its potential for research and development into learning experiences.

(continues)

Figure 8.3
(continued)

June 24, 2007 Page 5

Although the cabin might provide a more permanent learning space, it will not be portable. The other option of the more traditional portable classroom does not align well with the alternative values and educational philosophy espoused by the OEE courses and program.

I am sure that Dr. Koutsouvanos and the other OEE faculty will work with us enthusiastically and effectively in the fundraising drive that I have outlined.

CONCLUSION

The structure for reports is a versatile and flexible pattern for organizing information. Many specialized variations of this structure have evolved. Reference Guide 3, which follows this chapter, describes three of the most common ones: empirical research reports, feasibility reports, and progress reports.

Reference Guide 3: Three Types of Special Reports

CONTENTS

This Reference Guide describes the structure for three types of specialized reports:

For additional resources, visit Reference Guide 3 at www.techcomm.nelson.com.

Specialized Report Structures

Each is a variation on the general structure for reports discussed in Chapter 8. If you have not already done so, you should read that discussion because it provides information not repeated here that will help you write these three types of reports effectively.

EMPIRICAL RESEARCH REPORTS

In empirical research, the investigators gather information through carefully planned, systematic observations or measurements. When scientists send a satellite to investigate the atmosphere of a distant planet, when engineers test jet engine parts made of various alloys, when pollsters ask mature adults what kinds of outdoor recreation they participate in, and when sociologists observe people's behaviour at baseball games, they all are conducting empirical research. In your career, you will almost certainly perform some type of empirical research—and report on it in writing. This section describes the structure you can use for those reports.

TYPICAL WRITING SITUATIONS

Research to support decision making.

Empirical research has two distinct purposes. Most research aims to help people make practical decisions. For example, the engineers who test jet engine parts are trying to help designers determine which alloy to use in a new engine. Similarly, the researchers who study mature adults' recreational activities are trying to help decision makers in Parks and Recreation departments determine what sorts of services and facilities to provide for those adults.

Research to extend human knowledge.

A smaller portion of empirical research aims not to support practical decisions but rather to extend human knowledge. Here, researchers set out to learn how fish remember, what the molten core of the earth is like, or why people fall in love. Such research is usually reported in scholarly journals such as the *Journal of Chemical Thermodynamics, Journal of Cell Biology,* and *Journal of Social Psychology,* whose readers are concerned not so much with making practical business decisions as with extending the frontiers of human understanding.

These two aims of research sometimes overlap. Some organizations sponsor basic research in the hope that what is learned can later be turned to practical use. Likewise, some practical research produces results that help explain something about the world in general.

THE QUESTIONS READERS ASK MOST OFTEN

Whether it aims to support practical decisions, extend human knowledge, or achieve some combination of these two purposes, almost all empirical research is reported in the same structure. That's largely because the readers of reports on all types of empirical research tend to ask the same seven general questions:

- **Why is your research important to us?** Readers concerned with solving specific practical problems want to know what problems your research will help them address. Readers concerned with extending human knowledge want to know what your research contributes to that pursuit.
- **What were you trying to find out?** A well-designed empirical research project is based on carefully formulated research questions that the project will try to answer. Readers want to know what those questions are so they can determine whether they are significant.
- **Was your research method sound?** Unless your method is appropriate to your research questions and unless it is intellectually sound, your readers will not place any faith in your results or in your conclusions and recommendations.
- **What results did your research produce?** Naturally, your readers will want to learn what results you obtained.
- **How do you interpret those results?** Your readers will want you to interpret your results in ways that are meaningful to them.
- **What is the significance of those results?** What answers do your results imply for your research questions, and how do your results relate to the problems your research was to help solve or to the area of knowledge it was meant to expand?
- **What do you think we should do?** Readers concerned with practical problems want to know what you advise them to do. Readers concerned with extending human knowledge want to know what you think your results imply for future research.

STRUCTURE FOR EMPIRICAL RESEARCH REPORTS

To answer their readers' questions, writers of empirical research reports use a structure that contains seven elements. The following table shows how each of these elements corresponds with one of the readers' seven questions.

Structure for Empirical Research Reports

Report Element	Readers' Question
Introduction	Why is your research important to us?
Objectives of the research	What were you trying to find out?
Method	Was your research method sound?
Results	What results did your research produce?
Discussion	How do you interpret those results?
Conclusions	What is the significance of those results?
Recommendations	What do you think we should do?

Remember that a structure is not an outline; you may combine the elements of a structure in many ways (see page 185).

Much of the following advice about using this structure is illustrated through the use of two sample reports. The aim of the first report is practical. It is written by forestry consultants who evaluate woodland caribou winter ranges in Mount Robson Provincial Park. In their report, the consultants inform decision makers in BC parks about the use of results of an aerial survey and ground checks, which they use to make practical recommendations regarding habitat areas and fire management.

The aim of the second sample report is to extend human knowledge. In it, principal author Anita Kozyrskyi (2005) describes her study of the high-cost users of pharmaceuticals. Selected passages of this report, written for the Manitoba Centre for Health Policy, are quoted throughout this section.

Introduction

For additional advice on writing an introduction, see Chapter 3.

In the introduction to an empirical research report, you should seek to answer the readers' question, "Why is this research important to us?" Typically, writers answer that question in two steps: They announce the topic of their research and then explain the importance of the topic to their readers.

Announcing the Topic You can often announce the topic of your research simply by including it as the key phrase in your opening sentence. For example, here is the first sentence of the report on the woodland caribou:

Topic of report

> The Mount Robson Provincial Park Ecosystem Management Plan (Oikos et al., 1996) identified woodland caribou as a key wildlife management priority.

Here is the first sentence of the report on the high-cost users of pharmaceuticals:

Topic of report

> We have known for a long time that health care resources and costs are concentrated on a relatively small proportion of the population.

Explaining the Importance of the Research To explain the importance of your research to your readers, you can use either or both of the following methods: tell how your research is relevant to your organization's goals or tell how it expands existing knowledge on the subject.

- **Relevance to organizational goals.** In reports written to readers in organizations (whether your own or a client's), you can explain the relevance of your research by relating it to some organizational goal or problem. Sometimes, in fact, the importance of your research to the organization's needs will be so obvious to your readers that merely naming your topic will be sufficient. At other times, you will need to discuss at length the relevance of your research to the organization. In the second paragraph of the caribou report, for instance, the writers mention the potential refinement of the caribou winter range. That is, they explain the importance of their research by saying that it can lead to a conservation of biodiversity. For detailed advice about how to explain the importance of your research to readers in organizations, see Guideline 1 of Chapter 3.

- **Expansion of existing knowledge.** A second way to establish the importance of your research is to show the gap in current knowledge that it will fill. The following passage from the opening of the report on the pharmaceutical study illustrates this strategy.

The writer tells what is known about her topic

The writer identifies the gaps in knowledge that her research will fill

> Much of the available literature on heavy users of prescription medications originates from studies of elderly Americans with prescription insurance (Wrobel et al., 2003; Thomas et al., 2001; Mueller et al., 1997; Coulson and Stuart, 1992). A few studies on whole populations come to us from Denmark and Sweden (Halls and Nissen, 1994; Isacson and Haglund, 1989). Very little is known about high-cost users of pharmaceuticals among a general Canadian population in the context of public prescription insurance. This study provides a description, within the Province of Manitoba, of high-costs users of prescription medications compared with the rest of the population.

The writer provides this discussion of published research for three paragraphs. Each follows the same pattern: it identifies an area of research, tells what is known about that area, and further narrows the topic while identifying gaps in knowledge—gaps that will be filled by the research the writer has conducted. These paragraphs serve an important additional function performed by many literature reviews: they introduce the established facts and theories that are relevant to the writer's work and necessary to an understanding of the report.

Objectives of the Research

Every empirical research project has carefully constructed objectives. These objectives define the focus of your project, influence the choice of research method, and shape the way you interpret your results. Readers of empirical research reports want and need to know what the objectives are.

The following example from the caribou report shows one way you can inform your readers about your objectives:

Statement of research objectives

> Consistent with the Management Objectives and Actions outlined in the Ecosystem Management Plan to maintain woodland caribou habitat, the purpose of this evaluation was to:
> - Further refine woodland caribou ranges within the Prescription Ecosystem Management Zone (EMZ2) identified in the Mount Robson Provincial Park.
> - Determine the extent of winter use by caribou, particularly in subzones 5 (known caribou winter range) and 2 (suspected winter range).
> - Attempt to identify whether the winter range is being used by terrestrial lichen feeding caribou originating from Jasper National Park, arboreal lichen feeding caribou (i.e., mountain caribou) considered resident in B.C., or both.
> - Clarify provincial fire and forest health management approaches in Alberta and British Columbia to ensure consistent management direction with respect to woodland caribou habitat.

When reporting on research that involves the use of statistics, you can usually state your objectives by stating the hypotheses you were testing. Where appropriate, you can explain these hypotheses in terms of existing theory, again citing previous publications on the subject. The following passage shows how the writer who studied pharmaceutical use explains some of her hypotheses. Notice how the author begins with a statement of the overall goal of the research:

Overall goal

> This study provides a description, within the Province of Manitoba, of high-cost users of prescription medications compared with the rest of the population. The intent of this study is to provide a detailed characterization of this population so as to clarify whether its costs can be reduced or whether other interventions are needed. Answers are sought to the following questions:
> - What drug categories account for the high prescription costs?
> - Do differences in disease prevention explain the high prescription costs?
> - Are there other explanations for high-cost users? Do they use more expensive drugs? Are they taking too many drugs?
> - Is it possible to predict transitioning to high prescription drug use?

Method

Readers of your empirical research reports will look for precise details concerning your method. Those details serve three purposes. First, they let your readers assess the soundness of your research design and its appropriateness for the problems you are investigating. Second, the details enable your readers to determine the limitations that your method might place on the conclusions you draw. Third, they provide information that will help your readers repeat your experiment if they wish to verify your results or conduct similar research projects of their own.

The nature of the information you should provide about your method depends on the nature of your research. The writers of the caribou report provided two paragraphs reviewing existing information and explaining their equipment (truck radios and satellite), six paragraphs describing aerial border surveillance and ground search, and three paragraphs describing their data analysis.

The writer of the pharmaceutical study used much simpler methods and therefore could describe them quite briefly.

How can you decide which details to include? The most obvious way is to follow the general reporting practices in your field. Find some research reports that use a method similar to yours and see what they report. Depending on the needs of your readers, you may need to include some or all of the following:

Elements of your method to describe

- Every aspect of your procedure that you made a decision about when planning your research.
- Every aspect of your method that your readers might ask about.
- Any aspect of your method that might limit the conclusions you can draw from your results.
- Every procedure that other researchers would need to understand in order to design a similar study.

Results

The results of empirical research are the data you obtain. Although the results are the heart of any empirical research report, they may take up a very small portion of it. Generally, results are presented in one of two ways:

For information on using tables and other graphics for reporting numerical data, see pages 432–43.

- **Tables and graphs.** The report on pharmaceutical use contains 25 tables and 10 graphs.
- **Sentences.** When placed in sentences, results are often woven into a discussion that combines data and interpretation, as the next paragraphs explain.

Discussion

Sometimes writers briefly present all their results in one section and then discuss them in a separate section. Sometimes they combine the two in a single, integrated section. Whichever method you use, your discussion must link your interpretative comments with the specific results you are interpreting.

One way of making that link is to refer to the key results shown in a table or other graphic and then comment on them as appropriate. The following passage shows how the writer of the pharmaceutical use report did that with some of the results she presented in one of her tables:

> However, high-cost users have been increasingly taking more prescription medications. From 1997/98 to 2000/01, the average number of major conditions in high-cost users increased by 13%, while the average number of different medications increased by 14% (Figure 9).

When you present your results in text only (rather than in tables and graphs), you can weave them into your discussion by beginning your paragraphs with general statements that serve as interpretations of your data. Then you can cite the relevant results as evidence in support of the interpretations. Here is an example from the caribou habitat report:

> No evidence of caribou use (sign or sightings) was observed in either subzone 5 (caribou winter range) or subzone 2 (suspected winter range).

In a single report, you may use both of these methods of combining the presentation and discussion of your results.

Conclusions

Besides interpreting the results of your research, you need to explain what those results mean in terms of the original research questions and the general problem you set out to investigate. Your explanations of these matters are your conclusions.

For example, if your research project is sharply focused on a single hypothesis, your conclusion can be very brief, perhaps only a restatement of your chief results. However, if your research is more complex, your conclusion should draw all the strands together.

Link your conclusions to your objectives.

In either case, the presentation of your conclusions should correspond very closely to the objectives you identified toward the beginning of your report. Consider, for instance, the correspondence between objectives and conclusions in the caribou study. The first objective was to refine woodland caribou ranges within the Prescription Ecosystem Management Zone (EMZ2) identified in the Mount Robson Provincial Park. The first of the conclusions addresses that objective:

> Based upon the aerial survey and ground checks of the ecosystem units we suggest the area identified as caribou winter ranges within EMZ2 could be redefined to encompass a smaller area and more accurately reflect caribou winter habitat.

The other three objectives outlined in this report are addressed by conclusions on winter usage, point of caribou origin, and fire and forest health management approaches.

The caribou research concerns a practical question. Hence, its objectives and conclusions address practical concerns of particular individuals—in this case, the decision makers in BC Parks. In contrast, research that aims primarily to extend human knowledge often has objectives and conclusions that focus on theoretical issues.

For example, at the beginning of the pharmaceutical report, the researcher identifies several questions that her research investigated, and she states what answers she predicted her research would produce. Then, in her conclusion, she systematically addresses those same questions in terms of the results her research produced. Here is a summary of some of her objectives and conclusions.

Objective	Conclusion
What drug categories account for the higher prescription costs?	Nervous system medications and medications to treat gastrointestinal disease and diabetes accounted for a greater share of expenditures in high-cost than non-high-cost users.
Do differences in disease prevalence explain the higher prescription costs?	The prescription costs for some conditions such as cystic fibrosis, multiple sclerosis, HIV/AIDS and cancer were disproportionately higher than the prevalence of these conditions.
Is it possible to predict transitioning to high prescription cost use?	The study located transition points in the receipt of health care that increased the likelihood of a person becoming a high-cost user.

Typically, in presenting the conclusions of an empirical research project directed at extending human knowledge, writers discuss the relationship of their findings to the findings of other researchers and to various theories that have been advanced concerning their subject. The writer of the article on high-cost pharmaceutical users did that. The table you have just read presents only a portion of her overall discussion.

Recommendations

The readers of some empirical research reports want to know what, based on the research, the writer thinks should be done. This is especially true when the research is directed at solving a practical problem. Consequently, such reports usually include recommendations.

For example, the caribou report contains four recommendations. The first is the general recommendation that changes be made to the extent of caribou winter range. The other three involve specific recommendations regarding fire management strategies, zone management, and boundary redefinitions. As is common in research addressed to readers in organizations, these recommendations concern practical decisions. These decisions are often of a business, engineering, or scientific nature.

Even in reports designed to extend human knowledge, writers often include recommendations. These may convey their ideas about policy that should be instituted, future studies that should be made, adjustments in methodology that seem called for, and the like. In the final section of the pharmaceuticals report, for instance, the writer discusses the policy implications of her findings and how her findings can be acted on by provincial health care decision makers, analysts, and providers.

SAMPLE EMPIRICAL RESEARCH REPORTS

Figure SR.1 shows the contents list or the pharmaceutical report. For additional examples, see this book's website. To see examples of empirical research reports presented as journal articles, consult journals in your field.

FIGURE SR.1

Empirical Research Report Contents List

CONTENTS

REVISION CHECKLIST, PLANNING GUIDE, AND OTHER RESOURCES

WWW To obtain a revision checklist and planning guide you can download, visit Reference Guide 3 at www.techcomm.nelson.com.

A "Revision Checklist for Empirical Research Reports" that you can use in your course and on the job is available on the website for this book. The website also provides a "Planning Guide for Empirical Research Reports."

FEASIBILITY REPORTS

A feasibility report is an evaluation of the practicality and desirability of pursuing some course of action. Imagine, for instance, that you work for a company that designs and builds sailboats. The company thinks it might reduce manufacturing costs without hurting sales if it uses high-strength plastics to manufacture some parts traditionally made from metal. Before making such a change, however, the company wants you to answer several questions: Would plastic parts be as strong, durable, and attractive as metal ones? Is there a supplier who would make the plastic parts? Would the plastic parts really be less expensive than the metal ones? Would boat buyers accept the change? The company will use the information and analyses you provide in your report as the primary basis for deciding whether to pursue this course of action.

In this section, you will learn how to prepare effective feasibility reports.

TYPICAL WRITING SITUATION

All feasibility reports are written to help decision makers choose between two or more courses of action. Even when a feasibility report focuses on a single course of action, the readers are always considering a second course—namely, to leave things the way they are. Often, however, your readers will already have decided that some change is necessary and will be choosing between two or more alternatives to the status quo.

THE QUESTIONS READERS ASK MOST OFTEN

As decision makers think about the choices they must make, they ask many questions. From situation to situation, their basic questions remain the same, and the structure for feasibility reports is a widely used way of providing the answers. Here are the readers' typical questions:

- **Why is it important for us to consider these alternatives?** Decision makers ask this question because they want to know why they have to make any choice in the first place. They may need a detailed explanation of the problem before they appreciate the urgency of considering alternative courses of action. On the other hand, if they are already familiar with the problem, they may need only to be reminded of what it is.
- **Are your criteria reasonable and appropriate?** To help your readers choose between alternative courses of action, you must evaluate the alternatives in terms of specific criteria. At work, people want these criteria to reflect the needs and aims of their organization. And they want you to tell them explicitly what the criteria are so they can judge them.

- **Are your facts reliable?** Decision makers want to be sure that your facts are reliable before they take any action based on those facts.
- **What are the important features of the alternatives?** So that they can understand your detailed discussion of the alternatives, readers want you to highlight the key features of each alternative.
- **How do the alternatives stack up against your criteria?** The heart of a feasibility study is your evaluation of the alternatives in terms of your criteria. Your readers want to know the results.
- **What overall conclusions do you draw about the alternatives?** Based on your detailed evaluation of the alternatives, you will reach some general conclusions about the merits of each. Decision makers need to know your conclusions because these overall judgments form the basis for decision making.
- **What do you think we should do?** In the end, your readers must choose one of the alternative courses of action. Because of your expertise on the subject, they want you to help them by telling what you recommend.

STRUCTURE FOR FEASIBILITY REPORTS

The structure for feasibility reports provides a framework that writers and readers both have found effective for answering these questions. Each element corresponds with one of the questions decision makers ask when trying to decide about future action:

Structure for Feasibility Reports

Report Element	Readers' Question
Introduction	Why is it important for us to consider these alternatives?
Criteria	Are your criteria reasonable and appropriate?
Method	Are your facts reliable?
Overview of alternatives	What are the important features of the alternatives?
Evaluation	How do the alternatives stack up against your criteria?
Conclusions	What overall conclusions do you draw about the alternatives?
Recommendations	What do you think we should do?

Remember that a structure is not an outline; you may combine the elements of a structure in many ways (see page 185).

Of course, you may combine the elements in different ways, depending on the situation. For instance, you may integrate your conclusions into your evaluation, or you may omit a separate discussion of your criteria if they need no special explanation. But when preparing any feasibility report, you should consciously determine whether to include each of the seven elements, based on your understanding of your purpose, audience, and situation.

Introduction

For additional advice on writing an introduction, see Chapter 3.

In the introduction to a feasibility report, you should answer your readers' question, "Why is it important for us to consider these alternatives?" The most persuasive way to answer this question is to identify the problem your feasibility report will help your readers solve or the goal it will help them achieve: to reduce the number of rejected parts, to increase productivity, and so on. Beyond that, your introduction should announce the alternative courses of action you studied and tell generally how you investigated them.

Consider, for example, the way Phil, a process engineer in a paper mill, wrote the introduction of a feasibility report he prepared. (Phil's entire report appears in Figure SR.2 on pages 209–14.) Phil was asked to study the feasibility of substituting one ingredient for another in the furnish for one of the papers the mill produces (*furnish* is the combination of ingredients used to make the pulp for paper):

Problem

Possible solution

What Phil did to investigate the possible solution

> At present we rely on the titanium dioxide (TiO_2) in our furnish to provide the high brightness and opacity we desire in our paper. However, the price of TiO_2 has been rising steadily and rapidly for several years. We now pay roughly $1400 per ton for TiO_2, or about 70¢ per pound.
>
> Some mills are now replacing some of the TiO_2 in their furnish with silicate extenders. Because the average price for silicate extenders is only $500 per ton, well under half the cost of TiO_2, the savings are very great.
>
> To determine whether we could enjoy a similar savings for our 30-pound book paper, I have studied the physical properties, material handling requirements, and cost of two silicate extenders, Tri-Sil 606 and Zenolux 26 T.

Generally, the introduction to a long feasibility report (and most short ones) should also include a preview of the main conclusions and, perhaps, the major recommendations. Phil included his major conclusion:

Phil's main point

> I conclude that one of the silicate extenders, Zenolux 26 T, looks promising enough to be tested in a mill run.

The introduction of a feasibility report is often combined with one or more of the other six elements, such as a description of the criteria, a discussion of the method of obtaining facts, or an overview of the alternatives. It may also include the various kinds of background, explanatory, and forecasting information that may be found in the beginning of any technical communication (see Chapter 3).

Criteria

Criteria are the standards of evaluation.

Criteria are the standards that you apply in a feasibility study to evaluate the alternative courses of action you are considering. Phil evaluated the two silicate extenders by applying several specific criteria.

Two Ways of Presenting Criteria There are two common ways of telling your readers what your criteria are:

- **Devote a separate section to identifying and explaining them.** Writers often do this in long reports or in reports in which the criteria themselves require extended explanation.

- **Integrate your presentation of them into other elements of the report.** Phil did this in the following sentence from the third paragraph of his introduction:

Phil names his three criteria

> To determine whether we could enjoy a similar savings for our 30-pound book paper, I have studied the physical properties, material handling requirements, and cost of two silicate extenders, Tri-Sil 606 and Zenolux 26 T.

For each of the general criteria named in this sentence, Phil had some more specific criteria, which he described when he discussed his methods and results. For instance, at the beginning of his discussion of the physical properties of the two extenders, he named the three properties he evaluated.

Importance of Presenting Criteria Early Whether you present your criteria in a separate section or integrate them into other sections, you should introduce them early in your report. There are three good reasons for doing this. First, because your readers know that the validity of your conclusion depends on the criteria you use to evaluate the alternatives, they will want to evaluate the criteria themselves. They will ask, "Did you take into account all the considerations relevant to this decision?" and "Are the standards you are applying reasonable in these circumstances?"

Second, your discussion of the criteria tells readers a great deal about the scope of your report. Did you restrict yourself to technical questions, for instance, or did you also consider relevant organizational issues such as profitability and management strategies?

Third, presenting your criteria early means that your discussion of the alternative courses of action will make much more sense to your readers since they know in advance the criteria by which you evaluated the alternatives.

Sources of Your Criteria You may wonder how to come up with the criteria you will use in your study and report. Often, the person who asks you to undertake a study will tell you what criteria to apply. In other situations, particularly when you are conducting a feasibility study that requires technical knowledge that you have but your readers don't, your readers may expect you to identify the relevant criteria for them.

You may refine your criteria while you are writing your report.

In either case, you are likely to refine your criteria as you conduct your study. The writing process itself can help you refine your criteria because as you compose you must think in detail about the information you have obtained and decide how best to evaluate it.

Four Common Types of Criteria As you develop your criteria, you may find it helpful to know that, at work, criteria often address one or more of the following questions:

- **Will this course of action really do what's wanted?** This question is especially common when the problem is a technical one: Will this reorganization of the department really improve the speed with which we can process loan applications? Will the new type of programming really reduce computer time?
- **Can we implement this course of action?** Even though a particular course of action may work technically, it may not be practical. For example, it may require overly extensive changes in operations, equipment, or materials that are not readily available, or special skills that employees do not possess.

- **Can we afford it?** Cost can be treated in several ways. You may seek an alternative that costs less than some fixed amount or one that will save enough to pay for itself in a fixed period (for example, two years). Or you may simply be asked to determine whether the costs are "reasonable."
- **Is it desirable?** Sometimes a solution must be more than effective, implementable, and affordable. Many otherwise feasible courses of action are rejected because they create undesirable side effects. For example, a company might reject a plan for increasing productivity because it would impair employee morale.

Ultimately, your selection of criteria for a particular feasibility study will depend on the problem at hand and on the professional responsibilities, goals, and values of the people who will use your report. In some instances, you will need to deal only with criteria related to the question, "Does it work?" At other times, you might need to deal with all the criteria mentioned above, plus others. No matter what your criteria, however, announce them to your readers before you discuss your evaluation.

Method

By explaining how you obtained your facts, you answer your readers' question, "Are your facts reliable?" That is, by showing that you used reliable methods, you assure your readers that your facts form a sound basis for decision making.

The source of your facts will depend on the nature of your study—research, calls to manufacturers, interviews, meetings with other experts in your organization, surveys, laboratory research, and the like.

The amount of detail to provide depends on the situation.

How much detail should you provide about your methods? That depends on your readers and the situation, but in every case your goal is to say enough to satisfy your readers that your information is trustworthy. If the procedures are standard in your industry and well known to your readers, you do not need to explain them in detail.

In contrast, Clark needed to provide a great deal of information about his methods in a feasibility report that he prepared for his employer, a manufacturer of packaged mixes for cakes, breads, and cookies. His report is outlined in Figure SR.3 (page 215). Clark's assignment was to determine the feasibility of making a new, low-gluten muffin mix for a new line of products for people with special health conditions. To evaluate the alternative muffin mixes, Clark used several test procedures common in the food industry. However, to persuade his very demanding readers that his results were valid, he had to explain in detail how he obtained them. Readers in another organization might have required less detail to accept the validity of his findings.

Descriptions of methods are often combined with other sections.

Where is the best place to describe your methods? The best place for describing your methods depends partly on how many techniques you used. If you used only one or two techniques—say, research and interviews—you might describe each in a separate paragraph or section near the beginning of your report, perhaps in the introduction. On the other hand, if you used several different techniques, each pertaining to a different part of your analysis, you might describe each of them at the point where you present and discuss the results you obtained.

Of course, if your methods are obvious, you may not need to describe them at all. You must always be sure, however, that your readers know enough about your methods to accept your facts as reliable.

Overview of Alternatives

Ensure that your readers understand the alternatives before you begin evaluating them.

Before you begin your detailed evaluation of the alternatives, you must be sure that your readers understand what the alternatives are. Sometimes you need to devote only a few words to that task. Imagine, for instance, that you work for a chain of convenience stores that has asked you to investigate the feasibility of increasing starting salaries for store managers as a way of attracting stronger applicants for job openings. Surely your readers will not require any special explanation to understand the course of action you are assessing.

However, you may sometimes need to provide extensive background information or otherwise explain the alternatives to your readers. Your job is to ensure that your readers understand the alternatives before you begin evaluating them, whether it takes just a few words or many pages of explanation.

Evaluation

The heart of a feasibility report is the detailed evaluation of the course or courses of action you studied. Most writers organize their evaluation sections around their criteria.

For a full discussion of the divided and alternating patterns of organization, see pages 50–54.

Clark used the divided pattern in his evaluation of the alternative recipes for low-gluten muffins. Each individual test on each muffin mix produced a single, specific result that could be communicated in only a few words. He judged that his readers would have difficulty building a coherent understanding of each muffin's strengths and weaknesses if he used an alternating pattern based on criteria. By using the divided pattern, he presented all the results for a particular mix in one brief passage. Then, in his conclusions, he was able to compare the mixes quickly by citing the most significant differences among them.

Dismiss Obviously Unsuitable Alternatives Sometimes you will want to mention several alternatives but treat only one or a few thoroughly. Perhaps your investigation showed that the other alternatives failed to meet one or more of the critical criteria so that they should not be considered seriously.

Usually, it makes no sense to discuss obviously unsuitable alternatives at length. Instead, you might explain briefly the alternatives that you have dismissed and tell why you dismissed them. This entire discussion might take only a sentence or a paragraph. You might include it in the introduction (when you are talking about the scope of your report) or in your overview of the alternatives.

It is usually not wise to postpone your mention of unsuitable alternatives until the end of your report. If you do, some of your readers may keep asking, "But why didn't you consider so and so?" That question may distract them as they read the important sections of your report.

Put Your Most Important Points First The final piece of advice for presenting your evaluation of alternatives is to examine Guideline 3 in Chapter 14, which urges you to begin each segment of your communications with the most important information. This advice applies to the entire segment in which you evaluate the alternatives, as well as to its parts. By presenting the most important information first, you save your readers the trouble of trying to figure out what generalizations they should draw from the details you are presenting.

Conclusions

Your conclusions constitute your overall assessment of the feasibility of the alternative courses of action you studied. You might present your conclusions in two or three places in your report. You should certainly mention them in summary form near the beginning. If your report is long (say, several pages), you might also remind your readers of your conclusions at the beginning of the evaluation segment. Finally, you should provide detailed discussion of your conclusions in a separate section following your evaluation of the alternatives.

Recommendations

Tell what you think your readers should do, based on your study.

It is customary to end a feasibility report by answering the decision makers' question, "What do you think we should do?" Because you have investigated and thought about the alternatives so thoroughly, your readers will place special value on your recommendations. Depending on the situation, you might need to take only a single sentence or many pages to present your recommendations.

Sometimes your recommendations will pertain directly to the course of action you studied: "Do this," or "Don't do this." At other times you may perform a preliminary feasibility study to determine whether a certain course of action is promising enough to warrant a more thorough investigation. In that case, your recommendation would focus not on whether to pursue a certain course of action but on whether to continue studying it.

Sometimes you may discover that you were unable to gather all the information you needed to make a firm recommendation. Perhaps your deadline was too short or your funds too limited. Perhaps you uncovered an unexpected question that needs further investigation. In such situations, you should point out the limitations of your report and let your readers know what else they should find out so they can make a well-informed decision.

SAMPLE OUTLINE AND FEASIBILITY REPORT

The seven elements of the structure for feasibility reports can be incorporated into many different organizational patterns. Figures SR.3 shows you the outline for Clark's report on the muffin mixes (page 215).

Clark uses only four sections in his report. He does this by combining some of the structure's elements: In the opening section, he presents both the introduction and the overview; in the next section, he describes both his method of obtaining facts and his criteria; in the third section, he includes his results and analysis of his country; and in the final section, he intertwines his conclusions and recommendations.

You will find other differences between the outlines of Phil's and Clark's reports. The most important points to note are that (1) both reports contain all seven elements of a feasibility report and (2) the structure can be applied in many different ways to suit the purpose and readers of a particular report.

FIGURE SR.2
Feasibility Report

REGENCY INTERNATIONAL PAPER COMPANY

MEMORANDUM

FROM: Phil Hines, Process Engineer
TO: Jim Shulmann, Senior Engineer
DATE: December 13, 2007

SUBJECT: FEASIBILITY OF USING SILICATE EXTENDERS FOR 30-POUND BOOK PAPER

Summary

Entire report is summarized in 125 words.

I have investigated the feasibility of using a silicate extender to replace some of the TiO_2 in the furnish for our 30-pound book paper. Because the cost of the extenders is less than half the cost of TiO_2, we could enjoy a considerable savings through such a substitution.

The tests show that either one of the two extenders tested can save us money. In terms of retention, opacity, and brightness, Zenolux is more effective than Tri-Sil. Consequently, it can be used in smaller amounts to achieve a given opacity or brightness. Furthermore, because of its better retention, it will place less of a burden on our water system. With respect to handling and cost, the two are roughly the same.

I recommend a trial run with Zenolux.

Summary emphasizes conclusions and recommendations.

Introduction

At present we rely on the titanium dioxide (TiO_2) in our furnish to provide the high brightness and opacity we desire in our paper. However, the price of TiO_2 has been rising steadily and rapidly for several years. We now pay roughly $1400 per ton for TiO_2, or about 70¢ per pound.

Problem

(continues)

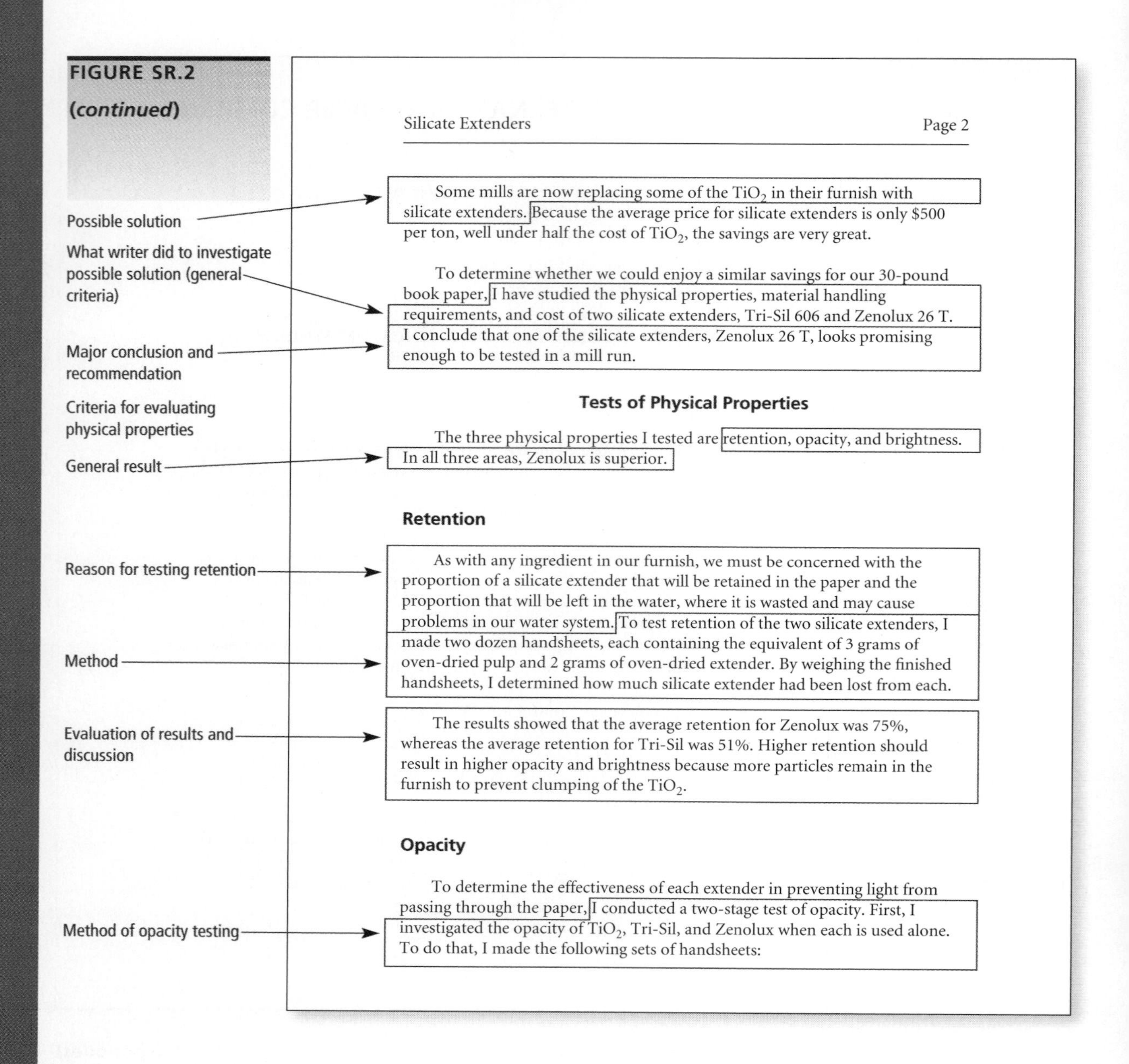

FIGURE SR.2
(continued)

Silicate Extenders Page 2

Some mills are now replacing some of the TiO_2 in their furnish with silicate extenders. Because the average price for silicate extenders is only $500 per ton, well under half the cost of TiO_2, the savings are very great.

To determine whether we could enjoy a similar savings for our 30-pound book paper, I have studied the physical properties, material handling requirements, and cost of two silicate extenders, Tri-Sil 606 and Zenolux 26 T. I conclude that one of the silicate extenders, Zenolux 26 T, looks promising enough to be tested in a mill run.

Tests of Physical Properties

The three physical properties I tested are retention, opacity, and brightness. In all three areas, Zenolux is superior.

Retention

As with any ingredient in our furnish, we must be concerned with the proportion of a silicate extender that will be retained in the paper and the proportion that will be left in the water, where it is wasted and may cause problems in our water system. To test retention of the two silicate extenders, I made two dozen handsheets, each containing the equivalent of 3 grams of oven-dried pulp and 2 grams of oven-dried extender. By weighing the finished handsheets, I determined how much silicate extender had been lost from each.

The results showed that the average retention for Zenolux was 75%, whereas the average retention for Tri-Sil was 51%. Higher retention should result in higher opacity and brightness because more particles remain in the furnish to prevent clumping of the TiO_2.

Opacity

To determine the effectiveness of each extender in preventing light from passing through the paper, I conducted a two-stage test of opacity. First, I investigated the opacity of TiO_2, Tri-Sil, and Zenolux when each is used alone. To do that, I made the following sets of handsheets:

FIGURE SR.2

(continued)

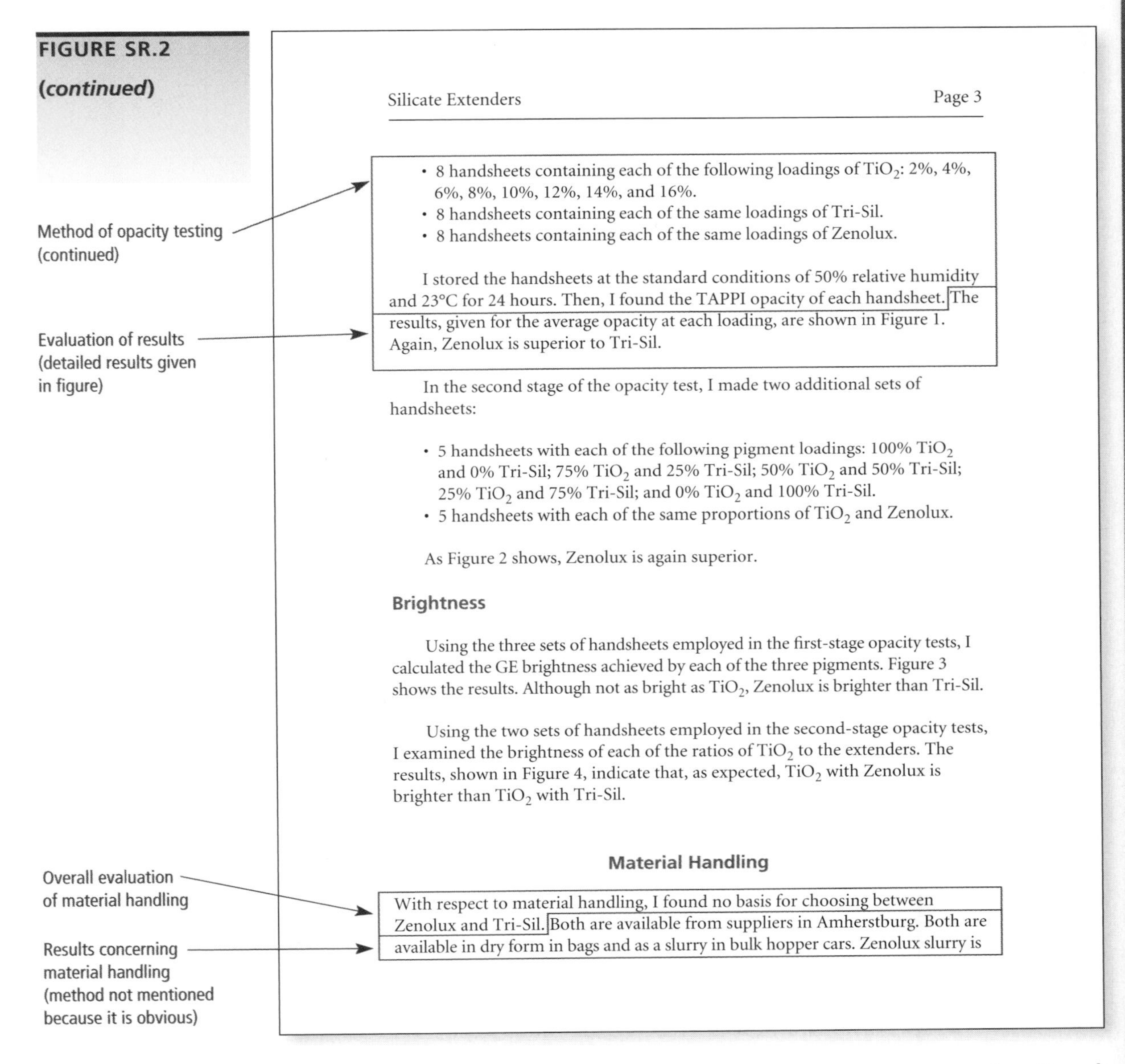

Silicate Extenders Page 3

- 8 handsheets containing each of the following loadings of TiO_2: 2%, 4%, 6%, 8%, 10%, 12%, 14%, and 16%.
- 8 handsheets containing each of the same loadings of Tri-Sil.
- 8 handsheets containing each of the same loadings of Zenolux.

I stored the handsheets at the standard conditions of 50% relative humidity and 23°C for 24 hours. Then, I found the TAPPI opacity of each handsheet. The results, given for the average opacity at each loading, are shown in Figure 1. Again, Zenolux is superior to Tri-Sil.

In the second stage of the opacity test, I made two additional sets of handsheets:

- 5 handsheets with each of the following pigment loadings: 100% TiO_2 and 0% Tri-Sil; 75% TiO_2 and 25% Tri-Sil; 50% TiO_2 and 50% Tri-Sil; 25% TiO_2 and 75% Tri-Sil; and 0% TiO_2 and 100% Tri-Sil.
- 5 handsheets with each of the same proportions of TiO_2 and Zenolux.

As Figure 2 shows, Zenolux is again superior.

Brightness

Using the three sets of handsheets employed in the first-stage opacity tests, I calculated the GE brightness achieved by each of the three pigments. Figure 3 shows the results. Although not as bright as TiO_2, Zenolux is brighter than Tri-Sil.

Using the two sets of handsheets employed in the second-stage opacity tests, I examined the brightness of each of the ratios of TiO_2 to the extenders. The results, shown in Figure 4, indicate that, as expected, TiO_2 with Zenolux is brighter than TiO_2 with Tri-Sil.

Material Handling

With respect to material handling, I found no basis for choosing between Zenolux and Tri-Sil. Both are available from suppliers in Amherstburg. Both are available in dry form in bags and as a slurry in bulk hopper cars. Zenolux slurry is

(continues)

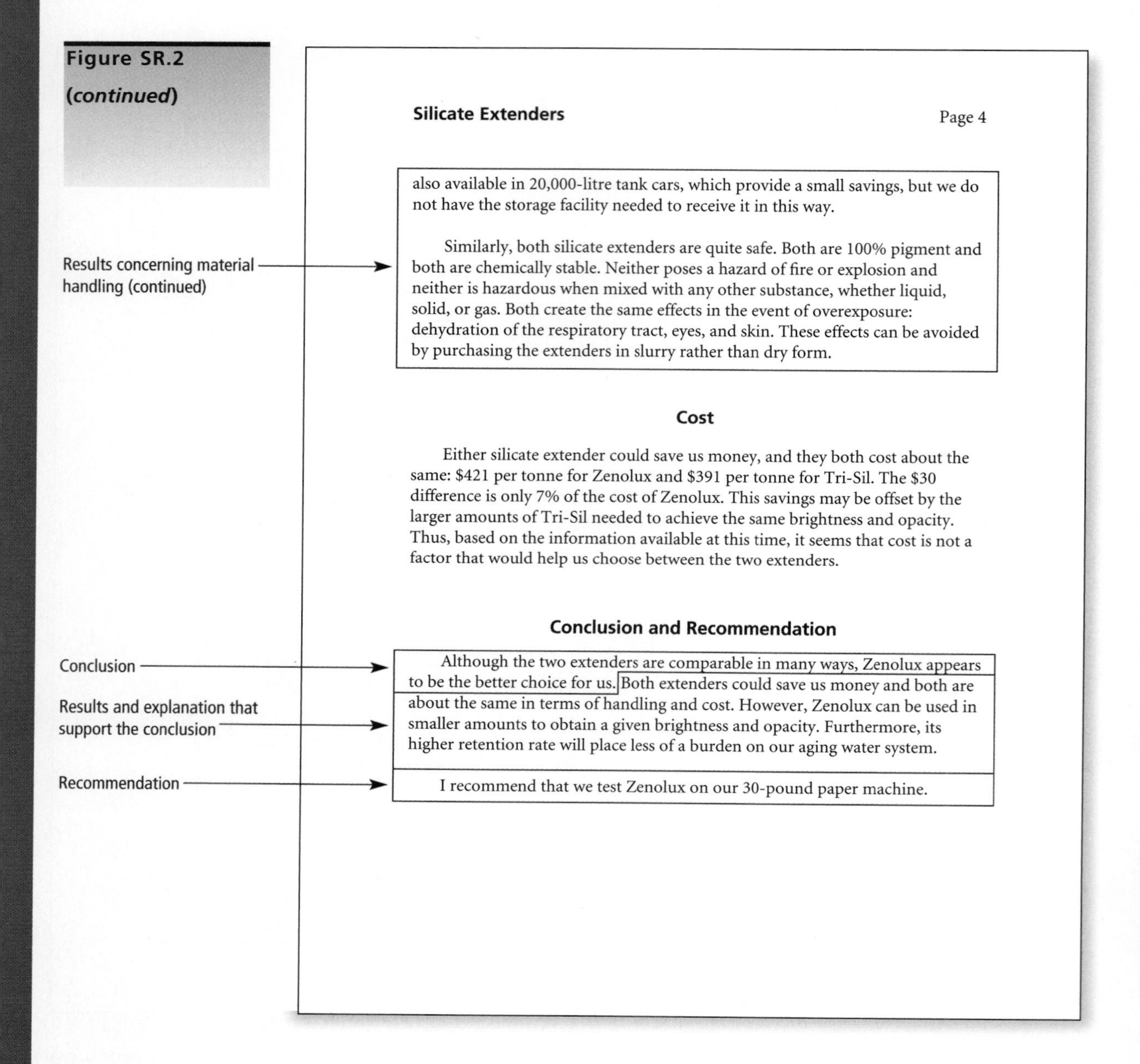

Figure SR.2 (*continued*)

FIGURE SR.2

(continued)

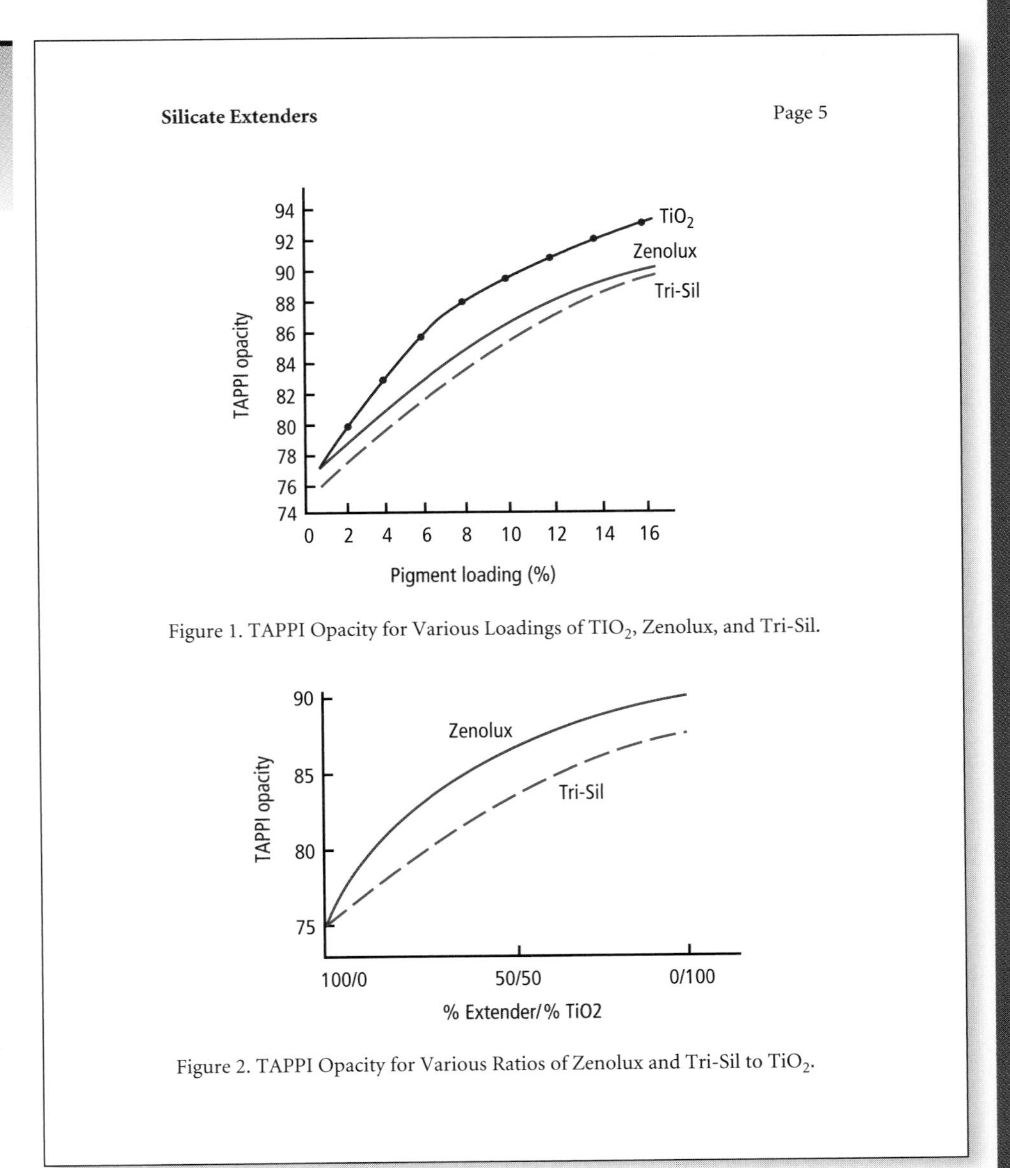

Figure 1. TAPPI Opacity for Various Loadings of TIO_2, Zenolux, and Tri-Sil.

Figure 2. TAPPI Opacity for Various Ratios of Zenolux and Tri-Sil to TiO_2.

(continues)

FIGURE SR.2

(continued)

Silicate Extenders

Page 6

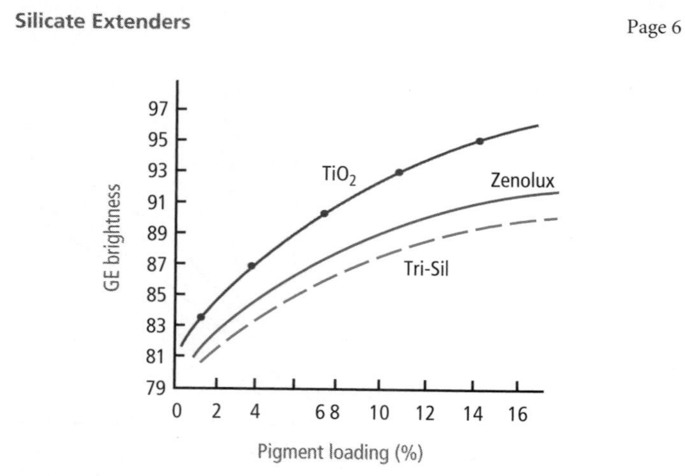

Figure 3. GE Brightness for Various Loadings of TiO_2, Zenolux, and Tri-Sil.

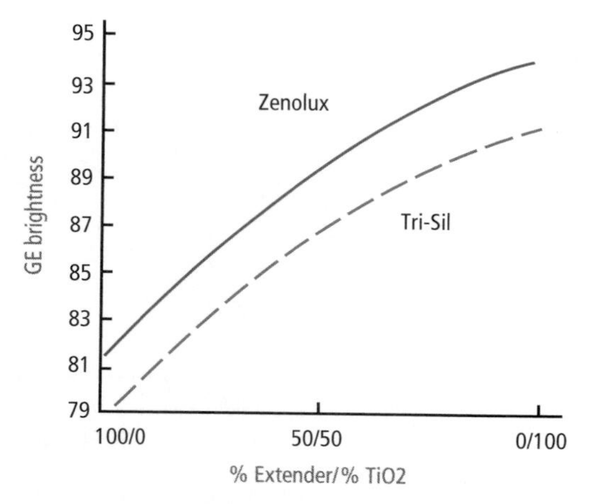

Figure 4. GE Brightness for Various Ratios of Zenolux and Tri-Sil to TiO_2.

FIGURE SR.3

Outline for Clark's Report

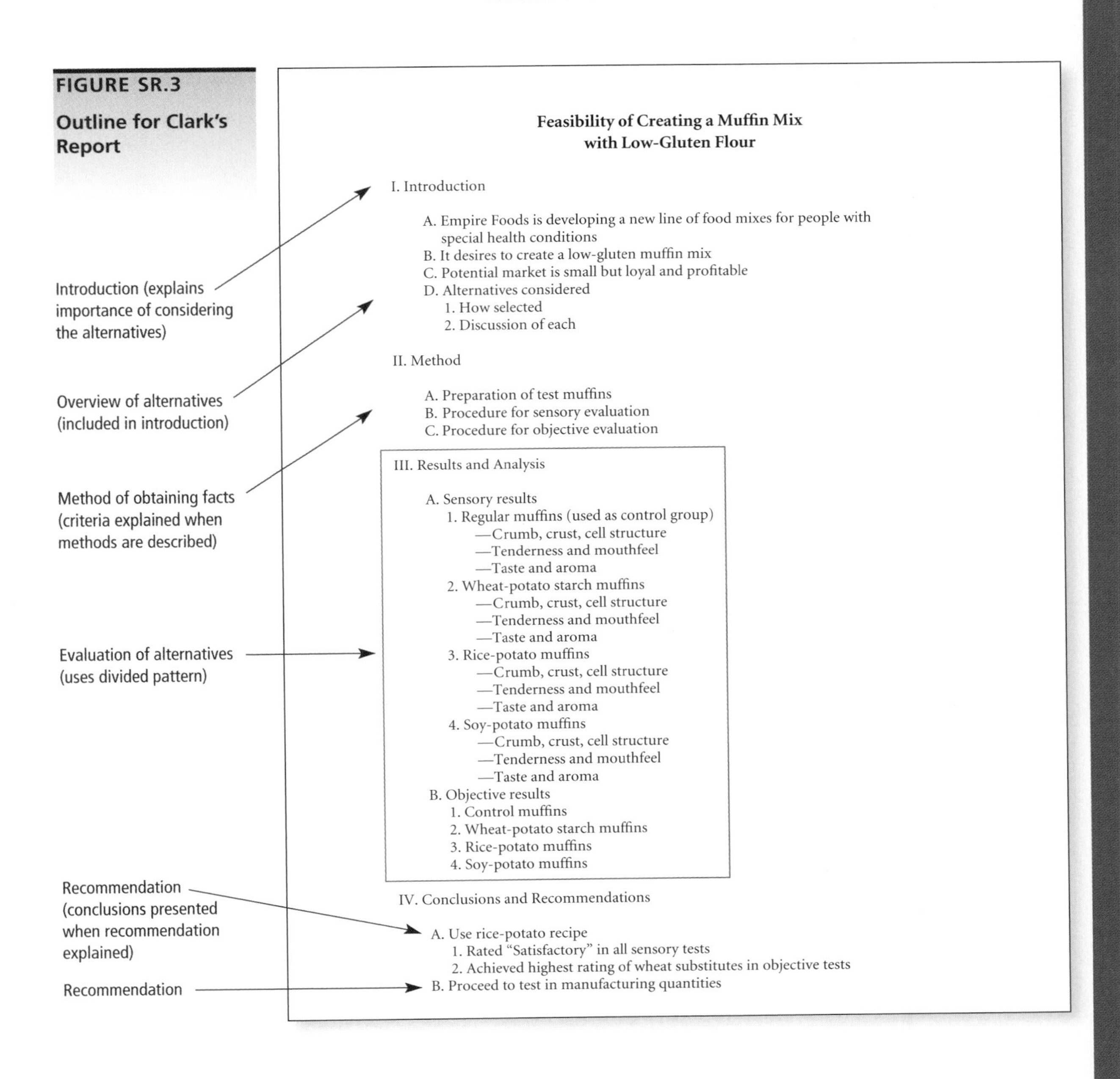

REVISION CHECKLIST, PLANNING GUIDE, AND OTHER RESOURCES

WWW To download a revision checklist and planning guide, visit Reference Guide 3 at www.techcomm.nelson.com.

A "Revision Checklist for Feasibility Reports" that you can use in your course and on the job is available at this book's website. The website also provides a "Planning Guide for Feasibility Reports."

PROGRESS REPORTS

A progress report is a report on work you have begun but not yet completed. The typical progress report is one of a series submitted at regular intervals, such as every week or month.

TYPICAL WRITING SITUATIONS

Some progress reports concern a single project.

Progress reports are prepared in two types of situations. In the first, you tell your readers about your progress on *one* project. Lee is a geologist employed by an engineering consulting firm. Her employer has assigned her to study the site that a large city would like to use for a civic centre and large office building. The city is worried that the site might not be geologically suited to such construction. Every two weeks, Lee must submit a progress report to her supervisor and to the city engineer. Her supervisor uses the report to make sure that Lee is conducting the study in a rapid and technically sound manner. The city engineer uses the report to see that Lee is proceeding according to schedule. She also looks for indications about the likely outcome of the study. Other work could be accelerated or halted as a result of Lee's preliminary findings.

Some progress reports concern all projects worked on during a single period.

In the second type of situation, you prepare progress reports that tell about your work on *all* your projects. Many employers require their workers to report on their activities at regular intervals all year round, year in and year out. Kayla is a person who must write such progress reports (often called *periodic reports*). She works in the research division of a large manufacturer of consumer products, where she manages a department that is responsible for improving the formulas for the company's laundry detergents—making them clean better and smell better, making them less expensive to manufacture, and making them safer for the environment. At any one time, Kayla's staff is working on between ten and twenty different projects.

As part of her regular responsibilities, Kayla must write a report every two weeks to summarize the progress on each of the projects. Her reports go to many readers, including the following: her immediate superiors, who want to be sure that her department's work is proceeding satisfactorily; researchers in other departments, who want to see whether her staff have made discoveries that they can use in the products they are responsible for (for example, dishwashing detergents); and corporate planners, who want to anticipate changes in formulas that will require alterations in production lines, advertising, and so on.

As the examples of Lee and Kayla indicate, progress reports may differ in many ways: They may cover one project or many; they may be addressed to people inside the writer's organization or outside it; and they may be used by people who have a variety of reasons for reading them, such as learning what they need to know to manage and to make decisions.

READERS' CONCERN WITH THE FUTURE

Although progress reports talk about the past, they are used to make decisions about the future.

Despite their diversity, however, almost all progress reports have this in common: Their readers are primarily concerned with the *future*. That is, even though most progress reports talk primarily about what has happened in the past, their readers usually want that information so that they can plan for the future.

For example, from your report they may be trying to learn the things they need to know in order to manage *your* project. They will want to know, for instance, what they should do (if anything) to keep your project going smoothly or to get it back on track. The progress reports written by Lee and Kayla are used for this purpose by some of their readers.

Other readers may be reading your progress reports to learn what they need to know in order to manage *other* projects. Almost all projects in an organization are interdependent with other projects. For instance, suppose you are conducting a marketing survey whose results will be used by another group as it designs an advertising campaign. If you have fallen behind your schedule, the other group's schedule may need to be adjusted.

Your readers may also be interested in the preliminary results of your work. Suppose, for instance, that you complete one part of a research project before you complete the others. Your readers may very well be able to use the results of that part immediately. The city engineer who reads Lee's reports about the possible building site wishes to use each of Lee's results as soon as it is available.

THE QUESTIONS READERS ASK MOST OFTEN

Your readers' concern with the implications of your progress for their future work and decisions leads them to want you to answer the following questions in your progress report. If your report describes more than one project, your readers will ask these questions about each of them:

- **What work does your report cover?** To be able to understand anything else in a progress report, readers must know what project or projects and what time period the report covers.
- **What is the purpose of the work?** Readers need to know the purpose of your work to see how it relates to their own responsibilities and to the other work, present and future, of the organization.
- **Is your work progressing as planned or expected?** Your readers will want to determine whether adjustments are needed in the schedule, budget, or number of people assigned to the project or projects on which you are working.
- **What results have you produced?** The results you produce in one reporting period may influence the shape of work in future periods. Also, even when you are still in the midst of a project, readers will want to know about any results they can use in other projects now, before you finish your overall work.
- **What progress do you expect during the next reporting period?** Again, your readers' interests will focus on such management concerns as schedule and budget and on the kinds of results they can expect.
- **How do things stand overall?** This question arises especially in long reports. Readers want to know what the overall status of your work is, something they may not be able to tell readily from all the details you provide.

- **What do you think we should do?** If you are experiencing or expecting problems, your readers will want your recommendations about what should be done. If you have other ideas about how the project could be improved, they, too, will probably be welcomed.

STRUCTURE FOR PROGRESS REPORTS

Remember that a structure is not an outline; you may combine the elements of a structure in many ways (see pages 185).

The structure for progress reports provides a very effective framework for answering your readers' questions about your projects:

Structure for Progress Reports

Report Element	Readers' Question
Introduction	What work does your report cover? What is the purpose of the work?
Facts and discussion	
Past work	Is your work progressing as planned or expected? What results have you produced?
Future work	What progress do you expect during the next reporting period?
Conclusions	In long reports—How do things stand overall?
Recommendations	What do you think we should do?

Introduction

In the introduction to a progress report, you can address the readers' first two questions. You can usually answer the question, "What work does your report cover?" by opening with a sentence that identifies the project or projects your report concerns and what time period it covers.

For additional advice about writing an introduction, see Chapter 3.

Sometimes you will not need to answer the second question—"What is the purpose of the work?"—because all your readers will already be quite familiar with its purpose. At other times, however, it will be crucial for you to tell your work's purpose because your readers will include people who don't know or may have forgotten it. You are especially likely to have such readers when your progress reports are widely circulated in your own organization or when you are reporting to another organization that has hired your employer to do the work you describe. You can usually explain purpose most helpfully by describing the problem that your project will help your readers solve.

The following sentences show how one writer answered the readers' first two questions:

> This is the fourth annual report submitted to Energy and Environment Ministers on *The Canada-Wide Acid Rain Strategy for Post-2000. The Strategy* provides a framework for resolving the acid rain problem in eastern Canada and preventing one in western and northern Canada. Annual reporting is a commitment under *The Strategy*, aimed at keeping decision makers and the public informed on the following: current and projected sulphur dioxide (SO_2) and nitrogen oxides (NO_x) emission levels in Canada, compliance with international emission reduction commitments, and progress in implementing the commitments in *The Strategy*.

Of course, your introduction should also provide any background information your readers will need in order to understand the rest of your report.

Facts and Discussion

In the discussion section of your progress report, you should answer these readers' questions: "Is your work progressing as planned or as expected?" "What results have you produced?" and "What progress do you expect during the next reporting period?"

Answering Your Readers' Questions Because the work to be accomplished during each reporting period is usually planned in advance, you can indicate your progress by comparing what happened with what was planned. Where there are significant discrepancies between the two, your readers will want to know why. The information you provide about the causes of problems will help your readers decide how to remedy them. It will also help you to explain any recommendations you make later in your report.

When you are discussing preliminary results, be sure to explain them in terms that will enable your readers to recognize their significance. Let your readers know whether the preliminary results are tentative or certain. That information will help them decide whether they can use the results.

Providing the Appropriate Amount of Information How much information should you include in your progress reports? Generally, readers prefer brief reports. Although you need to provide your readers with specific information about your work, don't include details unless they will help readers to decide how to manage your project or unless you believe readers will be able to make immediate use of them. As you work on a project, many minor events will occur, and you will have lots of small setbacks and triumphs along the way. Avoid discussing such matters. No matter how important these details may be to you, they are not likely to be interesting to your readers. Stick to the information your readers can use.

Organizing the Discussion You can organize your discussion segment in many ways. One way is to arrange it around time periods:

I. What happened during the most recent time period
II. What's expected to happen during the next time period

You will find that this organization is especially well suited for progress reports dealing with a single project that has distinct stages. However, you can also expand this structure for reports that cover either several projects or one project in which several tasks are performed simultaneously:

I. What happened during the most recent time period
  A. Project A (or Task A)
  B. Project B
II. What's expected to happen during the next time period
  A. Project A
  B. Project B

When you prepare a report that covers more than one project or more than one task, you might consider organizing it around those projects or tasks:

I. Work on Project A (or Task A)
  A. What happened during the last time period
  B. What's expected to happen during the next time period
II. Work on Project B
  A. What happened during the last time period
  B. What's expected to happen during the next time period

This way of organizing works very well in reports that are more than a few paragraphs long because it keeps all the information on each project together, making the report easy for readers to follow.

Emphasizing Important Findings and Problems As mentioned, your findings and problems are important to your readers. Be sure to highlight them and discuss them in sufficient detail to satisfy your readers' needs and desires for information.

Conclusions

Your conclusions are your overall views on the progress of your work. In short progress reports, there may be no need to include any conclusions, but if your report covers several projects or tasks, conclusions may help your readers understand the general state of your progress.

Recommendations

If you have any ideas about how to improve the project or increase the value of its results, your readers will want you to include them. Your recommendations might be directed at overcoming some difficulty that you have experienced or anticipate encountering in the future. Or they might be directed at refocusing or otherwise altering your project.

A Note on the Location of Conclusions and Recommendations

For most of your readers, your conclusions and recommendations are the most important information in your progress report. Therefore, you should usually include them at the beginning, either in the introduction or at the head of your discussion section. If they are brief, this may be the only place you need to state them. If they are long or if your readers will be able to understand them only after reading your discussion section, it's best to present your conclusions and recommendations at the end of your report, while still including a summary of them at the beginning.

TONE IN PROGRESS REPORTS

You may wonder what tone to use in the progress reports you prepare. Generally, you will want to persuade your readers that you are doing a good job. That is especially likely when you are new on the job and when your readers might discontinue a project if they feel that it isn't progressing satisfactorily.

Because of this strong persuasive element, some people adopt an inflated or highly optimistic tone. This sort of tone, however, can lead to difficulties. It might lead you to make statements that sound more like an advertising claim than a professional communication. Such a tone is more likely to make your readers suspicious than agreeable. Also, if you present overly optimistic accounts of what can be expected, you risk creating an unnecessary disappointment if things don't turn out the way you seem to be promising. And if you consistently turn in overly optimistic progress reports, your credibility with your readers will quickly vanish.

In progress reports, it's best to be straightforward about problems so that your readers can take appropriate measures to overcome them and so that they can adjust their expectations realistically. You can sound pleased and proud of your accomplishments without exaggerating them.

SAMPLE OUTLINES AND PROGRESS REPORT

The outlines of two progress reports prepared by Erin and Lloyd will suggest how the four elements of the structure for progress reports can be adapted to different situations. Erin is reporting on her investigation of problems with a hot saw, the machine that cuts white-hot ingots of metal in a steel mill (Figure SR.4, page 222), and Lloyd is submitting a weekly update on a project to introduce a new line of high-fashion women's clothes (Figure SR.5, page 223).

Notice that Erin organizes her report around time periods, devoting one section to her past work and a separate section to her future work. In contrast, Lloyd organizes his report around major activities: manufacturing, marketing, and so on. Within the section on manufacturing, he talks about past work (emphasizing a problem) and about future work; he even makes a recommendation there, rather than at the end.

When using the structure as a guide in preparing progress reports, try to be as responsive to your purpose and your readers as Erin and Lloyd have been to theirs.

REVISION CHECKLIST, PLANNING GUIDE, AND OTHER RESOURCES

WWW To obtain a revision checklist and planning guide you can download, visit Reference Guide 3 at www.techcomm.nelson.com.

A "Revision Checklist for Progress Reports" that you can use in your course and on the job is available at this book's website. The website also provides a "Planning Guide for Progress Reports."

Figure SR.4

Outline for Erin's Progress Report on Increasing Efficiency

Introduction (Erin explains work covered and its purpose; she tells the period covered)

Past work (Erin emphasizes her accomplishments)

Future work (she tells specifically what she will do)

Conclusion (Erin indicates that she's behind schedule, she asks for assistance, and she describes actions that can be taken immediately)

Progress on Efficiency Project

I. Introduction
- A. I am reporting on the efficiency project
- B. This report covers my work for the past two weeks
- C. I focused on the hot saw

II. Past Work
- A. I discovered a problem with the location of the tools
- B. I discovered a problem with the blades our supplier provides

III. Future Work
- A. During the next two weeks, I will visit Winnipeg to check on their hot-saw process
- B. I will also check on their loading dock procedures

IV. Conclusions
- A. I'm behind schedule; please let Zack help me for a few days
- B. I have two recommendations
    1. Build a tool stand next to the hot saw
    2. Look for a new supplier or get ours to meet our specifications

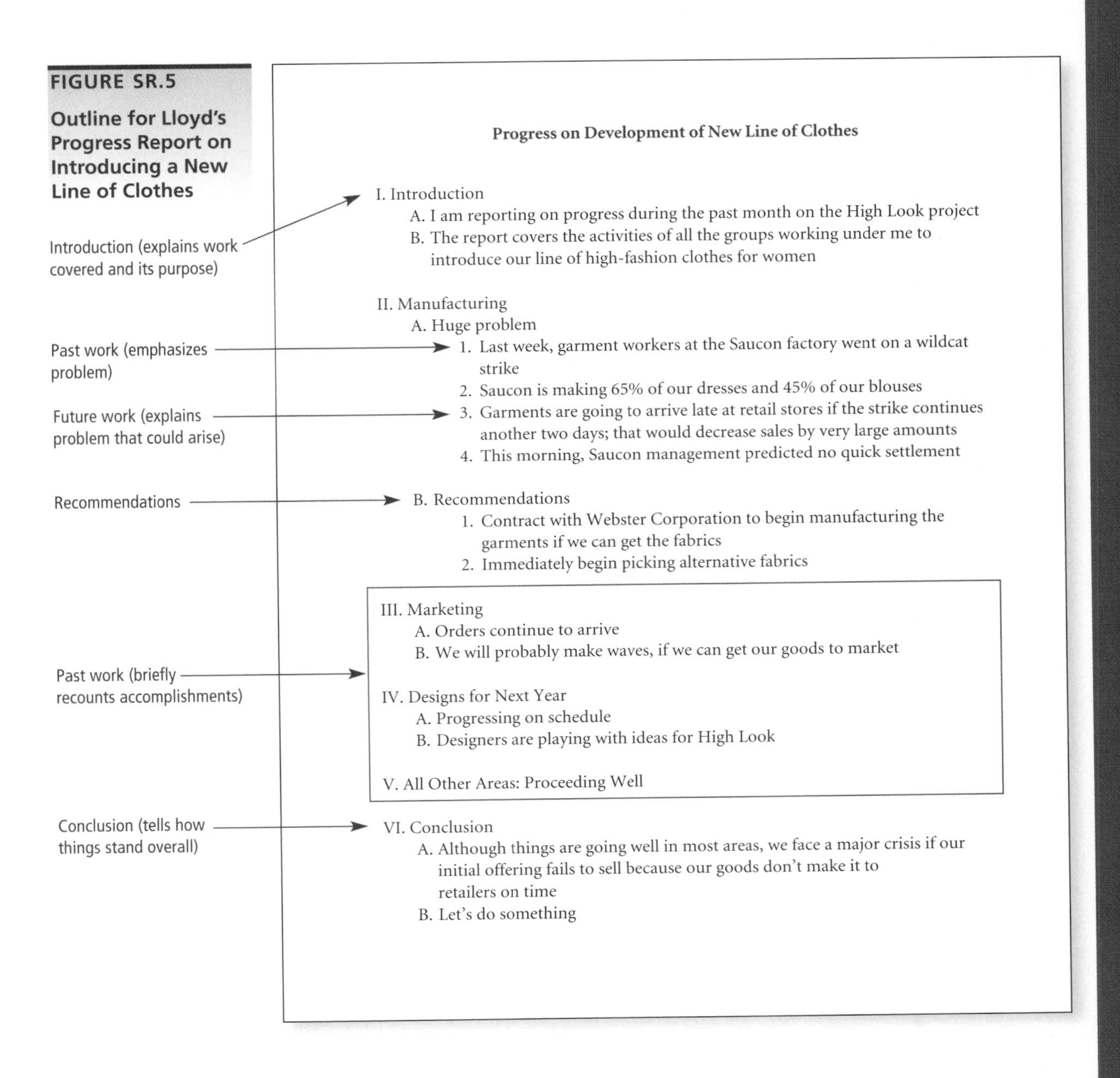

FIGURE SR.5

Outline for Lloyd's Progress Report on Introducing a New Line of Clothes

CASE

CONTAINING COSTS

Getting the job at Tumbler Mountain Property Management after graduation was a lucky break. The lifestyle in the village is fantastic all year round. There is mountain biking and snowboarding just minutes from your office, and you enjoy your position as a financial analyst. Today, though, it looks like you will probably have to stay late. You'll have to take on the switchback trail tomorrow.

Your manager has asked you to look at reducing costs in the Reservations Group. With Tumbler Mountain becoming a more popular destination, more and more properties and resorts are being booked through your firm. The number of booking agents in the Reservations Group has grown, and the staff often have to be paid overtime to keep up with the volume of calls. It's expensive to live in or even within commuting distance of Tumbler Mountain, and the salaries and benefits to compensate staff have to reflect that. Unfortunately, the Reservations Group, like Human Resources and Information Systems, is considered a non-essential service. Tumbler Mountain Property Management is about managing properties, not about hiring and paying people.

YOUR ASSIGNMENT

Your manager has asked you to prepare a feasibility report comparing options for containing costs in the reservations group. He has suggested that you look at ideas such as hiring more part-time staff (who don't need to be paid benefits), or perhaps outsourcing the function. There may be other options that he hasn't considered. He's not looking for a complete report this time—just an outline. Use the Revision Checklist for Feasibility Reports to ensure that you include each of the elements needed to create a report that your reader will find usable and persuasive.

Planning a Communication Strategy

CHAPTER 9 Planning for Usability

GUIDELINES

1. Identify the information your readers need
2. Organize around your readers' tasks
3. Identify ways to help readers quickly find what they want
4. For a complex audience, plan a modular communication
5. Look for a technical writing structure you can adapt
6. Plan your graphics
7. Outline, if this would be helpful
8. Check your plans with your readers
9. Ethics Guideline: Investigate stakeholder impacts

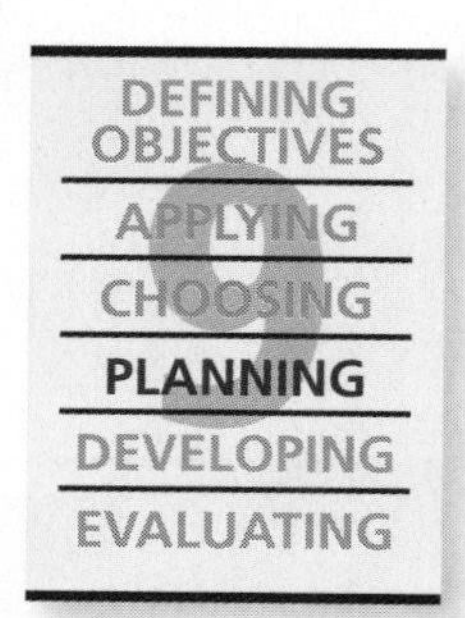

This chapter develops your expertise at translating your objectives into a successful plan of action.

Planning is an inevitable part of writing. Whenever you write to a friend, professor, coworker, or client, you begin with at least *some* idea of what you will say and how you will say it. However, there is a great difference between expert planning and poor planning. By developing your expertise in planning, you increase your ability to identify the content, organization, style, graphics, and other features that are most likely to achieve the results you desire. In addition, by developing your planning expertise you become a more efficient writer, one who can avoid many of the false starts, inconsistencies, and other difficulties that would require rework before the communication could be sent to its readers.

WWW For additional chapter resources, visit Chapter 9 at www.techcomm.nelson.com.

Overall, this chapter shows how to follow the same reader-centred approach that is explained throughout this book. Its guidelines tell how to plan communications that possess the two qualities that are indispensable for successful workplace communication: usability and persuasiveness *in your readers' eyes.*

YOUR GOAL WHEN PLANNING FOR USABILITY

As explained in Chapter 1, a usable communication enables its readers to find and use information easily as they perform some practical task. Their task might be purely mental, as when a manager at the Canadian Space Agency compares alternative designs for a new component for the international space station, or it might mix mental and physical activities, as when an astronaut follows a set of instructions for installing the component in outer space.

Research indicates that readers find communications usable if they are:

Qualities of a usable communication

- **Complete—from your readers' perspective.** A usable communication contains all the information its readers require.
- **Task-oriented.** A usable communication is organized around its readers' tasks and supports the readers' performance of these tasks as fully as possible.
- **Accessible.** A usable communication enables its readers to locate quickly the information they want.

This chapter's first eight guidelines identify the key strategies and resources for planning complete, task-oriented, and accessible communications. The ninth guideline describes important steps for assuring that the communications you plan are ethical.

GUIDELINE 1 Identify the Information Your Readers Need

Readers can't use what isn't there. Consequently, your first job when planning for usability is to ensure that your communication will be *complete,* that it will include all the information your readers need in order to perform their tasks.

To begin identifying the information your readers will need, list the questions they will bring to your communication or ask while reading it. As Chapter 2 suggests, you can develop this list by imagining your readers as they pick up and then use your communication. To gain additional insights into what your communication must say in order to meet your readers' needs, consider your readers' characteristics that will influence the kinds of explanations and background facts that will contribute to making your communication understandable and useful to them. As Chapter 2 suggests, these characteristics are your readers' professional specialty, organizational role, and cultural background; their familiarity with your topic, specialty, and situation; their personal preferences about the communications they read; and their relationship with you.

Research often turns up facts that are important to readers even though the readers would not know to ask for them.

To identify all the information your readers will need, however, you must go beyond the questions you can predict they will ask and the kinds of answers you believe will be most helpful to them. You may know important things that your readers don't realize they should ask about, or you may learn such things through research you do while preparing the communication. For example, Toni, a recently graduated systems analyst, has been assigned to research and write a report comparing several software packages that her employer, a large civil engineering firm, plans to distribute to all of its professional employees. By following Chapter 2's advice, she has identified many questions that her complex audience will ask. She knows, for instance, that the heads of the four engineering departments will each ask, "How efficiently can each program perform each of the most important design functions performed by my department?" And she knows that the head of the purchasing department will inquire, "What about licensing options?" By thinking about her readers' questions, Toni can identify many of the details she must discuss in her report. However, Toni has also learned that the company that makes one of the software programs she's evaluating is in financial trouble. If it goes out of business, Toni's company will be unable to obtain the assistance and upgrades it would expect to receive. Even though Toni's readers are not likely to ask, "Are any of the programs made by companies that appear to be on the verge of bankruptcy?" Toni should tell her readers about the company's financial problems.

On the other hand, Toni should avoid including information just because she finds it interesting or wants to demonstrate how much she knows. Such information only makes it more difficult for readers to locate and use the information they require. In a reader-centred communication, it's just as important to omit information your readers don't need as it is to provide all that they do.

GUIDELINE 2 Organize around Your Readers' Tasks

Once identified, a communication's content must be organized. As you learned in Chapter 1, reading involves a moment-by-moment interaction between reader and text. In this interaction, readers perform a sequence of mental tasks. In some cases, these mental tasks are mixed with physical tasks, as when people read instructions. Your communications will be most usable if you organize them to mirror and support the readers' tasks.

Use your mental portrait of your readers to envision the details of the reading tasks your readers want your communication to help them perform.

We all intuitively employ a task-oriented organization when we tell someone how to walk or ride to a particular destination. We mentally trace the route they should follow, and we organize our directions in a way that mirrors their progress: "Go to the third set of lights, turn left, and go for two blocks." In the same way, you can use a mental portrait of your readers (see Guideline 1 in Chapter 2) to create a task-oriented organization that guides them smoothly through a physical process or assists them in making a decision, planning future action, or performing other mental activities.

Imagine, for example, that you work in the dean's office of a small school. One of the computer's databases has been corrupted by a virus. You have been asked to compile a list with the names, email addresses, home addresses, phone numbers, programs, and academic levels of all students enrolled in an online technical communication course. Now, with the completed sheets in front of you, you must write a single list that includes all the information.

Organize in the way your readers will find most helpful when they try to use your information to fulfill their own goals.

How will you organize the list? If you followed the usual advice about organizing, you would organize it "logically." However, there are many logical ways to organize the list: by program, academic level, city, and so on. Although any of these choices would be logical, they would not be equally usable. To make a truly usable list, you would need to identify the task the dean will use the list to perform and then organize the list to help her perform this specific task. If the dean wants to use your list to address letters sent to students who are graduating, another letter to those who are in second year, and so on, you would organize the list according to academic level. If the dean will use your list to determine which departments enroll the most students in online technical communication classes, you would arrange the list according to the students' programs. For other uses, still other organizational patterns would be effective in helping the dean perform the tasks she wants to accomplish while reading your list.

Of course, readers' tasks vary greatly from situation to situation. Nevertheless, some tasks are common enough to make it possible to identify three strategies for organizing messages that almost always enhance usability:

- Organize hierarchically.
- Group together the items your readers will use together.
- Give the bottom line first.

These strategies are described in the following paragraphs.

Organize Hierarchically

Readers build mental hierarchies.

When we read, we encounter small bits of information one at a time: first what we find in this sentence, then what we find in the next sentence, and so on. One of our major tasks is to build these small bits of information into larger structures of meaning that we can store and work with in our own minds. Represented on paper, these mental structures are hierarchical. They look like outlines (Figure 9.1) or tree diagrams (Figure 9.2, page 232), with the overall topic divided into subtopics and some or all of the subtopics broken down into still smaller units. For readers, building these hierarchies can be hard work, as we've all experienced when we've had to reread a passage because we can't figure out how its sentences fit together.

Help readers by organizing hierarchically.

The easiest and surest way to help your readers build mental hierarchies is simply to present your information already organized this way. This doesn't mean that you need to start every writing project by making an outline (see Guideline 7, pages 236–37). Often, you may be able to achieve a hierarchical organization without outlining. But you should always organize hierarchically.

Group Together the Items Your Readers Will Use Together

Organize to support your readers' way of using your communication.

Creating a hierarchy involves grouping facts. You assemble individual facts into small groups and then gather these groups into larger ones. When grouping your facts, be sure to group together the information your readers will use together. As the example of the list

FIGURE 9.1

Outline Showing the Hierarchical Organization of a Report

	Parking Recommendations to Greenwood Area Council
Topic	I. Introduction
Topic	II. Problems
Subtopic	A. Insufficient parking
Smaller parts	1. Shoppers complain about inconvenience 2. Store owners complain about lost sales 3. Police report many parking violations
Subtopic	B. Traffic congestion
Smaller parts	1. People circle through the area waiting for a parking space to open up 2. Additional on-street parking impedes the free flow of traffic 3. Though illegal, some double-parking occurs, clogging streets
Topic	III. Possible solutions
Subtopic	A. Build a parking garage in the centre of the shopping area
Smaller parts	1. Advantages 2. Disadvantages 3. Cost
Subtopic	B. Build a parking lot on the edge of the shopping area and run shuttle buses to major stores
Smaller parts	1. Advantages 2. Disadvantages 3. Cost
Topic	IV. Recommendations

of technical writing students illustrates, you increase your communication's usability when your grouping matches your readers' tasks. Daniel's situation provides another example.

Daniel groups together the studies that he knows his readers would want to read about in one place.

Daniel is preparing a report that summarizes hundreds of research studies on the effects of sulphur dioxide (SO_2) emissions from smelters and utilities. Daniel could group his information in several ways. For instance, he could group together all the research published in a given year or span of years. Or he could discuss all the studies conducted in Europe in one place, all conducted in North America in another place, and so on. However, Daniel is employed by a provincial task force. His report will be read by members of the Manitoba Ministry of Conservation as they decide how to structure legislation on SO_2 emissions. Daniel knows that these readers will want to read about the impacts of SO_2 emissions on human health separately from the effects on the environment. Therefore, he organizes around kinds of impacts, not around the dates or locations of the studies.

Like Daniel, you can organize your communications in a usable, task-oriented manner by grouping together the information your readers will use together.

FIGURE 9.2

Tree Diagram Showing the Hierarchical Organization of a Report

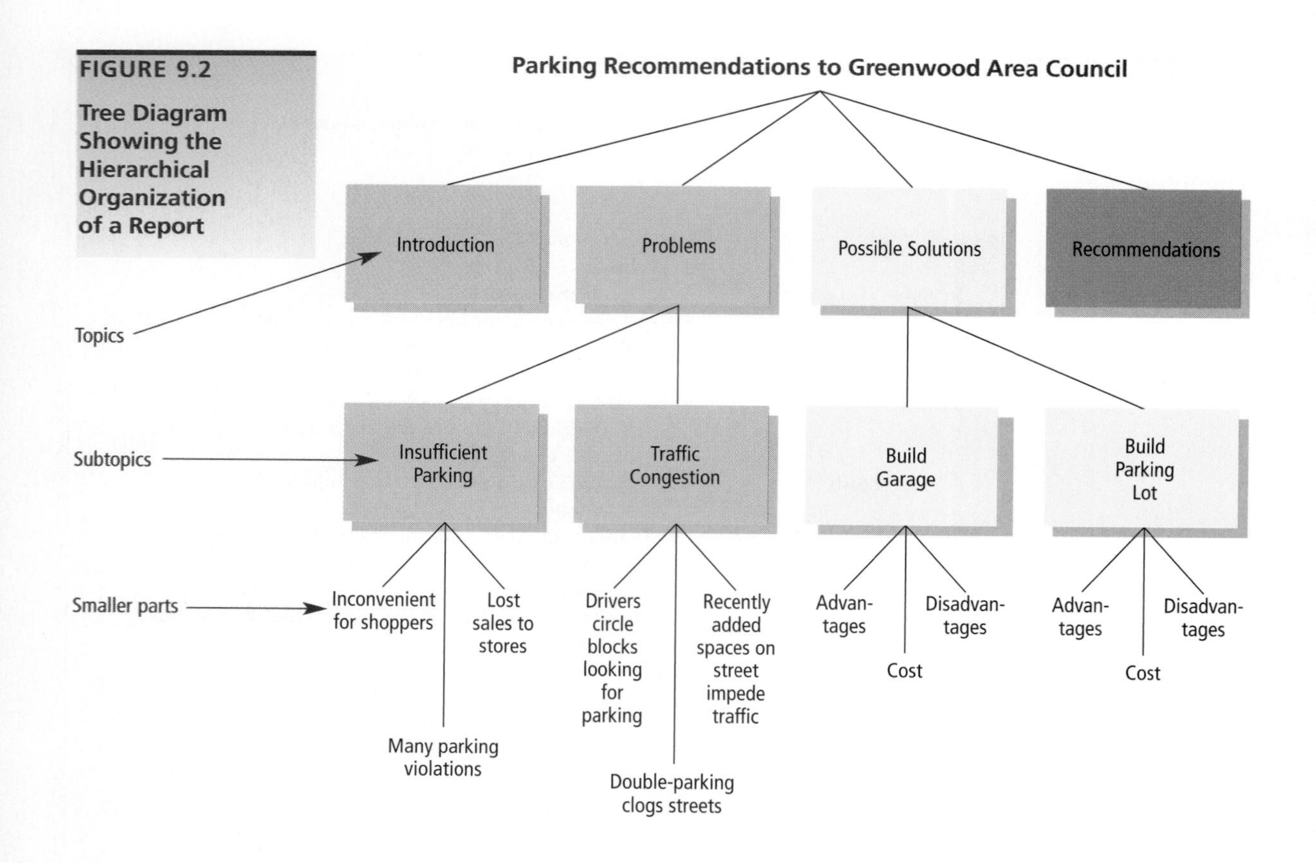

Give the Bottom Line First

Readers at work often say that their most urgent reading task is to find the writer's main point. The most obvious way to make it easy for them to find your main point is to put it first. Indeed, this strategy for task-oriented organization is the subject of one of the most common pieces of advice given in the workplace: "Put the bottom line first." The bottom line, of course, is the last line of a financial statement. Literally, the writers are being told, "Before you swamp me with details on expenditures and sources of income, tell me whether we made a profit or took a loss." However, this advice is applied figuratively to many kinds of communication prepared at work.

Readers want the main point quickly.

Of course, you don't always need to place your main point in the very first sentence, although this is sometimes the most helpful place for it. When appropriate, you can always provide relevant background information before stating your main point. But don't keep your readers in suspense. As soon as possible, get it out: Is the project on schedule, or must we take special action to meet the deadline? Will the proposed design for our product work, or must it be modified?

Focus on use, not logic, when you organize.

To some writers, it seems illogical to put the most important information first. They reason that the most important information is generally some conclusion they reached fairly late in their thinking about their subject, and they think it is logical to describe the process that led up to the conclusion before presenting this result. However, such a view is writer-centred. It assumes that information should be presented in the order in which the writer acquired it. However, such an organization runs counter to the sequence that readers would find most helpful.

In some circumstances, you should delay the main point.

There are, nevertheless, some situations in which it really is best to withhold the bottom line until later in a communication. You will find a discussion of such situations in Guideline 4 of Chapter 10 (pages 252–255). As a general rule, however, you can increase your readers' satisfaction and reading efficiency by creating a task-oriented organization in which you give the bottom line first.

GUIDELINE 3 Identify Ways to Help Readers Quickly Find What They Want

At work, readers often want to find a particular piece of information without reading the entire document that contains it. When you are planning, identify ways to make all information your readers might seek in this way readily accessible to them. Along with completeness and task-orientation, accessibility is one of the essential qualities of a usable communication.

To devise your accessibility strategies, turn once again to your mental portrait of your readers. Picture the various circumstances under which they would want to locate some particular subpart of your overall message. What will they be looking for? How will they search for it? With the answers to these two questions, you can plan the pathways that will guide your readers through your communication to the information they want.

For printed documents, you can construct these pathways with headings, topic sentences, tables of contents, and other devices.

For advice about creating site maps, see Chapter 5.

For websites, you can use menus, links within the text, and clickable images and icons. In website design, accessibility also depends substantially on the *site map,* which describes the way the parts of the site are related to one another and defines the routes by which visitors gain access to the information they want. It's best to plan the site so that they can get there in three or four clicks rather than five or six.

All of these strategies for accessibility are discussed in detail elsewhere in this book. The important point now is that when planning your communication, even before you begin drafting, you should identify the ways you will make your communication accessible to your readers. Note, too, that all three of Guideline 2's strategies for organizing around your readers' tasks also increase the readers' ability to quickly access the information they are seeking when they want to use some—but not all—of your communication's content.

GUIDELINE 4 For a Complex Audience, Plan a Modular Communication

Modular designs enable you to create a single communication that addresses readers with different sets of questions.

As explained in Chapter 2, at work you will sometimes need to write a single communication that must meet the needs of several readers or groups of readers, each with a different professional role, different background, and hence different set of questions. In the workplace, writers often meet the needs of these diverse audiences by creating modular communications in which different parts address different readers.

For example, many reports and proposals are written for both decision makers and advisers. Usually such reports have two parts: (1) a very brief summary—called an *executive summary* or *abstract*—at the beginning of the report, designed for decision makers who want only the key information; and (2) the body of the report, designed for advisers, who need the details. Typically, the executive summary is only a page or a few pages long, whereas the body may exceed a hundred pages. The body of the report might be divided into still other modules, one addressed to technical experts, one to accountants, and so on. Reports and proposals often include appendixes that present technical details.

www.techcomm.nelson.com

Websites are often designed using a similar strategy, with different areas addressed to different groups of users. Figure 9.3 shows an example.

By creating a modular design, you make your communication accessible. Each reader can go directly to the section or sections that are most relevant to him or her. A modular design also allows you to treat each topic at the level that is appropriate for a specific group of readers. Doing this frees you from the difficulties of trying to discuss a topic at a sufficiently detailed level for technical readers while also making it simple enough for non-specialists.

Researcher James W. Souther (1985) confirmed the usefulness of such a design from the decision makers' point of view. He asked a large group of managers how often they read each part of a long report. As Figure 9.4 indicates, they reported that they read the summary 100 percent of the time, but the body only 15 percent of the time. When decision makers want more information than the summary provides, they usually go to the introduction (which provides background information) and to the recommendations (where the writers explain their suggestions fully). The rest of the report is read by advisers and implementers.

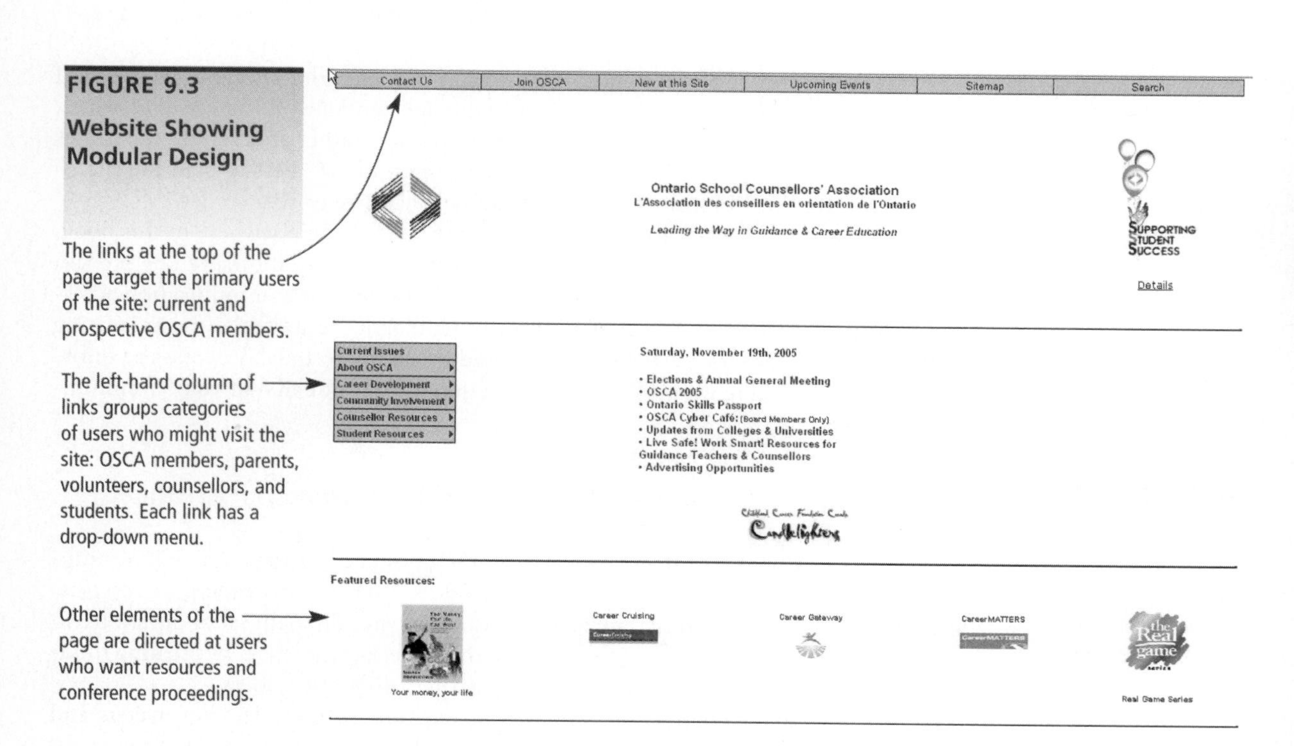

FIGURE 9.3

Website Showing Modular Design

The links at the top of the page target the primary users of the site: current and prospective OSCA members.

The left-hand column of links groups categories of users who might visit the site: OSCA members, parents, volunteers, counsellors, and students. Each link has a drop-down menu.

Other elements of the page are directed at users who want resources and conference proceedings.

FIGURE 9.4

How Decision Makers Read Reports

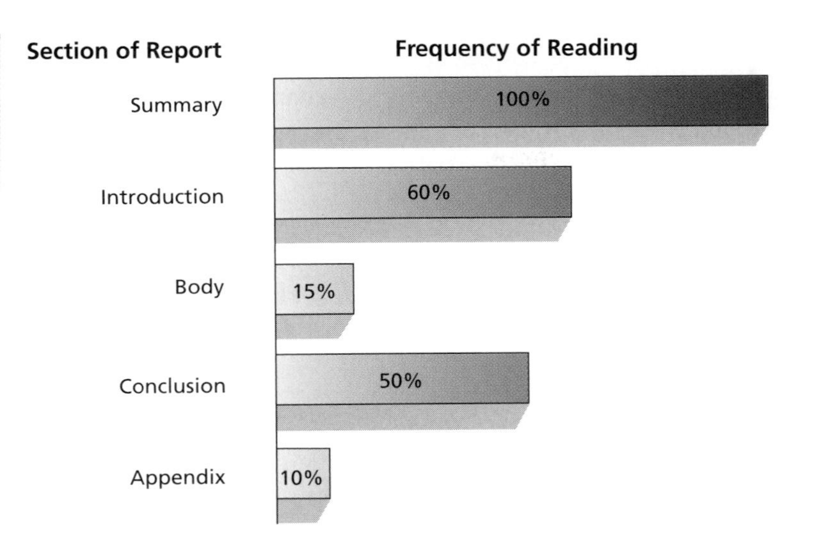

GUIDELINE 5 Look for a Technical Writing Structure You Can Adapt

At work, you will often write in a situation that closely resembles circumstances encountered by other people many times before. Like other people, you may need to report on a business trip, tell someone how to operate a piece of equipment, or request funds for a project you would like to conduct. For many of these recurring situations, writers employ conventional patterns for constructing their communications. These patterns are sometimes called *genres.*

At work, you will encounter many genres or structures: the business proposal, budget report, computer manual, feasibility report, project proposal, and environmental impact statement, to name a few. Carolyn R. Miller (1984), Graham Smart (1993), and other researchers suggest that each structure exists because writers and readers agree that it provides an effective pattern for meeting some particular communication need that occurs repeatedly. Figure 9.5 (page 236) shows how the structure for proposals accommodates readers (by answering their questions) and also writers (by providing them with a framework for presenting their persuasive claims).

Structures assist both readers and writers.

Because structures serve the needs of both readers and writers, they are ideal for writing reader-centred communications. They can be especially helpful at the planning stage because they suggest the kinds of information your readers probably want and the manner in which the information should be organized and presented. However, structures are not surefire recipes for success. Each represents a general framework for constructing messages in a typical situation. But no two situations are exactly alike. Moreover, for many situations no structure exists. To use structures effectively, look for an appropriate one; if you find one, adapt it to your particular purpose and readers.

Chapters 6 to 8 describe the general structures for instructions, proposals, and reports. These chapters also describe ways of adapting the structures to specific readers and purposes. At work, you may encounter more specialized structures developed within your profession or industry—or even within your own organization.

FIGURE 9.5

Structure for Proposal

Structure for Proposals

Topic	Readers' Question	Writer's Persuasive Point
Introduction	What is this communication about?	Briefly, I propose to do the following.
Problem	Why is the proposed project needed?	The proposed project addresses a problem, need, or goal that is important to you.
Objectives	What features will a solution to this problem need in order to be successful?	A successful solution can be achieved if it has these features.
Product or Outcome	How do you propose to do those things?	Here's what I plan to produce and how it has the features necessary for success.
Method	Are you going to be able to deliver what you describe here?	Yes, because I have a good plan of action (method), the necessary facilities, equipment, and other resources, a workable schedule, appropriate qualifications, and a sound management plan.
Costs	What will it cost?	The cost is reasonable.

GUIDELINE 6 Plan Your Graphics

Graphics are very common in on-the-job writing.

On-the-job communications often use graphics such as charts, drawings, and photographs rather than text to convey key points and information. Graphics convey certain kinds of information more clearly, succinctly, and forcefully than words. When planning a communication, look for places where graphics provide the best way for you to show how something looks (in drawings or photographs), explain a process (flowcharts), make detailed information readily accessible (tables), or clarify the relationship among groups of data (graphs).

Chapter 15 provides detailed advice about where to use graphics and how to construct them effectively. However, don't wait until you read that chapter to begin planning ways to increase your communication's usability with graphics.

GUIDELINE 7 Outline, If This Would Be Helpful

When they talk about planning a communication, many people mention outlining—and ask, "Is outlining worth the work it requires?" No single answer to this question is valid for all writers and all situations. The following general observations about outlining practices at work may help you determine when (and if) outlining might be worthwhile for you.

Outlining is rare for some types of communication.

On the job, outlining is rarely used for short or routine messages. However, many writers outline longer, more complex communications, especially if they expect to have difficulties in organizing. They use outlining as a way of experimenting with alternative ways of structuring their message before they begin to invest time in drafting. Similarly, if they encounter problems when drafting, some writers will try to outline the troublesome passage.

Outlining is common for reviews and team writing.

Also, writers sometimes wish to—or are required to—share their organizational plans with a manager or coworker. Outlining provides them with a convenient way of explaining their plans to such individuals.

Finally, outlining can help writing teams negotiate the structure of a communication they must create together (see Chapter 12).

When you create an outline, remember that the kind of tidy structure shown in Figure 9.1 (page 231) represents the product of the outlining process. Many writers begin organizing by making lists or diagrams, drawing arrows and pictures, and using many other techniques to think through possible ways of putting their information together for their readers.

Using Software to Outline

Most word processing programs include special tools for outlining. With these tools, you can, for instance, make an outline and then convert it immediately into the headings for your document. These tools will also allow you to convert from the normal view of your document to an outline view, so you can review the organizational structure that is evolving as you write. When you move material in the outline, the program automatically moves the corresponding parts of your full text, which can sometimes make for an extremely efficient way to revise a draft. Figure 9.6 (page 238) shows some of the features of one widely used word processing software outlining tool.

GUIDELINE 8 Check Your Plans with Your Readers

At the core of the seven guidelines you have just read is one common strategy: Focus on your readers as you plan. For that reason, the guidelines urged you to refer continuously to various sources of insight into what will make your communication usable in your readers' eyes. These include the understanding of your readers that you gained when defining your objectives and the structures that others have found successful in similar circumstances. Nevertheless, such planning involves guesswork about what will really work with your readers. Therefore, whenever you have the opportunity, bring your plans to your readers. Ask for their responses and requests. If it isn't feasible to check your plans with your readers, share them with someone who understands your readers well enough to help you find possible improvements. The better your plans, the better your final communication.

GUIDELINE 9 Ethics Guideline: Investigate Stakeholder Impacts

As Chapter 2 explained, the first step in writing ethically is to identify your communication's stakeholders—the people who will be affected by what you say and how you say it. Next, learn how your communication will impact these people and how they feel about these potential effects.

FIGURE 9.6

Outlining Functions of Word Processing Software

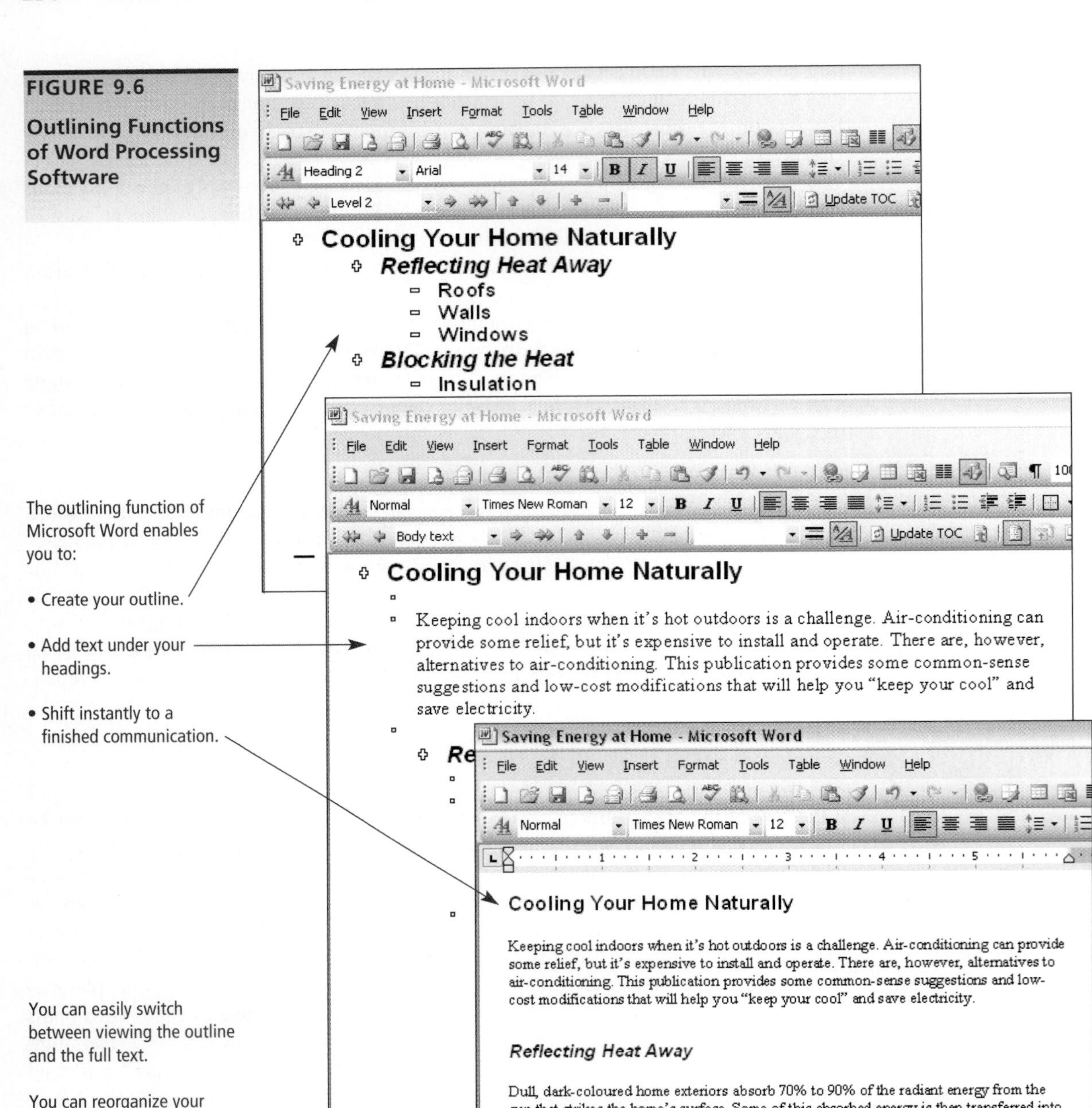

The outlining function of Microsoft Word enables you to:

- Create your outline.
- Add text under your headings.
- Shift instantly to a finished communication.

You can easily switch between viewing the outline and the full text.

You can reorganize your communication simply by moving a heading; the subheadings and text automatically move also.

Asking Stakeholders Directly

The best way to learn how your communication will affect its stakeholders is to talk to these individuals directly. In many organizations, such discussions are a regular step in the decision-making process that accompanies the writing of reports and proposals. When an action is considered, representatives of the various groups or divisions that might be affected meet together to discuss the action's potential impacts on each of them.

Similarly, government agencies often solicit the views of stakeholders. For example, the provincial ministries that write environmental impact statements are required to share drafts of these documents with the public so that concerned citizens can express their reactions. The final draft must respond to the public's comments.

Even in situations that are traditionally viewed as one-way communications, many managers seek stakeholder inputs. For instance, when they conduct annual employee evaluations, some managers draft their evaluation, then discuss it with the employee before preparing the final version.

Action You Can Take

If you are writing a communication for which there is no established process for soliciting stakeholders' views, you can initiate such a process on your own. To hear from stakeholders in your own organization, you can probably just visit or call. To contact stakeholders outside your organization, you may need to use more creativity and also consult with your coworkers.

If you already know the stakeholders well, you may be able to find out their views very quickly. At other times, an almost impossibly large amount of time would be needed to thoroughly investigate stakeholders' views. When that happens, the decision about how much time to spend can itself become an ethical decision. The crucial thing is to make as serious an attempt as circumstances will allow. There's all the difference in the world between saying that you can spend only a limited amount of time investigating stakeholders' views and saying that you just don't have *any* time to find out what the stakeholders are thinking.

Speaking for Others

To save time, you might try to imagine what the stakeholders would say if you spoke to them directly. Clearly this course of action is superior to ignoring stakeholders altogether. But you can never know exactly what others are thinking. The greater the difference between you and the stakeholders—in job title, education, and background—the less likely you are to guess correctly. So it's always best to let stakeholders speak for themselves. When you do this, avoid the mistake of assuming that all the stakeholders will hold the same view. The persons affected by a communication may belong to many different groups, and opinions can vary even within a single group. When letting stakeholders speak for themselves, seek out a variety of people.

For an additional discussion of ways to address an organization's reluctance to consider ethical issues, see the ethics guideline, page 81.

At work, you might need authorization from your managers to seek stakeholder views. If you encounter reluctance to grant you this permission, you have an opportunity to open a conversation with your managers about the ethical dimensions of the communication you are writing. Even if they aren't swayed this first time, you will have introduced the issue into the conversation at your workplace.

Seeking Stakeholder Views

In sum, on many issues, different people hold different views about what is the most ethical course of action. Asking stakeholders for their input does not guarantee that they all will be happy with what you ultimately write. But it does guarantee that you have heard their opinions. Without this step, you can scarcely attempt to take their needs and concerns into account.

CONCLUSION

WWW To download a planning guide for usability strategies, visit Chapter 9 at www.techcomm.nelson.com.

This chapter has described the reader-centred strategies that successful workplace communicators use in order to plan communications that are highly usable. As you have seen, to develop expertise in using these strategies you must possess the detailed knowledge of your readers that you gained by following Chapter 2's reader-centred advice for defining your communication's objectives.

EXERCISES

For additional exercises, visit www.techcomm.nelson.com.

Expertise

Imagine that you are employed full-time and have decided to take a course at a local educational institution. First, name a course you might like to take. Next, by following Guideline 1, list the information you would include in a memo in which you ask your employer to pay your tuition and permit you to leave work early two days a week to attend class.

Online

Develop a list of questions that high school students might ask about taking your program. Then try to answer these questions by going to your school's website. How accessible are the answers? How complete are they from the viewpoint of high school students? How could the website be made more usable for these students?

Collaboration

Imagine that you have been hired to create a brochure or website that presents your department in a favourable light to first-year students and students who are thinking of changing to your program. Working with another student and following the guidelines in this chapter, generate a list of things that you and your partner would want to say. How would the two of you group and order this information?

Ethics

Think of a policy of your school or employer that you would like to change. Imagine that you are going to write a report recommending this change. After identifying all the stakeholders, tell how you could gain a complete understanding of the relevant concerns and values of each stakeholder or stakeholder group. Which stakeholders' concerns and values would be most difficult for you to learn about? Why? Would you have more difficulty listening sympathetically to some stakeholders than others? If so, who and why? How would you ensure that you treat these stakeholders ethically?

CHAPTER 10 Planning Your Persuasive Strategy

GUIDELINES

1. Learn—and focus on—your readers' goals and values
2. Address your readers' concerns and counterarguments
3. Show that your reasoning is sound
4. Organize to create a favourable response
5. Build an effective relationship with your readers
6. Adapt your persuasive strategies to your readers' cultural background
7. Ethics Guideline: Employ ethical persuasive techniques

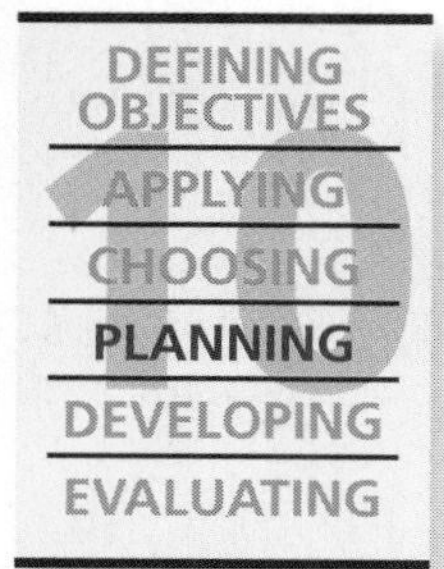

As explained in Chapter 1, a goal shared by all on-the-job writing is persuasiveness. In some communications, such as requests to management and proposals to clients, persuasion is of paramount importance. In other communications, such as instructions and technical reports, the persuasive aim is less obvious but still present. At the very least, writers want to create a favourable impression of themselves, their departments, and their companies.

Whether in the foreground or the background, your persuasive aims are important enough for you to develop specific plans for achieving them. To develop your expertise at planning persuasive strategies, this chapter begins with a practical discussion of how persuasion works and then presents seven guidelines for influencing your readers' thoughts, feelings, and actions in an effective and ethical manner.

HOW PERSUASION WORKS

WWW For additional chapter resources, visit Chapter 10 at www.techcomm.nelson.com.

According to researchers, the best way to influence how people think, feel, or act is to concentrate on shaping their attitudes (Petty & Cacioppo, 1986). At work, you will be concerned with your readers' attitudes toward a wide variety of subjects, such as products, policies, actions, and other people. As Chapter 2 explained, you may use your persuasive powers to change your readers' attitudes in any of the following ways:

Ways to change readers' attitudes

- **Reverse** an attitude you want your readers to abandon.
- **Reinforce** an attitude you want them to hold even more firmly.
- **Shape** their attitude on a subject about which they currently have no opinion.

Attitude is determined by the sum of thoughts about a topic.

What determines a person's attitude toward something? Researchers have found that it isn't a single thought or argument, but the *sum* of the various thoughts the person associates with the person or thing under consideration (Petty & Cacioppo, 1981). This is a critical point to remember as you plan your persuasive strategies. Consider the following example.

Jake's overall attitude is determined by his many thoughts related to the purchase.

The sum of his positive and negative thoughts will determine whether his overall attitude is positive or negative.

Jake is thinking about purchasing a particular bottled water system for his department. As he deliberates, a variety of thoughts cross his mind: the amount of money in his department's budget, the system's special features, the purity of the water, its appearance, his experience with another system provided by the same company, and his impression of the representative who sells the system. Each of these thoughts will make Jake's attitude more positive or more negative. He may think his budget has enough money (positive) or too little (negative); he may like the salesperson (positive) or detest the salesperson (negative); and so forth.

Some of these thoughts will probably influence Jake's attitude more than others. For instance, he may like the system's special features but that factor may be outweighed by reports that the system's upgrading method is very poor. If the sum of Jake's thoughts associated with the system is positive, he will have a favourable attitude toward it and may buy or lease it. If the sum is negative, he will have an unfavourable attitude and probably will not buy it.

Based on research on the psychology of persuasion and successful persuasive practices in the workplace, the following guidelines describe the major strategies for shaping readers' attitudes through your on-the-job writing.

GUIDELINE 1 Learn—and Focus on—Your Readers' Goals and Values

A powerful way to prompt your readers to experience favourable thoughts is to tell them how the idea, action, process, or product you advocate will help them accomplish some goal they want to achieve. For instance, if Jake's goal is to be sure that his department provides clients with great-tasting water, a writer could influence him to experience favourable thoughts toward a particular system by pointing out that it provides water with noticeably superior taste. If his goal is to keep costs low, a writer could influence him to view the system favourably by emphasizing that it is inexpensive to rent and operate.

Similarly, at work you can increase the persuasiveness of communications on any topic by focusing on the goals that guide your readers' decisions and actions. Here is a three-step process for doing so:

Process for developing persuasive strategies based on your readers' goals.

1. Identify your readers' goals.
2. Identify the ways that the ideas, actions, or other things you are recommending can assist your readers in achieving their goals.
3. Craft your communication to highlight the ways that the thing you are advocating would help your readers achieve some or all of their goals.

To identify your readers' goals, consider the following sources:

Sources of readers' goals

- Objectives of the organizations that employ your readers
- Values your readers derive from their organizations or from their personal lives
- Your readers' desires for achievement and growth

Organizational Goals

On the job, you will often write to coworkers in your company and to employees in other organizations. Because they are hired to advance their employer's interests, employees usually adopt their employer's goals as their own. Consequently, one way to identify your readers' goals is to examine the goals of the organization for which they work.

Look for specific objectives rather than general ones.

In any organization, some goals are general and some specific. At a general level, many organizations share the same goals: to increase revenue, operate efficiently, retain employees, and so on. However, each company also has its own unique goals. A company might be seeking to control 50 percent of its market, to have the best safety record in its industry, or to expand into another country in the next five years. Moreover, each department within a company has specific objectives. Thus, the research department aims to develop new and improved products, the marketing department seeks to identify new markets, and the accounting department strives to manage financial resources prudently. When identifying goals to serve as the basis for persuasion, choose specific goals over general ones.

Also, as you uncover an organization's various goals, focus on the ones that are most closely related to the specific idea, action, or process you are advocating. Then craft your communication to emphasize the ways that doing what you recommend will help your readers achieve the specific outcomes they desire.

Figure 10.1 shows a marketing brochure that uses organizational objectives to persuade its intended readers (beekeepers). Notice the boldfaced statements that proclaim how the system will increase profit potential, reduce costs, and reduce labour.

FIGURE 10.1

Brochure That Stresses Organizational Objectives

The Evolution of Beekeeping

BEE O SPHERE Technologies is a company that is advancing beekeeping practices through technology improvements and innovation. After years of studying bee behavior related to honey comb production, BEE O SPHERE is launching an innovative comb honey package BEE O PAC

This impressive little package gives the beekeeper compelling reasons to change. BEE O PAC's consumer-friendly size and fancy presentation attracts new customers to test comb honey and encourages repeat business.

BEE O PAC is made of food grade P.E.T. plastic formed to fit a standard 6 5/8" super.

There is a honey comb pattern embossed on the bottom of the pacs, eliminating the need for foundation wax.

The bees build their comb honey in the cavities. Look at the filled BEE O PAC from the front view (left) and back view (below)

Front Side

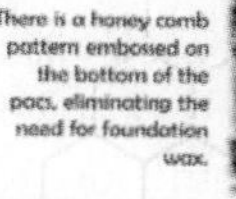

At harvest, the frames are separated into sixteen marketable units along preformed perforations.

Back Side

A lid is snapped on, and the combs can be sold without further handling. Wrap around labels provide additional security, and complete the package.

For the Bee, Beekeeper, Connoisseur

BEE O PAC

Reduce costs through this Impressive Packaging Innovation

Reduce labour up to 75% from hand cut method

No adaptation costs

No wax foundation

No messy cutting

No frames to clean

"THERE'S A BUZZ IN THE INDUSTRY ABOUT BEE O PAC!"

Values-Based Goals

Most companies have goals that involve social, ethical, and aesthetic values not directly related to profit and productivity. Some companies spell out these values in corporate credos; Figure 10.2 is an example. Even in companies that do not have an official credo, broad human and social values may provide an effective foundation for persuasion, especially when you are advocating a course of action that seems contrary to narrow business interests. For example, by pointing to the benefits to be enjoyed by a nearby community, you might be able to persuade your employer to strengthen its water pollution controls beyond what is required by environmental laws.

In addition, individual employees bring to work personal values that can influence their decisions and actions. Especially when writing to only one or a few readers, learn whether they have personal values that are related to the thing you are recommending.

Achievement and Growth Goals

Studies of employee motivation by Abraham Maslow (1970) and by Frederick Herzberg (1968) highlight another type of benefit you can use in your persuasive strategies. Both researchers confirmed that, as everyone knew, employees are motivated

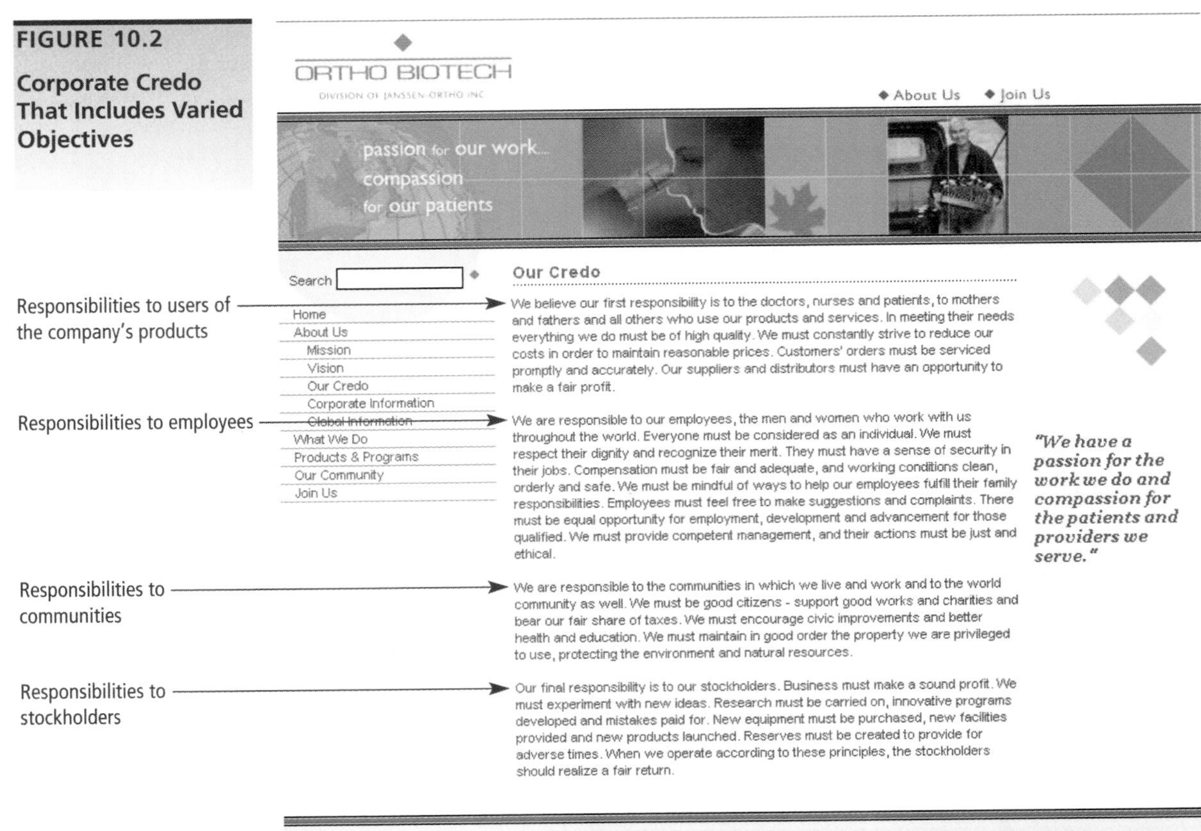

FIGURE 10.2

Corporate Credo That Includes Varied Objectives

by such considerations as pay and safe working conditions. However, the studies also found that after most people feel they have an adequate income and safe working conditions, they become less easily motivated by these factors. Consequently, such factors are called *deficiency needs:* they motivate principally when they are absent.

People seek more than pay from their jobs.

Once their deficiency needs are met—and even before—most people are motivated by so-called *growth needs,* including the desire for recognition, good relationships at work, a sense of achievement, personal development, and the enjoyment of work itself.

Many other studies have confirmed these findings. For instance, a study of supervisors showed that achievement, recognition, and personal relationships are much more powerful motivators than money (Munter, 1987). Another found that computer professionals are less motivated by salary or status symbols than by the opportunity to gain new skills and knowledge in their field (Warren, Roth, & Devanna, 1984).

These studies demonstrate that one powerful persuasive strategy is to show how the decisions or actions you are advocating will help your readers satisfy their desires for growth and achievement. When you request information or cooperation from a coworker, mention how much you value his or her assistance. When you evaluate a subordinate's performance, discuss accomplishments as well as shortcomings. When you ask someone to take on additional duties, emphasize the challenge and opportunities for achievement that lie ahead.

Recruiting materials often appeal to the readers' desires for achievement and growth.

Figure 10.3 shows a Procter & Gamble website designed to recruit new employees. Notice the many ways it appeals to graduates by emphasizing the opportunities they would have with the company to assume responsibility, take on challenges, and grow professionally in this promotion-from-within organization.

GUIDELINE 2 Address Your Readers' Concerns and Counterarguments

When people read, they not only pay attention to the writer's statements but also generate their own thoughts. For example, as Jake reads a brochure about a water system he may purchase, he responds to a description of some special feature by thinking, "Hey, that sounds useful" (positive) or "I don't see what good that will do me" (negative). Some of a reader's thoughts even arise independently of any specific statement the writer makes. For instance, while Jake is reading the brochure, he may remember a comment about the machine made by a manager whose department already owns one.

Readers' self-generated thoughts are very influential.

Whether readers' self-generated thoughts are favourable or unfavourable, research shows that they can have a greater influence on readers' attitudes than any point made in the communication itself (Petty & Cacioppo, 1986). Consequently, when you are trying to persuade, avoid saying anything that might prompt negative thoughts. And try to counteract any negative thoughts that seem likely to arise in your readers' minds despite your best efforts.

Avoid inspiring negative thoughts.

One way to avoid arousing negative thoughts is to answer all the important questions your readers are likely to ask while reading your communication. Most of these questions will simply reflect the readers' efforts to understand your position or proposal thoroughly. "What are the costs of doing as you suggest?" "What do other people who have looked into this matter think?" "Do we have employees with the education, experience, or talent to do what you recommend?" If you answer such questions satisfactorily, they

FIGURE 10.3

Careers Section of a Corporate Website That Stresses Growth Needs

This Careers section of a corporate website appeals to the desire for achievement and growth of potential employees.

Opportunities for professional development:
- Individual coaching and training.
- Development of general management and leadership skills.
- Chance to work within different business units, functions, and countries.

Opportunities for achievement:
- Promise of substantial responsibility.
- Policy of promoting from within.

The display of two rankings from national periodicals highlights the employer's reputation as a top employer in Canada.

create no problem whatever. But if you ignore them, they may prompt negative thoughts. Readers may think you have overlooked some important consideration or have decided to ignore evidence that might weaken your position. The remedy, quite simply, is to anticipate these questions and answer them satisfactorily.

Address objections your readers might raise.

In addition to asking questions, readers may generate arguments against your position. For example, if you say there is a serious risk of injury to workers in a building used by your company, your readers may think to themselves that the building has been used for three years without an accident. If you say, a new procedure will increase productivity by at least 10 percent, your readers may think to themselves that some of the data on which you are basing your estimate are inaccurate. Readers are especially likely to generate counterarguments when you attempt to reverse their attitudes. Research shows that people resist efforts to persuade them that their attitudes are incorrect (Petty & Cacioppo, 1986).

To deal effectively with counterarguments, you must offer some reason for relying on your position rather than on the opposing position. For example, imagine that you are proposing the purchase of a certain piece of equipment and that you predict your readers will object because it is more expensive than a competitor's product that they believe to be its equal. You might explain that the competitor's product, though less expensive to purchase, is more expensive to operate and maintain. Or you might point out that the equipment you are recommending has certain capabilities that make it well worth the additional cost.

Imagine that you are addressing highly skeptical readers.

To anticipate the questions and counterarguments you must address, follow Chapter 2's advice for identifying readers' questions, but focus specifically on questions that skeptical readers might pose. Another way of anticipating readers' counterarguments is to learn their reasons for holding their present attitudes. Why do they do things the way they do? What arguments do they feel most strongly support their present position? At times, you may have to show the weaknesses of your readers' reasons in order to demonstrate the superiority of your own. But do so diplomatically. The last thing you want to do is arouse the defensiveness of the people you are trying to persuade.

In your efforts to address your readers' concerns and counterarguments, be careful not to mention objections your readers are unlikely to raise on their own. There's always the chance that they will find the objection more persuasive than your rebuttal.

Figure 10.4 shows a letter in which the marketing director of a radio station attempts to persuade an advertising agency to switch some of its ads to her station. Notice how she anticipates and addresses possible counterarguments.

GUIDELINE 3 Show That Your Reasoning Is Sound

A third strategy for writing persuasively is to show that your reasoning is sound. In most working situations, one of the most favourable thoughts your readers could possibly have is, "Yeah, that makes sense"—and one of the most unfavourable is, "Hey, there's a flaw in your reasoning."

Sound reasoning is especially important when you are trying to influence your readers' decisions and actions. In addition to identifying potential benefits that will appeal to your readers, you also must persuade them that the decision or action you advocate will actually bring about these benefits—that the proposed new equipment really will reduce costs enough to pay for itself in just eighteen months or that the product modification you are recommending really will boost sales 10 percent in the first year.

Sound reasoning is also essential when you are describing conclusions you have reached after studying a group of facts, such as the results of a laboratory experiment or

FIGURE 10.4

Letter That Addresses Counterarguments

LIFE 101.1 CLFN

545 Falls Road, Ste. 1100
Barrie, ON L4M 2B2

August 17, 2007

Mr. David Nordstrom
Vice President
Focus Marketing
4840 Church St., Ste. 200
Peterborough, ON
K9J 6P2

Dear Mr. Nordstrom:

What a lucky coincidence that we should have met at Chances R last week. As I told you then, I believe you can improve the advertising service that you provide many of your clients by switching to our station some of the radio advertising that you presently place for your clients with 88.1 Fun (CFUN) and 106.9 The Fox (CFOX). Since we met, I have put together some figures.

The primary advantage of advertising with us is that you will extend the size of your audience considerably. As an example, consider the reach of the campaign you are running for Fuller's Furniture. You currently buy 45 spots over four weeks on the other two stations. According to Arbitron market analysis, you thereby reach 63% of your target market—women in the 25–54 demographic group—an average of 16 times per week. However, there is a lot of overlap in the audiences of those two adult contemporary stations: many listeners switch back and forth between them.

In contrast, we serve a very distinct audience with our contemporary Christian offering at LIFE 101.1 (CLFN). By using only 30 spots on those two stations and giving 15 to us, you would increase your reach to 73% of the 25–54 group. It's true, of course, that this change would reduce from 16 to 13 the number of times the average listener hears your message. However, this is an increase of 10% in the portion of your target audience that you would reach.

Writer states a counter-argument that might occur to the reader.

Writer specifies a benefit to the reader that will outweigh the objection.

(continues)

consumer survey. In such cases, you must persuade your readers that your conclusions are firmly based on the facts.

Notice that in each of the situations just mentioned (as well as in any other you might encounter on the job), you must not only use sound reasoning, but also convince your readers that your reasoning is sound. The ability to do so is one of the most valuable writing skills you can develop. The following discussion, based largely

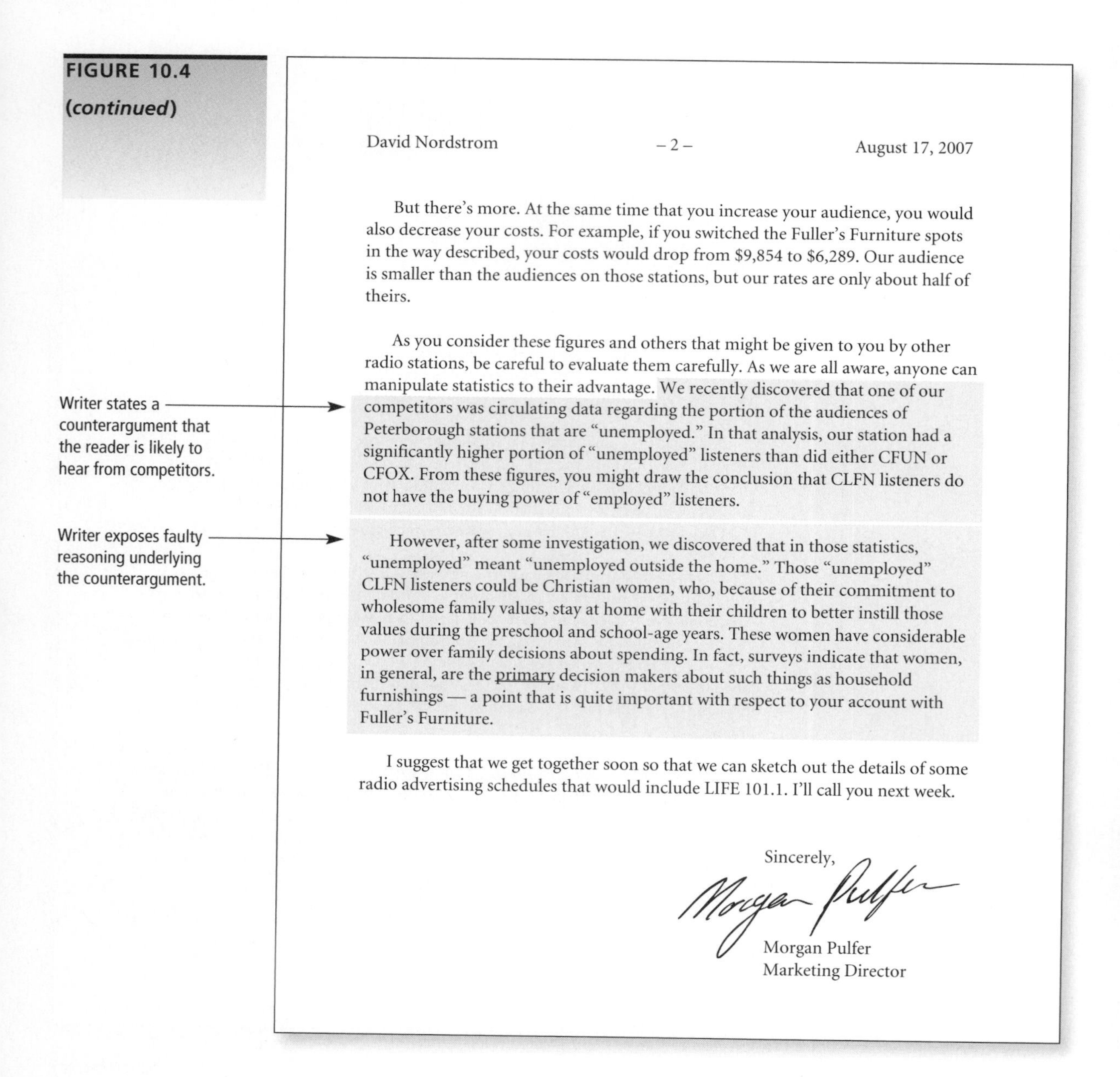

FIGURE 10.4 (*continued*)

David Nordstrom –2– August 17, 2007

But there's more. At the same time that you increase your audience, you would also decrease your costs. For example, if you switched the Fuller's Furniture spots in the way described, your costs would drop from $9,854 to $6,289. Our audience is smaller than the audiences on those stations, but our rates are only about half of theirs.

As you consider these figures and others that might be given to you by other radio stations, be careful to evaluate them carefully. As we are all aware, anyone can manipulate statistics to their advantage. We recently discovered that one of our competitors was circulating data regarding the portion of the audiences of Peterborough stations that are "unemployed." In that analysis, our station had a significantly higher portion of "unemployed" listeners than did either CFUN or CFOX. From these figures, you might draw the conclusion that CLFN listeners do not have the buying power of "employed" listeners.

However, after some investigation, we discovered that in those statistics, "unemployed" meant "unemployed outside the home." Those "unemployed" CLFN listeners could be Christian women, who, because of their commitment to wholesome family values, stay at home with their children to better instill those values during the preschool and school-age years. These women have considerable power over family decisions about spending. In fact, surveys indicate that women, in general, are the primary decision makers about such things as household furnishings — a point that is quite important with respect to your account with Fuller's Furniture.

I suggest that we get together soon so that we can sketch out the details of some radio advertising schedules that would include LIFE 101.1. I'll call you next week.

Sincerely,

Morgan Pulfer
Marketing Director

on the work of Stephen Toulmin, will help you master this skill (Toulmin, Rieke, & Janik, 1984).

How Reasoning Works

To accept your reasoning as sound, your readers must feel that they understand the following elements: your *claim,* your *evidence,* and your *line of reasoning.* The following diagram illustrates the relationship among these three elements:

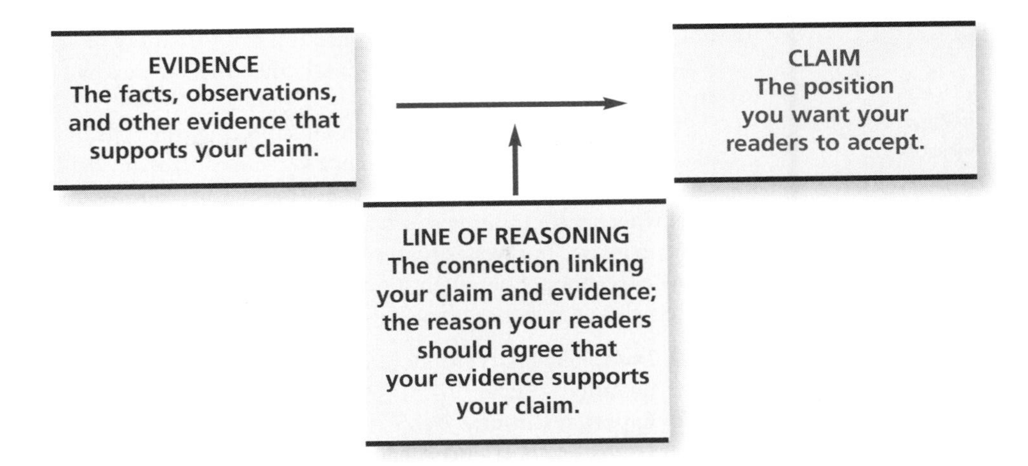

Imagine that you work for a company that manufactures fabric. You have found out that one of your employer's competitors recently increased its productivity by using software to automate some of its manufacturing processes. If you were to recommend that your employer adopt similar technology, your argument could be diagrammed as follows:

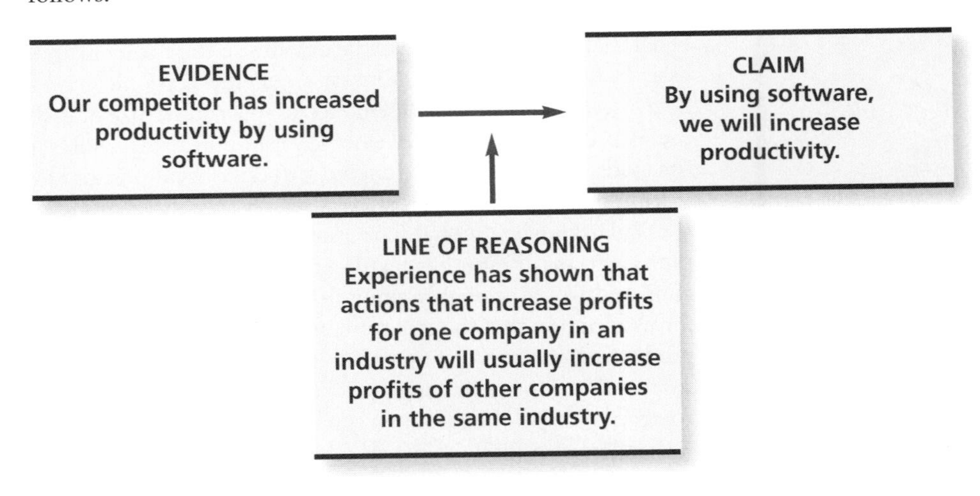

To accept your claim, readers must be willing to place their faith in both your evidence and your line of reasoning. The next two sections offer advice about how to persuade them to do so.

Present Sufficient and Reliable Evidence

First you must convince your readers that your evidence is both sufficient and reliable.

How to provide sufficient evidence.

To provide *sufficient* evidence, you must furnish all the details your readers are likely to want. For instance, in the example of the textile mills, your readers would probably regard your evidence as skimpy if you produced only a vague report that the other company had somehow used technology and saved some money. They would want to know how the company had used hardware and software; how much money had been saved; what processes had been adopted; whether the savings had justified the cost of the equipment; and so forth.

How to provide reliable evidence.

To provide *reliable* evidence, you must produce the type of evidence your readers are likely to accept. The type of evidence varies greatly from field to field. For instance, in science and engineering, certain experimental procedures are widely accepted as reliable, whereas common wisdom and unsystematic observation usually are not. In contrast, in many business situations, personal observations and anecdotes provided by knowledgeable people often are accepted as reliable evidence. However, the following three types of evidence are widely accepted:

- **Data.** Readers typically respond very favourably to claims that are supported by quantitative data.
- **Expert testimony.** People with advanced education, first-hand knowledge, or extensive experience related to a topic are often credited with special understanding and insight.
- **Examples.** Specific instances can effectively support general claims.

Explicitly Justify Your Line of Reasoning Where Necessary

To argue persuasively, you must not only present sufficient and reliable evidence, but also convince your readers that you are using a valid line of reasoning to link your evidence to your claim. Writers often omit any justification of their line of reasoning in the belief that the justification will be obvious to their readers. In fact, that is sometimes the case. In the construction industry, for example, people generally agree that if an engineer uses the appropriate formulas to analyze the size and shape of a bridge, the formulas will accurately predict whether or not the bridge will be strong enough to support the loads it must carry. The engineer doesn't need to justify the formulas themselves.

Avoid false assumptions.

In many cases, however, readers search aggressively for a weak line of reasoning. In particular, they are wary of arguments based on false assumptions. For example, they may agree that if another textile mill like yours saved money by computerizing, then your mill would probably enjoy the same result; your readers, however, may question the assumption that the other mill is truly like yours. Maybe it makes a different kind of product or employs a different manufacturing process. If you think your readers will suspect that you are making a false assumption, offer whatever evidence or explanation you can to dispel their doubt.

Avoid overgeneralizing.

Readers also look for places where writers have overgeneralized by drawing broad conclusions from too few specific instances. If you think your readers will raise such an objection to your argument, mention additional cases. Or, narrow your conclusion to better match the evidence you have gathered. For example, instead of asserting that your claim applies to all textile companies in all situations, argue that it applies to specific companies or specific situations.

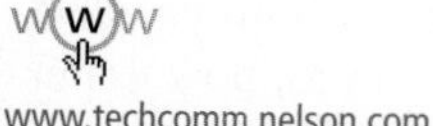

www.techcomm.nelson.com

GUIDELINE 4 Organize to Create a Favourable Response

Organization can affect persuasiveness.

The way you organize a communication may have almost as much effect on its power to persuade as what you say. That point was demonstrated by researchers Sternthal, Dholakia, and Leavitt (1978), who presented two groups of people with different versions of a talk urging that a federal consumer protection agency be established. One version *began* by saying that the speaker was a highly credible source (a lawyer who graduated from Harvard and had extensive experience with consumer issues); the other version *ended* with that information.

Among people initially opposed to the speaker's recommendation, those who learned about his credentials at the beginning responded more favourably to his arguments than did those who learned about his credentials at the end. Why? This outcome can be explained in terms of two principles you learned in Chapter 1. First, people react to persuasive messages moment by moment. Second (and here's the key point), their reactions in one moment will affect their reactions in subsequent moments. In this experiment, those who learned of the speaker's credentials before hearing his arguments were relatively open to what he had to say. But those who learned of the speaker's credentials only at the end worked more vigorously at creating counterarguments as they heard each of his points. After those counterarguments had been recorded in memory, they could not be erased simply by adding information about the speaker's credibility.

Thus, it's not only the array of information that is critical in persuasion, but also the way readers process the information. The following sections suggest two strategies for organizing to elicit a favourable response: Choose carefully between direct and indirect organizational patterns, and create a tight fit among the parts of your communication.

Choose Carefully between Direct and Indirect Organizational Patterns

The direct pattern goes directly to the main point.

As you learned in Chapter 9, the most common organizational pattern at work begins by stating the bottom line—the writer's main point. Communications organized this way are said to use a *direct pattern* of organization because they go directly to the main point and only afterward present the evidence and other information related to it. For example, in a memo recommending the purchase of a new ergonomic chair, you might begin with the recommendation and then explain why you think the new chair is desirable.

The indirect pattern delays the main point.

The alternative is to postpone presenting your main point until you have presented your evidence or other related information. This is called the *indirect pattern* of organization. For example, in a memo recommending the purchase of a new ergonomic chair, you might first explain the health problems created by the current office seating, withholding until later your recommendation that new seating be purchased.

To choose between the direct and indirect organizational patterns, focus your attention on your readers' all-important initial response to your message. The direct pattern will start your readers off on the right foot when you have good news to convey: "You're hired," "I've figured out a solution to your problem," or something similar. By starting with the good news, you put your readers in a favourable frame of mind as they read the rest of your message.

Use the direct pattern when expecting a favourable response.

The direct pattern also works well when you are offering an analysis or recommending a course of action that you expect your readers to view favourably—or at least objectively—from the start. Leah is about to write such a memo, in which she will recommend a new system for managing the warehouses for her employer, a company that manufactures hundreds of parts used to drill oil and gas wells. Leah has chosen to use the direct pattern shown in the left-hand column of Figure 10.5 (page 254). This is an appropriate choice because her readers (upper management) have expressed

FIGURE 10.5 **Comparison of Direct and Indirect Organizational Patterns for Organizing Leah's Memo**

The direct pattern presents the recommendation first.

Direct Pattern	Indirect Pattern
I. Leah presents her recommended strategy	I. Leah discusses the goals of the present system from the *reader's point of view*
II. Leah explains why her way of warehousing is superior to the present way	II. Leah discusses the ways in which the present system does and does not achieve the readers' goals
III. Leah explains in detail how to implement her system	III. Leah presents her recommended strategy for achieving those goals more effectively, focusing on the ways her recommendation can overcome the shortcomings of the present system
	IV. Leah explains in detail how to implement her system

dissatisfaction with the present warehousing system. Consequently, she can expect a favourable reaction to her initial announcement that she has designed a better system.

The indirect pattern delays recommendation; it's for use where the reader may react unfavourably.

The direct pattern is less effective when you are conveying information your readers might view as bad, alarming, or threatening. Imagine, for example, that Leah's readers are the people who set up the present system and that they believe it to be working well. If Leah begins her memo by recommending a new system, she might put her readers immediately on the defensive because they might feel she is criticizing their competence. Then she would have little hope of receiving an open and objective reading of her supporting information. By using an indirect pattern, however, Leah can *prepare* her readers for her recommendation by first getting them to agree that it might be possible to improve on the present system. The right-hand column of Figure 10.5 shows an indirect pattern she might use.

You may wonder why you shouldn't simply use the indirect pattern all the time. It presents the same information as the direct organization (plus some more), and it avoids the risk of inciting a negative reaction at the outset. The trouble is that this pattern frustrates the readers' desire to learn the main point first.

In sum, the choice between direct and indirect patterns of organization can greatly affect the persuasiveness of your communications. To choose, you need to follow the basic strategy suggested throughout this book: Think about your readers' moment-by-moment reactions to your message.

Create a Tight Fit among the Parts of Your Communication

When you organize a communication, you can also strengthen its persuasiveness by ensuring that the parts fit together tightly. This advice applies particularly to longer communications, where the overall argument often consists of two or more subordinate arguments. The way to do this is to review side by side the claims made in the various parts of a communication.

Every part must match up with all the other parts.

For example, imagine that you are writing a proposal. In an early section, you describe a problem your proposed project will solve. Here, your persuasive points are that a problem exists and that the readers should view it as serious. In a later section, you describe the project you propose. Check to see whether this description tells how the project will address each aspect of the problem that you described. If it doesn't, either the discussion of the problem or the discussion of the project needs to be revised so the two match up. Similarly, your budget should include expenses that are clearly related to the project you describe, and your schedule should show when you will carry out each activity necessary to complete the project successfully.

Of course, the need for a tight fit applies not only to proposals but also to any communication whose various parts work together to affect your readers' attitudes. Whenever you write, think about ways to make the parts work harmoniously together in mutual support of your overall position.

GUIDELINE 5 Build an Effective Relationship with Your Readers

One of the most important factors influencing the success of a persuasive communication is how your readers feel about you. You will remember from Chapter 1 that you should think of the communications you prepare at work as interpersonal interactions. If your readers feel well disposed toward you, they are likely to consider your points openly and without bias. If they feel irritated, angry, or otherwise unfriendly toward you, they may immediately raise counterarguments to every point you present, making it extremely unlikely that you will elicit a favourable reaction, even if all your points are clear, valid, and substantiated. Good points rarely win the day in the face of bad feelings.

The following sections describe two ways you can present yourself to obtain a fair—or even a favourable—hearing from your readers at work: Present yourself as a credible person, and present yourself as a friend, not a foe.

Present Yourself as a Credible Person

The more credible you are, the more persuasive you are.

Your credibility is your readers' beliefs about whether or not you are a good source for information and ideas. If people believe you are credible, they will be relatively open to what you say. They may even accept your judgments and recommendations without inquiring very deeply into your reasons for making them. If people do not find you credible, they may refuse to give you a fair hearing no matter how soundly you state your case.

To see how much your readers' perceptions of you can affect the way they respond to your message, consider the results of the following experiment. Researchers H. C. Kelman and C. I. Hovland (1953) recorded a speech advocating lenient treatment of young offenders. Before playing their tape for one group, they identified the speaker as an ex-delinquent out on bail. They told another group that the speaker was a judge. Only 27 percent of those who heard the "delinquent's" talk responded favourably to it, while 73 percent of those who heard the "judge's" talk responded favourably.

Several factors that influence credibility.

Researchers have conducted many studies on the factors that affect an audience's impression of a person's credibility. In summarizing this research, Robert Bostrom (1981) has identified five key factors.

Strategies for Building Credibility

- **Expertise.** Expertise is the knowledge and experience that readers believe you possess that is relevant to the topic. *Strategies you can use:*
  - Mention your credentials.
  - Demonstrate a command of the facts.
  - Avoid oversimplifying.
  - Mention or quote experts so their expertise supports your position.
- **Trustworthiness.** Trustworthiness depends largely on your readers' perceptions of your motives. If you seem to be acting from self-interest, your credibility is low; if you seem to be acting objectively or for goals shared by your readers, your credibility is high. *Strategies you can use:*
  - Stress values and objectives that are important to your readers.
  - Avoid drawing attention to personal advantages to you.
  - Demonstrate knowledge of the concerns and perspectives of others.
- **Group membership.** You will have more credibility if you are a member of the readers' own group or a group admired by your readers. *Strategies you can use:*
  - If you are associated with a group admired by your readers, allude to that relationship.
  - If you are addressing members of your own organization, affirm that relationship by showing that you share the group's objectives, methods, and values.
  - Use terms that are commonly employed in your organization.
- **Dynamic appeal.** An energetic, enthusiastic person has much higher credibility than a passive, guarded one. *Strategies you can use:*
  - State your message confidently and directly.
  - Show enthusiasm for your ideas and subject.
- **Power.** For example, simply by virtue of her position, a boss acquires some credibility with subordinates. *Strategies you can use:*
  - If you are in a position of authority, identify your position if your readers don't know it.
  - If you are not in a position of authority, associate yourself with a powerful person by quoting the person or saying that you consulted with him or her or were assigned the job by that individual.

Figure 10.6 shows a letter that puts many of these strategies into action. In it, a supplier attempts to persuade a consultant to let it make a sales presentation.

FIGURE 10.6

Sales Letter That Relies Heavily on Building Credibility

Embassy Meats Inc.

92254 Kingston Road • Scarborough, ON • M4K 2L2

30 November 2007

Ms. Nazira Kabbari
President and Owner
The Wedding Planners
1850 Montreal Road
Vanier, ON K7L 9Y3

Dear Ms. Kabbari:

Our firm is very interested in presenting its Halal product credentials to you for consideration.

While you may never have heard about Embassy Meats, chances are you're familiar with our products. We supply Halal products to the catering services of the Capital University cafeterias and to the Muslim Cultural Centre in Centretown. We are also the supplier to the Embassy of the Islamic Republic of Iran.

Writer emphasizes experience in reader's city.

But more important to you as President of The Wedding Planners is Embassy Meats' experience as a Halal supplier to special events catering firms. That experience includes work for such clients as Weddings International (Toronto), Your Special Day Planners (Kingston), and Muslim Events Planning (Oshawa).

Writer demonstrates experience with reader's type of company.

Our mission is to expand our client base in the Ottawa market. To that end, our firm is interested in presenting our customer list, testimonials, product lists, and product samples for your review. Our products include a broad range of Halal options, including chicken, turkey, and beef hand-slaughtered by Muslims using the Zabiha method. As a result, you can trust our certification from the Islamic Food and Nutrition Council of America (IFANCA).

Writer cites evidence of his company's standing.

We're confident that you'll like what you see and taste. But we're also confident you'll recognize the possibilities that an Embassy Meats/The Wedding Planners relationship would afford.

Writer expresses enthusiasm and confidence.

(continues)

Present Yourself as a Friend, Not a Foe

Readers may see you as an adversary even when you wish to help.

Psychologist Carl Rogers (1952) identified another important strategy in fostering open, unbiased communication: Reduce the sense of threat that people often feel when others are presenting ideas to them. According to Rogers, people are likely to feel threatened even when you make *helpful* suggestions. As a result, your readers may see you as an adversary even though you don't intend at all to be one.

Ms. Nazira Kabbari 30 November 2007 – 2 –

Writer closes on goal-oriented energetic note.

Our goal is to apply our capabilities to the catering challenges in the Ottawa market in the upcoming year. Please let me know how I might proceed in presenting our qualifications and products to you. I can be contacted at (416) 885-2798.

Thank you for your consideration of Embassy Meats Inc.

Sincerely,

Mahmoud El-Fahmawi

Mahmoud El-Fahmawi
President and Owner

Here are four methods, based on the work of Rogers, that you can use to present yourself as non-threatening to your readers.

Strategies for Presenting Yourself as a Friend, Not a Foe

- Praise Your Readers
  - When writing to an individual, mention one of his or her recent accomplishments.
  - When writing to another organization, mention something it prides itself on.
  - When praising, be sure to mention specifics. General praise sounds insincere.
- Present Yourself as Your Readers' Partner
  - Identify some personal or organizational goal of your readers that you will help them attain.
  - If you are already your readers' partner, mention that fact and emphasize the goals you share.
- Show That You Understand Your Readers
  - Even if you disagree with your readers, state their case fairly.
  - Focus on areas of agreement.
- Maintain a Positive and Helpful Stance
  - Present your suggestions as ways of helping your readers do an even better job.
  - Avoid criticizing or blaming.

To see how to apply this advice, consider the following situation. Marg Lakwurtz is a regional manager for a company that employs nearly 1,000 students in many cities during the summer months to paint houses and other large structures. Each student is required at the beginning of the summer to make a $100 deposit, which is returned at the end of the summer. Students complain that the equipment they work with isn't worth anywhere near $100 so the deposit is unjustified. Furthermore, they believe that the company is simply requiring the deposit so it can invest the money during the summer. The company keeps all the interest on the deposits, not giving any to the students when it returns their deposits at the end of the summer.

Although she is presenting complaints from student employees, Marg wisely frames her memo in a non-threatening way.

Marg agrees that the students have a good case, so she wants to suggest a change. She could do that in a negative way by telling the company that it should be ashamed of its greedy plot to profit by investing other people's money. Or she could take a positive approach, presenting her suggested change as a way the company can better achieve its own goals. In the memo shown in Figure 10.7 (page 260), she takes the latter course. Notice how skillfully she presents herself as her readers' partner.

GUIDELINE 6 Adapt Your Persuasive Strategies to Your Readers' Cultural Background

What's persuasive in one culture may not be in another.

The persuasive strategies you've just read about may need to be adjusted in light of the cultural background of your readers. For example, readers' notions of benefits vary from country to country. In Canada, companies selling facial moisturizer emphasize the product's anti-aging properties and may even allude to the increased self-esteem that will result from its use. In contrast, companies selling the same facial moisturizer in China promote the fair skin that their product will yield in its customers. However, to sell the same moisturizer in India, the company must persuade its readers that the product will not only yield fair skin but also keep a user's skin smooth the entire day.

Also, people from different cultures have different views about what constitutes a good reason for taking a particular action. Where Canadians base their arguments on facts, a Chinese audience might consider a personal appeal by a leader, such as a celebrity or a political personage, to be more persuasive. Moreover, an Indian audience would consider a religious appeal, or an appeal based on superstition, to be a more valid justification for action.

Similarly, what readers consider an appropriate role for a writer differs from culture to culture. For example, in Canada and in India, writers often "tell it like it is." But in China, a writer would avoid causing the reader or anyone else to lose face by such an approach. Instead, a writer might use an analogy to get the same message across to the audience.

These cultural differences mean that you must study the cultural norms of your readers whenever you hope to persuade someone from another culture through your communication.

GUIDELINE 7 Ethics Guideline: Employ Ethical Persuasive Techniques

Nowhere are the ethical dimensions of on-the-job writing more evident than when you are trying to persuade other people to take a certain action or adopt a certain attitude. Here are four guidelines for ethical persuasion that you might want to keep in mind: Don't mislead, don't manipulate, open yourself to your readers' viewpoint, and argue from human values.

FIGURE 10.7

Memo in Which the Writer Establishes a Partnership with the Reader

PREMIUM PAINTING COMPANY

February 12, 2007

TO: Martin Sneed

FROM: Marg Lakwurtz

RE: Improving Worker Morale

Marg opens by praising the reader.

Now that I have completed my first year with the company, I have been thinking over my experiences here. I have enjoyed the challenges that this unique company offers, and I've been very favourably impressed with our ability to find nearly 1,000 temporary workers who are willing to work so energetically and diligently for us.

In fact, it's occurred to me that the good attitude of the students who work for us is one of the indispensable ingredients in our success. If they slack off or become careless, our profits could drop precipitously. If they become careless, splattering and spilling paint for instance, we would lose much of our profits in cleaning up the mess.

Marg presents herself as the reader's partner in the mutual goal of raising worker morale.

Because student morale is so crucial to us, I would like to suggest a way of raising it even further—and of ensuring that it won't droop. From many different students in several cities, I have heard complaints about the $100 security deposit we require them to pay before they begin work. They feel that the equipment they work with is worth much less than that, so that the amount of this security deposit is very steep. Furthermore, there is a widespread belief that the company is actually cheating them by taking their money, investing it for the summer, and keeping the profits for itself.

Marg uses "we" to reinforce the sense of partnership.

Because the complaints are so widespread and because they can directly affect the students' sense of obligation to the company, I think we should try to do something about the security deposit. First we could reduce the deposit. If you think that would be unwise, we could give the students the interest earned on their money when we return the deposits to them at the end of the summer.

FIGURE 10.7
(continued)

Martin Sneed
February 12, 2007
Page 2

Marg shows that she understands the reader's perspective.

I realize that it would be difficult to calculate the precise amount of interest earned by each student. They begin work at different times in May and June, and they end at different times in August and September. However, we could establish a flat amount to be paid each student, perhaps basing it on the average interest earned by the deposits over a three-month period.

Marg counters a possible objection by making a positive suggestion.

Marg states and addresses another counterargument.

I'm sure that it has been somewhat beneficial for the company to have the extra income produced by the deposits. But the amount earned is still rather modest. The additional productivity we might enjoy from our students by removing this irritant to them is likely to increase our profits by much more than the interest paid.

Marg closes on a positive note, re-emphasizing her desire to work as her reader's partner.

If you would like to talk to me about this idea or about the feelings expressed by the students, I would be happy to meet with you.

Don't Mislead

When you are writing persuasively, respect your readers' right to evaluate your arguments in an informed and independent way. If you mislead your readers by misstating facts, using intentionally ambiguous expressions, or arguing from false premises, you deprive your readers of their rights.

Don't Manipulate

The philosopher Immanuel Kant originated the enduring ethical principle that we should never use other people merely to get what we want. Whenever we try to influence our readers, the action we advocate should advance their goals as well as our own.

Under Kant's principle, for instance, it would be unethical to persuade readers to do something that would benefit us but harm them. High-pressure sales techniques are unethical because their purpose is to persuade consumers to purchase something they may not need or even want. Persuasion is ethical only if it will lead our readers to get something they truly desire.

Open Yourself to Your Readers' Viewpoint

To keep your readers' goals and interests in mind, you must be open to their viewpoint. Instead of regarding their counterarguments as objections you must overcome, try to understand what lies behind their concerns. Consider ways of modifying your original ideas to take your readers' perspective into account.

In this way, rather than treating your readers as adversaries, accept them as your partners in a search for a course of action acceptable to you all (Lauer, 1994). Management experts call this the search for a "win-win" situation—a situation in which all parties benefit (Covey, 1989).

Argue from Human Values

Whenever you feel that human values are relevant, don't hesitate to introduce them when you are writing to persuade. Many organizations realize the need to consider these values when making decisions. In other organizations, human values are sometimes overlooked—or even considered to be inappropriate topics—because the employees focus too sharply on business objectives. However, even if your arguments based on human values do not prevail, you will have succeeded in introducing a consideration of these values into your working environment. Your action may even encourage others to follow your lead.

CONCLUSION

WWW To download a planning guide for persuasive strategies, visit Chapter 10 at www.techcomm.nelson.com.

This chapter has focused on writing persuasively. As you can see, nearly every aspect of your communication affects your ability to influence your readers' attitudes and actions. Although the guidelines in this chapter will help you write persuasively, the most important persuasive strategy of all is to keep in mind your readers' needs, concerns, values, and preferences whenever you write.

A guide you can use when planning your persuasive strategies is available on this book's website.

EXERCISES

For additional exercises, visit www.techcomm.nelson.com.

Expertise

1. Find a persuasive communication that contains at least 25 words of text. It may be an advertisement, marketing letter, memo from school or work, or a webpage. What persuasive strategies are used in the text and any images that are included? Are the strategies effective for the intended audience? Are all of them ethical? What other strategies might have been used? Present your responses in the way requested by your professor.

2. Do the "Unsolicited Recommendation" assignment given in Appendix C.

Online

Study the website of a company or organization related to your career. What attitudes toward the organization and its products or services does the site promote? What persuasive strategies are used in the text and images in order to persuade readers to adopt the desired attitudes? How effective are these strategies? Are all of them ethical?

Collaboration

Working with one or two other students, analyze the letter shown in Figure 10.8 on pages 264–265, identifying its strengths and weaknesses. Relate your points to the guidelines in this chapter. The following paragraphs describe the situation in which the writer, Scott Houck, is writing.

Before going to college, Scott worked for a few years at Thompson Textiles. In his letter, he addresses Thompson's Executive Vice President, Georgiana Stroh. He is writing because in college he is learning many things that made him think Thompson would benefit if its managers were better educated in management techniques. Thompson Textiles could enjoy these benefits, Scott believes, if it would offer management courses to its employees and if it would fill job openings at the managerial level with postsecondary graduates. However, if Thompson were to follow Scott's recommendations, it would have to change its practices considerably. Thompson has never offered courses for its employees and has long sought to keep payroll expenses low by employing people without a postsecondary education, even in management positions. (In a rare exception to this practice, the company has guaranteed Scott a position after he graduates.)

To attempt to change the company's policies, Scott decided to write a letter to one of the most influential people on its staff, Ms. Stroh. Unfortunately for Scott, throughout the three decades that Stroh has served as an executive officer at Thompson, she has consistently opposed company-sponsored education and the hiring of postsecondary graduates. Consequently, she has an especially strong motive for rejecting Scott's advice: She is likely to feel that, if she agreed that Thompson's educational and hiring policies should be changed, she would be admitting that she had been wrong all along.

Ethics

Find an online or print communication you feel is unethical. Identify the specific elements in the text images that you feel are untrue, misleading, or manipulative. Present your analysis in the way your professor requests.

FIGURE 10.8

Letter for Collaboration Exercise

58043 Residence Levoux
College Royale
Lennoxville, QC H4Z 1X4
April 16, 2007

Georgiana Stroh
Executive Vice President
Thompson Textiles Incorporated
1010 rue Notre Dame
Hull, QC H3T 1V8

Dear Ms. Stroh:

As my second year draws to a close, I am more and more eager to return to our company, where I can apply my new knowledge and skills. Since our recent talk about the increasingly stiff competition in the textile industry, I have thought quite a bit about what I can do to help Thompson continue to prosper. I have been going over some notes I have made on the subject, and I am struck by how many of the ideas stemmed directly from the courses I have taken here at College Royale.

Almost all of the notes featured suggestions or thoughts I simply didn't have the knowledge to consider before I went to college! Before I enrolled, I, like many people, presumed that operating a business required only a certain measure of commonsense ability—that almost anyone could learn to guide a business down the right path with a little experience. However, I have come to realize that this belief is far from the truth. It is true that many decisions are common sense, but decisions often only appear to be simple because the entire scope of the problem or the full ramifications of a particular alternative are not well understood. A path is always chosen, but often is it the best path for the company as a whole?

In retrospect, I appreciate the year I spent supervising the Eaton Boulevard Plant because the experience has been an impetus to actually learn from my classes instead of just receiving grades. But I look back in embarassment upon some of the decisions I made and the methods I used then. I now see that my previous work in our factories and my experience in cadets did not prepare me as well for that position as I thought they did. My mistakes were not so often a poor selection among known alternatives, but were more often sins of omission. For example, you may remember that we were constantly running low on packing cartons and that we sometimes ran completely out, causing the entire line to shut down. Now I know that instead of haphazardly placing orders for a different amount every time, we should have used a forecasting model to determine and establish a reorder point and a reorder quantity. But I was simply unaware of many of the techniques available to me as a manager.

I respectfully submit that many of our supervisory staff are in a similar situation. This is not to downplay the many contributions they have made to the company. Thompson can directly attribute its prominent position in the industry to the devotion and hard work of these people. But very few of them have more than a high school education or have read even a single text on management skills. We have always counted on our supervisors to pick up their management

FIGURE 10.8

Letter for Collaboration Exercise (*continued*)

Georgiana Stroh Page 2

skills on the job without any additional training. Although I recognize that I owe my own opportunities to this approach, this comes too close to the commonsense theory I mentioned earlier.

The success of Thompson depends on the abilities of our managers relative to the abilities of our competition. In the past, every company used this commonsense approach, and Thompson prospered because of the natural talent of people like you. But in the last decade, many new managerial techniques have been developed that are too complex for the average employee to just "figure out" on his or her own. For example, people had been doing business for several thousand years before developing the Linear Programming Model for transportation and resource allocation problem-solving. It is not reasonable to expect a high school graduate to recognize that his or her particular distribution problem could be solved by a mathematical model and then to develop the LP from scratch. But as the global marketplace grows more vast, competition will stiffen as others take advantage of these innovations. I fear that what has worked in the past will not necessarily work in the future: we may find out that what our managers don't know can hurt us. Our managers must be made aware of advances in technology, management theory, and operations innovations, and they must be able to use them to transform our business as changing market conditions demand.

I would like to suggest that you consider the value of investing in an in-house training program dealing with relevant topics to augment the practical experience our employees are gaining. In addition, when management or other fast-track administrative positions must be filled, it may be worth the investment to hire postsecondary graduates whose coursework has prepared them to use techniques to help us remain competitive. Of course, these programs will initially show up on the bottom line as increased expenses, but it is reasonable to expect that, in the not-so-long run, profits will be boosted by newfound efficiencies. Most important, we must recognize the danger of adopting a wait-and-see attitude. Our competitors are now making this same decision; hesitation on our part may leave us playing catch-up.

In conclusion, I believe I will be a valuable asset to the company, in large part because of the education I am now receiving. I hope you agree that a higher education level in our employees is a cause worthy of our most sincere efforts. I will contact your office next week to find out if you are interested in meeting to discuss questions you may have or to review possible implementation strategies.

Sincerely,

Scott Houck

CHAPTER 11 Obtaining a Job

CHAPTER OVERVIEW

Central Principles of the Reader-Centred Approach

Writing Your Résumé

Electronic Résumés: Special Considerations

Writing Your Job Application Letter

Ethical Issues in the Job Search

Writing for Employment in Other Countries

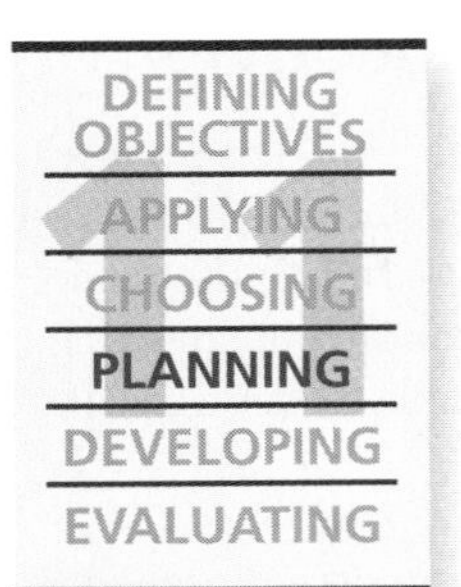

As you learned in Chapter 1, this book aims to help you develop your communication expertise by teaching you the strategies used by successful workplace communicators. At the heart of these communicators' success is the reader-centred way they approach their communications.

The purpose of this chapter is to show you how to take a reader-centred approach in preparing two communications that will play important roles in your career: your résumé and job application letter.

For additional chapter resources, visit Chapter 11 at www.techcomm.nelson.com.

The résumé and job application letter illustrate with special clarity that the success of your work-related communications depends on your readers' response to them. If your résumé and letter persuade employers to invite you for an interview, they have succeeded. If they don't produce such a response in their readers, they haven't succeeded. Thus, in addition to helping you understand how to employ the reader-centred approach to your own on-the-job writing, this chapter will help you see why it is so important for you to do so.

CENTRAL PRINCIPLES OF THE READER-CENTRED APPROACH

Think constantly about your readers.

To gain the maximum value from this chapter, take a moment to review the central principles of the reader-centred approach that underlie the advice given here and in the rest of this book. As you learned in Chapter 1, this approach's key strategy is very simple: Throughout all your work on a communication, think constantly about your readers.

- As you make each writing decision, consider your readers' characteristics, situation, and goals—the needs, desires, interests, preferences, and other factors that will shape their response to what you say.
- Concentrate on crafting a communication that will be persuasive and usable in your readers' eyes. Persuasiveness and usability are indispensable qualities of all successful work-related communications.
- Focus specifically on the ways your readers will respond, moment by moment, while they are reading your communication. These moments are the only opportunity your communication has to influence your readers directly.

WRITING YOUR RÉSUMÉ

The following sections explain how to write highly effective résumés by taking a reader-centred approach to all the activities of the communication process:

- **Defining objectives** for your résumé
- **Planning** your résumé
- **Drafting** your résumé's text, graphics, and visual design
- **Evaluating** your draft
- **Revising** your draft based on the results of your evaluation

A section provides advice for preparing Web and scannable résumés, which many employers request.

DEFINING YOUR RÉSUMÉ'S OBJECTIVES

Begin work on your résumé in the same way you'd begin work on any other work-related communication: by carefully defining its objectives. In the reader-centred approach, this crucial activity, like all others, is approached from the readers' perspective. First, determine who your readers will be, and then conduct the research necessary to pinpoint what will make your résumé persuasive and usable to these specific individuals.

Identifying Employers

www.techcomm.nelson.com

If you haven't identified the jobs or co-op positions you'd like to apply for, you might talk with professors in your department as well as staff at the career placement office. Another way to identify openings is networking, in which you ask friends, family members, former employers, and others to help you spot opportunities. As you speak with each new person, you can also ask him or her to suggest other people who can assist. Newspapers, professional organizations, and professional publications directed to people who have graduated in your field are also good sources for leads. And you can use the Web, consulting such sites as Job Bank (www.jobbank.gc.ca), sponsored by the Government of Canada; commercial boards such as Monster (www.monster.ca) and Workopolis (www.workopolis.com); and provincial and local Web listings.

Determining Your Résumé's Persuasive Objectives

Employers seek different qualifications for different jobs. To persuade the employers you've identified that you possess the specific qualifications they are seeking, you must understand as precisely as possible what they will be looking for as they read your résumé.

At a general level, these qualifications fall into the following three categories:

What Employers Seek in New Employees

- **Technical expertise.** Employers want to hire people who can perform their jobs adeptly with a minimum of on-the-job training.
- **Supporting abilities.** Most jobs require a wide range of abilities and employability skills beyond the purely technical ones of your field. Among others, these often include communication, numeracy, critical thinking and problem solving, information management, and interpersonal and personal skills.
- **Favourable personal attributes.** Employers want to hire motivated, self-directed, and responsible individuals. They also want people who will work well with their other employees as well as with customers, suppliers, and others outside the organization.

Try to learn *exactly* what the employer wants.

Of course, for any particular job, employers seek specific technical know-how, supporting abilities, and personal attributes. Instead of asking, "Does this applicant have technical expertise?" an employer will ask, "Can this person analyze geological samples?" "Write computer programs in object-oriented C++?" "Manage a consumer electronics store?" Your task is to find out the precise set of technical know-how, supporting abilities, and personal attributes that is sought by the employers to whom you are applying.

To identify these specific qualifications, learn as much as you can about the daily work of people in the position you would like to hold. Interview someone who holds such a position, talk with professors in your department, or read materials about careers at your library or campus employment centre, or online. If you have an advertisement or job description from the employer to whom you are applying, study it carefully. The more fully you understand what your readers want in a new employee, the better prepared you will be to write a résumé that demonstrates you have the knowledge, skills, and personal qualities that employers are seeking.

Understanding Your Résumé's Usability Objectives

To be successful, your résumé must not only highlight persuasive facts about you, but also present these facts in a highly usable manner. For résumé readers, the key factor in usability is the speed with which they can obtain the information they want about you. Many employers receive hundreds, even thousands, of applications a week. Consequently, the first person to read these résumés may spend a very brief time looking at each one, hoping to spot promising candidates by quickly scanning through a tall stack of résumés. Often these initial readers are human resources staff, not specialists in the area with the opening. Résumés that pass the initial screening are forwarded to managers and others in the department that wishes to hire a new employee. These readers know exactly what qualifications they desire, and they look specifically for them. However, many other responsibilities clamour for their attention so they, too, read quickly.

Thus, your résumé's usability objectives are to satisfy the needs of two equally important but different groups of speed readers: one group that is unfamiliar with the details of your specialty and the job for which you've applied, and another group that is very knowledgeable about both.

Dealing with the Complexity of Objectives

When defining objectives, you can uncover the full complexity of a communication situation. You may have noticed, for example, a tension between the usability and persuasive objectives of your résumé. To be persuasive, it must be packed with favourable information about your qualifications. To be usable for hurried readers, it must be something that can be read quickly. Expertise in on-the-job writing involves the ability to recognize such complexities when you define objectives and the ability to craft a communication that resolves these complexities in a manner that satisfies and persuades its readers. The following sections provide advice for planning and drafting a résumé that meets both of these conflicting objectives.

PLANNING

In the reader-centred writing process, you first put your objectives to use when planning the overall strategies you will use in your communication. In this planning, you decide what you will include and how, in general, you will organize and present this content.

As you plan, you can often obtain valuable assistance by looking for a conventional pattern that others have used for achieving communication objectives similar to yours. There are two major structures for organizing information on résumés.

- **Chronological résumé.** In a chronological (or experiential) résumé, you organize information about yourself around your experiences, grouping them under such headings as "Education," "Employment," and "Activities." Under these headings, you describe your experiences using reverse chronological order, in ways that demonstrate that you possess the qualifications employers want. A chronological résumé is the best choice for most students and people new in their careers. Figure 11.1 shows an example.

FIGURE 11.1

Chronological Résumé

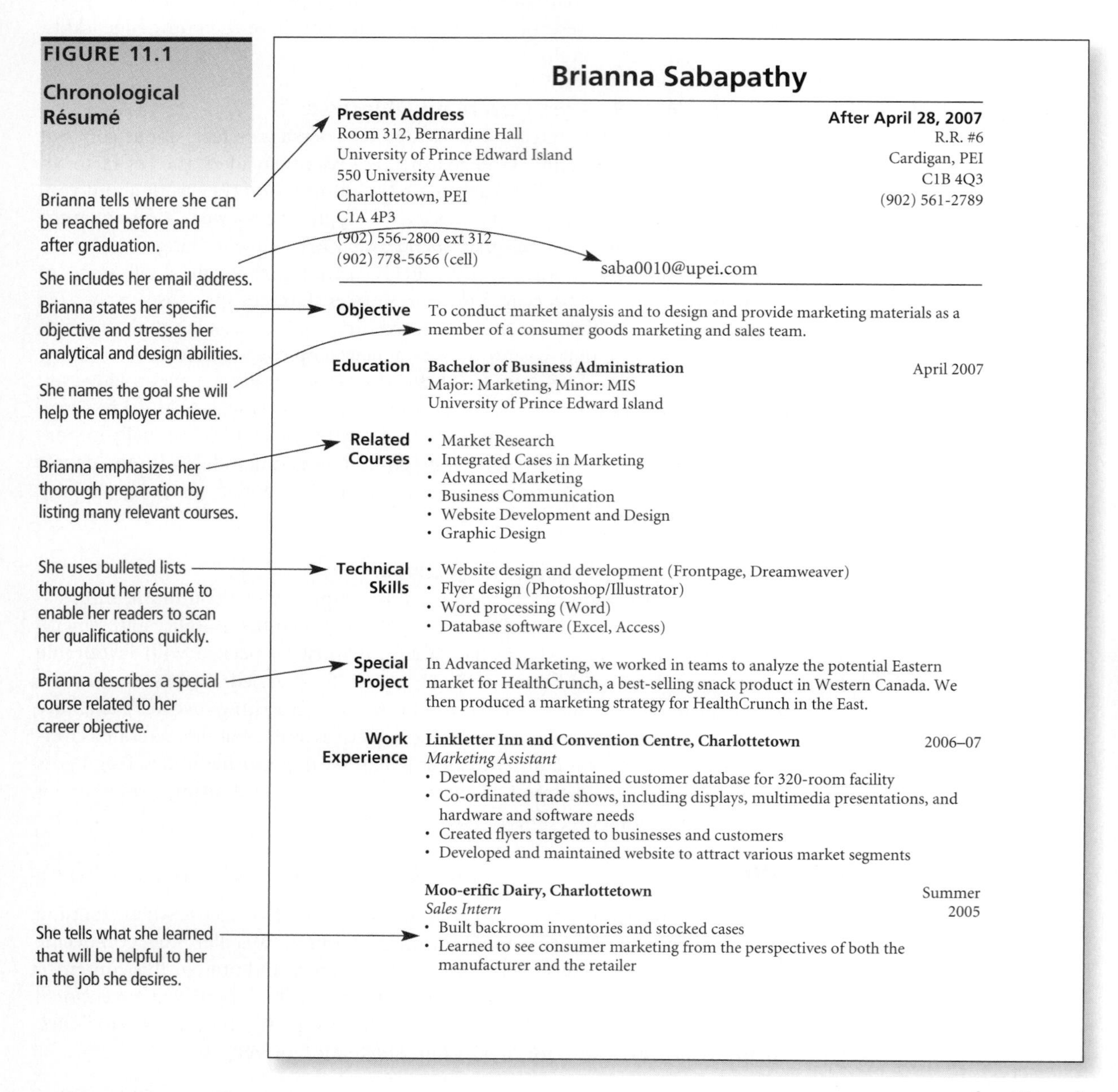

(continues)

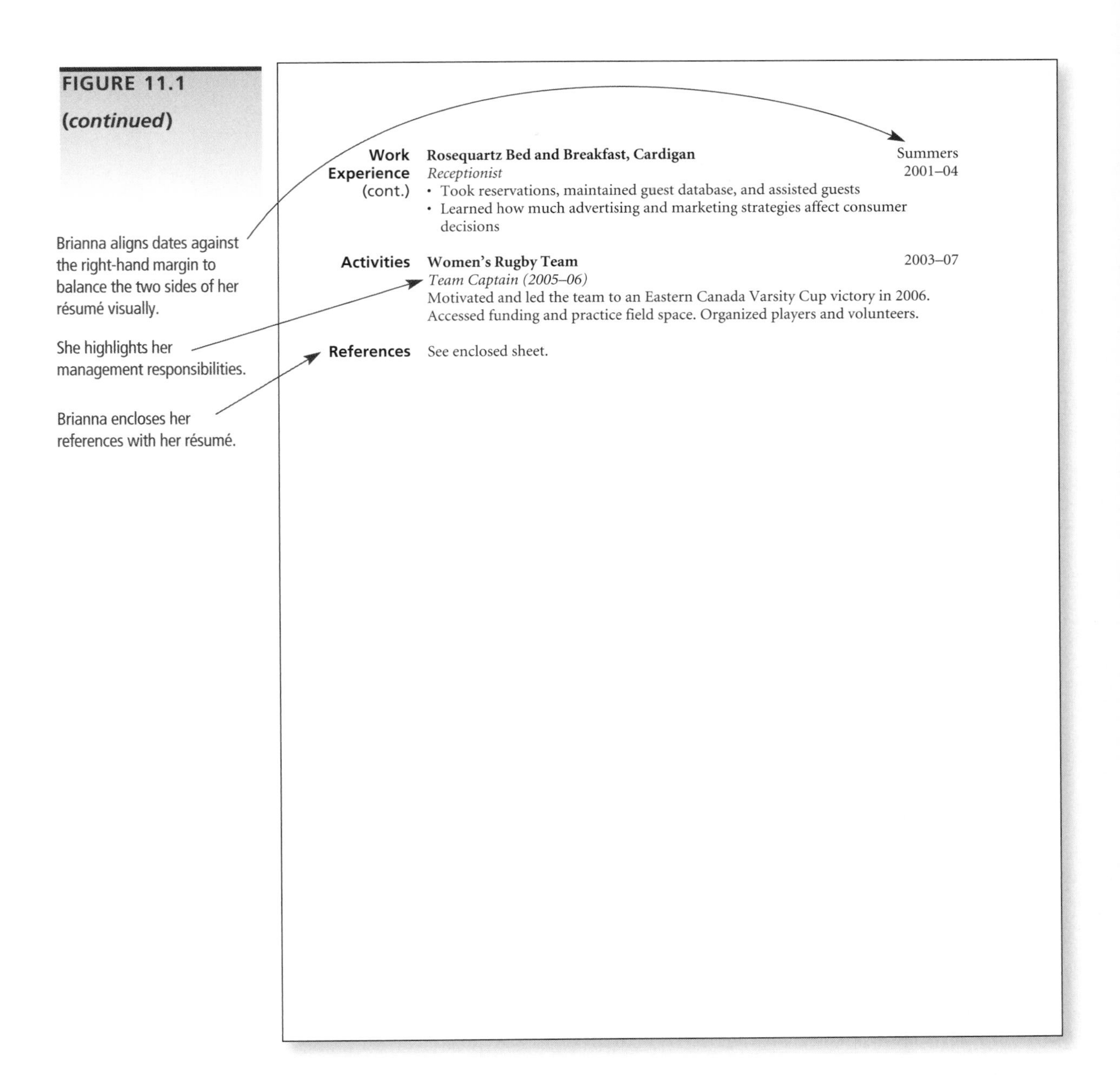

- **Functional résumé.** In a functional résumé, you organize key sections around your abilities and accomplishments, using such headings as "Technical Abilities," "Management Experience," and "Communication Skills." Later, you list the school you attended and the jobs you've held. Functional résumés usually work best for individuals with enough professional experience to be able to list several on-the-job responsibilities and accomplishments in each of several categories. Figure 11.2 (page 272) shows a functional résumé for a project manager with over ten years of experience.

FIGURE 11.2
Functional Résumé

Jan names the goal that he will help the employer achieve.

He highlights his special qualifications in a separate section.

Jan uses the present tense in his "Management" and "Budget" sections because they are continuing duties; he uses the past tense in his entry about "Innovation" because it describes a completed project.

He describes a major accomplishment.

Jan provides information about the budget's size.

Because he presented information about substantial on-the-job responsibilities and achievements above, he does not elaborate on his jobs here.

Because he has substantial professional experience, Jan de-emphasizes his university experience by giving only basic facts.

Jan shows commitment to continued professional development.

JAN PIKOLSKI

Objective

A project management position where I can lead a team to assist a telecommunications manufacturer in achieving high quality and productivity.

Skills and Accomplishments

Management	Supervise a team of six specialists who design and develop HiPath Wireless Switch and Location Server software product versions.
Innovation	Developed and oversaw new methods of testing the Serving Mobile Location Centre (SMLC) network entity.
Technical Expertise	Produced Java-based bridges to Wireless Network Services, provided sales engineer support to GUI-based business intelligence software, and developed a high level of familiarity with IP 3G networks, IP based multimedia messaging, and IP systems security.
Project Responsibility	Restructured and integrated the IT infrastructure to implement a project-based reporting structure so that project status, priority, budget, and visibility were defined.
Budget Responsibility	Manage an annual budget of $2 million.

Employment History

R&D Manager, Siemens Canada, Ottawa, Ontario, 2001–present
Development Manager, Saraide, Ottawa, Ontario, 1999–2001
Customer Training Manager, Jetform, Ottawa, Ontario, 1996–1999

Education

B.A. in Computer Science, Queen's University, 1993
Professional development courses in Project Management, Java, SQL, UNIX

Professional Societies

Canadian Advanced Technology Association (Chapter Treasurer, 2005–06)
Canadian Society of Telecommunications Professionals

Languages

Fluent in German and French, spoken and written

REFERENCES AVAILABLE UPON REQUEST

634 Clover Ave., Smiths Falls, ON K4L 5T3 (613) 354-8992
jpikolski@magma.ca

Used skillfully, either structure can help to resolve the tension, noted above, between a résumé's usability and persuasive objectives. For example, both use a tablelike structure. The headings enable employers to scan quickly for the information they want. Lists and sentence fragments pare away unnecessary words to make room for an abundance of persuasive facts.

To use these or any other structure effectively, you must adapt their helpful features to your specific readers and situation. For instance, rather than copying the résumé headings someone else has used, you must create the headings that show most persuasively how your qualifications match the ones sought by the employers to whom you are applying. Similarly, you must choose the specific facts, the order in which to present them, and the details of your page design that your readers will find most persuasive and usable. This chapter's advice for drafting, given below, will help you adapt the structures for résumés to the specific jobs you want.

How Long Should Your Résumé Be?

Length depends on qualifications.

In planning, people often ask, "How long should my résumé be?" The answer is that it should be as short as possible while still presenting the facts about you that employers will find most persuasive. For some students, this is one page. However, some students have extensive qualifications that justify a second page.

DRAFTING

When you draft, you transform your plans into a finished communication by creating the text and constructing its visual design. As when planning, base all your decisions on the knowledge of your readers' characteristics, situation, and goals that you developed while defining your résumé's objectives. The following five strategies will help you use this knowledge to create a draft that employers will find highly persuasive and usable.

General Strategies for Drafting Résumés

- **Take an employer's perspective.** Make all decisions concerning your résumé by thinking about the qualities an employer will hope to find in a job applicant.
- **Think creatively about your qualifications.** Think beyond course titles and job titles. Your challenge is to describe the knowledge and experience you possess that relates to the job you want.
- **Be specific.** Don't say that you are "proficient on the computer" or that you've "taken several advanced courses." Name the software you've mastered or the projects you've completed using the programming language.
- **Write succinctly.** Trim your prose to the bare bones, cutting unnecessary words without sacrificing clarity and impact.
- **Eliminate irrelevant information.** No matter how proud you are of some fact, leave it out if an employer won't be impressed by it. Don't bury your good qualifications among items that are unimportant to your readers.

Drafting the Text of a Chronological Résumé

The following sections detail ways to apply the six general strategies in an experiential résumé.

Name and Contact Information Help your readers locate your résumé in a stack of applications by placing your name prominently at the top of your page. Enable them to contact you quickly for an interview or with a job offer by including your postal address, email address, and phone number(s). If you will be living at another address during the summer or other part of the year, give the relevant information.

Most employers want you to state your objectives.

Professional Objective Most employers want you to tell them your objectives. This surprises some students, who think, "After I present my qualifications, shouldn't an employer be able to match me to an appropriate opening?" Harried staff trying to work their way through piles of applications don't have time to do that.

Tell what you will give, not what you want to get.

Make it your objective to achieve results that the employer desires.

When drafting your objective, take the reader-centred approach of telling what you will give to your future employer rather than the writer-centred approach of telling what you want to gain from the employer. Here's a simple, two-step procedure for doing this:

1. **Identify the results sought by the department or unit in which you wish to work.** If you don't know what they are, ask a professor, talk with someone in the kind of job you want, or contact your campus employment centre for assistance.
2. **State that your objective is to help achieve those results.**

Eric, a student in sports studies, and Jena, a computer science major, used this process to improve the first drafts of their objectives. Their initial drafts read as follows:

Writer-centred, vague objectives

> **Eric:** A position where I can extend my knowledge of nutrition and sports studies.
>
> **Jena:** A position as a computer systems analyst in the field of software development with an innovative and growing company.

When Eric and Jena considered their statements from their readers' perspective, they realized that they were merely stating what they wanted to gain from their jobs. They said nothing about how they would use their knowledge to benefit their employers. Eric and Jena revised their objectives to read as follows:

Reader-centred, results-oriented revisions

> **Eric:** A position in which I develop individualized nutrition and exercise programs that enable competitive athletes to achieve peak performance.
>
> **Jena:** To develop and maintain software systems that provide efficient, easy-to-use control of inventory, ordering, and billing for a mail-order business.

If you are applying for a co-op or summer job, tell what you will "help" or "assist" the employer to do.

Objective for internship

> A co-op placement in which I help aerospace engineers design high-quality components for commercial aircraft.

Your goal in the rest of your résumé is to make a compelling case that you have the qualifications necessary to do an outstanding job in the position you describe in your objective.

Education Place your education section immediately after your identifying information, unless you have had enough work experience to make the knowledge you gained at work more impressive to employers than the knowledge you gained in classes. Name your school, diploma or degree, and actual or anticipated graduation date. If your grade point average is good, include it. If your average in your field is higher, give it. But don't stop there. Provide additional details about your education that employers will see as relevant to the job you want. Here are some examples.

You can highlight any of these credentials by giving it a separate heading (for example, "Honours" or "Co-op Terms").

Special Educational Qualifications to Highlight

- Advanced courses directly relevant to the job you want (give titles, not course numbers)
- Courses outside your field of study that broaden the range of abilities you would bring to an employer
- Co-op assignments or other on-the-job academic experiences
- Special projects, such as a thesis or a design project in an advanced course
- Academic honours and scholarships
- Study abroad
- Training programs provided by employers

Work Experience When drafting your work experience section, list the employers, their cities, your job titles, and your employment dates. Then present facts about your work that will impress future employers.

Facts to Highlight about Your Work Experience

- **Your accomplishments.** Describe projects you worked on, problems you addressed, goals you pursued, products you designed, and reports you helped write. Where possible, emphasize specific results—number of dollars saved, additional units produced, or extra customers served.
- **Knowledge gained.** Be resourceful in highlighting things you learned that increase your ability to contribute to your future employer. Realizing that her duties of stacking ice cream packages neatly in supermarket freezers might not seem relevant to a job in marketing, Brianna describes the insight she gained in that job: "Learned to see consumer marketing from the perspectives of both the manufacturer and the retailer" (Figure 11.1).
- **Responsibilities given.** If you supervised others, say how many. If you controlled a budget, say how large it was. Employers will be impressed that others have entrusted you with significant responsibility.

When organizing and describing your work experience, follow these guidelines for achieving high impact.

For more information about putting actions in verbs, see page 92.

Organizing and Describing Your Work Experience

- **Put your actions in verbs, not nouns.** Verbs portray you in action. Don't say you were responsible for the "analysis of test data" but that you "analyzed test data." Avoid such weak phrases as "responsibilities included," "duties were," or "other duties as required."
- **Use strong verbs.** When choosing your verbs, choose specific, lively verbs, not vague, lifeless ones. Avoid saying simply that you "made conceptual engineering models." Say that you "designed" or "created" the models. Don't say that you "interacted with clients" but that you "responded to client concerns."
- **Use parallel constructions.** When making parallel statements, use a grammatically correct parallel construction. Non-parallel constructions slow reading and indicate a lack of writing skill.

Changing *correspondence* to *corresponded* makes it parallel with *trained* and *prepared*.

Not Parallel	Parallel
• Trained new employees	• Trained new employees
• Correspondence with customers	• Corresponded with customers
• Prepared loan forms	• Prepared loan forms

Activities At the very least, participation in group activities indicates that you are a pleasant person who gets along with others. Beyond that, it may show that you have acquired certain abilities that are important in the job you want.

Special Abilities Let employers know about exceptional achievements and abilities of any sort, using such headings as "Languages" or "Certifications."

Interests If you have interests such as golf or skiing that could help you build relationships with coworkers and clients, you may wish to mention them, though a separate section for interests is unnecessary if the information is provided in your activities section.

Personal Data Employment law prohibits employers from discriminating on the basis of gender, sexual orientation, religion, race, colour, age, place of national origin, or disability. It also prohibits employers from inquiring about matters unrelated to the job for which a person has applied. For instance, employers cannot ask if you are married or plan to marry. Most job applicants welcome these restrictions because they consider such questions to be personal or irrelevant. Other information that should be left off a résumé includes your social insurance number, data regarding your physical appearance (such as height or weight), and date of birth.

References You may prefer to omit your references from your résumé, including instead a line that reads, "References available upon request." By including your references, you increase your résumé's usability by enabling employers to contact them immediately and directly, unless you prefer that the employer not do so. You also increase your résumé's persuasiveness by letting readers see the names and titles of the impressive people who will speak favourably about you. If you prefer to omit your references from your résumé, create a separate page entitled "References" and be prepared to take it with you to the interview.

Employers expect three to five references, so include this many. Select a mix of people who, taken together, can describe the range of your qualifications. As appropriate, choose professors in your field and other key courses and former employers. Avoid listing your parents and their friends, who (an employer might feel) are going to say nice things about you no matter what. Provide titles, business addresses, phone numbers, and email addresses. If one of your references has changed jobs so that his or her business address doesn't indicate how the person knows you, provide the needed information: "My supervisor while at Sondid Company." Obtain permission from the people you want to list as references so they aren't taken by surprise by a phone call. Give your résumé to your references so they can quickly review your qualifications when they receive an inquiry from an employer.

Drafting the Text for a Functional Résumé

A functional résumé has the same aims as a chronological one. However, in a functional résumé, you consolidate the presentation of your accomplishments and experience in a special section located near the beginning, rather than weaving this information into your sections on education, work experience, and activities. Figure 11.2 shows an example.

Headings for functional résumés

For this special section, use a title that emphasizes its contents, such as "Skills" or "Skills and Achievements." Within the section, use subheadings that identify the major areas of ability and experience you would bring to an employer. Typical headings include "Technical," "Management," "Financial," and "Communication." However, the specific headings that will work best for you depend on what employers seek when recruiting people for the kind of job you want. For example, the headings in Jan Pikolski's résumé (Figure 11.2) focus specifically on skills required of project managers.

Because you aggregate your skills and accomplishments in a special "Skills" section, the other sections of your functional résumé should be brief to avoid redundancy.

Designing Your Résumé's Appearance

At work, good visual design is crucial to achieving your usability and persuasive goals. Nowhere is that more true than with a résumé, whose design must support rapid reading, emphasize your most impressive qualifications, and look attractive. The résumés shown in this chapter achieve these objectives through a variety of methods you can use.

Designing Your Résumé's Visual Appearance

- Short, informative headings
- Lists
- Bullets
- Italics
- Variety of type sizes
- Different typefaces for headings than for text
- White space to separate sections
- Ample margins (¾" to 1")
- Visual balance

To check visual balance, fold your résumé vertically. Both sides should have a substantial amount of type. Neither should be primarily blank.

Multicolumn résumés are easiest to create with the "Tables" feature of a word processing program. Create a narrower left-hand column for headings and a wider right-hand column for facts.

EVALUATING

When you evaluate a draft of any communication, you look it over for potential problems and possible improvements.

Ask someone else to review your résumé.

Unfortunately, we all have difficulty seeing our own work the way our readers will. We often see what we intended—which is not always what we accomplished. Show your draft to other people. Tell them about the job you want so they can read from your readers' perspective. Ask whether each feature will contribute to your communication's usability and persuasiveness. Good reviewers for your résumé include classmates, your writing professor, and professors in your major department. Campus employment centre staff can also help with résumé review.

Also, proofread carefully yourself. Employers say repeatedly that even a single error in spelling or grammar can eliminate a résumé from further consideration. Check for consistency in the use of italics, boldface, periods, bullets, and abbreviations as well as in the format for dates and other parallel items. Finally, check for consistency in the vertical alignment of information that isn't flush against the margin.

REVISING

Revising would be easy if it involved only making corrections in matters where there is a clear right and wrong. However, the review of your résumé may turn up problems for which there's no immediately obvious answer. You may even get contradictory advice from different reviewers. Here, too, let your judgment and creativity be guided by your knowledge of your readers. After all, their responses are what really matter.

Tailoring Your Résumé to Specific Employers

Don't prepare a "generic" résumé that vaguely addresses a variety of jobs and therefore fails to target any of them with the specificity employers want.

The reader-centred process emphasizes the importance of tailoring every feature of a communication to the needs and goals of the specific readers you are addressing. This approach is the key to developing expertise at communicating successfully in your career. If you are applying for more than one type of job, prepare more than one résumé. In each, tailor every section to the specific kind of position you are seeking, starting with your statement of objectives and ending with your selection of references. The more closely your résumé matches the employer's needs, the more likely you will be to get the interview.

WWW To download a checklist for revising résumés, visit www.techcomm.nelson.com.

A revision checklist you can use as you prepare your résumé is available on this text's website.

ELECTRONIC RÉSUMÉS: SPECIAL CONSIDERATIONS

www.techcomm.nelson.com

Increasingly, job applicants and employers are using electronic résumés rather than paper ones. The following sections provide advice about three types: scannable résumés, résumés submitted via the Internet, and webpage résumés.

SCANNABLE RÉSUMÉS

Many large and high-tech employers have software, not people, as the first readers of your résumé. If it arrives in paper form, it is fed into a scanner. Résumés that arrive by email or fax enter the firm's database directly. To find applicants who might be invited for a job interview, the employer searches its database for résumés that have words—*keywords*—that the employer believes would appear in the résumés of good candidates for the opening to be filled. The electronic résumé management (ERM) system displays a list of the résumés with the most matches, called *hits*. These are the only résumés a person would read.

If you suspect that your résumé might be scanned, call the employer to ask or else enclose both a regular résumé and a scannable one. To increase the number of hits your résumé will receive, make a list of keywords to include in it that employers are likely to ask their computers to look for. Here are some suggestions.

Keywords for Résumés

- Words in the employer's ad or job description
- All degrees, certifications, and licences you've earned: B.A., B.S., R.N. (registered nurse), P.E. (professional engineer licence), C.P.A. (certified public accountant)
- Advanced topics you've studied
- Computer hardware, software, programming languages, and operating systems you've mastered: Excel, AutoCAD II, C++, Linux
- Specialized equipment and techniques you've used in school or at work: X-ray machine, PK/PD analysis, ladderlogic
- Job titles and the specialized tasks you performed
- Buzzwords in your field: client server, LAN, TQM (total quality management), B2B (business-to-business), VOIP
- Names of professional societies to which you belong (including student chapters)
- Other qualifications an employer would desire: leadership, writing ability, interpersonal skills, etc.

To increase the chances that the database query will select your résumé as one a human should read, do the following.

Increasing the Number of Hits Your Résumé Receives

- Put your keywords in nouns, even if your scannable résumé becomes wordy as a result.
- Be redundant. For instance, write "Used Excel spreadsheet program" rather than "Used spreadsheet program" or "Used Excel." You don't know which term an employer will use, and if both terms are used, you will have two hits.
- Create a "Keywords" section for keywords you can't easily work into any of your résumé's other sections.
- Proofread carefully, as software queries don't match misspellings.
- Use 12-point type.
- Use a standard typeface (e.g., Times, Arial, or Helvetica).
- Avoid italics, underlining, and decorative elements such as vertical lines, borders, and shading. Boldface, all caps, and bullets are okay.
- Use blank lines and boldface headings to separate sections.
- Use a single column of text.
- Put your name at the top of every page, on a line of its own.
- Use laser printing or high-quality photocopying on white paper.
- Mail your résumé flat and without staples.

RÉSUMÉS SUBMITTED VIA THE INTERNET

Some employers ask applicants to submit their résumés via the Internet, either as email attachments or by completing online forms at the employer's website. In either case, your résumé will almost surely be managed from an ERM, so you should follow the advice for scannable résumés. To complete an online application form, draft your text in a word processing program, save it as an unformatted text or ASCII file, then review and proofread it carefully before pasting it into the employer's form.

WEBPAGE RÉSUMÉS

Although most employers do not read webpage résumés, a few do. If you create a résumé accessible via the World Wide Web, consider the following advice:

- **Link to samples of your work and fuller descriptions of your experiences and capabilities.**
- **Keep your design simple, uncluttered.** Don't let fancy embellishments distract employers' attention from your qualifications.
- **Include a "mail-to" link to your email address.**
- **Post your online résumé only at secure services.** Some services have no security, making it possible for other people to alter your résumé.

WRITING YOUR JOB APPLICATION LETTER

Importance of a job application letter

www.techcomm.nelson.com

Your job application letter can be a powerfully persuasive part of the package you submit to employers. You need to work through the activities of the communication process to ensure that your job application letter meets its own, yet complementary, objectives.

DEFINING YOUR LETTER'S OBJECTIVES

Your job application letter has somewhat different work to do than does your résumé. Although the readers are the same, the usability and persuasive objectives differ.

The usability objectives are different because readers look for different information in a letter than in a résumé. Their major letter-reading questions are as follows:

Employer's questions while reading a job application letter.

- **Why do you want to work for me instead of someone else?** You should explain why you chose to apply to this specific company.
- **How will you contribute to my organization's success?** You need to make explicit in your job application letter how your qualifications match the organization's needs and the requirements of the position. Your letter acts as a bridge between what the organization needs (the job ad) and what you have to offer (your résumé).
- **Will you work well with my other employees and the people with whom we do business?** Because every job involves extensive interactions with coworkers, employers want to hire people who excel in these relationships. Of course, interactions with clients, suppliers, and others outside the employer's organization are also important.

The persuasive objectives of your letter also differ from those of your résumé:

Your job application letter's persuasive objectives

- To respond to the employer's questions, just listed, in ways that make the employer want to hire you.
- To convey a favourable sense of your enthusiasm, creativity, commitment, and other attributes that employers value but can't be communicated easily in your résumé.

PLANNING

To begin planning your job application letter, conduct the research necessary to answer the reader's first question, "Why do you want to work for me instead of someone else?" Seek things you can praise. You'll find it helpful to categorize the facts you discover as either writer-centred or reader-centred. Examples of *writer-centred facts* are the benefits the organization gives employees or the appealing features of its co-op program. You won't gain anything by mentioning writer-centred facts in your letter because whenever you talk about them you are saying, in essence, "I want to work for you because of what you'll give me." When employers are hiring, they are not looking for people to give things to.

Look for information about things the organization is proud of.

In contrast, *reader-centred facts* concern things the organization is proud of—a specific innovation it has created, a novel process it uses, a goal it has achieved. These are facts you can build on to create a reader-centred letter.

In addition to looking for things you can praise, research the goals and activities of people who hold the job you want. The more you know about the specific job you want, the more persuasively you will be able to answer the reader's second question, "How will you contribute to my organization's success?"

Here are several ways to obtain specific, reader-centred information about an employer.

For more detailed advice about how to conduct research, see Chapter 13 and Reference Guide 4.

Learning about Employers

- **Draw on your own knowledge.** If you've already worked for the company, you may already know all you need to know. Remember, however, that you need information related to the specific area of activity in which you'd like to work.
- **Ask an employee, professor, or other knowledgeable person.** People often have information that isn't available in print or online.
- **Contact the company.** It might send publications about itself.
- **Search the Web.** Start with the employer's own website. Also use resources such as Industry Canada's Strategis (www.strategis.ic.gc.ca), which provides detailed information about thousands of Canadian businesses on the Canadian Company Capabilities link.
- **Consult your campus's placement office.**
- **Visit the library.** Business newspapers and magazines, as well as trade and professional journals, are excellent sources of information.

DRAFTING

When drafting your letter, think of it as having three parts: introduction, qualifications section, and conclusion.

Introduction

The major thing to accomplish in your introduction is to indicate the job you want and why. Here's a weak, but very common, way to do that.

Weak introduction

> I wish to apply for a position as ______. I am a student at ______ majoring in ______.

You can begin your letter much more effectively by using the reader-centred facts you discovered through your research:

- Praise the accomplishment, project, or activity you learned about. Praise is almost always welcomed by a reader, provided that it seems sincere.
- State that your reason for applying for the job is your desire to contribute to the success of these or similar accomplishments, projects, or activities.

Here's an introduction in which the writer uses this strategy.

Shawana gives specific praise related to the job she wants.

She introduces herself and expresses her desire to contribute.

She identifies the job she wants.

> While reading the August issue of *Automotive Week,* I learned that you needed to shut down your assembly line for only 45 minutes when switching from making last year's car model to this year's model. This 500 percent reduction in shutdown time over last year is a remarkable accomplishment. As a graduate in manufacturing engineering at Western University, I would welcome a chance to contribute to further improvements in the production processes at your plant. Please consider me for the opening in the Production Design Group that you advertised through the University's Career Services Centre.

Note that Shawana isn't praising the features of the car or the huge profit the company made. That would be superficial praise anyone could give without having any real understanding of the organization. Instead, Shawana focuses on a specific accomplishment, discovered through her research, that is directly related to her own specialty.

Qualifications Section

In your letter's qualifications section, explain how your knowledge and experience prepare you to contribute significantly to the employer's organization. Some job applicants divide this section into two parts: one on their education and the other on their work experience or personal qualities. Other applicants devote a paragraph to their knowledge and skills in one area, such as engineering, and a second paragraph to their knowledge and skills in another area, such as communication. Many other methods of organizing are also possible.

No matter how you organize this section, indicate how the specific facts you convey about yourself relate to the demands of the position you want. Don't merely repeat information from your résumé. Explain its relevance to the job you want. Don't merely list courses you've taken. Tell also how the knowledge you gained will help you to do a good job for the employer. Don't merely list previous job titles and areas of responsibility. Indicate how the skills you gained will enable you to succeed in the job you are seeking.

What the student learned

How this will enable the student to contribute

> In my advanced physical chemistry course, I learned to conduct fluoroscopic and gas chromatographic analyses similar to those used in your laboratory to detect contaminants in the materials provided by your vendors.
>
> In one course, we designed a computer simulation of the transportation between three manufacturing facilities, two warehouses, and seventeen retail outlets. Through this class, I gained substantial experience in designing the kinds of systems used by your company.

Conclusion

In the conclusion of your letter, look ahead to the next step. If you are planning to follow up your letter with a phone call, indicate that. In some situations, such a call can be helpful in focusing an employer's attention on your résumé. If you are planning to wait to be contacted by the employer, indicate where you can be reached and when.

Using a Conventional Format

Finally, when you draft a letter of application, be sure to use a conventional format. Standard formats are described in Appendix A and in Figure 11.3, page 285.

EVALUATING

Don't skip this step when preparing your job application letter. This letter is an especially challenging communication to write. Look it over very carefully yourself, and ask others to help you determine whether your letter is coming across in the way you intend. The following paragraphs discuss some of the most important things to look for.

Review the Personality You Project

Employers will examine everything you say for clues to your personality. When you say what appeals to you about the organization and explain why you are well suited for the job, you are revealing important things about yourself. Notice, for instance, how the first sentences of the second and third paragraphs of Brianna's letter (Figure 11.3) show her to be a goal-oriented person who plans her work purposefully. When reviewing your own draft, pay special attention to the personality you project. An attribute employers especially value is enthusiasm for their organization and for the work you would do there.

Review Your Tone

Some people have difficulty indicating an appropriate level of self-confidence. You want the tone of your letter to suggest to employers that you are self-assured but not brash or overconfident. Avoid statements like this:

Overconfident tone

> I am sure you will agree that my excellent education qualifies me for a position in your Design Department.

The phrase "I am sure" will offend some readers. And the sentence as a whole pushes the readers out of consideration by asserting that the writer has already performed for them their job of evaluating the writer's qualifications. The following sentence is more likely to generate a favourable response:

More effective tone

> I hope you will find that my education qualifies me for a position in your Design Department.

Achieving just the right tone in the conclusion of a letter is also rather tricky. You should avoid ending your letter like this:

Ineffective, demanding tone

> I would like to meet with you at your earliest convenience. Please let me know when this is possible.

To sound less demanding, the writer might revise the second sentence to read, "Please let me know *whether* this is possible" and add "I *look forward* to hearing from you."

Finally, remember that you must ensure that each sentence states your meaning clearly and precisely. And you must eradicate *all* spelling and grammatical errors.

REVISING

When revising your application letter, you may encounter the same difficulties you encounter when revising your résumé. The people who help you evaluate your letter may identify problems but not tell you how to solve them, or they may give you conflicting advice. To revise well, you have to use your own creativity and good judgment. And you need to think constantly about how your readers will react, moment by moment, to what you have written.

FIGURE 11.3

Letter of Application to Accompany the Résumé Shown in Figure 11.1

Room 312, Bernardine Hall
University of Prince Edward Island
550 University Avenue
Charlottetown, PEI C1A 4P3
March 8, 2007

Jonathan Ganz
Vice President, Sales and Marketing
Burdick Marketing
993 Princess St.
Charlottetown, PEI C1A 1T2

Dear Mr. Ganz:

RE: Marketing and Communication Co-ordinator Position

I was very impressed last month when I read in *Retail News* that your firm added five large accounts to its client list in 2006. I was especially interested to learn that this success is built upon the power and rapid response of software-based statistical analysis tools developed by your staff.

As a marketing student soon to graduate from UPEI, I would very much like to work for Burdick as a Marketing and Communication Co-ordinator dedicated to developing effective marketing strategies for clients.

To prepare myself for employment in marketing and sales, I have taken several classes in market research and advanced marketing (B2B, B2C). I have also excelled in communication classes such as technical communication, website development and design, and graphic design.

To learn the practical applications of concepts I've studied in school, I have worked in selling to consumers and business. My jobs taught me how unreliable our commonsense predictions about consumer behaviour can be and how important it is to research the market before introducing a new product.

In addition, my position as team captain of the Women's Rugby Team helped me develop my leadership and communication skills that will assist me in working effectively with other Burdick employees and with Burdick's clients.

I'm sure you realize that a letter and a résumé (which I've enclosed) can convey only a limited sense of a person's motivation and qualifications. I would welcome the opportunity to meet with you to discuss my credentials. Next week, I will call your assistant to see if you are able to grant me an interview. I hope that I may look forward to speaking with you.

Sincerely yours,

Brianna Sabapathy

Brianna Sabapathy
saba0010@upei.com
(902) 556-2800 ext 312

Enclosure

Annotations:

- Brianna addresses a specific person.
- She states the position for which she is applying.
- Brianna shows specific knowledge of the company and offers praise related to it.
- She states that she wishes to pursue a goal that the employer pursues.
- Brianna lists specific courses that have prepared her for the specific job she wants.
- She describes one way in which her job experience will benefit the employer.
- Brianna tells how abilities she developed in school will benefit the employer.
- She requests an interview.

WWW To download a checklist for application letters, visit www.techcomm.nelson.com.

A checklist you can use as you prepare your job application letters is available on this text's website.

ETHICAL ISSUES IN THE JOB SEARCH

One of the most interesting—and perplexing—features of workplace writing is that the ethical standards differ from one situation to another. The résumé and job application letter illustrate this point because the expectations that apply to them are significantly different from those that apply to other common types of on-the-job writing.

For instance, when writing a résumé you are permitted to present facts about yourself in a very selective way that would be considered unethical in some other kinds of workplace writing. For example, imagine that you have suggested that your company reorganize its system for keeping track of inventory. Several managers have asked you to investigate this possibility further, then write a report about it. In this report, your readers will expect you to include unfavourable information about your system as well as favourable information. If you omit the unfavourable information, your employer will judge that you have behaved unethically in order to win approval of your idea.

If you were to submit your résumé to these same readers, however, they would not expect you to include unfavourable information about yourself. In fact, they would be surprised if you did.

Furthermore, these readers would expect you to present the favourable information about yourself in language as impressive as possible, even though they might feel you were ethically bound to use cooler, more objective writing in your proposal about the inventory system. Here are the ethical guidelines that do apply.

Guidelines for an Ethical Résumé and Job Application Letter

- Don't list diplomas or degrees you haven't earned, or jobs you haven't held.
- Don't list awards or other recognition you haven't actually received.
- Don't take sole credit for things you have done as a team member.
- Don't give yourself a job title you haven't had.
- Don't phrase your statements in a way that is intended to mislead your readers.
- Don't list references who haven't agreed to serve as references for you.

If you are unsure whether you are writing ethically in some part of your résumé, ask your professor or someone else who is familiar with workplace expectations about this type of communication. More generally, you might also ask for advice from more experienced people whenever you are unsure about the ethical expectations that apply to any communication you write at work.

WRITING FOR EMPLOYMENT IN OTHER COUNTRIES

In many other countries, résumés and job application letters look very different from those used in Canada. In China, for instance, a lengthy résumé, including personal data, academic credentials, and evidence of effort (particularly academic) is expected when you are applying for employment. A letter of application is not required; rather, a phone call or another method of personal contact is used to introduce the résumé. Moreover, the résumé is not tailored to the position to which you are applying.

Similarly, in India, employment applicants prepare a single comprehensive and lengthy résumé-like document called the biodata. Educational attainment is emphasized by placement at the beginning of this document, and the information is organized in chronological rather than reverse chronological order. A letter of application requesting an opportunity to put one's education into practice traditionally accompanies the biodata, and a list of references is always included.

If you apply for a job in another country you will need to produce a résumé and job application letter that might differ significantly from the ones described in this chapter. However, you will still be able to succeed if you follow the reader-centred writing process described here—one in which you use your understanding of your readers' needs, situation, preferences, and other important characteristics to guide your writing.

CONCLUSION

www.techcomm.nelson.com

This chapter has demonstrated how the reader-centred approach used by successful workplace communicators can help you create a highly effective résumé and letter of application. The strategies you saw in action here are ones you can use for all your work-related writing:

- Think continuously about your readers.
- Use your knowledge of your readers to guide all your writing decisions.

The rest of this book is devoted to developing your communication expertise by providing detailed reader-centred advice you can use whenever you write at work.

EXERCISES

For additional exercises, visit www.techcomm.nelson.com.

Expertise

1. Find a sample résumé at your school's career services centre or in a book about résumé writing. Evaluate it from the point of view of its intended reader. How could it be improved?

2. Using the Web, newspapers, or journal articles, locate four or more openings that appeal to you. Create a unified list of the qualifications they specify. Then identify your skills and experiences that match each item on the employers' list.

3. Complete the assignment in Appendix C on writing a résumé and a letter of application.

Online

Using the Web, find an employer or online job board that asks you to fill out an online résumé form. Evaluate the extent to which the form helps and hinders you in presenting your qualifications in the most persuasive manner.

Collaboration

Collaborating with another student in your class, work together on developing a keyword list for each of you that you could use when creating a scannable résumé.

Ethics

Using the library or the Web, read an article that discusses the attitudes of employers toward unethical résumés. Take notes you can share with your class.

CHAPTER 12

Creating Communications with a Team

GUIDELINES

1. Select the most effective structure for your team's collaboration
2. Create a consensus on the communication's objectives
3. Involve the whole team in planning
4. Make a project schedule
5. Share leadership responsibilities
6. Make meetings efficient
7. Encourage discussion, debate, and diversity of ideas
8. Be sensitive to possible cultural and gender differences in team interactions
9. Use computer support for collaboration when it's available

For additional chapter resources, visit Chapter 12 at www.techcomm.nelson.com.

With satisfaction and pride, Jack sets his copy of the report on his desk. For three months, he and four coworkers have laboured together to evaluate three possible plans for improving performance of the network at their employer's headquarters building. Together they researched, together they evaluated, and together they wrote the sixty-page report with which Jack is now pleased. For Jack, as for most professionals, the ability to work effectively on communication teams is an indispensable skill. The recommendation report was Jack's fourth collaborative project in the past twelve months, not counting three oral presentations he made with others.

Advantages of creating communications with a team

Employees work together on communication teams for several reasons. Many projects benefit greatly from the expertise of people specializing in several fields. Other projects are too large to be completed on time without the combined efforts of several people. Just as important, experience has shown that groups can generate a larger and more creative pool of ideas than can one person working alone. For this last reason, management theory and practice increasingly emphasize teamwork over individual efforts in all areas, including writing and speaking.

Despite these good reasons for establishing communication teams, the thought of working on one can create anxiety. Jack was apprehensive the first time his boss assigned him to a team. Because Jack hadn't worked in writing groups in school, he didn't know what to expect. He feared that others in the group might overrule his good ideas or that his ideas might be accepted but blended in with the others so that he wouldn't receive the recognition he deserved. He also feared that he might be left with a huge portion of the work because others would not feel as committed to the project as he did. In fact, Jack has encountered some of these problems while working on teams, but overall he has found teamwork to be quite satisfying. And he has earned a reputation for being an effective writing-team member, a quality he hadn't previously realized that he possessed or that his employer valued.

This chapter's nine guidelines will help you develop similar expertise and enjoy similar success when preparing either written or oral communications with a team. You can also find additional advice for collaborating on oral communications in Chapter 17.

GUIDELINE 1 Select the Most Effective Structure for Your Team's Collaboration

A communication team performs the same essential activities as does an individual writer: defining objectives, planning, drafting, evaluating the draft, and revising it. The team may organize its work on these activities in a variety of ways, assigning each to either one team member, selected members, or the whole team.

Teams can assign each writing activity to one person, selected team members, or the entire team.

	Defining Objectives	Planning	Drafting	Evaluating	Revising
One Member	One defines	One plans	One drafts	One evaluates	One revises
Some Members	Some define	Some plan	Some draft	Some evaluate	Some revise
Whole Team	Whole team defines	Whole team plans	Whole team drafts	Whole team evaluates	Whole team revises

Three general ways of organizing a team's work

Here are three general patterns that teams use to organize their work together.

- **Maximum dialogue patterns.** In this model of team organization, the entire team performs all functions together. Because it involves contributions by all team members, this model can provide a very thorough and original result.
- **Maximum specialization patterns.** In this model, each activity is performed by a single team member, who passes the results of his or her work to someone else. For instance, the team leader may define the communication's objectives, a second person may plan and draft it, a third person may evaluate the draft, and a fourth person (perhaps a technical editor) may revise it. In one variation of this model, two or more people may draft, each working independently of the others. For example, one person may draft the text and another may draft the charts, drawings, and other graphics. Alternatively, people from different technical areas may each draft the sections for which they are the experts.
- **Hybrid patterns.** A hybrid structure involves some mixture of individual, small group, and all-team writing activities. In one common variation, the whole team defines the communication's objectives and plans the overall strategy for communicating it. Individuals then draft different sections, and the whole team evaluates the results. Finally, a single individual or a selected subset of team members revise, based on the whole team's suggestions. Hybrid structures are the most common for longer communications.

No pattern is best for all situations. Your team's organizational plan should respond to the nature of the communication you are preparing and the talents and areas of expertise of your team members. For many projects, however, a hybrid pattern produces the best result by combining the benefits of team discussions at some points with the efficiency of individual work at other points. In school, student teams often obtain the best results from defining objectives and planning together, drafting individually, and then evaluating and revising together.

All of the remaining guidelines in this chapter focus on the interactions that occur when two or more team members are working together at some stage of the communication process.

GUIDELINE 2 Create a Consensus on the Communication's Objectives

The first activity of the communication process—defining objectives—is even more important when you are participating on a team than when you are preparing a communication individually. Because each member brings a unique perspective to the task, the team can build on one another's ideas and insights to achieve a particularly comprehensive view of the communication's readers and purpose. This kind of collaboration is especially helpful with long, complex communications, such as formal reports and proposals.

Failure to achieve consensus about objectives can cause many problems.

However, differences among the team members' perspectives may also create problems. If each member insists on emphasizing his or her own perspective on the communication, the team may have difficulty determining what's important to the people whose perspective matters most: the readers.

Moreover, even when the team agrees on a specific set of objectives, team members may interpret them differently. For example, the members of a writing team might agree that the readers of their report will want to know the research procedures the team used. But some team members might think the readers want a non-technical explanation, while others may think a technical one is appropriate.

No matter who drafts the description of the research procedures, some of the team members are likely to say it doesn't achieve the communication's objectives. The work may need to be redone, and feelings are likely to be strained as a result.

To avoid such problems, urge members of your communication teams to follow these four suggestions.

Defining Objectives as a Team

- Take time to explore the diverse views of all team members regarding your communication's objectives.
- Keep talking until consensus is reached.
- Discuss communication strategies the team might use to achieve its objectives. Such discussion may reveal unexpected differences among team members that can then be addressed openly.
- Remain open to new insights about your readers and purposes as your team's work progresses.

GUIDELINE 3 Involve the Whole Team in Planning

One major advantage of planning as a team is identical with a major advantage of defining objectives as a team: Many minds working together will almost always generate better results than could one person working alone. Another major advantage stems from the way teams often organize their work. On almost all large projects and on many small ones, teams often divide research, drafting, and other tasks among their members. When this delegating of tasks occurs, each team member must be clear about what the team wants him or her to produce. Otherwise, the person may spend hours on work the team will ask to have redone. Such misunderstandings bruise feelings and waste time. Here are four suggestions for avoiding such problems.

Planning Team Projects

- **Discuss plans in detail.** Vague plans leave the door wide open for problems. For each part of a communication, team members should discuss what topics are to be covered, what sources should be consulted, what main points are to be highlighted, and what graphics are to be included.
- **Write an outline.** One way to focus your team's planning efforts on the details of your report is to write an outline together. Outlining also gives everyone a written record of what the team decided, so no one forgets while doing independent work.
- **Create a storyboard.** Your team can also plan together by using the storyboard technique devised by James R. Tracey and now widely employed in

For information about a different variation of storyboarding used with oral presentations, see Chapter 17.

industry. After the team has prepared a written outline, it assigns each section to one team member. Before the team's next meeting, the member responsible for a given section fills out a set of standard one-page forms, preparing one form for each segment of the section. On the form, shown in Figure 12.1, the writer indicates the segment's title or topic, its main point (or thesis), its subpoints, and the figures it will include.

The team then meets to review the results. Some teams tape the storyboards on the walls in the order in which they will appear in the finished communication. Team members look for overlap, omissions, and inconsistencies. Typically, this review produces new ideas as well. After revision, the storyboards can serve as guides to drafting.

- **Use a style guide.** A style guide tells how to handle various details of writing that team members might otherwise handle differently from one another. For instance, it might tell what margins and typefaces to use, how tables and graphs should look, and what abbreviations should be used. Some employers publish their own style guides. Others ask their employees to use an appropriate general one, such as the *Chicago Manual of Style* or the *APA Style Manual.* If your employer doesn't use a specific style guide, your team can create one, perhaps by delegating this responsibility to one member. By distributing a style guide before your team begins drafting, you can greatly reduce the editing required to achieve consistency in your final communication.

FIGURE 12.1

Storyboard for Use by Writing Teams

Before anyone begins drafting, team members can use a storyboard to decide together what the content of each section will be.

This process helps each team member draft sections that will fit together tightly and require a minimum of revision.

Storyboard

Project ____________________ Writer ____________________

Section ____________________ Subsection ____________________

TOPIC:

Thesis Statement:

Main Subpoints to Be Made — Figure

1.

2.

3.

Figure Title

GUIDELINE 4 Make a Project Schedule

Schedules are helpful for almost any team project that requires more than one meeting, but especially for projects in which some of the tasks will be performed by individual team members working independently. The schedule lets each person know exactly when his or her work must be completed, and it enables every team member to see the adverse consequences of missing any deadline along the way. When you create a schedule, include the following elements.

Creating a Project Schedule

- **Include time to define the project's objectives.** If the project leader has already defined the objectives, he or she should take time to discuss them with the team. Alternatively, the entire team can define the communication's objectives by discussing its purpose and readers. In either case, this definition is needed to guide group discussions and individual work throughout the project.
- **Establish frequent checkpoints.** The team needs to meet often enough to see that the work that individual members are doing independently is proceeding in the way the team wants. Early discovery of problems helps the individual as much as it helps the team because the individual is saved from investing additional time doing work that will have to be redone. Frequent meetings also give the team ample opportunity to refine or alter its plans before team members have devoted overly large amounts of energy and creativity to work that later would have to be changed substantially.
- **Include time for editing.** Even if the team uses an outline, storyboards, and a style guide, sharp differences may appear among sections. These may involve writing style, or they may involve development of ideas or overlap of content. The editing required to smooth out these differences might be assigned to one team member, or it might be undertaken by the whole team acting as an editorial committee. Schedule the time needed for this work.

Create a team schedule as soon as possible.

The best time to make a schedule is at the very beginning of the specialized, technical work that will produce much of the content of the communication. For instance, Jack's team made its schedule when it first received its assignment to investigate the networking alternatives. This enabled the team to create a schedule that integrated writing activities with the research and other work that the team was to conduct. When teams instead plan to do all of the writing work at the end of the project, they frequently leave too little time for writing. Consequently, their communication is either done poorly or late—outcomes that you and your teammates will surely want to avoid.

GUIDELINE 5 Share Leadership Responsibilities

Communication experts Kenneth D. Benne and Paul Sheats (2000) have identified a wide range of roles that team members must play if they are going to maximize their productivity. The roles fall into two groups: *task roles,* which keep the team moving toward its goal, and *group maintenance* roles, which assure good working relationships among the team members. The roles most essential for successfully completing team-written communications are described here.

Roles	Contributions
Task roles	
Initiators	offer new ideas, propose new solutions, and restate old issues in a novel way. They provide creativity and direction as the team explores its subject matter and communication strategies.
Information seekers	request clarification and additional information. They ensure that the team members understand all relevant factors—including their subject matter, readers, and communication alternatives.
Information givers	furnish the facts needed by the team, sometimes on their own initiative, sometimes in response to information seekers.
Opinion seekers	ask others to express their judgments, values, and opinions. They also share their views with the team.
Clarifiers	clear up misunderstanding or confusion by explaining points or providing additional information.
Summarizers	consolidate the team's deliberations by stating concisely what has been said or decided. They help team members see what has been accomplished so the team can proceed to the next task.
Energizers	motivate the team to take action, often by communicating a sense of enthusiasm or by emphasizing its commitment to its goals.
Group maintenance roles	
Encouragers	offer warmth, praise, and recognition during team discussions. They support quieter team members, whom they gently encourage to join in.
Harmonizers	help team members explore differences of opinion without hurting one another's feelings. They detect and reduce friction by helping the team to focus on ideas rather than on personalities.
Feeling expressers	share their own feelings or vocalize those of the team, thereby enabling members to deal with emotions that might interfere with the team's ability to work together productively.
Compromisers	volunteer concessions of their own positions on controversial issues and suggest a middle ground when other team members seem stuck in opposing positions. They help all team members realize that they are contributing even when their ideas are altered.
Gatekeepers	encourage all team members to participate, and they create opportunities for silent members to speak.

Teams often work best when all members share leadership responsibilities.

The most important thing for you and your teammates to remember is that these roles are all leadership roles for which team members all should take responsibility when the need arises. In this way, your whole team shares leadership responsibility. Teams are usually stronger when their members share leadership rather than rely on one person to provide it all. Of course, fulfilling leadership roles also involves taking full responsibility for your own assignments, completing them thoroughly, on time, and with high quality.

GUIDELINE 6 Make Meetings Efficient

Nothing is more precious to a communication team than time. Because it's usually so hard to find moments when all members are free, every one of these moments should be used well. Moreover, every member usually has so many other responsibilities that meetings must be productive to justify the minutes and hours taken away from other duties. The following strategies help make meetings productive.

Conducting Efficient Meetings

- **Prepare an agenda.** Before the meeting, have someone prepare an agenda by listing the major topics to be discussed. Open the meeting by having the team review the agenda to ensure that everyone agrees on what is to be accomplished.
- **Discuss each item on the agenda.** As the meeting proceeds, team members can keep the discussion on track by referring to the agenda.
- **Bring discussion on each topic to a close.** When a discussion becomes repetitious, you can focus the team's attention on the decision to be made by saying something like, "We seem to have explored the options pretty thoroughly. Let's make a decision." If team members continue to talk about a topic even when they have reached consensus, you can formalize the agreement by saying, "I think I hear everyone agreeing that . . ." If someone objects, the team can clear up any points that still need to be resolved.
- **Make notes.** After the discussion on a topic has been ended, record what has been decided, who will be responsible for taking action on the matter, and when that action will be taken. The team can then move on to the next item.
- **Sum up.** After all the topics have been covered to everyone's satisfaction, sum up the results of the meeting. Such a summation consolidates what the team has accomplished and reinforces its decisions.
- **Set goals for the next meeting.** Before the meeting breaks up, make sure that all team members know what they have to do before the next meeting. Doing this helps ensure that when you meet again, you will have new ideas and material to discuss.
- **Set a date, time, and place for the next meeting.** If necessary, distribute the notes taken immediately after the meeting to confirm responsibilities.

In your quest for efficiency, remember that even at work a writing team is a social group in which friendliness and interest in one another are natural and desirable. In your attempts to be efficient, don't take the fun out of the occasion. On the other hand, don't let the fun take up so much of the meeting that people feel exasperated because so little was accomplished.

GUIDELINE 7 Encourage Discussion, Debate, and Diversity of Ideas

As mentioned above, one of the chief benefits of group work is that many people bring their expertise and creativity to a project. To take full advantage of that benefit, all team members must offer their ideas freely—even if the ideas conflict with one another. In fact, debate and disagreement can be very useful if carried out in a courteous and non-threatening way. Debate ensures that the team won't settle for the first or most obvious suggestion. It also enables your team to avoid *groupthink,* a condition in which everyone uncritically agrees at a time when critical thinking is really what's needed.

Encouraging debate and diversity of ideas can be difficult. Some people are naturally shy about speaking, and many avoid disagreeing, especially if they fear that their ideas will be treated with hostility rather than openness and politeness. To promote healthy debate and the consideration of a rich diversity of ideas, you and your teammates can use these four strategies:

Ways to encourage debate and diversity of ideas

- **Invite everyone to speak.** The quiet members of a team often have good ideas but are timid about offering them. They may need to be invited before they will speak up. You might simply say, "What do you think of that idea, Abdul?" or "What have you been thinking about this topic, Maria?" If some team members are very talkative, you may have to create a place for the quiet person in the team's conversation by interrupting the more talkative people. You might say, "I've liked the things Pei-ju has had to say, but I wonder what Ashwin's ideas are," or "Let's take a minute to hear what Tomas feels about the ideas we've been discussing." Another way of inviting everyone to speak is to establish an operating procedure that gives each person an equal amount of time to talk about every major new topic before the floor is opened to a more general discussion.
- **Listen with interest and respect.** When a team member does speak, it is critical that others respond with interest and respect—even if they disagree with what the person is saying. If someone feels that his or her ideas have been treated rudely or harshly, that person is unlikely to contribute openly again.

  Show that you welcome each person's contributions.

  One simple way to indicate that you welcome someone's contribution is to make remarks that show you are paying attention. If you agree with the person, say such things as "That's interesting," "I hadn't thought of that," or even simply "Uh-huh."

  If you disagree, indicate that you welcome the person's contribution by showing that you want to understand the person's position fully. Ask questions if you are unclear about anything. Or paraphrase the person's position, perhaps saying something like, "What I think I hear you saying is. . . ." If you have misunderstood, the speaker has the opportunity to correct you. Maybe you don't disagree after all. But even if you do, hear the person out. Don't cut the speaker off. If you won't listen, others won't speak, and then your team will lose its good ideas as well as its not-so-good ones.

  Maintaining eye contact makes people feel you are listening attentively to them.

  In addition to sounding interested, look interested in what others are saying. That will encourage them to share their ideas. Start by maintaining eye contact. Researchers have discovered that when speakers think they are receiving more eye contact, they also think their listeners are more attentive (Kleck & Nuessle, 1967). Researchers have also found that people believe that listeners who maintain lots of eye contact are friendlier than those who maintain less eye contact (Kleinke, Bustos, Meeker, & Staneski, 1973).

Another way to show your interest is to refrain from nervous gestures. Researchers have found that speakers feel uncomfortable if their listeners engage in such nervous gestures as cleaning their fingernails, drumming their fingers, or holding their hands over their mouths (Mehrabian, 1972). Replace such gestures with notetaking or assume a relaxed but alert posture.

Listen actively.

The listening strategies just described are sometimes referred to as *active listening skills* because they involve positive actions you can take to understand other people and to show them you want to hear their message accurately. The term *active listening* is particularly appropriate because it can remind you that in team meetings you should put as much effort into listening as you put into speaking.

- **Be considerate when you discuss other members' drafts.** Team meetings can become particularly awkward when drafts produced by members working independently are reviewed by the group as a whole. Many people almost automatically resist any suggestion for change in their drafts. This resistance is entirely understandable. The writers have invested considerable personal creativity and effort in the project and so are reluctant to see that work undone. Furthermore, writers sometimes feel that the team is criticizing their overall writing ability when it requests changes. The resulting defensive responses by the writers can prevent needed improvements and undermine the mutual good will that is essential to effective teamwork.

  On the other side, the people reviewing another team member's work often avoid suggesting changes because they sympathize with the drafter's anxiety and fear a conflict. "We just can't think of any way to make this better," they say—no matter how badly improvements are needed.

  Such resistance by the writer and other team members can be very counterproductive. Most drafts can be improved, and open discussion of them can discover how that can be done.

Open your comments on drafts with a positive statement.

  The most effective way to promote an open discussion is to be considerate of one another's feelings. Present your ideas for changing a draft as suggestions or options: "Here's another way you could say that." When discussing options, focus on the positive reasons for choosing one option over another. Avoid making statements that sound like criticism of the writer's work. Also, accompany suggestions with praise for what is strong in the writer's draft. Indicate that you are offering ideas for improving a draft that is basically good, not corrections for a draft that is fundamentally bad.

  You can also be considerate of others' feelings when your draft is being reviewed by the team. Help the other team members feel comfortable by being open, not defensive. Make them feel that their ideas are welcomed rather than resented.

- **Treat drafts as team property, not individual property.** Another way to promote open discussion is to encourage team members to give up their sense of personal "ownership" of the material they draft. While still taking pride in your contributions, you and your teammates can try to see your drafts as something you've created for the team so that the drafts are the group's property, not your personal property. This doesn't mean that you and your teammates should give up supporting the strong points of your own drafts if you think they have been underestimated. It means that when the team is reviewing your draft, you should join in a reasoned discussion of the best way to write "our" communication rather than struggling to protect "my" writing from assaults by others. Your openness about your draft will encourage others to take the same attitude concerning theirs.

Swapping responsibilities can diminish the sense of personal ownership.

Your team can also reduce the troublesome sense of personal ownership by agreeing to swap responsibilities at a certain point. For example, after second drafts are written, the first section might become the responsibility of the person who drafted the second section, the second section might become the responsibility of the person who drafted the third section, and so on. Also, the team members can agree that at some point everyone's drafts will be combined for editing and polishing by the whole group or by an individual the group designates. At that point, the draft clearly becomes group property, not individual property.

GUIDELINE 8 Be Sensitive to Possible Cultural and Gender Differences in Team Interactions

Different people approach team projects in different ways. It's very helpful to any team for all members to be aware of their own predispositions and to be sensitive to those of others. The team's goal should be to interact in ways that allow each member to make his or her maximum contribution. For this outcome to happen, various individuals may need to adapt their customary ways of interacting. For example, because they are eager to contribute, some members may speak immediately on each issue that arises. An advantage to the team of this strategy is that these individuals quickly put ideas on the table for others to consider and, perhaps, build on. However, this strategy also has the potential to discourage other members from saying anything. Perhaps these other members fear that the matter has already been settled or are hesitant to contradict the first speaker. The challenge faced by a team is to develop an approach to conducting its interactions that gets the fullest possible contribution from all members.

To succeed in encouraging maximum participation by all team members, each person must be sensitive and responsive to the individual styles of everyone else. It can be helpful to remember that a person's style may be significantly influenced by gender and cultural background.

Gender and Collaboration

For example, research shows that many men state their ideas and opinions as assertions of fact. When exploring ideas, they may argue over them in a competitive manner. In contrast, many women offer their ideas tentatively, introducing them with statements such as, "I think," or, "I'm not sure about this, but. . . ." If there is disagreement, they may support part or all of the other person's ideas and seek to reach consensus (Lay, 1989).

Although it would be counterproductive to assume that all men or all women behave in these ways, knowing about such general patterns can enable you to increase your sensitivity to your own habits and to recognize the different practices of others. A key point is to avoid valuing your own mode of interacting as necessarily the best one. As explained above, you can help to create successful team interactions by adapting your own modes and supporting people with different modes so that all can contribute to the maximum of their potential.

Culture and Collaboration

On the job, you will almost certainly find yourself engaged in some form of international or intercultural collaboration. In companies with international operations, collaboration between offices can also mean collaboration between residents of different countries. Moreover, many employees in Canada were born and raised elsewhere in the world or have grown up in families or neighbourhoods deeply influenced by other cultures.

Recognize cultural differences.

Writing researcher Deborah S. Bosley (1993) has identified some differences in the ways that people from various cultures may interact on writing teams. Whatever your cultural background, you can contribute to your team's success by responding sensitively to people raised in cultures that have different styles. Here are some differences to be alert to.

- **Expressing disagreement.** In some cultures, individuals typically express disagreement directly. In some other cultures, people say "no" only indirectly in order to save face for themselves and for the other person.
- **Making suggestions.** People from some cultures offer suggestions freely. To avoid embarrassing others, people from some other cultures avoid saying anything that might be interpreted as disagreement.
- **Requesting clarification.** In some cultures, individuals frequently ask others to explain themselves more clearly. People from some other cultures feel that it is rude to ask for clarification because doing so would imply that the speaker doesn't know what he or she is talking about or hasn't succeeded in explaining things clearly.
- **Debating ideas.** Whereas members of work teams in some cultures debate ideas vigorously as a way of exploring ideas, people from some other cultures regard such behaviour as disloyal and unacceptable.

Responding to possible cultural differences

When you work on an international or intercultural writing team, avoid using your own culture's standards to interpret and evaluate the ways people interact. If you are from a culture where people freely express disagreement and offer suggestions, don't automatically conclude that a quiet team member is indifferent or bored. If you are from a culture where people express themselves indirectly, don't automatically judge a person negatively because he or she interacts more bluntly. In either case, remind yourself that the other person may be interacting in a way unfamiliar to you because that is the accepted mode in his or her culture. In a non-confrontational way, perhaps outside a team meeting, ask the person about his or her expectations about team interactions and share your own. Part of the experience of working on intercultural writing teams is learning about the customs of people in other cultures—and sharing with people from other cultures the customs in your own. To increase your team's effectiveness, you must also adjust your expectations, interpretations and interactions based on what you learn.

GUIDELINE 9 Use Computer Support for Collaboration When It's Available

Computer technology offers a fascinating variety of support for teams that are creating communications together. Here are some examples:

Examples of computer support for team projects

- Email allows team members to share ideas and drafts without meeting. Using a program called a *listserv*, an individual can send an email message to a single address from which it will be forwarded automatically to all other team members.

- Some software, including Web-based course management software used at educational institutions, allows all team members to participate simultaneously in an online discussion even though each person is in a different location.
- Some corporations use teleconferencing rooms so that team members in locations distant from one another can converse even as they use software to draft or edit a document together.
- Word processing software allows all team members to work on a single draft. Each person's suggested revisions are shown in a different colour, so team members can identify the revisions' sources and track these changes. These programs also allow an individual to attach a written or voice annotation to a specific place in a communication, thereby enabling other team members to read or hear the person's comments while simultaneously looking at the passage under discussion.

The advantages of these electronic tools can be enormous. Some let you "converse" with one another even if you are not working at the same time. Others allow you to "meet" together even if you aren't in the same location. Although some of these tools are associated with expensive software programs, others are included in standard products such as Microsoft Word. When working with a team, explore the electronic tools that can increase the efficiency of your team.

Because many of these tools rely on email or similar capabilities for conveying messages electronically, follow Chapter 5's guidelines for using email. These will help you communicate effectively and avoid inadvertently offending other team members.

CONCLUSION

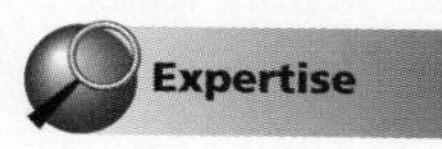

To download a collaboration feedback and evaluation form, visit Chapter 12 at www.techcomm.nelson.com.

Team writing is very common in the workplace. Although all the guidelines in this book apply to team writing as well as to individual writing, team projects require some additional skills. The nine guidelines presented in this chapter will help you create team efforts in which the team members work together productively and enjoy their mutual effort. A feedback and evaluation form team members can use to help one another develop their collaboration skills is available on this text's website.

EXERCISES

For additional exercises, visit www.techcomm.nelson.com.

Expertise

1. Write a memo to your professor in which you describe difficulties encountered by a team on which you have worked. Describe strategies presented in this chapter you employed in an attempt to address these problems. Also reflect on things that you could have done differently that might have overcome the problems.
2. Write a memo to your professor describing the style and strategies you have brought to team projects in the past. First, identify the strengths of your approach, referring to specific advice given in this chapter. Second, pinpoint ways in which you can achieve an even higher level of effectiveness.

Online

1. Explore the word processing or other software you are using to identify features that enable you to collaborate over the Internet through such means as emailing files in which you have embedded review comments or sharing files stored on a server that you and your teammates can access.

2. Explore your Internet browser or Web-based course management software to determine what capabilities it has that would enable you to collaborate more effectively on a team communication project.

Collaboration

1. If you are currently involved in a team project, fill out the collaboration feedback and evaluation form for each of the other team members.

2. If you are about to begin a group project, write a memo to your professor detailing the plans you have made in accordance with the advice given in Guidelines 3 and 4.

Ethics

Various team members may bring different talents, commitments, and values to collaborative projects. What are your ethical obligations if one team member doesn't contribute, persistently completes work late, or resists compromise? What if one member is plagiarizing? Who are the stakeholders in these cases? What are your options? What is an ethical course of action? What are your obligations if one or more members of your team ask you to contribute more or in a manner that others consider more productive? Would any of your responses be different if the project were assigned at work rather than in school? Present your results in the way your professor requests.

CHAPTER 13

Conducting Research

GUIDELINES

1. Define your research objectives
2. Create an efficient and productive research plan
3. Check each source for leads to other sources
4. Carefully evaluate what you find
5. Begin interpreting your research results even as you obtain them
6. Take careful notes
7. Ethics Guideline: Observe copyright law and intellectual property rights
8. Ethics Guideline: Document your sources

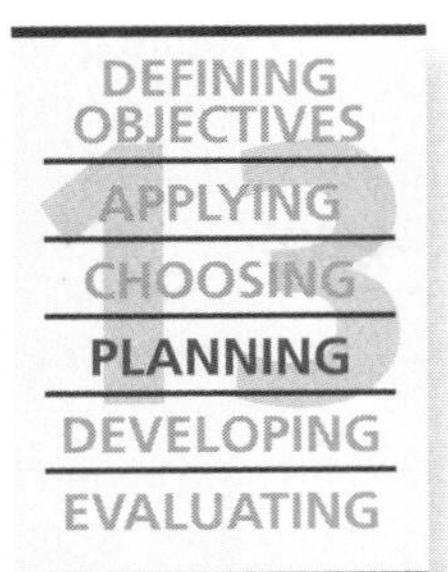

To succeed in the writing you do in your career, you must provide your readers with information and ideas they will find useful and persuasive. Sometimes, you will already possess this knowledge before you begin to plan your communication. Often, however, you will need to conduct research to discover and develop your communication's content. This chapter will help you develop your expertise at conducting reader-centred research on the job.

SPECIAL CHARACTERISTICS OF ON-THE-JOB RESEARCH

Workplace research focuses on finding the information, ideas, and arguments that will meet your readers' needs.

The initial step in developing this expertise is understanding that research at work differs significantly from research in school. First, the purposes are very different. As explained in Chapter 1, your professors assign writing projects in order to advance your personal, intellectual, and professional development. Typically, your research goal is to gain either a general overview or a comprehensive understanding of a topic that will be useful *to you* sometime *in the future.* In contrast, at work you will write for practical purposes, such as helping managers, coworkers, or clients perform practical tasks and make good decisions on issues that confront them right now. Your research goal will be to develop ideas, information, and arguments that *your readers* will find to be valuable *right now.*

At work, you will need to research very efficiently.

Second, in the workplace it is much more important to be able to conduct research efficiently, without taking extra time to travel down avoidable dead ends or study material that will be irrelevant to your readers' current situation. You will need to produce your results quickly because your readers will have an immediate need for your results, and because you will also have many other responsibilities and tasks to complete.

Third, in the workplace some (but not all) of the ethical principles concerning research differ from those that apply in school.

WWW For additional chapter resources, visit Chapter 13 at www.techcomm.nelson.com.

This chapter's first six guidelines will help you quickly and efficiently gather the information, ideas, and arguments your readers need from your communications. Two other guidelines focus on important ethical issues in workplace research. In addition, Reference Guide 4, which follows this chapter, offers detailed advice for skillfully using five research methods often employed on the job.

GUIDELINE 1 Define Your Research Objectives

You can streamline your research by defining in advance what you want to find. After all, you are not trying to dig up everything that is known about your subject. You are seeking only information and ideas that will help you achieve your communication's objectives.

Base your research objectives on your communication's usability and persuasive goals.

In fact, the most productive way to begin your research is by reviewing your communication's usability and persuasive goals (see Chapter 2). Your research objectives should be built squarely on them.

First, identify the information you need to write a communication your readers will find to be highly usable. Review the questions your readers will ask about your topic. The questions you can't answer immediately are the ones you need to research. Also,

Determine what you need to research in order to make your communication usable in your readers' eyes.

review the kinds of answers your readers will want you to provide. For example, remind yourself whether they will want general information or specific details, introductory overviews or technical explanations. In addition, consider the point of view from which your readers will read your communication. Are they going to want it to meet their needs as engineers or accountants, consumers or producers? The answers to these questions will help you determine not only the kinds of information you must obtain but also where you should look for it.

Figure out what you need to research in order to make your communication persuasive to your readers.

Second, identify the kinds of information and arguments that will make your communication most persuasive in your readers' eyes. Given your readers' goals, values, and preferences, what kinds of information and arguments are most likely to influence their attitudes and actions? For example, are they primarily interested in efficiency or profitability, safety or consumer acceptance? Consider also the types of evidence and kinds of sources your readers will find most compelling. For instance, are they more likely to be swayed by quantitative data or by testimonials from leaders in their field?

Be prepared to modify your objectives as you proceed.

Although you should define your research objectives at the outset, be prepared to revise them as you proceed. Research is all about learning. One thing you may learn along the way is that you need to investigate something you hadn't originally thought important—or even thought about at all.

GUIDELINE 2 Create an Efficient and Productive Research Plan

Many people conduct research haphazardly. They dash off to the library or log onto the Internet in the hopes of quickly finding just the right book or website. If the first source fails, they scoot off to another one. Such an approach can waste time and cause you to miss very helpful sources, including (perhaps) the one that includes exactly the information you need.

You will conduct your research most efficiently and productively if you begin by making a plan.

Making a Research Plan

- **Identify the sources and methods most likely to help you write an effective, reader-centred communication.** Consider the full range of information you must locate in order to assist and persuade your readers. Here are some strengths and weaknesses of several useful sources and methods:
  - **Books.** Broad coverage of established topics; often reviewed for accuracy in the publication process; quickly become out of date on rapidly developing topics.
  - **Research journals.** Up-to-date discussions that have been judged valid by specialists in the writer's field; often very sharply focused.
  - **Trade journals.** Solutions to practical problems encountered by many organizations in an industry or field; sometimes lack balance and depth.
  - **Popular periodicals.** General introductions to topics from the perspective of the non-specialist; not generally considered authoritative on technical matters.
  - **Internet searches.** Very current information whose quality can sometimes be difficult to assess.

Guideline 4 provides detailed advice about evaluating information you find in these and other sources.

Some kinds of research sources are more likely than others to enable you to meet your readers' needs and influence their attitudes and actions.

- **Interviews and surveys.** Opportunity to gather exactly the information you need to assist and persuade your readers; not suited to all topics.
- **Specialized methods in your field.** Ideal for answering the questions you are learning to address through classes in your field.
- **Your memory and creativity.** Invaluable resources; may need to be tested against other people's thoughts and experiences.

- **Consult general sources first.** By gaining a general view of your subject, you increase the ease with which you can locate, comprehend, and interpret the more detailed facts you are seeking. Useful general sources include encyclopedias, review articles that summarize research on a particular subject, and articles in popular magazines.
- **Conduct preliminary research when appropriate.** For example, before interviewing a technical specialist or upper-level manager, conduct the background research that will enable you to focus the interview exclusively on facts this person alone can supply. Similarly, before conducting a survey, determine what other surveys have learned and study the techniques they used.
- **Make a schedule.** Establish a deadline for completing all your research that leaves adequate time for you to draft, evaluate, and revise your communication. Then set dates for finishing the subparts of your investigation, remembering to complete general and preliminary research before proceeding with other sources and methods.
- **Study the research methods you are going to use.** In your schedule, include time to study research methods you haven't used before. Also, provide time to learn advanced techniques for methods, such as searching the Internet, whose basics you already know. The Reference Guide that follows this chapter provides detailed advice for skillfully using five research methods that are very helpful on the job.

Expertise in using research sources and methods is as important as expertise in any other writing activity.

While planning your research, you may find it helpful to use a planning guide like this one:

WWW For a copy of this planning guide that you can fill in, visit Chapter 13 at www.techcomm.nelson.com.

Planning Guide

Readers' Question	Possible Sources	Assessment of Each Source	When to Consult
Are our competitors developing this technology more rapidly than we are?	Competitor reports to stockholders	Biased	Next week
	Trade journals	Probably reliable	Immediately
When will our design be ready?	Kami Mason, Project Manager	Objective, informed	Close to completion of report

GUIDELINE 3 Check Each Source for Leads to Other Sources

One good source can lead you to others.

Conducting research is often like solving a crime. You don't know exactly what the outcome will be—or where to find the clues. Consequently, it makes sense to check each source for leads to other sources. Scrutinize the footnotes and bibliographies of every book, article, and report you consult. When you locate a book in the library stacks, browse through books nearby. When you interview people, ask them to suggest additional places to look and people to contact. Be sure to schedule time to follow up on promising leads.

GUIDELINE 4 Carefully Evaluate What You Find

You have no use for information that your readers won't find useful or persuasive or that you yourself don't believe to be credible. Consequently, you should evaluate continuously the facts and ideas you discover. If you discover that your readers will perceive a particular person to be biased, move on to someone with more credibility. If you find that a book or webpage about your topic treats it at the wrong level or with the wrong focus, close it and move on to something else.

In evaluating, be as alert to your own biases as you are to those of your sources.

When evaluating sources, be as cautious about your own biases as you are about any biases your sources may possess. Don't dismiss a source simply because it contradicts your views or presents data that fail to support your conclusions. Your readers depend on your thoroughness and integrity.

The following questions can help you evaluate the sources you consult in your research.

Special considerations for evaluating Internet sources are described on pages 321–23.

Questions for Evaluating a Research Source

- Is it accurate?
- Is it up to date?
- Is it supported by evidence my readers will find compelling?
- Is it clearly relevant to my readers' situation?
- Is it complete?
- Is it unbiased?
- Does it conflict with other evidence?

GUIDELINE 5 Begin Interpreting Your Research Results Even as You Obtain Them

Consider all of your research results from your readers' perspective.

Research involves more than just amassing information. To make your results truly useful and persuasive to your readers, you must also interpret them in light of your readers' desires, needs, and situation. For example, imagine that you have been asked to study two expensive pieces of equipment used for laboratory analyses needed by your employer. One piece, you discover, performs a certain function 9 percent more rapidly than the other. This fact alone would not be sufficient for a decision-making reader. You need to interpret the fact by telling the reader whether the greater speed would improve operations significantly enough to justify the added cost. And answering this secondary—but crucial—question may require additional research. The following questions will help you interpret your research results.

Questions for Interpreting Research Results

- What do I conclude from these research results—and what else might I need to learn to test my conclusions?
- Are other interpretations possible—and do I need to explore them?
- What does this mean my readers should do—and what do I need to investigate to enssure that they have the information necessary to enable them to carry it out?
- What must my readers do to carry out my recommendation—and what else must I learn so I can write a communication that will enable them to do that?
- What are the implications for stakeholders who are not my readers—and what else can I learn that will enable me to suggest ways to avoid undesirable consequences for them?

GUIDELINE 6 Take Careful Notes

Careful notes can prevent wasteful backtracking.

A simple but critical technique for conducting productive, efficient research is to take careful notes every step of the way. When recording the facts and opinions you discover, be sure to distinguish quotations from paraphrases so you can properly identify quoted statements in your communication. Also, clearly differentiate ideas you obtain from your sources and your own ideas in response to what you find there.

In addition, make careful bibliographic notes about your sources. Include all the details you will need when documenting your sources (see Guideline 8, page 309). For books and articles, record the following details.

Information to Record about Your Sources

Books

- Author's or editor's full name
- Exact title
- City of publication
- Year of publication
- Edition
- Page numbers

Articles

- Author's or editor's full name
- Exact title
- Journal title
- Volume (and issue unless pages are numbered consecutively throughout the volume)
- Year of publication
- Page numbers

For interviews, record the person's full name (verify the spelling!), title, and employer, if different from your own. Special considerations apply when your sources are on the Internet; they are described in Appendix B.

It is equally important for you to record the information you will need if you later find that you need to consult this source again. For instance, when you are working in a library, jot down the call number of each book; when interviewing someone, get the person's phone number or email address; and when using an Internet site, copy the universal resource locator (URL).

As you proceed, be sure to keep a list of sources that you checked but found to be useless. Otherwise, you may find a later reference to the same sources but be unable to remember that you have already examined them.

GUIDELINE 7

Ethics Guideline: Observe Copyright Law and Intellectual Property Rights

The individuals and organizations that produced the resources you encounter in your research have legal rights that it is your legal and ethical obligation to honour. Broadly speaking, these rights are provided by laws concerning intellectual property, which include the following areas:

Three areas of intellectual property law

- **Patent law.** Governs such things as inventions and novel manufacturing processes.
- **Trademark law.** Pertains to such things as company and product names (Microsoft, Pentium), slogans ("Drivers Wanted"), and symbols (the Nike "swoosh").
- **Copyright law.** Deals with such things as written works, images, performances, and computer software.

Works are automatically copyrighted as soon as they are created.

When you are writing at work, copyright law will probably be the most important to you. Copyright law was created to encourage creativity while also providing the public with an abundant source of information and ideas. To achieve these goals, copyright law enables the creators of a work to profit from it while also allowing others to use the work in limited ways without cost.

The rights of copyright owners

Any communication, such as a report, letter, email, photograph, or diagram, is copyrighted as soon as it is created. If the creator generated the work on his or her own, that individual owns the copyright to it. If the creator made the work while employed by someone else, the copyright probably belongs to the employer. Whether the copyright owner is an individual or an organization, the owner has the legal right to prohibit others from copying the work, distributing it, displaying it in a public forum, or creating a derivative work based on it. When copyright owners grant others permission to do any of these things, they may charge a fee or make other contractual demands. The copyright owner has these rights even if the work does not include the copyright notation or the copyright symbol: ©.

Note, however, that no one can copyright an idea, only the particular expression of an idea in words, images, or some other tangible form. Thus, if you paraphrase someone else's text, you are not violating copyright—though you would still have an ethical obligation to acknowledge your source (see Guideline 8).

www.techcomm.nelson.com

It is also both legal and ethical for you to use other people's work without their permission if the work is in the *public domain.* Such works include federal and provincial statutes and judicial decisions, as well as works whose copyright has expired. Also, private individuals and organizations sometimes put their work in the public domain. The owners of websites that offer free use of clip art are one example; authors who dedicate their software to the public domain are another.

Employees can use work owned by their employers without permission.

Finally, you can generally use work that other people working for your employer created as part of their job responsibilities. In fact, in the workplace it is very common for employees to use substantial parts of communications created by other employees. For instance, when you are creating the final report on a project, you may incorporate portions of the proposal written to obtain the original authorization for the project as well as parts of progress reports written during the project. Similarly, when you are creating or updating a website, you may use text and images that your employer created for print communications.

Copyright law and other intellectual property law is complicated enough that whenever you have questions about its application to something you are writing, you should consult your professor, your employer's legal department, or the resources available at the website for this book.

GUIDELINE 8 Ethics Guideline: Document Your Sources

On the job, as in school, you have an ethical obligation to credit the sources of your ideas and information by listing those sources in a reference list, footnotes, or bibliography. However, the question of whether you need to document a source is different from the question of whether you need permission from the copyright owner. Moreover, standards for deciding exactly what sources need to be listed at work differ considerably from the standards that apply at school.

To determine whether you need to document a particular source at work, answer the following questions.

You will find information about how to write bibliographic citations in Appendix B.

Determining Whether You Need to Document a Source at Work

- **Is the information I obtained from this source common knowledge?** Both in school and at work, you must indicate the source of ideas and information that (1) you have derived from someone else, and (2) are not common knowledge.

  However, what's considered common knowledge at work is different from what's considered common knowledge at school. At school, it's knowledge every person possesses without doing any special reading. Thus, you must document any material you find in print.

  At work, however, common knowledge is knowledge that is possessed by or readily available to people in your field. Thus, you do not need to acknowledge material you obtained through your courses, your textbooks, the standard reference works in your field, or similar sources.

- **Does my employer own it?** As explained above, employers own the writing done at work by their employees. Consequently, it is usually considered perfectly ethical to incorporate information from one proposal or report into another without acknowledging the source.

- **Am I taking credit for someone else's work?** On the other hand, you must be careful to avoid taking credit for ideas that aren't your own. In one case, an engineer was fired for unethical conduct because he pretended that he had devised a solution to a technical problem when he had actually copied the solution from a published article.

Research journals have expectations about documentation that closely resemble those in school.

- **Am I writing for a research journal?** In articles to be published in scientific or scholarly journals, ethical standards for documentation are far more stringent than for on-the-job reports and proposals. In such articles, thorough documentation is required even for ideas based on a single sentence in another source. Thus, you must document any information you find in print or online. In research labs where employees customarily publish their results in scientific or scholarly journals, even information drawn from internal communications may need to be thoroughly documented.
- **Whom can I ask for advice?** Because expectations about documentation can vary from company to company and from situation to situation, the surest way to identify your ethical obligations is to determine what your readers and employer expect. Consult your manager and coworkers, and examine communications similar to the one you are preparing.

CONCLUSION

To download a planning guide for conducting research, visit Chapter 13 at ww.techcomm.nelson.com.

This chapter's eight reader-centred guidelines apply to all your research efforts regardless of the research method you employ. A planning guide for this chapter is available on this text's website. Following this chapter is Reference Guide 4, which provides additional advice for using five research techniques that are frequently employed in the workplace.

EXERCISES

For additional exercises, visit www.techcomm.nelson.com.

Expertise

1. Choose a concept, process, or procedure that is important in your field. Imagine that one of your professors has asked you to explain it to first-year students in your program. (See Reference Guide 4 for guidelines for using each of the following research methods.)
   a. Use brainstorming or freewriting to generate a list of things you might say in your talk.
   b. Use a flowchart, matrix, cluster sketch, or table to generate a list of things you might say.
   c. Compare your two lists. What inferences can you draw about the strengths and limitations of each technique?
2. Imagine that a friend wants to purchase some item about which you are knowledgeable (for example, a cell phone or an MP3 player). The friend has asked your advice about which brand to buy. Design a matrix in which you list two or three brands and also at least six criteria you recommend your friend use to compare them. Fill in the matrix as completely as you can. Each box you can't fill indicates an area you must research. Describe the methods you would use to gather the additional information. (See Reference Guide 4 for advice about using a matrix as a research tool.)

3. Imagine that you have been asked by the chair of your department to study student satisfaction with its program offerings. Devise a set of six or more closed questions and four open-ended questions you could use in a survey or interviews. (For information about survey and interview questions, see Reference Guide 4, pages 330–335.

4. Create a research plan for a project you are preparing for your technical communication course.

Online

1. Use two search engines and an Internet directory to look for websites on a topic related to your field of study. How many hits does each produce? Compare the first ten results from each search in terms of the quality of the sites and the amount and kind of information the search engine or Internet directory provides about each one. (For information about using search engines and Internet directories, see Reference Guide 4, pages 318–323.)

2. Using a search engine and online library resources, identify three websites, two books, and two articles on a topic related to your field. Which would you find most interesting? Which would be most helpful if you were writing a paper on the topic for a class? Which would be most helpful if you were writing a report on the topic for your employer? (For information about using search engines and online library resources, see Reference Guide 4, pages 324–330.

Collaboration

Working with another student, choose a topic that interests you both. Find five websites that provide substantial information on your topic. Which sites are most appealing to you initially? Following the advice on page 306, evaluate each site, and then compare the results with your initial impression of it.

Ethics

Create a bibliography of sources concerning an ethical issue related to your field of study or career. Include four websites, one book, and two journal articles that you believe would help you understand various approaches to this issue.

Reference Guide 4: Five Research Methods

CONTENTS

WWW For additional resources, visit Reference Guide 4 at www.techcomm.nelson.com.

This Reference Guide tells how to use five research methods commonly employed on the job. Often, they are used in combination with one another. Each method is discussed separately to create a reference source that allows you quickly to access the advice about the particular method you need to use. Note, however, that in many writing situations you will need to use two or more of these.

Research Methods

EXPLORING YOUR OWN MEMORY AND CREATIVITY

Almost always, your best research aids will include your own memory and creativity. The following sections discuss four methods of exploiting the power of these mental resources. Each method can be useful at the beginning of your research and at many points along the way:

- Brainstorming
- Freewriting
- Drawing a picture of your topic
- Creating and studying a table or graph of your data

BRAINSTORMING

When you brainstorm, you generate thoughts about your subject as rapidly as you can through the spontaneous association of ideas, writing down whatever thoughts occur to you.

Brainstorming lets your thoughts run free.

The power of brainstorming arises from the way it unleashes your natural creativity. By freeing you from the confines imposed by outlines or other highly structured ways of organizing your ideas, brainstorming lets you follow your own creative lines of thought.

Brainstorming is especially helpful in enabling you to focus on the core ideas you want to communicate to your readers, whether in an overall message or in one part of a longer message. It also works well in group writing projects: When the members brainstorm aloud, the ideas offered by one person often spark ideas for the others.

Brainstorming Procedure

1. Review your knowledge of your readers and your communication situation.
2. Ask yourself, "What do I know about my subject that might help me achieve my communication purpose?"
3. As ideas come, write them down as fast as you can, using single words or short phrases. As soon as you list one idea, move on to the next thought that comes to you.
4. When your stream of ideas runs dry, read back through your list to see if your previous entries suggest any new ideas.
5. When you no longer have any new ideas, gather related items in your list into groups to see if this activity inspires new thoughts.

When brainstorming, avoid evaluating your ideas.

The key to brainstorming is to record ideas quickly without evaluating any of the thoughts that come to mind. Record everything. If you shift your task from generating ideas to evaluating them, you will disrupt the free flow of associations on which brainstorming thrives.

Here's the first part of a brainstorming session by Nicole. She wanted to write to her boss about ways to improve the quality control procedures at her company, which makes machines that keep patients alive during organ transplant operations. Sitting at her computer, Nicole began by simply keying in her initial thoughts on her subject.

While she was brainstorming, Nicole sometimes moved from one idea to a related one. At other times, she jumped to a completely unrelated thought.

Ideas for Quality Control Recommendations

Problem: Present system is unreliable
Everyone is "supposed" to be responsible for quality
No one has specific responsibilities
Workers feel rushed, sometimes ignore quality checks
People's lives are at stake
Near-fatal failure last year in Toronto
Need procedures to test critical components when they arrive from supplier
People follow their own shortcuts, using personal assembly techniques
Don't realize harm they could do
Product's overall record of performance is excellent

After running out of ideas, Nicole grouped related items under common headings. Note how organizing her notes around her three main topics spurred her to develop additional thoughts.

Ideas for Quality Control Recommendations

Importance of Quality Control
- Lives depend on it (near-fatal failure in Toronto)
- Avoid product liability lawsuits
- Keep sales up

New ideas suggested by outlining

Present System
- Overall record is excellent but could easily fail
- Everyone is supposed to be responsible for quality, but no one is
- Workers follow their own shortcuts and personal assembly techniques
- Workers feel rushed, sometimes ignore quality control checks
- Only consistent test is after the machine is built (some parts may cause the machine to fail due to flaws that are not detectable in the whole-machine test)

New idea

Strategies for a Better System
- Test critical parts as they are delivered from the supplier
- Insist that standard assembly procedures be followed
- Assign specific responsibilities—and rewards

New idea

New idea

FREEWRITING

Freewriting is very much like brainstorming. Here, too, you set your natural creativity free from the confines of structured thought. You rapidly record your ideas as they pop into your mind. Only this time, you write text rather than a list. The goal is to keep your ideas flowing.

Freewriting Procedure

1. Review your knowledge of your readers and your communication situation.
2. Ask yourself, "What do I know about my subject that will help me achieve my communication purpose?"
3. As ideas come, write them down as sentences. Follow each line of thought until you come to the end of it, then immediately pick up the next line of thought that suggests itself.
4. Write rapidly without making corrections or refining your prose. If you think of a better way to say something, start the sentence anew.
5. Don't stop for gaps in your knowledge. If you discover that you need some information you don't possess, note that fact, then keep on writing.
6. When you finally do run out of ideas, read back through your material to select the ideas worth telling your readers. Then discard the rest.

Freewriting can help you develop your main ideas.

Freewriting is especially helpful when you are trying to develop your main points. Use it when writing brief communications or parts of a long communication when you still need to interpret your data or formulate your recommendations.

Below is a sample of freewriting done by Michel, an employee of a company that makes precision instruments. He spent two weeks investigating technologies to be placed aboard airplanes for detecting microbursts and wind shear, two dangerous atmospheric conditions that have caused several crashes. He wrote this freewriting draft when deciding what to say in the opening paragraph of his report.

In Michel's freewriting, ideas are jumbled together, but they are now recorded so he can sort through and organize them.

> Wind shear and microbursts have been blamed for several recent airline crashes (find out which ones—St. John's?). People are studying several technologies for detecting these conditions. Then pilots can fly around them. The technologies might be used on the ground or in planes. Onboard devices would be much more helpful to pilots. Several technologies are being studied. The equipment would need to be approved by Transport Canada. It is eager for such devices to be developed. There would be a high demand for onboard devices from airlines. Many companies are working on them. The key point is we could make a lot of profit if we develop the right instrument first. We need to pick the most promising technology and develop it.

Freewriting produces ideas, not a draft.

The following passage shows the paragraph Michel wrote after freewriting. Note that it further develops some ideas from his freewriting, and it omits others. It is a fresh start—but one built on the ideas Michel generated while freewriting.

Through freewriting, Michel realized that profitability was a key point and so placed it first in his subsequent draft.

His ideas are now organized and clearly developed.

We have a substantial opportunity to develop and successfully market instruments that can be placed aboard airplanes to detect dangerous wind conditions called wind shear and microbursts. The conditions have been blamed for several recent air crashes, including one of a Lockheed L-1011 that killed 133 people. Because of the increasing awareness of the danger of these wind conditions, Transport Canada is encouraging research into a variety of technologies for detecting them. Most concern systems placed at airports or in large airplanes. We could establish a highly profitable niche by developing a system for use in small, private aircraft. In this report, I will review four major technologies, assessing the suitability of each for development by our company.

DRAWING A PICTURE OF YOUR TOPIC

Try thinking visually about your topic.

Another effective strategy for exploiting your memory and creativity is to explore your topic visually. Here are four kinds of diagrams that writers at work have found to be useful.

Flowchart

When you are writing about a process or procedure, try drawing a flowchart of it. Leave lots of space around each box in the flowchart so you can write notes next to it. Here's a flowchart that Nicole used to generate ideas for the report recommending improved quality control procedures in the manufacturing and delivery of the medical equipment sold by her employer.

Above the flowchart, Nicole wrote the ideas that occurred to her as she studied the chart.

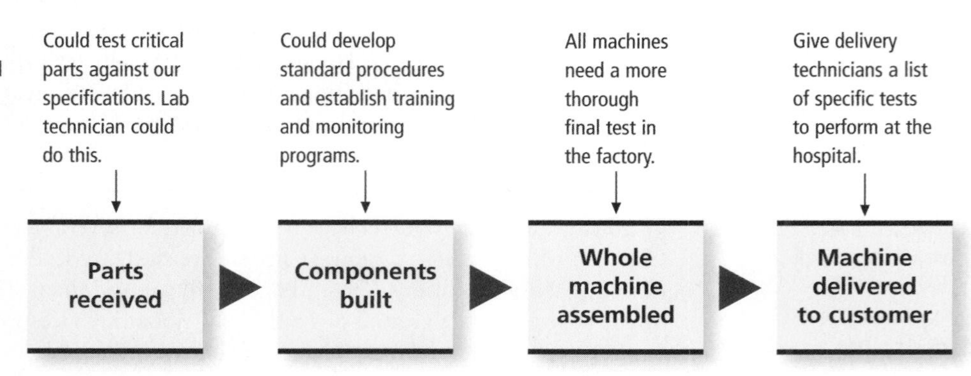

Matrix

A matrix is a table used to generate and organize ideas.

When you are comparing two or more alternatives in terms of a common set of criteria, drawing a matrix can aid you in systematically identifying the key features of each item being compared. Make a table in which you list the alternatives down the left-hand side and write the topics or issues to be covered across the top. Then, fill in each cell in the resulting table by brainstorming. Blank boxes indicate information you need to obtain. Michel created the following matrix on his computer, but he could also have made it with pencil and paper.

Matrix

System	How it Works	Limitations	Potential Competition
Doppler radar	Detects rapidly rotating air masses, like those found in wind shear	Technology still being researched	General Dynamics Hughes
Infrared detector	Detects slight increases in temperature that often accompany wind shear	Temperature doesn't always rise	CMC Electronics
Laser sensor	Sudden wind shifts affect reflectivity of air that lasers can detect	Provides only a 20-second warning for jets travelling at a typical speed	Marconi Exel Tech

Cluster Sketch

Your ideas radiate from the centre of the page.

Creating a cluster sketch is a simple, powerful technique for exploring a topic visually. Write your overall topic in a circle at the centre of a piece of paper, then add circles around the perimeter that identify the major issues or subtopics, joining them with lines to the main topic. Continue adding satellite notes, expanding outward as far as you find productive. Figure RM.1 shows a cluster sketch created by Ali, an engineer who is leading a team assigned to help a small city locate places where it can drill new wells for its municipal water supply.

A variation of the cluster sketch is the idea tree, shown in Figure RM.2 (page 318). At the top of a sheet of paper, write your main topic. Then list the main subtopics or issues horizontally below, joining them to the main topic with lines. Continue this branching as long as it is fruitful.

FIGURE RM.1
Cluster Sketch

Cluster sketches allow you to draw a map of your topics.

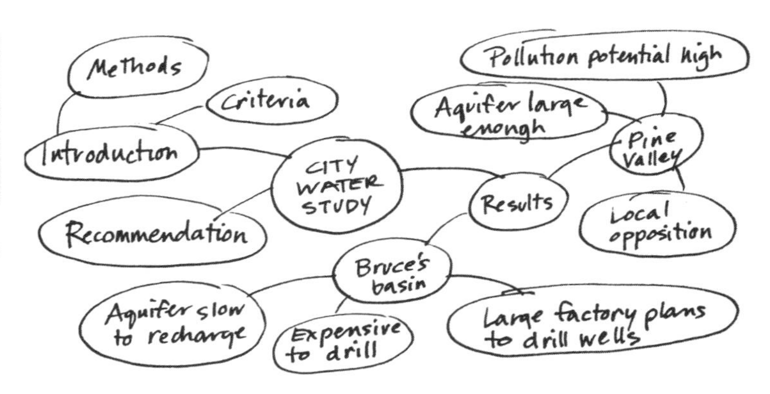

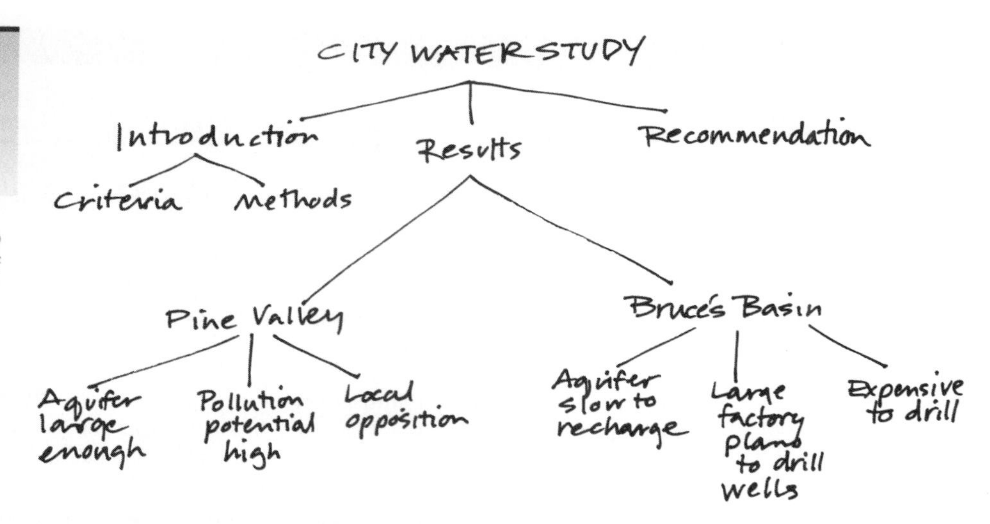

FIGURE RM.2
Idea Tree

Idea trees help you develop the hierarchical structure of your communications.

CREATING AND STUDYING A TABLE OR GRAPH OF YOUR DATA

Often at work you will need to write communications about data, such as the results of a test you have run, costs you have calculated, or production figures you have gathered. In such cases, many people find it helpful to begin their writing process by making the tables or graphs that they will include in their communication. Then they can begin to interpret the data arrayed before them, making notes about the data's meaning and significance to their readers.

SEARCHING THE INTERNET

The Internet has created a rich and constantly evolving aid to researchers. From your computer at home, school, or work, the Internet lets you read technical reports from companies such as IBM, download software, view pictures taken by NASA spacecraft in remote areas of the solar system, or join online discussions on an astonishing array of topics. Figure RM.3 lists just a few of the resources the Internet makes available to you.

Two types of tools can assist you in your Internet search. As a researcher, it's important that you understand the differences between the two.

WWW For links to a variety of search engines and subject directories, visit Reference Guide 4 at www.techcomm.nelson.com.

- **Search Engines.** Search engines allow you to locate websites that include one or more words that you enter into the search engine. You can also use them to search for images, PowerPoint presentations, MP3s, and other types of files posted on the Web. Although it might seem that a search engine is scouring the entire Web each time you submit a request, the engine actually operates much differently. At regular intervals, perhaps every few days or weeks, the search engine uses a computer program called a spider to crawl across the Web, extracting the words, Internet address, and other information from each location it visits. The spider delivers these results to a database, which indexes and stores the information. When you order a search, the search engine looks through its database. There are single-source search engines such as AltaVista and Google, but there are also meta search engines such as MetaCrawler and Dogpile, which search several single-source engines simultaneously.

FIGURE RM.3

Some Major Internet Resources for Research

Internet Resource	Examples
Corporate reports and information	IBM posts technical documents, Microsoft offers detailed information on its products, and the Canadian Wildlife Federation reports on its environmental projects. Thousands of other for-profit and non-profit organizations do the same. Examples IBM Research Papers on Networking http://www.research.ibm.com Microsoft product information and downloading http://www.microsoft.com Canadian Wildlife Federation http://www.cwf-fcf.org
Technical and scientific journals	Many technical and scientific journals are available online, though often only to people or through libraries that pay an online subscription fee. Examples *Journal of Cell Biology* http://www.jcb.org *IEEE Transactions on Software Engineering* http://www.computer.org/tse
Government agencies	Many government agencies have websites at which they provide reports, regulations, forms, and similar resources. Examples Canadian Space Agency http://www.space.gc.ca Health Canada http://www.hc-sc.ca Parks Canada http://www.pc.gc.ca
Interest group discussions	Internet sites called *newsgroups* enable people to ask questions, offer advice, and share insights online. Newsgroups exist on thousands of topics, many organized around professional or research interests. Browser to try Google http://google.com (click on "Groups")

- **Subject Directories.** These directories are created by people rather than computers. These people search the Internet for sites that are likely to be of interest to their users. Then they place each site in a highly structured framework that enables you to locate sites on a particular topic by browsing through a series of hierarchically organized menus. For instance, using the popular directory Yahoo to find information on basking sharks (the second largest of all sharks), you would first choose the category "Science," then "Biology," then "Zoology," then "Animals, Insects, and Pets," then "Fish," then "Saltwater," then "Species," then "Basking Shark."

CHOOSING THE SEARCH TOOLS TO USE

Advantages and disadvantages of search engines and subject directories.

Search engines and subject directories provide distinctly different kinds of support for your research. Which is better? There's no absolute answer. Because a search engine's spider returns information about every site it visits, a search engine does not distinguish between sites created by experts and those created by second graders. Consequently, searching through the hundreds or thousands of sites identified by a search engine can be tedious and time consuming. A subject directory can simplify your search because its database includes only a limited number of sites screened and selected by the people who created the directory. Thus it may return only a handful of sites in response to your search. On the other hand, a directory may miss many sites that could be very useful to you.

Different search engines work differently and produce different results. The same is true for subject directories.

Even among search engines, there are substantial differences that affect the results they produce for you. For example, the spiders for different search engines follow different paths and therefore produce different results. As mentioned above, a search for "basking shark food" with Teoma produced 6,760 sites. The same search with Google produced five times as many: 31,100. In addition to the number of sites identified, the differences between search engines affect such things as the types of sites found, the types of information included with the search results, the order in which the sites are listed, and the freshness of the sites (how recently the spiders have updated the database).

For most research, it's best to use more than one Internet search tool.

Subject directories also differ from one another. Depending on your subject area, you may wish to consult specialized search tools, such as Biz/ed and SciNet, that specialize in topics such as business and economics, or science and technology. There are also search portals, such as Eureka and HotSheet, that provide extensive lists of search tools, and invisible Web tools, such as CompletePlanet and Turbo 10. Large portions of the Web are not retrieved by search engines, and many sites are not listed in regular subject directories, so invisible Web tools attempt to list some of those "deep Web" sites. Because of such differences in results obtained using these various methods, it's often worthwhile to use more than one Internet search tool to ensure that you are obtaining a thorough, balanced understanding of the topic you are researching on your readers' behalf.

CONDUCTING KEYWORD SEARCHES

Your success in keyword searches depends on your ability to use your vocabulary skillfully.

When the search tools look for websites for you, they look for sites that use the words, called keywords, that you type into your computer. Your skill at playing what is essentially a word game can greatly increase the efficiency and effectiveness of your search. Choose your keywords carefully and creatively. Your goal is to use the words that match the words used in the sites that will be most valuable to you. For example, if you search for "genetic engineering" you will undoubtedly find a multitude of websites. That's fine, except that many of the sites on this topic that are the most valuable for many research purposes use the term "gene splicing" instead. Because you can't predict in advance what terms others might use, you need to use your own vocabulary flexibly and creatively. If you don't locate exactly what you need with the first words you enter, brainstorm. Think of synonyms for the words you've listed, scan through an encyclopedia or reference book to see what terms it uses, or ask a reference librarian (reference librarians know a great deal about electronic resources as well as about printed ones).

Boolean operators enable you to sharpen your search.

Use Boolean operators to narrow, expand, or shape your search results. If you enter two or more words, most search engines will automatically look for all sites that have *any* of these words. With Boolean operators, however, you can define other relationships among your keywords.

Using Boolean Operators

When you use these words . . .	the computer searches for . . .
laser AND surgery	only sites that contain both words
laser OR surgery	all sites that contain either word
surgery NOT laser	only sites that contain the first word but not the second
Many search tools also include other operators	
"laser surgery"	only sites with this exact phrase: these words next to one another in this order
laser NEAR surgery	all sites that contain the terms within a certain number of words of one another (e.g., ten words or fifteen words)
surg* (a wildcard)	all sites that contain the root word within another word (e.g., *surgeon, surgery, surgical*)

Different search engines and subject directories use different sets of operators, and some use different symbols for the same operators. For instance, some use the plus sign (+) instead of AND, and the minus sign (−) instead of NOT. With some, you can click on a link for advanced search functions that are equivalent to the Boolean operators, as well as other features for sharpening your search, such as by indicating the language used in the site and whether the site has images or a movie. Figure RM.4 (page 322) shows an example. An especially helpful feature of some search tools allows you to search only certain domains.

EVALUATING YOUR SEARCH RESULTS

Use the list of results to target the most promising sites.

Whether you are using a search engine or a subject directory, you need to evaluate the results you receive. Start by evaluating the list of sites produced by your search tool. The information in this list can often tell you whether it would even be worth taking the time required to link to the site. Although different search tools display different sets of information, the following result from AltaVista illustrates the way you can use this information to evaluate sites without even going to them.

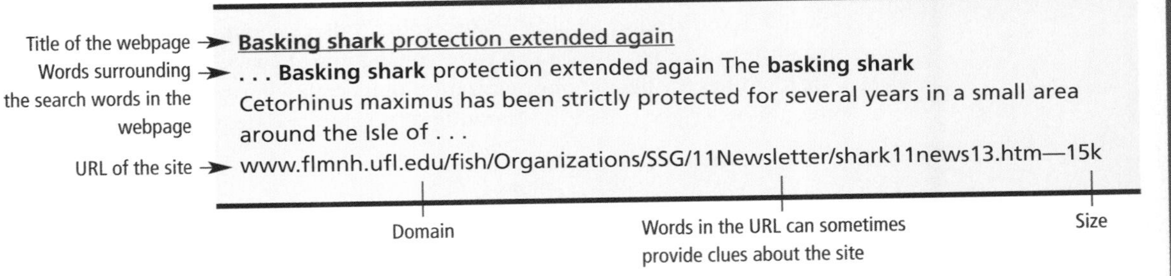

FIGURE RM.4

Advanced Search Page for an Internet Search Engine

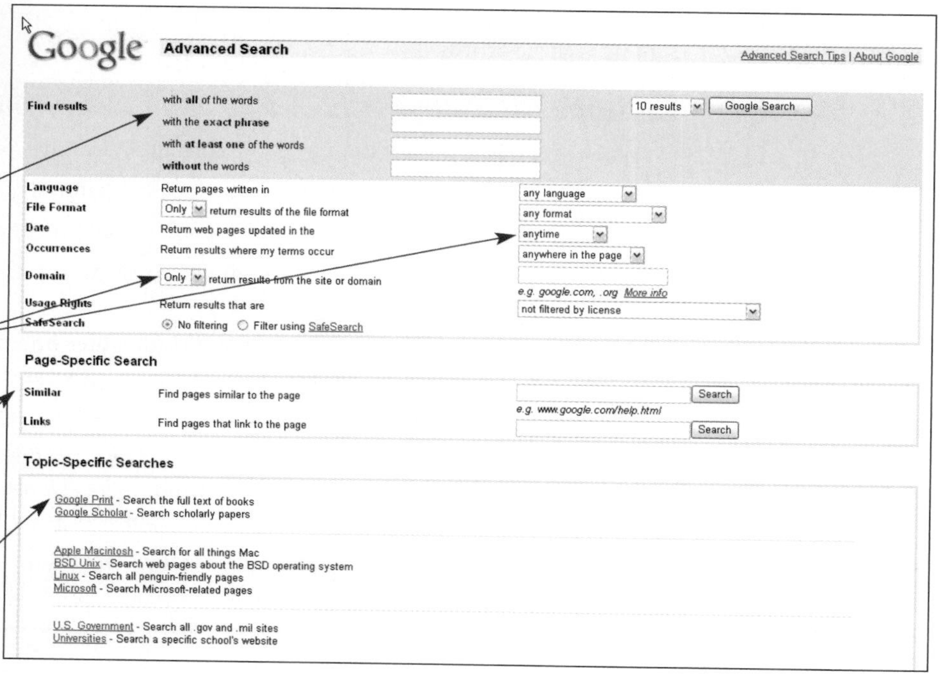

Google Advanced Search

Advanced Search Tips | About Google

Find results — with **all** of the words / with the **exact phrase** / with **at least one** of the words / **without** the words — 10 results — Google Search

Language — Return pages written in — any language

File Format — Only — return results of the file format — any format

Date — Return web pages updated in the — anytime

Occurrences — Return results where my terms occur — anywhere in the page

Domain — Only — return results from the site or domain — e.g. google.com, .org More info

Usage Rights — Return results that are — not filtered by license

SafeSearch — No filtering — Filter using SafeSearch

Page-Specific Search

Similar — Find pages similar to the page — Search — e.g. www.google.com/help.html

Links — Find pages that link to the page — Search

Topic-Specific Searches

Google Print - Search the full text of books
Google Scholar - Search scholarly papers

Apple Macintosh - Search for all things Mac
BSD Unix - Search web pages about the BSD operating system
Linux - Search all penguin-friendly pages
Microsoft - Search Microsoft-related pages

U.S. Government - Search all .gov and .mil sites
Universities - Search a specific school's website

These alternatives let you define the relationships among the keywords you enter.

The drop-down menus enable you to sharpen your search in several ways.

For these two features, you can follow the threads from a webpage you have found useful to other webpages that might also help.

These alternatives allow you to perform topic-specific searches on subsets of information posted on the Internet.

The site's domain can be especially helpful because, on the Internet, different types of organizations are assigned to different domains, which are included in the addresses of the sites created by those organizations. Here are a few.

.ca	Canadian generally, including educational and government
.com	Commercial (sites for businesses)
.edu	US educational (sites for four-year universities)
.gov	US government
.org	Not-for-profit organizations
.mil	US military
.pro	Certified professionals

Depending on what you are looking for, sites in any of these domains may be good sources or bad ones. For example, a .com site may provide useful details about the features of products, but biased information about product quality and lawsuits against a corporation.

Finally, if you decide to visit a site, evaluate its contents critically. In addition to applying the evaluative criteria described in Chapter 13 (page 306), determine whether the site identifies the person or organization that created it, whether you can contact the creator, and when the site was last modified (how up to date it is).

KEEPING RECORDS

Bookmark valuable sites and record their URLs.

Finally, when conducting Internet research, keep careful records of the sites you find valuable. It's easy to lose your way when searching the Internet, which can make it difficult to relocate a site you need to visit again. Most browsers provide a bookmark feature that lets you add any page you are visiting to a personalized menu of sites you can return to with a single click. Even so, it's best to write down the URL of any site whose information you believe you will provide to your readers.

Be sure to record the date you visit each site. Sites can change and even disappear suddenly, so this date is a crucial part of your bibliographic citation, as Appendix B explains.

USING THE LIBRARY

For many communications, the library, or learning resource centre, will be your best source of information and ideas.

Libraries are as much online as on the shelf.

The first step in using the library effectively is to discard the old image of it as a place that primarily houses books and periodicals. Although libraries still feature these publications, most are now as much online as on the shelf. In fact, many libraries are so computerized that you don't even need to enter the buildings to use their resources. You can access most of their resources online.

Library resources fall into two broad categories:

Major library resources.

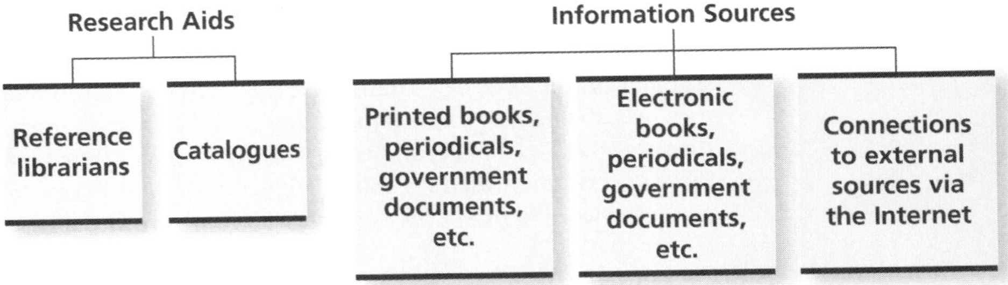

Generally, your excursions in library research will begin with one of the research aids, which can guide you to the most productive information sources. The following sections will help you use the research aids productively and also introduce some of the information sources with which you may not be familiar.

REFERENCE LIBRARIANS

You will rarely find any research aid more helpful than reference librarians. They can tell you about specialized resources that you may not be aware of, and they can explain how to use the time-saving features of these resources.

Tell the reference librarian your communication's objectives.

Reference librarians will be able to give you the best help if you indicate very specifically what you want. In addition to stating your topic, describe what the purpose of your communication is, who your readers are, and how your readers will use your communication.

LIBRARY CATALOGUE

The library catalogue lists the complete holdings of a library, including books, periodicals, journals, technical reports, music, and visual materials. In most libraries, the catalogue is computerized so that you can search for items in several ways. If you are looking for a particular book whose title you know or for work written or edited by a person whose name you know, library catalogues are very simple to use. However, when you begin by looking for information about a particular topic, your success may depend on your ingenuity and knowledge of how to use the online catalogue that most libraries have. Figure RM.5 shows the initial menu for a typical online catalogue.

To search for a specific topic, you have two choices:

Two kinds of library searches

- **Subject search.** To aid researchers, librarians include subject headings in the record for each library item. When you indicate that you want to do a subject search, the computer will prompt you to enter the words that identify the subject you are looking for. The computer will search through all items that have been tagged with the exact words you entered.
- **Keyword search.** When you indicate that you wish to conduct a keyword search, you will also be prompted to enter the words that identify your subject. This time, however, the computer will search the entire contents of all its records, including the title, author, and subject lines as well as contents lists and other information that particular records might have.

CONDUCTING SUBJECT AND KEYWORD SEARCHES

Subject and keyword searches in online library catalogues are very similar to keyword searches on the Internet. Therefore, all the advice given on page 320 applies. However, there is also one very important difference: When identifying the words used to describe the subject of a book, librarians use a formal and restricted set of terms that are defined in a large volume titled the *Library of Congress Subject Headings List*. In the discussion about keyword searches on the Internet, you learned that some websites might use the

FIGURE RM.5

Menu for Typical Online Library Catalogue

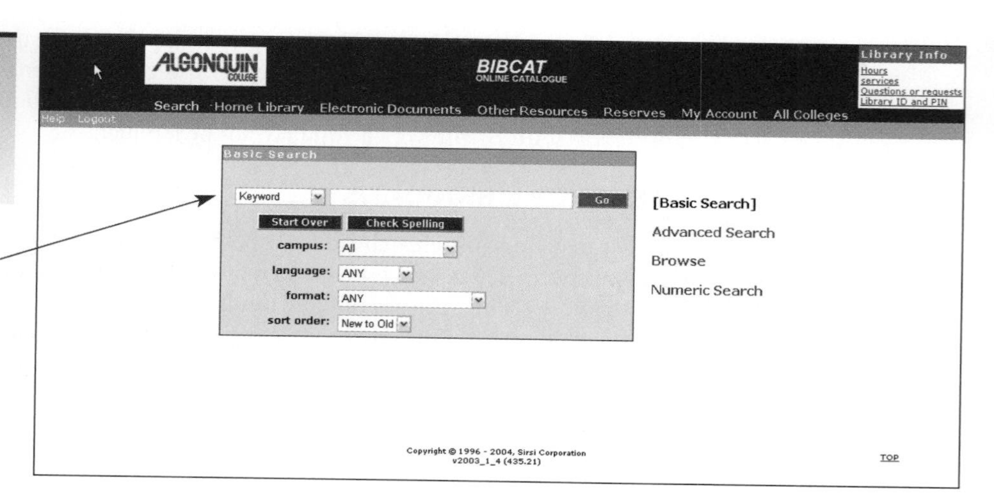

If you aren't looking for a specific author or title, use the *Keyword* or *Subject* search. You can also search under *Series* or *Periodical* Title.

term "gene splicing" and others might use "genetic engineering." In library subject headings, only "genetic engineering" is used. Consequently, while "gene splicing" will produce some results in an Internet search, it won't produce any in a subject search in a library. For help in determining the correct terms for subject searches, you have three resources:

- Many online library catalogues will tell you the correct term if you use an incorrect one that it recognizes as a synonym.
- The *Library of Congress Subject Headings List* is available at any library.
- Reference librarians are most willing to assist you.

Many online library catalogues will let you choose between abbreviated and extended displays of your search results. Extended displays are usually more helpful because they give you more information to consider as you decide whether or not to look at the entry for a particular item. Figures RM.6, RM.7, and RM.8 (pages 326–27) show the sequence of screens used to locate a book on genetic engineering.

REFINING AND EXTENDING YOUR SEARCH

If your initial effort produces too few results, an overwhelming number of them, or an inadequate quality or range of them, there are several ways to refine and extend your search:

- **Look in the catalogue entries of books you find for leads to other books.** Catalogue entries not only name the subject headings under which a book is catalogued, but they also provide links to lists of other works that also have those subject headings (see Figure RM.8, page 327).
- **Use Boolean operators and similar aids.** If you receive too many results, you may narrow the search in many of the ways described in the discussion of Internet searches (see pages 320–321).
- **Use other resources.** Don't limit yourself only to resources you locate through the library catalogue. Your best source of information may be a corporate publication or other item not listed there. A reference librarian can help you identify other aids to use.
- **When you go to the library shelves, browse.** Sometimes books that will assist you are located right next to books you found through the library catalogue. Don't miss the opportunity to discover them. Browse the shelves.

INDEXES

Indexes are research aids that focus on specific topics or specific types of publications. Most catalogue the contents of periodicals, but some include television programs, films, and similar items. Most indexes are available online or on CD-ROM. At some libraries, indexes can be accessed through the same computer screens that provide access to the catalogue.

To use indexes well, you need to select ones that cover the kinds of material you want:

- **General periodical indexes.** These index the contents of publications directed at a general audience. Examples include MasterFile and the Canadian Reference Centre.

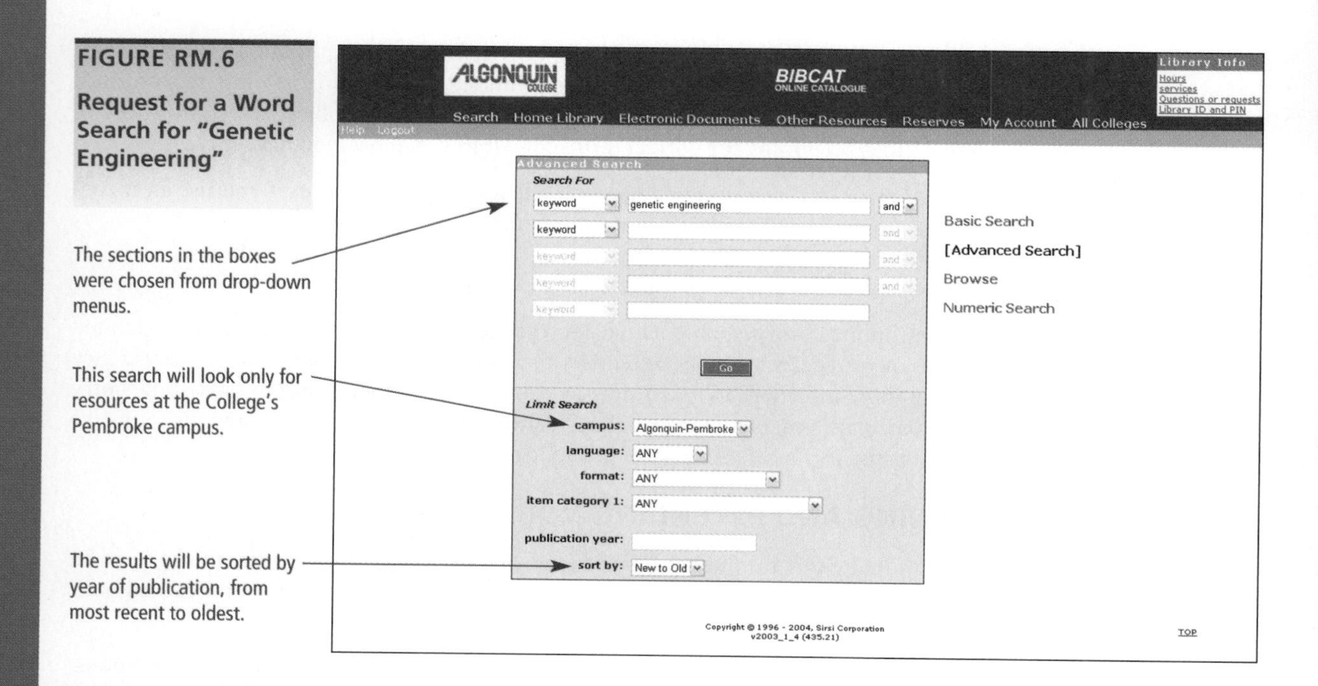

FIGURE RM.6

Request for a Word Search for "Genetic Engineering"

The sections in the boxes were chosen from drop-down menus.

This search will look only for resources at the College's Pembroke campus.

The results will be sorted by year of publication, from most recent to oldest.

FIGURE RM.7

Results of a Word Search for "Genetic Engineering"

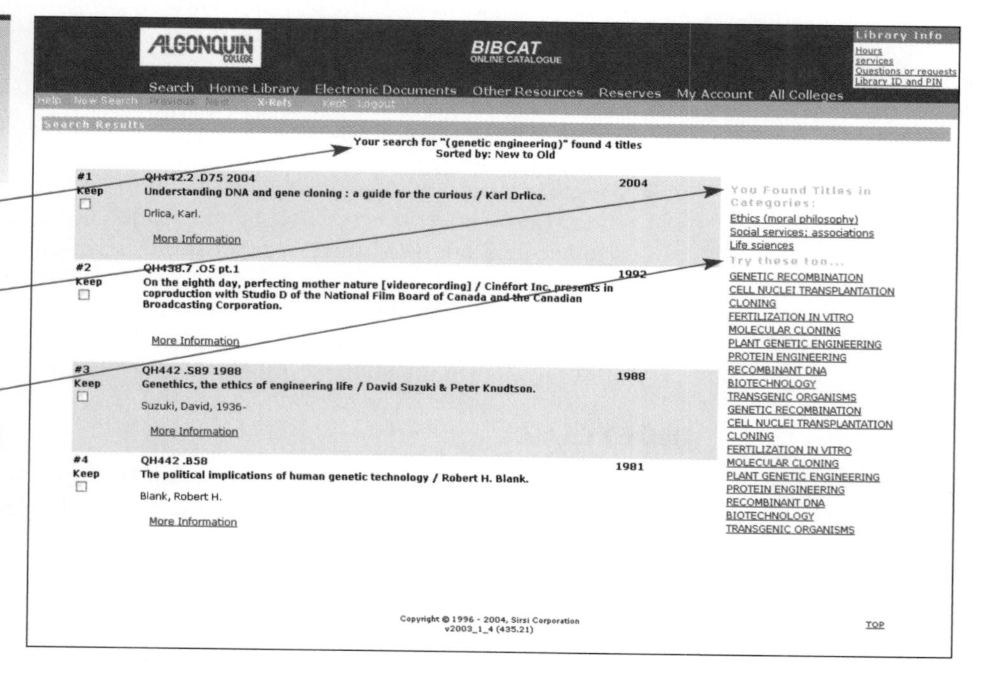

This line indicates the total number of items this search produced (4).

This section indicates the subject areas of the items that were found in this search.

This section indicates related subject areas that might prove helpful as keywords or subject terms for subsequent searches on this topic.

FIGURE RM.8

Catalogue Entry for One Book Listed in a Search for "Genetic Engineering"

Bibliographic information about the book.

Clicking on one of these options will allow you to keep this item information, and variations on this item information, for emailing or printing.

Clicking on these keywords can provide leads to more items.

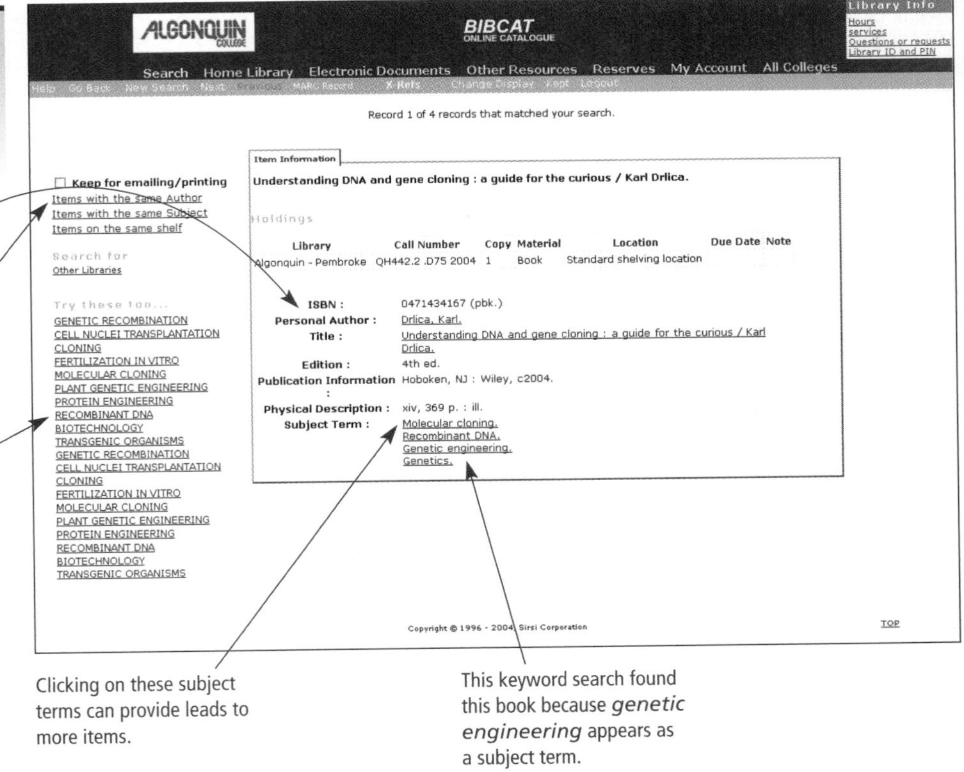

Clicking on these subject terms can provide leads to more items.

This keyword search found this book because *genetic engineering* appears as a subject term.

- **Specialized periodical indexes.** Almost every field has at least one. Examples are Canadian Business and Current Affairs, Medline Plus, and ERIC (education, psychology, and sociology).
- **Newspaper indexes,** such as Canadian Newsstand.

Some indexes include abstracts.

Some indexes not only list articles, but also provide an abstract (or summary) of each one. Figure RM.9 (page 328) shows an index entry that includes an abstract. By scanning through an abstract, you can usually tell whether reading the entire article would be worthwhile.

Find out what indexing terms are used.

Whether you are using a printed or an online index or database, you can often speed your search by looking at the index's thesaurus. Different indexes use different sets of terms. Most publish this list in a thesaurus that can be accessed through a menu selection in the online version or found in the front or back of the printed version.

Use "Help" to customize a search.

Online indexes work the same way that library catalogues do. You can search by the same variables: author, title, words. Searches can also be limited in similar fashion, although different indexes do this in different ways, so you should look at the "Help" feature for instructions.

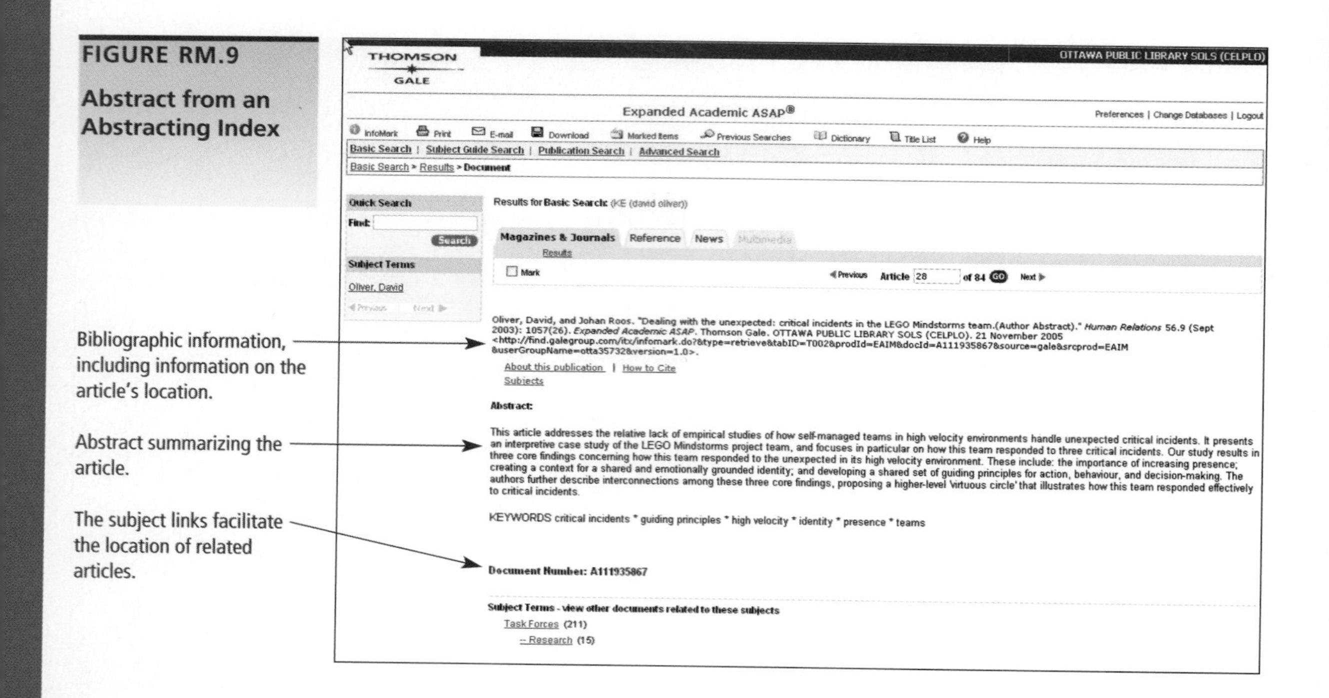

FIGURE RM.9

Abstract from an Abstracting Index

REFERENCE WORKS

When you hear the term *reference works*, you probably think immediately (and quite correctly) of encyclopedias, dictionaries, atlases, and similar storehouses of knowledge, thousands of pages long. What you may not realize is that many of these resources, such as the *Encyclopedia Britannica* and the *Canadian Encyclopedia*, are now available online or on CD-ROM, so that finding information in them can be very quick and easy.

In addition to such familiar reference works as the *Encyclopedia Britannica*, thousands of specialized reference works exist, some of which may relate to your area of study. For example, there are works such as *Gray's Anatomy*, *Robert's Rules of Order*, and the *Encyclopedia of World History*.

GOVERNMENT DOCUMENTS

Every year, the Government of Canada and provincial and regional governments publish thousands of documents, ranging from pamphlets and brochures to periodicals, technical reports, and books. Some are addressed to the general public, while others are addressed to specialists in various fields. Sample titles include *Aboriginal Diabetes Initiative*, *Anti-Personnel Land Mines*, and *Artists in the Labour Force*.

The following websites are especially helpful. A reference librarian can help you find many others.

Indexes to Canadian Government Publications

- Government of Canada Publications
  www.canada.gc.ca/publications
- Statistics Canada
  www.statcan.ca
  www.statcan.ca/english/Estat

COMPUTERIZED FULL-TEXT SOURCES

You can download full texts.

Many libraries also offer computer access to the full text of various sources. These include standard reference works, such as the *Encyclopedia Britannica,* and specialized publications such as scientific journals. These utilities allow you to search for topics in the same sorts of ways described in the discussion of the library catalogue. When you locate an article of interest, you can read the text, download it, or print it.

INTERVIEWING

At work, your best source of information will often be another person. In fact, people will sometimes be your only source of information because you'll be researching situations unique to your organization or its clients and customers. Or you may be asking a subject matter expert (SME) for information that this person possesses that is not yet available in print or from an online source.

The following advice focuses on face-to-face interviews, but it applies also to telephone interviews, which are quite common in the workplace.

PREPARING FOR AN INTERVIEW

Preparing for an interview involves three major activities:

Take a reader-centred approach to selecting your interviewee.

- **Choose the right person to interview.** Approach this selection from your readers' perspective. Pick someone you feel confident can answer the questions your readers are likely to ask in a way that your readers will find useful and credible. If you are seeking someone to interview who is outside your own organization, the directories of professional societies may help you identify an appropriate person.
- **Make arrangements.** If you expect the interview to take more than a few minutes, contact the person in advance to make an appointment. Let the person know the purpose of the interview. This will enable him or her to start thinking about how to assist you before you arrive. Be sure to say how long you think the interview will take. This will enable your interviewee to carve out time for you from all of his or her other responsibilities. Also, if you would like to record the interview, ask permission in advance.

Decide in advance what you hope to learn from the interview.

- **Plan the agenda.** As the interviewer, you will be the person who must identify the topics that need to be discussed. Often, it's best simply to generate a list of topics to inquire about. But if there are specific facts you need to obtain, identify them as well. To protect against forgetting during the interview, bring a written list of your topics and specific questions. For advice on phrasing questions, see the section on surveys (pages 332–33). If possible, provide your list of topics and specific questions to the interviewer beforehand.

CONDUCTING THE INTERVIEW

Do only 10 percent to 20 percent of the talking.

Unless you are seeking a simple list of facts, your goal in an interview should be to engage the other person in a conversation, not a question-and-answer session. In this conversation, your goal should be to have the other person do 80 percent to 90 percent of the talking—and to have him or her focus on the information you need. To achieve these goals, you will need to ask your questions well and maintain a productive interpersonal relationship with your interviewee. Figure RM.10 suggests practical steps that you can take.

It's especially important that you assume leadership for guiding the interview. You are the person who knows what information you need to obtain on your readers' behalf. Consequently, you may need to courteously redirect the conversation to your topics.

CONCLUDING THE INTERVIEW

End on time.

During the interview, keep your eye on the clock so that you don't take more of your interviewee's time than you requested. As the time limit approaches, do the following:

- **Check your list.** Make sure that all your key questions have been answered.
- **Invite a final thought.** One of the most productive questions that you can ask near the end of an interview is, "Can you think of anything else I should know?"
- **Open the door for follow-up.** Ask something like this: "If I find that I need to know a little more about something we've discussed, would it be okay if I contacted you?"
- **Thank your interviewee.** If appropriate, send a brief thank you note by letter, memo, or email.

CONDUCTING A SURVEY

While an interview enables you to gather information from one person, a survey enables you to gather information from *groups* of people.

Surveys support practical decision making.

On the job, surveys are almost always used as the basis for practical decision making. Manufacturers survey consumers when deciding how to market a new product, and employers survey employees when deciding how to modify human resources policies or benefit packages. Some surveys, such as those used to predict the outcomes of federal and provincial elections, require the use of specialized techniques that are beyond the scope of this book. However, in many situations, you will be able to conduct surveys that provide a solid basis for on-the-job decision making if you follow the suggestions given in the following sections.

FIGURE RM.10

Creating a Productive Conversation During an Interview

Goal	Guidelines
Establish rapport.	■ Arrive on time. ■ Thank the person for agreeing to meet with you.
Explain your goal.	■ Tell what you are writing and who your readers will be. ■ Explain the use your readers will make of your communication. ■ Describe the outcome you desire.
Ask questions that encourage discussion.	■ Use open questions that ask the interviewee to explain, describe, and discuss. They can elicit valuable information that you might not have thought to ask for. Avoid closed questions that request a yes/no or either/or response. Closed question Does the present policy create any problems? Open question What are your views of the present policy? ■ Use neutral, unbiased questions Biased question Don't you think we could improve operations by making this change? Neutral question If we made this change, what effect would it have on operations? ■ Begin with general questions, supplemented by more specific follow-up questions that seek additional details important to you. General question Please tell me the history of this policy. Follow-up question What role did the labour union play in formulating the policy?
Show that you are attentive and appreciative.	■ Maintain eye contact and lean forward. ■ Respond with an occasional "uh-huh" or "I see." ■ Comment favourably on the interviewee's statements. Examples "That's helpful." "I hadn't thought of that." "This will be useful to my readers."
Give your interviewee room to help you.	■ If the interviewee pauses, be patient. Don't jump in with another question. Assume that he or she is thinking of some additional point. Look at him or her to convey that you are waiting to hear whatever he or she will add. ■ If the interviewee begins to offer information out of the order you anticipated, adjust your expectations.
Keep the conversation on track.	■ If the interviewee strays seriously from the topic, find a moment to interrupt politely in order to ask another question. You might preface the question by saying something like this: "My readers will be very interested to know . . . "
Be sure you understand and remember.	■ If anything is unclear, ask for further explanation. ■ On complicated points, paraphrase what your interviewee has said and then ask, "Have I understood correctly?" ■ Take notes. Jot down key points. Don't try to write down everything because that would be distracting and would slow down the conversation. ■ Double-check the spelling of names, people's titles, and specific figures.

WRITING THE QUESTIONS

The first step in writing survey questions is to decide exactly what you want to learn. Begin by focusing on the decisions that your information will help your readers make. This will help you determine what sorts of information your survey must yield. Consider the following example.

Formulate survey questions based on the factors your readers will consider when making their decision.

> Kevin works for a small restaurant chain that has asked him to study the feasibility of opening a doughnut shop next to campus. He already has investigated possible sites and looked into licensing, insurance, wholesale suppliers, and related matters. Now he must find out if there would be enough business to make the shop profitable. As he thinks about the decisions that will be based on his report, he realizes that his employer will be interested not only in predicting the amount of sales but also in learning what kinds of doughnuts, pastries, coffee, and other products to offer; what prices to charge; and even what the opening and closing hours should be. Consequently, Kevin must design questions on each of these areas.

The following suggestions will help you create an effective questionnaire that provides useful information and elicits the cooperation of the people you ask to fill it out.

- **Mix closed and open questions.** *Closed questions* allow only a limited number of possible responses. They provide answers that are easy to tabulate. *Open questions* allow the respondent freedom in devising the answer. They provide respondents an opportunity to react to your subject matter in their own terms. See Figure RM.11.

  You may want to follow each of your closed questions with an open one that simply asks respondents to comment. A good way to conclude a survey is to invite additional comments.
- **Ask reliable questions.** A *reliable* question is one that every respondent will understand and interpret in the same way. For instance, if Kevin asked, "Do you like high-quality pastries?" different readers might interpret the term "high-quality" in different ways. Kevin might instead ask how much the respondents would be willing to pay for pastries or what kinds of snacks they like to eat with their coffee.
- **Ask valid questions.** A *valid* question is one that produces the information you are seeking. For example, to determine how much business the doughnut shop might attract, Kevin could ask either of these two questions:

Invalid

How much do you like doughnuts?

Valid

How many times a month would you visit a doughnut shop located within three blocks of campus?

The first question is invalid because the fact that students like doughnuts does not necessarily mean that they would patronize a doughnut shop. The second question is valid because it can help Kevin estimate how many customers the shop would have.

FIGURE RM.11

Closed and Open Questions for Surveys

Closed Questions	
Forced Choice	▪ Respondents must select one of two choices (yes/no, either/or). Example Would you buy doughnuts at a shop near campus, yes or no?
Multiple Choice	▪ Respondents select from several predefined alternatives. How many times a month would you visit the shop? _____ 1 to 2 _____ 3 to 4 _____ 5 or more
Ranking	▪ Respondents indicate an order of preference. Example Please rank the following types of doughnuts, using a 1 for your favourite, and so on.
Rating	▪ Respondents pick a number on a scale. Example Please circle the number on the following scale that best describes the importance of the following features of a doughnut shop: Music Unimportant 1 2 3 4 5 Important Tables Unimportant 1 2 3 4 5 Important
Open Questions	
Fill in the Blank	▪ Respondents complete a statement. Example When deciding where to eat a late-night snack, I usually base my choice on _________.
Essay	▪ Respondents can frame responses in any way they choose. Example Please suggest ways we could make a doughnut shop that would be appealing to you.

- **Avoid biased questions.** Don't phrase your questions in ways that seem to guide your respondents to give a particular response.

Biased | Wouldn't it be good to have a coffee shop near campus?

Unbiased | How much would you like to have a coffee shop near campus?

- **Place your most interesting questions first.** Save questions about the respondent's age or similar characteristics until the end.
- **Limit the number of questions.** If your questionnaire is lengthy, people may not complete it. Decide what you really need to know and ask only about that.
- **Test your questionnaire.** Even small changes in wording may have a substantial effect on the way people respond. Questions that seem perfectly clear to you may appear puzzling or ambiguous to others. Before completing your survey, try out your questions with a few people from your target group.

Be sure that your questionnaire works before you distribute it.

CONTACTING RESPONDENTS

There are three methods for presenting your survey to your respondents:

Don't bias your respondents' answers.

- **Face-to-face.** In this method, you read your questions aloud to each respondent and record his or her answers on a form. It's an effective method of contacting respondents because people are more willing to cooperate when someone asks for their help in person than they are when asked to fill out a printed questionnaire. The only risk is that your intonation, facial expressions, or body language may signal that you are hoping for a certain answer. Research shows that respondents tend to give answers that will please the questioner.
- **Telephone.** Telephone surveys are convenient for the writer. However, it can sometimes be difficult to use a directory to identify people who represent the group of people being studied.

Response rates can be very low.

- **Email or handout.** Emailing or handing your survey forms to people you hope will respond is less time consuming than conducting a survey face-to-face or by telephone. Generally, however, only a small portion of the people who receive survey forms in these ways actually fill them out and return them. Even professional survey specialists typically receive responses from only about 20 percent of the people they contact.

SELECTING YOUR RESPONDENTS

At work, writers sometimes present their survey questions to every person who belongs to the group whose attitudes or practices they want to learn about. For example, an employee assigned to learn what others in her company feel about a proposed change in health care benefits or a switch to flextime scheduling might send a survey questionnaire to every employee.

Your sample should reflect the composition of the overall group.

However, surveys are often designed to permit the writers to generalize about a large group of people (called a *population*) by surveying only a small portion of individuals in the group (called a *sample*). To ensure that the sample is truly representative of the population, you must select the sample carefully. Here are four types of samples you can use:

- **Simple random sample.** Here, every member of the population has an equal chance of being chosen for the sample. If the population is small, you could put the name of every person into a hat, then draw out the names to be included in your sample. If the population is large—all the students at a university, for example—the creation of a simple random sample can be difficult.
- **Systematic random sample.** To create a systematic random sample, you start with a list that includes every person in the population—perhaps by using a phone book or student directory. Then you devise some pattern or rule for choosing the people who will make up your sample. For instance, you might choose the fourteenth name on each page of the list.

Convenience samples can give unreliable results.

- **Convenience sample.** To set up a convenience sample, you select people who are handy and who resemble in some way the population you want to survey. For example, if your population is the student body, you might knock on every fifth door in your dormitory, or stop every fifth student who walks into the library. The weakness of such samples is obvious: From the point of view of the attitudes or behaviours you want to learn about, the students who live with their parents or in apartments may be significantly different from those who live in dorms, just as those who don't go to the library may differ in substantial ways from those who do.

- **Stratified sample.** Creating a stratified sample is one way to partially overcome the shortcomings of a convenience sample. For instance, if you know that 15 percent of the students in your population live at home, 25 percent live in apartments, and 60 percent live in dormitories, you would find enough representatives of each group so that they constituted 15 percent, 25 percent, and 60 percent of your sample. Even if you can't choose the people in each group randomly, you would have made some progress toward creating a sample that accurately represents your population.

Use enough respondents to persuade your readers.

When creating your sample, you must determine how many people to include. On one hand, you want a manageable number; on the other hand, however, you also want enough people to form the basis for valid generalizations. Statisticians use formulas to decide on the appropriate sample size, but in many on-the-job situations, writers rely on their common sense. One good way to decide is to ask what number of people your readers would consider to be sufficient.

CASE

ADVISING JP

This morning you stop outside the Bagel Bar to talk with Jean-Paul Leon, a third-year human resource administration student. JP tells you with a mixture of excitement and anxiety that he has finally decided to join the other graduating students who are busily looking for work. He has drafted a résumé and begun writing letters to employers listed on the Career Services website.

"Look!" he exclaims. "One of the aerospace companies I'm writing to is mentioned in this article that Professor Samhaber asked us to read." He holds out an article from *Human Resources Professional*. "They've moved to decentralized Human Resources. Each of their business units has its own HR Specialists who deal with all recruitment and selection, compensation and benefits, and training and development requirements. It's been a huge success—it works better than one impersonal Human Resources Department centralized in a corporate location. This sounds like such a great place to work—a big company that's based on teams." Then his excitement turns to anxiety. "I'm worried that they won't like my résumé and application, though. I've gone over both again and again, and my roommate has, too. But I'm still worried."

As you try to reassure him, he pulls out his drafts and places them into your hands. "Take a look at them and tell me what you really think. I need all the help I can get." You have to meet with your project group for another class, but you agree to look over his drafts and email him later.

Now it's late, and you've started to read JP's résumé and letter. As you do, you think back over some of the things you know about him. He's an active and energetic person, talkative and extroverted. He knew everybody in the program by the end of the first week. Throughout his time at school, he has spent lots of time with a group called Angel Flight, a volunteer organization that sponsors service activities on campus and off. In fact, this past year you've seen less of him outside class because he has spent so much time serving as Angel Flight's volunteer recruiter. "I have to ensure that I get good volunteers from the first-year students," he told you once, "so that there's a succession of people moving through the organization. If everyone graduates and leaves all at once, there's no one to take their place."

Of course, JP had to work throughout his time at school. His father and mother expected him to keep working in the family butcher shop, particularly on the weekends when his ability to speak English and French helped them keep their customers well served. But he still kept a good GPA of 3.6, even though he didn't have the free time that some other students had.

The ability you admire most in JP, though, is his skill as a DJ for student events on campus. He likes to play with Virtual DJ software, and he's really good at mixing. He seems to have a sense for what the crowd will react well to. He's a great guy to go clubbing with, because he's so familiar with the software the DJs use. He can often predict the type of beat, or even the track, that the VJ is going to move into next. And of course, JP always knows lots of people wherever he goes. You have met a lot of good people through JP.

YOUR ASSIGNMENT

Decide what you will say to JP about his letter and résumé. What strengths will you praise? What changes will you suggest? What questions will you ask to determine whether he might include additional information? Assume that his résumé will be read first by a person, not a scanner.

854 Graveline St.
Gatineau, QC
J8P 3H1

February 12, 2007

Judy Leger
Human Resources Office
CTI Aerospace
555 rue Barrage
Longueuil, QC
H4T 2P2

Dear Ms. Leger:

I saw your ad on the college website. I am very impressed with your company. I hope that you will consider me for an opening in Human Resources.

In May, I will graduate from Algonquin College's three-year business program, specializing in human resource administration. I learned a great deal about all of the human resources functions. Furthermore, I have gained a through overview of the Canadian business environment, and I have studied successful and unsuccessful recruitment campaigns through the case study method.

In addition to my educational qualifications, I have experience in customer service and recruitment. While working at our family business, I had many opportunities to apply the knowledge and skills that I learned in college. Likewise, in my extracurricular activities, I have gained experience working and communicating with people. For example, I acted as Angel Flight's volunteer recruiter during the past year. I also worked as a part-time DJ for student events.

I would like to talk to you in person about my qualifications. Please tell me how that can be arranged.

Sincerely,

JP Leon

JP Leon

JP LEON
854 Graveline St.
Gatineau, QC J8P 3H1
(819) 839-0009

PERSONAL

Born: March 17, 1985
Health: Excellent
Willing to relocated

PROFESSIONAL OBJECTIVE

To work in human resources.

EDUCATION

Algonquin College, Ottawa, Ontario.
Diploma in Human Resource Administration.
May 2005.

Courses included Business Administration, Management Information Systems, Software Applications, Global Business Practices, various Human Resources courses.

WORK EXPERIENCE

Boucherie Leon
2001–present

Served customers in the family butcher shop. Helped customers choose their purchases and listened politely to their complaints. Cash register operation. Stocked coolers and shelves.

JP's DJ Services
2002–present

Provided music for student events on campus.

ACTIVITIES

Angel Flight, Volunteer Recruiter

Human Resources Club, member

REFERENCES

Sebastian Leon
854 Graveline St
Gatineau, QC

Professor Samhaber
School of Business
Algonquin College
Ottawa, ON

Developing the Communication Elements

CHAPTER 14

Drafting Paragraphs, Sections, and Chapters

CHAPTER 15

Creating Reader-Centred Graphics

CHAPTER 16

Designing Pages and Screens

CHAPTER 17

Creating and Delivering Oral Presentations

REFERENCE GUIDE 5

Thirteen Types of Graphics

CHAPTER 14 Drafting Paragraphs, Sections, and Chapters

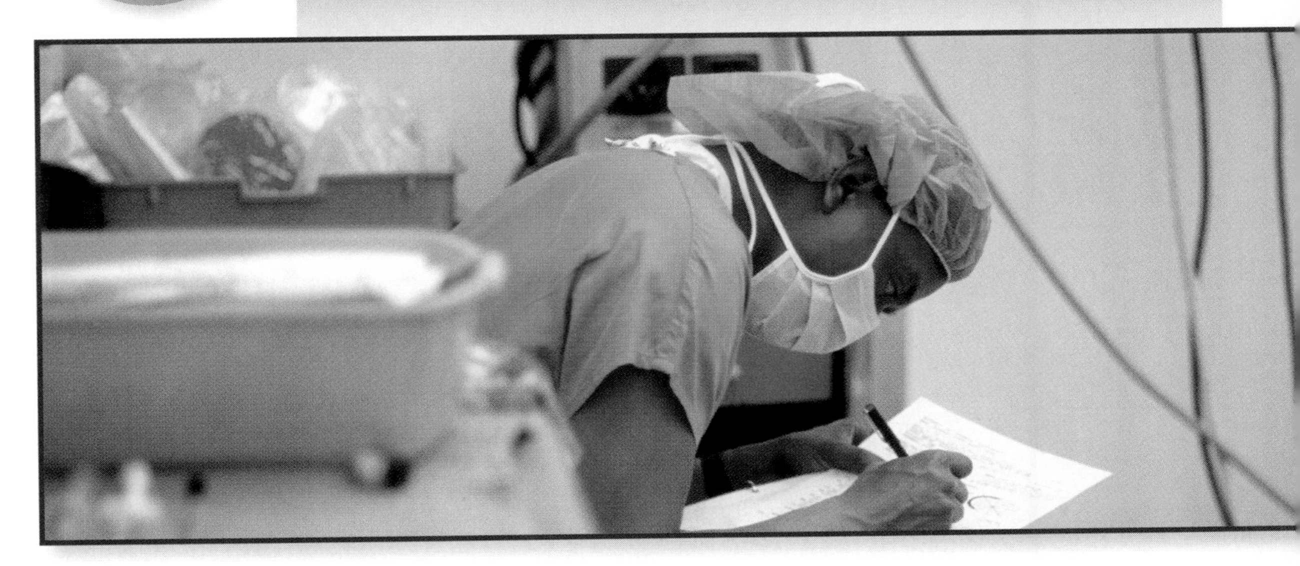

GUIDELINES

1. Begin by announcing your topic
2. Present your generalizations before your details
3. Move from most important to least important
4. Reveal your communication's organization
5. Consult conventional strategies when having difficulties organizing
6. Consider your readers' cultural background when organizing
7. Ethics Guideline: Remember the human consequences of what you're drafting

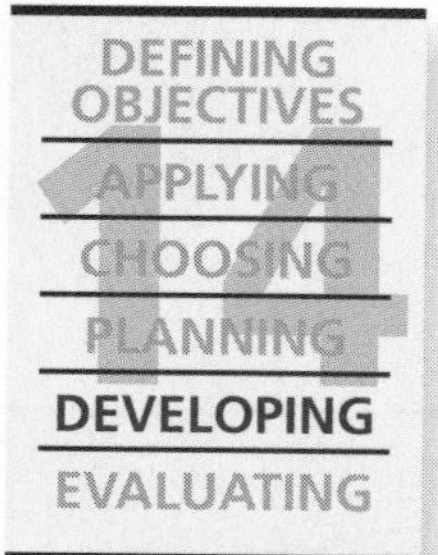

This is the first of four chapters that will help you develop expertise at transforming plans into action as you draft your communication's text, graphics, and visual design. All of the advice in these chapters is research-based. Also, all of it employs the reader-centred approach in which you keep your readers constantly in mind, using your understanding of them as the basis for all your writing decisions.

It's important for you to take a reader-centred approach while drafting because every decision you make while creating your text, graphics, and page design will affect the ways your readers respond to your message. And, as you learned in Chapter 1, their response will determine whether your communication succeeds or fails.

DRAFTING PARAGRAPHS, SECTIONS, AND CHAPTERS

For additional chapter resources, visit Chapter 14 at www.techcomm.nelson.com.

This chapter will help you draft text that has the two essential ingredients of successful on-the-job writing: usability and persuasiveness. Its guidelines apply to paragraphs, to the groups of paragraphs that make up the sections and chapters of longer communications, and even to communications in their entirety. For convenience's sake, this chapter uses the word *segments* to designate these variously sized text units.

How can this chapter's advice apply with equal validity to segments that range in size from a few sentences to an entire communication that may be tens or hundreds of pages long? There are two reasons, one related to what makes a communication *usable* and one related to what makes it *persuasive.*

- **Usability: Regardless of a segment's size, readers must perform the same set of mental tasks as they seek to understand and use its content.** You may have heard a paragraph defined as a group of sentences about the same subject. With only slight variation, that definition applies equally well to larger segments: A section is a group of paragraphs on the same subject, and an entire communication is a group of sections on the same topic. To understand and use any segment, whether a short paragraph or an entire communication, readers do the same things:

  1. Determine its topic.
  2. Figure out how its various parts (words, sentences, paragraphs, etc.) fit together to form a cohesive whole.

  Because all segments make the same mental demands on readers, you can increase the usability of every segment you write by using the same strategies.
- **Persuasiveness: Regardless of a segment's size, readers mentally process its persuasive claims and evidence in the same way.** As you learned in Chapter 1, readers perform a consistent set of mental activities when reading persuasive arguments. For instance, they look for benefits to their organizations and themselves, and they spontaneously raise counterarguments. Because these reading activities remain constant, you can enhance the persuasiveness of every segment you write by employing the same strategies.

This chapter's first six guidelines will help you draft highly usable, highly persuasive segments of all sizes. A seventh guideline discusses the ethical importance of remembering the human consequences of what you are drafting.

GUIDELINE 1 Begin by Announcing Your Topic

You have undoubtedly heard that you should begin your paragraphs with topic sentences. This guideline extends that advice to all your segments, large and small: To make your communications easier to understand and use, begin every segment with a topic statement that tells what the segment is about.

How Topic Statements Increase Usability

How do topic statements increase usability? A key element in usability is simply the ease with which readers can understand your message. As mentioned above, to understand a paragraph, readers must establish in their own minds the meaningful relationship among the sentences from which it is built. The same goes for larger segments. To understand a section or chapter, they must understand how its paragraphs fit together to construct its larger meaning.

Researchers have discovered that as people try to discern the relationships among the parts of a paragraph or longer segment, they engage in two kinds of mental processing:

- **Bottom-up processing.** In bottom-up processing, readers proceed in much the same way as people who are working a jigsaw puzzle *without* having seen a picture that tells them whether the finished puzzle will show a comic figure, a city street, or three cats. As they read, they try to guess how the small bits of information they gather from each sentence fit together with the information from the other sentences to form the segment's general meaning.
- **Top-down processing.** In top-down processing, readers proceed like people who have seen a picture of the finished jigsaw puzzle *before* they begin. Because they know the communication's overall structure in advance, they know immediately how the information they obtain from each sentence fits into the larger meaning of the segment.

Top-down processing is more efficient than bottom-up processing.

Although readers engage continuously in both processes, the more top-down processing they can perform, the more easily they can understand and remember the message. You can help your readers do more top-down processing by telling them explicitly what each segment is about. There are many ways to make these helpful statements. Here are three of the most common ones.

Ways to indicate your topic

Indicating the Topic of a Segment

- **Use a sentence.** An example is the sentence that introduced this list of strategies ("There are many ways to make these helpful statements").
- **Use a single word.** The word "First" that begins the third paragraph on page 25 tells the reader, "You are now going to read a segment that explains the 'four-step' procedure for defining objectives that was just mentioned."
- **Use a question.** The question ("How do topic statements increase usability?") that begins the second paragraph of the discussion of this guideline told you that you were about to read a segment explaining the ways topic statements contribute to usability.

Topic statements are especially helpful to readers when placed at the beginning of a segment.

Although topic statements help readers wherever they appear in a segment, they are especially helpful when placed at the beginning. To demonstrate this connection, researchers John D. Bransford and Marcia K. Johnson (1972) asked people to listen to the following passage being read aloud.

Passage used in an experiment that demonstrated the importance of top-down processing

> The procedure is actually quite simple. First you arrange things into different groups. Of course, one pile may be sufficient depending on how much there is to do. If you have to go somewhere else due to lack of facilities, that is the next step; otherwise you are pretty well set. It is important not to overdo things. That is, it is better to do too few things at once than too many. In the short run this may not seem important but complications can easily arise. A mistake can be expensive as well. At first the whole procedure will seem complicated. Soon, however, it will become just another facet of life. . . . After the procedure is completed, one arranges the materials into different groups again. Then they can be put into their appropriate places. Eventually they will be used once more and the whole cycle will then have to be repeated. However, that is part of life.

Initial topic statements help readers understand and remember.

The researchers told one group the topic of this passage in advance; they told the other group afterwards. Then they asked both groups to write down everything they remembered from what they had heard. People who had been told the topic (washing clothes) before hearing the passage remembered many more details than those who were told afterwards.

As Bransford and Johnson's study suggests, by stating the topic at the beginning of your segments, you increase the usability of your text for readers who want to understand your communication thoroughly and remember what you said. You also increase your communication's usability for readers who are skimming for particular facts. The first sentence of each segment tells them whether the segment is likely to contain the information they are seeking.

When you explicitly state the topic of each segment, small and large, you create a hierarchy of topic statements. Figure 14.1 shows the interlocking, hierarchically organized topic statements from a report in which an engineering firm is seeking to persuade a financially troubled zoo that it can overcome its difficulties by hiring the firm for several construction projects that will attract more visitors. The topic statements correspond to the following major points from one section of the report's outline.

This outline corresponds to the page shown in Figure 14.1.

II. Problem: Budget Crisis
   A. History of the Crisis
   B. Cause of the Crisis
      1. Rising costs are *not* the cause
      2. Declining revenues are the cause

GUIDELINE 2 Present Your Generalizations before Your Details

In many segments you write at work, you will be presenting detailed facts about your topic in order to explain or support a general point you want your readers to understand or accept. You can increase both the usability and the persuasiveness of most of your on-the-job writing not only by stating your topic at the beginning of each segment

FIGURE 14.1

Interlocking Topic Statements (see page 344 for the outline of this page)

Topic announced for chapter (whole discussion of the budget crisis).

Topic announced for first section (history of the crisis).

Topic announced for second section (causes of the crisis).

Topic announced for first subsection (one possible cause: rising costs).

Topic announced for second subsection (second possible cause: falling revenues).

Chapter 2

PROBLEM: BUDGET CRISIS

The Metropolitan Zoo faces a severe budget crisis. The crisis first surfaced last August, when the zoo discovered that operating expenses for the year were going to exceed income by $247,000. Emergency measures, including a reduction in working hours for some employees, lowered the actual loss by December 31 to $121,000. However, the zoo faces similar difficulties again this year—and in future years—unless effective measures are taken.

Causes of the Problem

What is causing this budget crisis, which first appeared in a year when the zoo thought it would enjoy a large profit? The crisis is *not* caused by rising costs. In fact, the zoo actually reduced operating costs by 3% last year. The greatest savings were related to energy expenses. The new power plant began operation, reducing fuel consumption by 15%. Also, design changes in the three largest animal houses conserved enough heat to reduce their heating expenses by 9%. Finally, a new method of ordering and paying for supplies lowered expenses enough to offset inflation.

The budget crisis is caused instead by declining revenues. During the past year income from admission fees, concession sales, and donations has dropped. Because of the decline in paid admissions, overall income from this source was $57,344 less last year than the year before. [This discussion of falling revenue continues for two pages.]

(Guideline 1), but also by stating your general point about your topic *before* you present your details.

How Initial Generalizations Make Writing Easier to Understand and Use

When you present your generalizations first, you save your readers the work of trying to figure out what your general point is. Imagine, for instance, that you are a manager who finds the following sentences in a report:

Details without an initial generalization

> Using the sampling technique just described, we passed a gas sample containing 500 micrograms of VCM through the tube in a test chamber set at 25°C. Afterwards, we divided the charcoal in the sampling tube into two equal parts. The front half of the tube contained approximately ⅔ of the charcoal while the back half contained the rest. Analysis of the back half of the tube revealed no VCM; the front half contained the entire 500 micrograms.

As you read these details, you probably find yourself asking, "What does the writer want me to get from this?" You would have been saved that labour if the writer had begun the segment with the following statement:

> We have conducted a test that demonstrates the ability of our sampling tube to absorb the necessary amount of VCM under the conditions specified.

By placing the generalization at the head of the paragraph, the writer would also help you use the segment more efficiently as you performed the managerial task of determining whether the writer's conclusion is valid. Because you would already know the writer's generalization, you could immediately assess whether each detail does, indeed, provide adequate support for the conclusion that the sample tube absorbs the necessary amount of VCM. In contrast, if the writer waited to reveal the conclusion after presenting the details, you would have to recall each detail from memory—or even reread the passage—in order to assess the strength of its support for the writer's conclusion.

How Initial Generalizations Make Writing More Persuasive

Left to themselves, readers are capable of deriving all sorts of generalizations from a passage. Consider the following sentences:

> Richard moved the gas chromatograph to the adjacent lab.
> He also moved the electronic balance.
> And he moved the experimental laser.

Left to themselves, different readers will draw different conclusions from the same facts.

One reader might note that everything Richard moved is a piece of laboratory equipment and consequently might generalize that "Richard moved some laboratory equipment from one place to another." Another reader might observe that everything Richard moved was heavy and therefore might generalize that "Richard is strong." A member of a labour union in Richard's organization might generalize that "Richard was doing work that should have been done by a union member, not by a manager" and might file a grievance. Different generalizations lead to different outcomes.

Present your generalization before your readers begin to formulate contradictory ones.

A key point is that readers naturally formulate generalizations even if none are provided. Of course, when you are writing persuasively you will want your readers to draw one particular conclusion—and not other possible ones. You can increase your chances of succeeding by stating your desired generalization explicitly and by placing that generalization ahead of your supporting details so that your readers encounter it before forming a different generalization on their own.

Sometimes You Shouldn't Present Your Generalizations First

Although you can usually strengthen your segments by stating your generalizations before your details, watch for situations where you need to follow a different strategy. If you launch a segment with a generalization that is likely to provoke a negative reaction from your readers, you may decrease your communication's persuasiveness. In such cases, you can usually increase your communication's persuasiveness by postponing your general points until *after* you've laid the relevant groundwork with your details by using the indirect organizational pattern (described in Chapter 10).

How Guideline 2 Relates to Guideline 1

Taken together, Guidelines 1 and 2 advise you to announce your topic and state your main point about it at the beginning of each segment. Often, you can make your writing more concise and forceful by doing both in a single sentence. Consider the opening for a two-page discussion of an engineer's experimental results.

Separate sentences state the topic and the writer's generalization

> We conducted tests to determine whether the plastic resins can be used to replace metal in the manufacture of the CV-200 housing. The tests showed that there are three shortcomings in plastic resins that make them unsuitable as a replacement for the metal.

The first sentence announces the topic of the segment (the tests), and the second states the writer's generalization about the topic (the resins are not a good substitute for metal). By combining both the topic and generalization into a single sentence, the engineer could save many words.

Topic and generalization are combined in one sentence

Our tests showed three shortcomings in plastic resins that make them unsuitable as a replacement for metal in the manufacture of the CV-200 housing.

GUIDELINE 3 Move from Most Important to Least Important

Guidelines 1 and 2 discussed the way you begin your segments. This guideline focuses on the way you order the material that follows your opening sentence or sentences. It applies particularly to passages in which you present parallel pieces of information, such as a list of five recommendations or an explanation of three causes of a problem. Whether you present each item in a single sentence or in several paragraphs, put the most important item first and then proceed in descending order of importance.

Initial placement makes your key information more accessible and adds persuasive emphasis.

By putting the most important information first, you increase your communication's usability by making it easier for readers who scan to locate your key points without reading your entire communication. Placing your most important information first also increases your communication's persuasiveness by presenting the strongest support for your arguments in the most prominent spot.

Determine what's most important by considering your readers' viewpoint.

To identify the most important information, consider your communication from your readers' viewpoint. What information will they be most interested in or find most persuasive? For example, in the segment on the three shortcomings of plastic resins, readers will certainly be more interested in the major shortcoming than the minor ones. Similarly, if the readers must be persuaded that plastic resins are not a good substitute for metal, they are sure to find the major shortcoming more compelling than minor ones.

Sometimes the most important information shouldn't be first.

Occasionally, you may encounter situations where you must ignore this guideline in order to present your overall message clearly and economically. For instance, to explain clearly the multiple causes of flooding along a river, you may need to describe events chronologically even though the event that occurred first was not the one with the greatest impact. In general, though, presenting the most important information first will be most helpful and persuasive to your readers.

GUIDELINE 4 Reveal Your Communication's Organization

Combined with strategies described in Chapters 9 and 10, the advice given in this chapter's first three guidelines enables you to organize communications in reader-centred ways that are highly usable and persuasive. To realize the full benefit of your organizing efforts, however, you must also "reveal" a communication's organization by drawing a "map" of it for your readers.

If it is well drawn, your map can increase your communication's usability for the two most common types of readers. For readers hoping to find a specific piece of information without reading other material, your map provides quick access to their destination. For readers who are reading the entire communication sequentially, the map explains how the parts fit together, thereby making your communication easier to understand. Even in the most carefully organized communication, the relationships among the parts may not be evident to readers unless the writer explicitly indicates them. That's because information about organization can be quite distinct from information about subject matter, as the following diagram indicates.

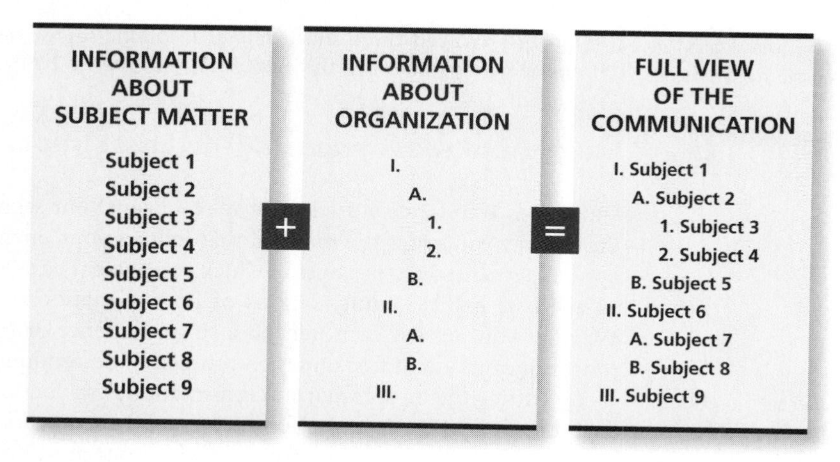

As another way to appreciate the distinction between information about subject matter and information about organization, look at Figure 14.2. Unless you can read Japanese, the text will be incomprehensible to you, but the organization of information is clear, thanks to the skillful use of several techniques described later in this chapter.

By providing a map that reveals the way your communication is organized, you also increase the communication's persuasiveness. The map allows your readers to expend less attention unraveling the communication's organization so they can devote more attention to the data and other evidence that supports your claims. By reducing the effort required to understand your communication, your map also increases the likelihood that readers will study your entire argument with care rather than skimming or stopping reading altogether.

The following sections discuss four techniques for revealing organization:

- Forecasting statements
- Transitions
- Headings
- Visual arrangement of your text on the page

Forecasting Statements

Forecasting statements tell readers what's coming up.

Forecasting statements tell the reader the organization of what lies ahead. Often, they appear along with a topic sentence, which they supplement. For instance, here is the topic sentence for a section of a brochure published by a large chain of garden nurseries:

Topic sentence | Our first topic is the trees found in southern Ontario.

Although this topic sentence tells what the section is about, it gives no hint of the way the section has been organized. The section might discuss evergreen trees first and then deciduous trees. It might discuss healthy trees and then diseased ones, or it might be

FIGURE 14.2

Page Printed in Japanese

The blue block gives primary emphasis to the page's title.

Large, bold type shows where major sections begin.

Indentation signals subordination.

A table provides another way of using visual design to signal the organization of information.

画面の調整

調整のしかた

1 色温度調整画面を操作する

色温度調整画面では色温度を変えることができます。画面の色を印刷したものと同じ色になるように変えたいときなどに使います。色温度は9300K（青みがかった白）から5000K（暖色の赤）まで調整できます。
この調整はすべての入力信号に対して有効です。

1 MENUボタンを押す。MENU画面が表示されます。
2 ブライトネスまたはコントラストボタンを押して 色温度調整を選び、MENU ボタンを押す。
3 ブライトネスボタン☼↓/↑を押して9300K、5000K、あるいは のオプションのどれかを選択します。
   - お買い上げ時は、２種類の色温度が設定されています。出荷時の設定は9300K、5000Kです。
   - のオプションでは、全色温度を使ってディスプレイの設定が可能です。コントラストボタンを押してご希望の色温度に調整してください。
4 MENUボタンを一度押すとMENU画面 に戻ります。二度押すと、通常の画面に戻ります。

ご注意
GEM（Graphic Enhancement Mode）を使用時の色温度調整は9300Kから11000Kの間での調整となります。GEMについて詳しくは、64ページの「Graphic Enhancement Mode（GEM）」の項を参照してください。

2 位置調整画面を操作する

位置調整画面では画像の位置を調整できます。この調整は現在受信している入力信号に対してのみ有効です。

1 MENUボタンを押す。MENU画面表示されます。
2 ブライトネスおよびコントラストボタンを押して、「位置調整」を選び、MENUボタンを押す。
3 垂直方向は、ブライトネスボタン☼↓/↑で調整する。
4 水平方向は、コントラストボタン◑←/→で調整する。
5 MENUボタンを一度押すとMENU画面 に戻ります。二度押すと、通常の画面に戻ります。

3 画調整画面を操作する

画調整画面では、コンバージェンスの調整とモアレ調整ができます。コンバージェンスが良い場合は画面上の赤、緑、青の電子ビームが正しく整列しています。コンバージェンスが正しく整列していない場合は、赤または青の影が見えること（特にテキストで）があります。これは画像の明瞭度、またはフォーカスに影響を与えます。モアレは画面上に現れる波模様や点状模様です。モアレを少なくするために、モアレのキャンセル調整を行います。
この調整はすべての入力信号に対して有効です。

1 MENUボタンを押す。MENU画面が表示されます。
2 ブライトネスまたはコントラストボタンを押して、「画調整」を選び、MENUボタンを押す。
3 ブライトネスボタン☼↓/↑を押して希望の調整項目を選びます。
4 コントラストボタン◑←/→を押して調整する。
5 MENUボタンを一度押すとMENU画面に戻ります。二度押すと、通常の画面に戻ります。

項目	機能
Hコンバージェンス	水平方向のコンバージェンスを調整します。
V コンバージェンス	垂直方向のコンバージェンスを調整します。
モアレキャンセル	画面上の点状あるいは波形の線を少なくします。
* モアレ調整	モアレキャンセルの効果を調整します。

- （モアレ調整）は、モアレキャンセルはが「オン」の状態のときに現れます。

J

61

organized in any of various other ways. Readers would have no way of knowing. A forecasting statement resolves that uncertainty.

Forecasting statement

Our first topic is the trees found in southern Ontario. Some of the trees are native, some imported.

Forecasting and topic statements are often combined in one sentence:

Forecasting statement

> Our first topic is the trees—both native and imported—found in southern Ontario.

Forecasting statements may vary greatly in the amount of detail they provide. The sample sentences above provide both the number and the names of the categories to be discussed. A more general preview is given in the next example, which tells its readers to expect a list of actions but not what these actions are or how many will be discussed:

Forecasting statement

> To solve this problem, the department must take the following actions.

When you are deciding how much detail to include in a forecasting statement, consider the following points.

Writing Forecasting Statements

- **Say something about the segment's arrangement that readers will find helpful.** Usually, the more complex the relationship among the parts, the greater the amount of detail that is needed.
- **Say only as much as readers can easily remember.** A forecasting statement should help readers, not test their memories. When forecasting a segment that will discuss a three-step process, you could name all the steps. If the process has eight steps, state the number without naming them.
- **Forecast only one level at a time.** Don't list all the contents of a communication at its outset. That will only confuse your readers. Tick off only the major divisions of a particular section. If those divisions are themselves divided, provide each of them with its own forecasting statement.

Transitions

As readers proceed from one part of a communication to another—from sentence to sentence, paragraph to paragraph, and section to section—they must figure out how these parts relate to one another. Chapter 4 ("Developing an Effective Style") tells how to lead readers smoothly from one sentence to the next. The discussion that follows explains how to construct transitional statements that help readers move between larger units of text—from one paragraph or group of paragraphs to the next.

How to Write a Transition Transitional statements indicate two things. First, they tell what the upcoming segment is about. In this way, they often serve as topic sentences. Second, they indicate the relationship of what is coming and what has just ended. You can accomplish these two tasks in separate sentences:

Two-sentence transitional statement

> In our guidelines for handling Exban, we have now completed our suggestions for guarding against accidental spills during transportation. We turn now to our guidelines for storing the product.

Or you can use a single sentence:

One-sentence transitional statement

> After we developed our hypothesis, we were ready to design our experiment.

These sample transitions all would appear at the opening of the new segment. Transitional statements can also occur at the end of the segment just being concluded. Here is the last sentence of a section in a report about the sales performance of a corporation's online store:

Transitional statement at the end of a segment

> In light of the marketing information just described, the performance of our online store can be interpreted more precisely.

The following section then begins with a topic statement:

Topic statement for next segment

> During the past twelve months, our online store performed extremely well in the parts of the market that were generally weak, but below average in those that were strong.

You can also make transitions through visual design.

Other Means of Making Transitions Sometimes you can signal transitions without using any words at all. For example, in a report that presents three brief recommendations, you might arrange the recommendations in a numbered list. The numbers themselves would provide the transition from one recommendation to the next. Similarly, in a memo covering several separate topics, the transition from one to the next might be provided by giving each topic a heading (see the discussion of headings that follows).

Headings

Headings are used extensively in on-the-job writing.

A third technique for revealing organization is to use headings. These serve as signposts that tell readers what the successive parts of a communication are about. At work, writers use headings not only in long documents, such as reports and manuals, but also in short ones, such as letters and memos. To see how effectively the insertion of headings can reveal organization, look at Figures 14.3 and 14.4 (pages 352 and 353), which show two versions of the same memo, one without headings and one with headings.

Avoid giving every paragraph its own heading.

Provide headings wherever there is a major shift in topic. In much on-the-job writing, such a shift occurs every few paragraphs. Avoid giving every paragraph its own heading, which would give your text a disjointed appearance rather than helping readers see how things fit together. An exception occurs in communications designed to provide readers quick access to specific pieces of information, as in a warranty, troubleshooting guide, reference manual, or fact sheet. In these documents, headings may even label sentence fragments or brief bits of data, turning the communication into something very much like a table of facts.

To be helpful, each heading must unambiguously indicate the kind of information that is included in the passage it labels.

Creating Text for Headings

- **Ask the question that the segment will answer for your readers.** Headings that ask questions such as, "What happens if I miss a payment on my loan?" or "Can I pay my loan off early?" are especially useful in communications designed to help readers decide what to do.
- **State the main idea of the segment.** For example, this strategy could be used in a brochure on bicycling safety. With headings such as "Ride with the Traffic" and "Use Streets with Parked Cars," the readers' attention is focused on the key points of a passage.

- **Use a key word or phrase.** This type of heading is especially effective when a full question or statement would be unnecessarily wordy. For instance, in a request for a high-end multimedia production system, the section that discusses prices might have a heading that reads, "How Much Will the System Cost?" However, the single word "Cost" would serve the same purpose.

FIGURE 14.3
Memo without Headings

The lack of visual cues hides the organization of this memo's contents.

Garibaldi Corporation
INTEROFFICE MEMORANDUM

June 15, 2007

TO Vice Presidents and Department Managers
FROM Mario Carutti, President
RE PURCHASES OF COMPUTER AND SCANNER EQUIPMENT

Three months ago, I appointed a task force to develop corporate-wide policies for the purchase of computers and scanner equipment. Based on the advice of the task force, I am establishing the following policies.

The task force was to balance two possibly conflicting objectives: (1) to ensure that each department purchases the equipment that best serves its special needs and (2) to ensure compatibility among the equipment purchased so the company can maintain an efficient network for all our computers and scanners.

I am designating one "preferred" vendor of computers and two "secondary" vendors.

The preferred vendor, YYY, is the vendor from which all purchases should be made unless there is a compelling reason for selecting other equipment. To encourage purchases from the preferred vendor, a special corporate fund will cover 30% of the purchase price so that individual departments need fund only 70%.

Two other vendors, AAA and MMM, offer computers already widely used in Garibaldi; both computers are compatible with our network. Therefore, the special corporate fund will support 10% of the purchase price of these machines.

We will select one preferred vendor and no secondary vendor for scanners. The task force will choose between two candidates: FFF and TTT. I will notify you when the choice is made early next month.

FIGURE 14.4

Memo with Headings

Garibaldi Corporation
INTEROFFICE MEMORANDUM

June 15, 2007

TO Vice Presidents and Department Managers
FROM Mario Carutti, President
RE PURCHASES OF COMPUTER AND SCANNER EQUIPMENT

Three months ago, I appointed a task force to develop corporate-wide policies for the purchase of computers and scanner equipment. Based on the advice of the task force, I am establishing the following policies.

Headings help reveal the memo's organization.

Objectives of Policies
The task force was to balance two possibly conflicting objectives: (1) to ensure that each department purchases the equipment that best serves its special needs and (2) to ensure compatibility among the equipment purchased so the company can maintain an efficient network for all our computers and scanners.

Computer Purchases
I am designating one "preferred" vendor of computers and two "secondary" vendors.

Indenting indicates that these are the two parts of the computer purchase policy.

> **Preferred Vendor:** The preferred vendor, YYY, is the vendor from which all purchases should be made unless there is a compelling reason for selecting other equipment. To encourage purchases from the preferred vendor, a special corporate fund will cover 30% of the purchase price so that individual departments need fund only 70%.
>
> **Secondary Vendor:** Two other vendors, AAA and MMM, offer computers already widely used in Garibaldi; both computers are compatible with our network. Therefore, the special corporate fund will support 10% of the purchase price of these machines.

Headings are bold to make them stand out.

Scanner Purchases
We will select one preferred vendor and no secondary vendor for scanners. The task force will choose between two candidates: FFF and TTT. I will notify you when the choice is made early next month.

Parallelism in headings is often desirable.

The different types of headings may be mixed. Often, however, parallel headings make a communication's content easier to access and understand. For instance, when you are describing a series of steps in a process, parallel phrasing cues readers that the segments being labelled are logically parallel: "Building a Database," "Entering Data," and so on.

A heading's visual design can indicate its place in the communication's hierarchy.

The visual appearance of headings is important to readers. To be useful, headings must stand out. If more than one level of heading is used, they must communicate their hierarchical relationships visually.

Designing Headings Visually

1. Make headings stand out from the text, perhaps using these strategies.
   - Use bold.
   - Use a different colour than is used for the text.
   - Place headings at the left-hand margin or centre them.
2. Make major headings more prominent than minor ones. Here are some strategies to consider.
   - Make major headings larger.
   - Centre the major headings and tuck the others against the left-hand margin.
   - Use all capital letters for the major headings and initial capital letters for the others.
   - Give the major headings a line of their own and put the others on the same line as the text that follows them.
3. Give the same visual treatment to headings at the same level in the hierarchy.

Usually, you can use two or more of these techniques together so that they reinforce one another. By doing so, you can readily create two or three easily distinguishable levels of headings that instantly convey the organizational hierarchy of your communication. Figure 14.5 shows an example.

Chapter titles are a form of heading.

If you feel that additional levels are needed, try making the most important sections into separate chapters, with the chapter titles serving as the highest level of heading (see Figure 14.5). Alternatively, you can reinforce the headings with letters and numbers in outline fashion, as shown in Figure 14.6 (page 356).

By convention, headings and the topic statements that follow them reinforce each other, with the topic statement repeating one or more key words from the heading. Topic statements do not contain pronouns that refer to the headings. For example, the heading "Research Method" would not be followed by a sentence that says, "Designing this was a great challenge for us." Instead, the sentence would read, "Designing the research method was a great challenge for us."

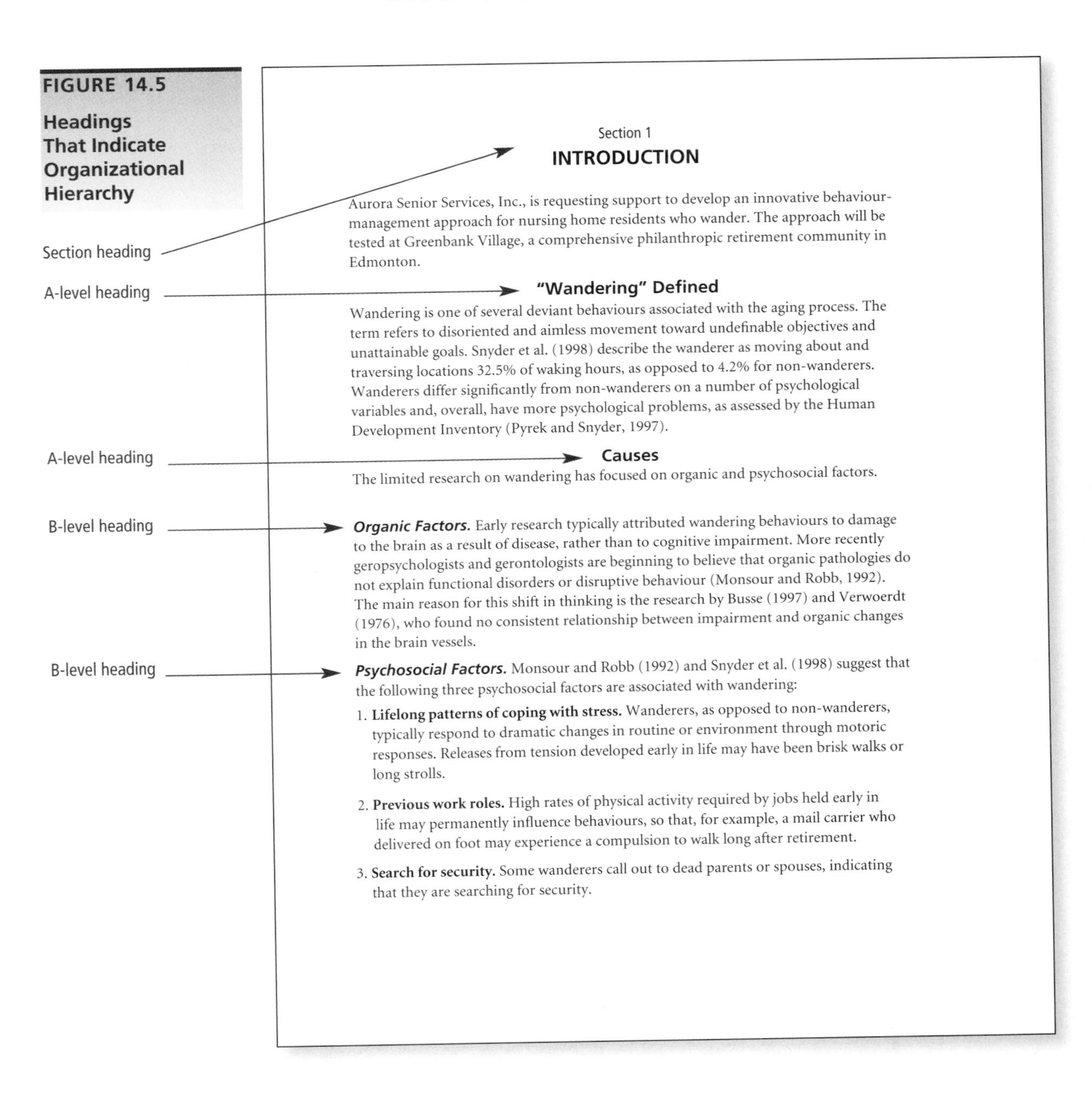

Section 1

INTRODUCTION

Aurora Senior Services, Inc., is requesting support to develop an innovative behaviour-management approach for nursing home residents who wander. The approach will be tested at Greenbank Village, a comprehensive philanthropic retirement community in Edmonton.

"Wandering" Defined

Wandering is one of several deviant behaviours associated with the aging process. The term refers to disoriented and aimless movement toward undefinable objectives and unattainable goals. Snyder et al. (1998) describe the wanderer as moving about and traversing locations 32.5% of waking hours, as opposed to 4.2% for non-wanderers. Wanderers differ significantly from non-wanderers on a number of psychological variables and, overall, have more psychological problems, as assessed by the Human Development Inventory (Pyrek and Snyder, 1997).

Causes

The limited research on wandering has focused on organic and psychosocial factors.

Organic Factors. Early research typically attributed wandering behaviours to damage to the brain as a result of disease, rather than to cognitive impairment. More recently geropsychologists and gerontologists are beginning to believe that organic pathologies do not explain functional disorders or disruptive behaviour (Monsour and Robb, 1992). The main reason for this shift in thinking is the research by Busse (1997) and Verwoerdt (1976), who found no consistent relationship between impairment and organic changes in the brain vessels.

Psychosocial Factors. Monsour and Robb (1992) and Snyder et al. (1998) suggest that the following three psychosocial factors are associated with wandering:

1. **Lifelong patterns of coping with stress.** Wanderers, as opposed to non-wanderers, typically respond to dramatic changes in routine or environment through motoric responses. Releases from tension developed early in life may have been brisk walks or long strolls.
2. **Previous work roles.** High rates of physical activity required by jobs held early in life may permanently influence behaviours, so that, for example, a mail carrier who delivered on foot may experience a compulsion to walk long after retirement.
3. **Search for security.** Some wanderers call out to dead parents or spouses, indicating that they are searching for security.

FIGURE 14.5
Headings That Indicate Organizational Hierarchy

FIGURE 14.6

Contents Page from a Research Report That Uses a Decimal Outlining System

CONTENTS

v

Visual Arrangement of Text

You can also reveal your communication's organization through the visual arrangement of your text on the page:

Arranging Text Visually to Reveal Organization

- **Adjust the location of your blocks of type.** Here are three adjustments you can make:
  - Indent paragraphs lower in the organizational hierarchy.
  - Leave extra space between the end of one major section and the beginning of the next.
  - In long communications, begin each new chapter or major section on its own page, regardless of where the preceding chapter or section ended.
- **Use lists.** By placing items in a list, you signal readers that items hold a parallel place in your organizational hierarchy. Usually, numbers are used for sequential steps. Bullets are often used where a specific order isn't required.

Number List	**Bullet List**
The three steps we should take are:	Three features of the product are:
1. _______	• _______
2. _______	• _______
3. _______	• _______

When you are constructing a list, give the entries a parallel grammatical construction: All the items should be nouns, all should be full sentences, or all should be questions, and so on. Mixing grammatical constructions distracts readers and sometimes indicates a shift in point of view that breaks the tight relationship that should exist among the items.

	Without Parallelism	**With Parallelism**
	Benefits	Benefits
	■ Increased sales	■ Increased sales
	■ Decreased production costs	■ Decreased production costs
Not parallel	■ Morale that is higher	■ Higher morale
	Recommendations	Recommendations
	1. Sell our Port Hope plant.	1. Sell our Port Hope plant.
	2. Place the two Chalk River plants under a single manager.	2. Place the two Chalk River plants under a single manager.
Not parallel	3. We should also seek a better location for our Pickering warehouse.	3. Seek a better location for our Pickering warehouse.

GUIDELINE 5 Consult Conventional Strategies When Having Difficulties Organizing

Every writer occasionally gets stumped when trying to organize a particular paragraph, section, or chapter. Often the problem is one that many others have faced, such as how to describe a certain process or how to explain the causes of a particular event. For many commonly encountered organizational problems, there are conventional strategies for arranging material in ways that will be understandable and useful to readers. By consulting these strategies, you will often find a quick and effective solution to your own problem.

Reference Guide 1 tells how to use six strategies that are especially useful on the job. As you study the strategies, remember to use them only as guides. To make them work in your particular context, you will need to adapt them to your purpose and the needs of your readers.

GUIDELINE 6 Consider Your Readers' Cultural Background When Organizing

The advice you have read so far in this chapter is based on the customs of readers in Canada and other countries where readers expect and value what might be called a *linear* organization. In this organizational pattern, writers express their main ideas explicitly and develop each one separately, carefully leading readers from one to another. As international communication experts Myron W. Lustig and Jolene Koester explain, this pattern can be visualized as "a series of steps or progressions that move in a straight line toward a particular goal or idea" (1993, p. 218).

Different cultures use different organizational patterns.

In other cultures, writers and readers are accustomed to less linear patterns of organization. For example, a Chinese writer introduces the topic of the communication before providing the background, the rationale for selection of the topic, an explanation of how the writer came across the topic, and a justification for its interest or value to the writer.

A less linear style is also used by the majority of writers in India. In Hindi, the main language of India, paragraphs do not stick to one unified idea or thought, as they do in Canada. Jamuma Kachru (1988) explains that in the preferred Hindi style, the writer may digress and introduce material related to many different ideas.

Use appropriate organizational patterns when addressing persons from other cultures.

Because of these cultural differences, serious misunderstandings can arise when your readers are employed by other companies in another country, work for your own company in another country, or even work in your own building but were raised observing the customs of another culture. Such misunderstandings cannot be avoided simply by translating the words of your communication: The whole message must be structured to suit the customs of your readers' culture.

GUIDELINE 7 Ethics Guideline: Remember the Human Consequences of What You're Drafting

When drafting their communications, employees sometimes become so engrossed in the technical aspects of their subject that they forget the human consequences of what they're writing. When this happens, they write communications in which their stakeholders are overlooked. Depending on the situation, the consequences can be quite harmful or relatively mild—but they can always lead to the unethical treatment of other people.

Mining Accidents

An example is provided by Beverly A. Sauer (1994), who studied the reports written by American federal employees investigating mining accidents in which miners were killed. In their reports, as Sauer points out, the investigators typically focused on technical information about the accidents without paying sufficient attention to the human tragedies caused by the accidents. In one report, for example, the investigators described the path of an underground explosion as it travelled through an intricate web of mineshafts and flamed out of various mine entrances. At one entrance, investigators wrote, "Debris blown by the explosion's forces damaged a jeep automobile parked near the drift openings." The investigators didn't mention in this passage that in addition to damaging the jeep, the explosion killed sixteen miners who were in the mineshafts.

Besides overlooking the victims of the disasters, the investigators' reports often failed to identify the human beings who created the conditions that caused the accidents. One report said, "The accident and resultant fatality occurred when the victim proceeded into an area of known loose roof before the roof was supported or taken down." This suggests that the miner was crushed to death by a falling mine roof because he was careless. Sauer's research showed, however, that in the same mines ten fatalities had occurred in five years—and seven had resulted from falling roofs. Managers of the mines were not following safety regulations, and mine safety inspectors were not enforcing the law.

Writing with Awareness of Human Consequences

Of course, there's nothing the inspectors' reports can do on behalf of the deceased miners or their families. However, the stakeholders in the inspectors' reports include other miners who continue to work in what is the most dangerous profession in the world. As Sauer pointed out, the investigators' readers included federal officials responsible for overseeing the American mining industry. If the inspectors wrote in ways that made these readers more aware of the human consequences of mining accidents—and of the human failings that often bring them about—those officials might be more willing to pass stricter laws and insist that existing regulations be strictly enforced.

As Sauer explains, in Europe, where government regulation of mining is much stronger, miners' deaths and injuries are much rarer events than in the United States. The high accident rate that makes mining so dangerous in the United States, she argues, results in part from the way mining investigators write their reports.

Of course, most people aren't in professions where lives are at stake. In any profession, however, it's possible to become so focused on your technical subject matter that you forget the human consequences of your writing.

To avoid accidentally treating others unethically, you can take the following steps.

Avoiding Accidentally Treating Others Unethically

- **When beginning work on a communication, identify its stakeholders (page 34).** Certainly, the stakeholders of the mining disaster reports included miners, whose lives are endangered when government officials fail to enact and enforce safety measures.
- **Determine how the stakeholders will be affected by your communication (page 237).**
- **Draft your communication in a way that reflects proper care for these individuals.** Be sure that all your decisions about what to say, what *not* to say, and how to present your message are consistent with your personal beliefs about how you should treat other people.

CONCLUSION

This chapter has suggested that you can increase the usability and persuasiveness of your communications if you begin your segments with topic statements; present generalizations before details; organize from most important to least important; reveal your communication's organization with headings, forecasting statements, and similar devices; and consider your readers' cultural background.

Remember that the first six guidelines in this chapter are suggestions, not rules. The only "rule" for writing segments is to be sure your readers know what you are talking about and how your various points relate to one another. Sometimes you will be able to do this without thinking consciously about your techniques. Often, however, you will be able to increase the clarity and persuasiveness of your segments by drawing on the advice presented here.

To ensure that you are always drafting ethically, the seventh guideline is something you should treat as a rule: Remember the human consequences of what you are writing.

EXERCISES

Expertise

1. Circle the various parts and subparts of the passage in Figure 14.7 to show how the smaller segments are contained within larger ones.

2. Identify the topic statements in Figure 14.7 by putting an asterisk before the first word of each sentence that indicates the topic of a segment.

3. Circle all the forecasting statements in Figure 14.5 (page 355). For those segments that lack explicit forecasting statements, explain how readers might figure out the way in which they are organized.

Online

Examine the ways that this chapter's guidelines are applied by a website that explains a technical or scientific topic. For example, study the explanation of a medical topic at the Canadian Cancer Society (www.cancer.ca), Parkinson Society Canada (www.parkinson.ca), or the Heart and Stroke Foundation (ww2.heartandstroke.ca). Present your analysis in the way your professor requests.

Collaboration

1. Figure 14.8 (page 362) shows an email whose contents have been scrambled. Each statement has been assigned a number. Working with one or two other students, do the following.

   a. Write the numbers of the statements in the order in which the statements would appear if the email were written in accordance with the guidelines in this chapter. Place an asterisk before each statement that would begin a segment. (When you order the statements, ignore their particular phrasing. Order them according to the information they provide the reader.)

   b. Using the list you just made, rewrite the email by rephrasing the sentences so that the finished message conforms with all the guidelines in this chapter.

Ethics

Using the Internet or library, find the ethics codes for three corporations. Examine the ways this chapter's guidelines help these corporations write codes that clearly convey their ethical commitments.

FIGURE 14.7

Passage Containing Several Levels of Segments

Importing Insects

By importing insects from other parts of the world, countries can sometimes increase the productivity of their agricultural sector, but they also risk hurting themselves. By importing insects, Australia controlled the infestation of its continent by the prickly pear, a cactus native to North and South America. The problem began when this plant, which has an edible fruit, was brought to Australia by early explorers. Because the explorers did not also bring its natural enemies, the prickly pear grew uncontrolled, eventually rendering large areas useless as grazing land, thereby harming the nation's farm economy. The problem was solved when scientists in Argentina found a small moth, *Cactoblastis cactorum*, whose larvae feed upon the prickly pear. The moth was imported to Australia, where its larvae, by eating the cactus, reopened thousands of acres of land.

In contrast, the importation of another insect, the Africanized honeybee (AHB) could threaten the well-being of Canada. It once appeared that the importation of the insect might bring tremendous benefits, since the AHB produces about twice as much honey as do the bees native to North America.

However, the Ministry of Agriculture and Agri-Food has determined that the introduction of the AHB into Canada would cause serious problems. The problems arise from the peculiar way the AHB swarms. When the honey bees native to North America swarm, about half the bees leave the hive with the queen, moving a short distance. The rest remain, choosing a new queen. In contrast, when the AHB swarms, the entire colony moves, sometimes up to 100 km away. If AHBs intermix with the domestic bee population, they might introduce their swarming traits. Beekeepers could be abandoned by their bees, and large areas of cropland could be left without the services of this pollinating insect.

Unfortunately, the AHB has moved northward to the USA from Sao Paulo, Brazil, where several years ago a researcher released 27 swarms of the AHB from an experiment. AHBs are now found in Texas, Nevada, Arizona, New Mexico, and California.

Thus, while the importation of insects can sometimes benefit a country, imported insects can also alter the country's ecological system, hereby harming its agricultural business.

FIGURE 14.8

Email for Collaboration Exercise

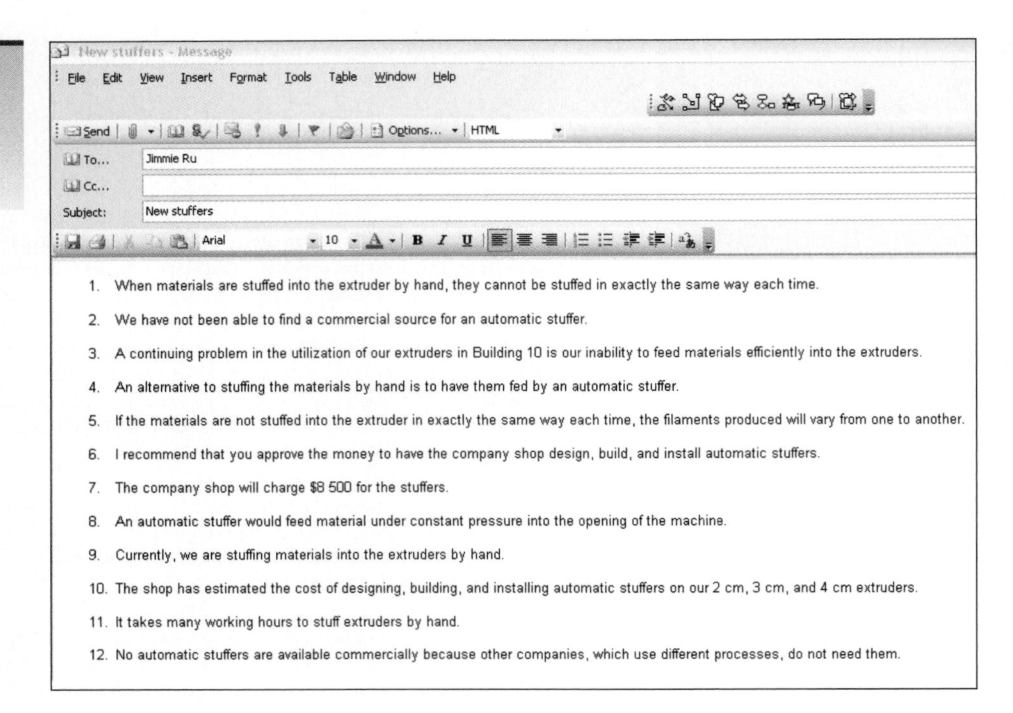

CHAPTER 15 Creating Reader-Centred Graphics

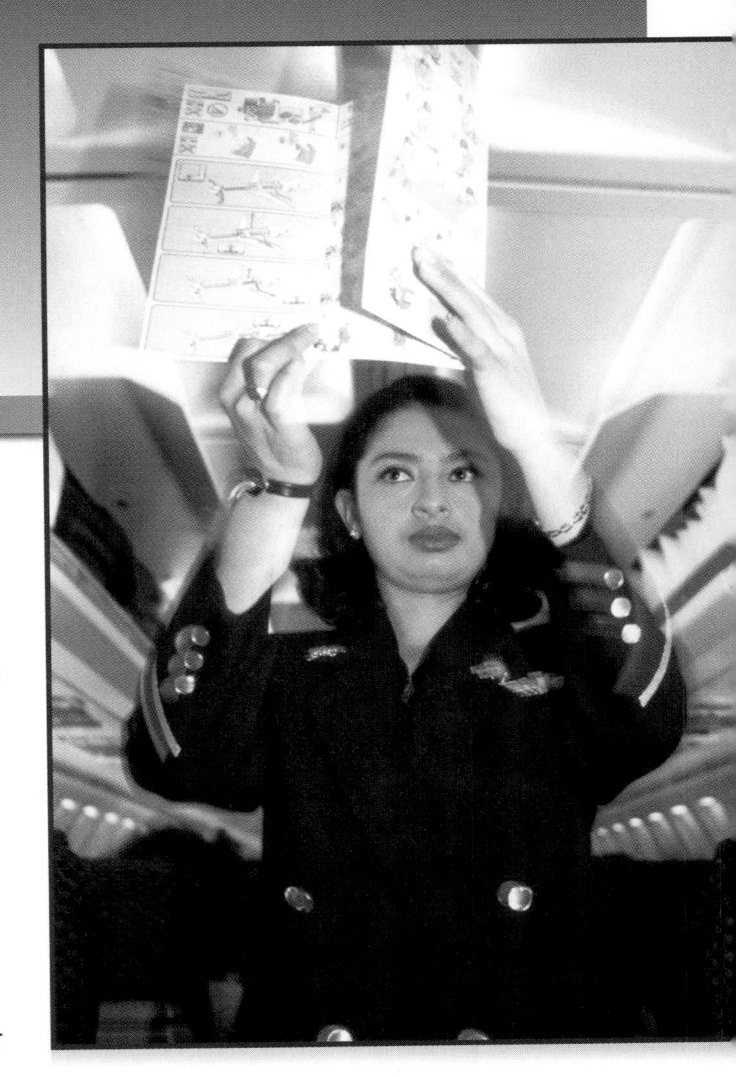

GUIDELINES

1. Look for places where graphics will increase your communication's usability
2. Look for places where graphics will increase your communication's persuasiveness
3. Choose the types of graphics that are best at achieving your objectives
4. Make each graphic easy to understand and use
5. Use colour to support your message
6. Adapt existing graphics to your readers and purpose
7. Integrate your graphics with your text
8. Before addressing an international audience, check your graphics with people from other cultures
9. Ethics Guideline: Avoid graphics that mislead

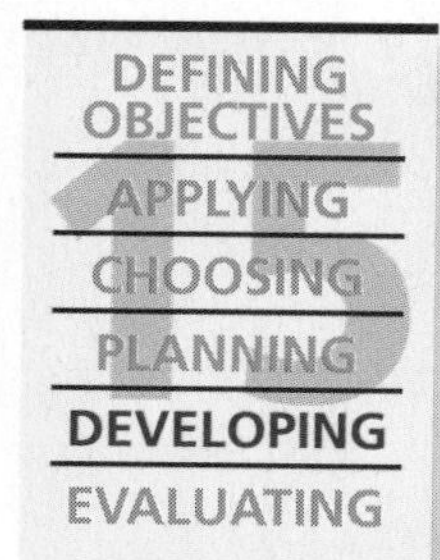

For additional chapter resources, visit Chapter 15 at www.techcomm.nelson.com.

Graphics are the tables, drawings, photos, flowcharts, and other visual representations that play such an important role in on-the-job writing. They often convey part of a message more clearly, usefully, and persuasively than words. Sometimes they even carry the entire message.

For instance, in 1994, HP used its first wordless manual, which served all of its customers around the globe, no matter what their native language. Figure 15.1 shows an example from an HP manual. Wordless instructions are used widely for products that are marketed internationally because the pictures can overcome language barriers.

Although you may never create a wordless communication, you will be able to increase the usability and persuasiveness of your communications for your readers by skillfully designing graphics and carefully integrating them with your text. To develop this expertise in using graphics, take the same reader-centred approach that you follow when drafting your text. Think constantly about your readers, the way you want them to respond to each graphic, and the strategies that will make each one most usable and persuasive in your readers' eyes.

This chapter's eight reader-centred guidelines will help you develop your expertise at using graphics effectively in written and online documents and in oral presentations. A ninth discusses several ethical considerations that apply specifically to graphics.

Reference Guide 5 for this section provides additional advice about how to construct thirteen types of graphics that are widely used in the workplace.

GUIDELINE 1 Look for Places Where Graphics Will Increase Your Communication's Usability

Paradoxically, although society has become increasingly visual when taking in information, most of us still think primarily of words when we want to convey information to others. Consequently, an important step in creating reader-centred communications is to search actively for places where visual representations can help you achieve your communication objectives by replacing, supplementing, or reinforcing your text (Schriver, 1997).

FIGURE 15.1

Wordless Instructions

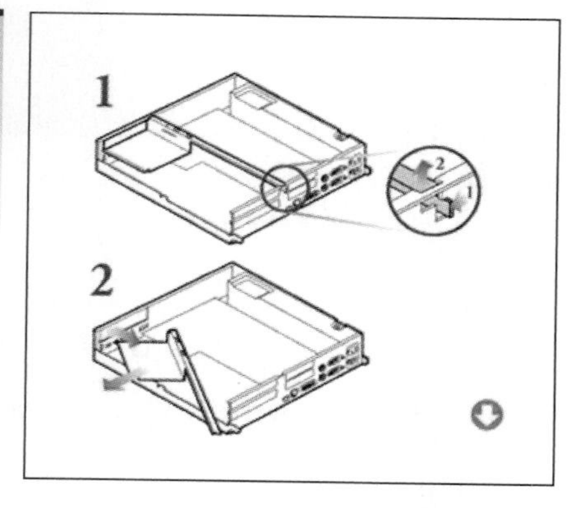

Wordless communications are especially useful when addressing international audiences who do not share a common language.

Search actively for places to use graphics.

Begin by taking a reader-centred look for places where graphics can increase your communication's usability. Here are some ways that graphics help your readers understand and use the information you provide.

Ways you can use graphics to help your readers

- **Show your readers how something looks or is constructed.** Drawings and photographs can often show readers the appearance of things with greater clarity than words could achieve. For example, an engineer used the diagram on the left-hand side of Figure 15.2 to show his readers the construction of the equipment he designed for using lasers to make computer chips. And the Canadian Space Agency used the photograph on the left-hand side of Figure 15.2 to help its readers envision Canadarm2 after it was installed on the International Space Station.
- **Show your readers how to do something.** Ideal for instructions, graphics explain operations that would be difficult to describe—and understand—in text. Figure 15.3 shows readers how to remove a protective cap from a print head.

FIGURE 15.2 Graphics That Show How Something Looks or Is Constructed

Ion source
Ion accelerator
Vacuum pump
Mass separation magnet
Beam sweeping electrodes
Ion beam
Multiple wafer target chamber for batch reduction
Rotary drive
Current integrator
Vacuum pump

FIGURE 15.3

Drawing That Shows How to Do Something

This figure shows the action from the same point of view that a person performing the task would have.

Note how the figure includes text that aids the reader.

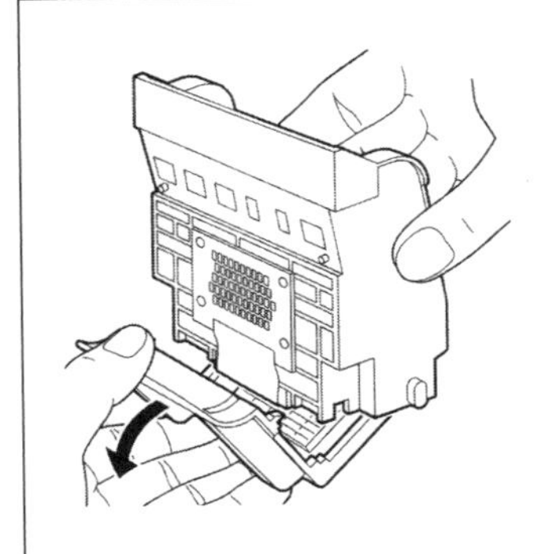

6 Remove the protective cap from the Print Head.

➡ ***Notes***

- Do not touch the Print Head Nozzles or electrical contacts.
- Do not attempt to reattach the protective cap once it has been removed.

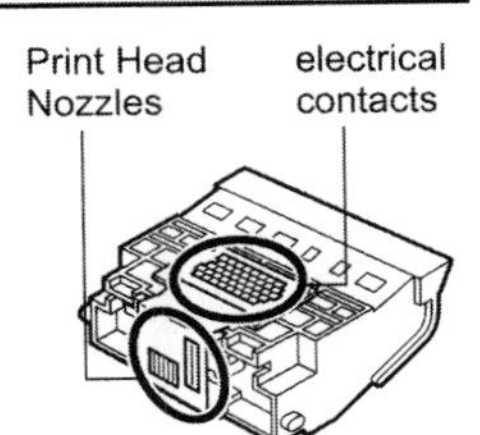

- **Explain a process.** Many processes are best understood visually. By reading Figure 15.4, people can quickly understand how seawater picks up chemicals as it loops through hot-spring systems under the ocean.
- **Show how something is organized.** The diagram shown in Figure 15.5 explains the complex relationships among the computer programs and physical parts inside a cell phone.

FIGURE 15.4

Diagram That Explains a Process

By incorporating text into this drawing, the writer increases the ease with which readers can understand the process it describes.

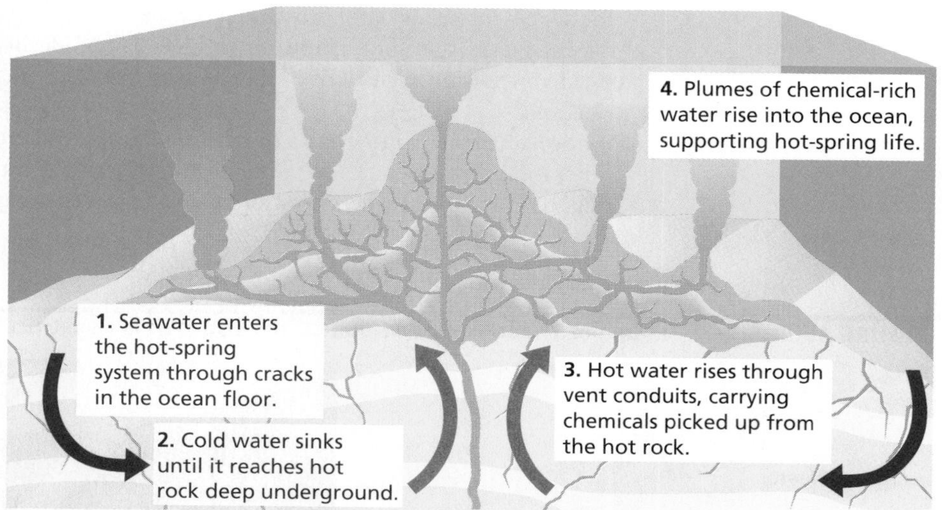

Hot rock deep underground causes seawater to loop through hot-springs, picking up heat and chemicals as it goes.

FIGURE 15.5 **Diagram That Shows How Something Is Organized**

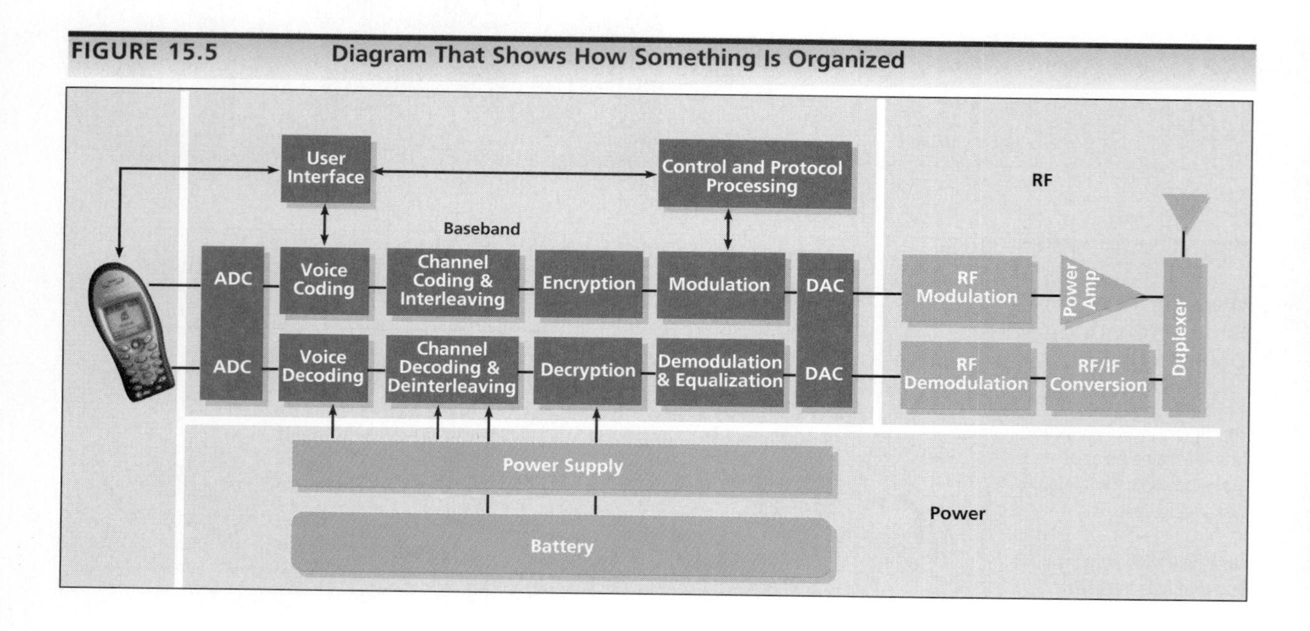

- **Display data to make particular facts easy to find.** Using the table shown in Figure 15.6, readers can quickly determine the time and temperature they should use when developing their own photographic film.
- **Show trends and other numerical relationships.** Various graphics such as line graphs, bar graphs, and pie charts enable readers to grasp trends and other numerical relationships much more quickly than they could from sentences. A team of biological researchers used the graph shown in Figure 15.7 to help their readers see the relationship between river water temperature and the abundance of a particular kind of shellfish.

FIGURE 15.6

Graphic That Helps Readers Find Particular Facts

Boldface is used in the two columns for 68°F (20°C) to highlight the temperature photographers prefer to use.

Kodak Developer	Developing Time (in Minutes)									
	SMALL TANK (Agitation at 30-Second Intervals)					LARGE TANK (Agitation at 1-Minute Intervals)				
	65°F (18° C)	**68°F (20° C)**	70°F (21° C)	72°F (22° C)	75°F (24° C)	65°F (18° C)	**68°F (20° C)**	70°F (21° C)	72°F (22° C)	75°F (24° C)
HC-110 (Dil B)	8 ½	**7 ½**	6 ½	6	5	9 ½	**8 ½**	8	7 ½	
D-76	9	**8**	7 ½	6 ½	5 ½	10	**9**	8	7	
D-76 (1:1)	11	**10**	9 ½	9	8	13	**12**	11	10	
MICRODOL-X	11	**10**	9 ½	9	8	13	**12**	11	10	
MICRODOL-X (1:3)*	—	—	15	14	13	—	—	17	16	
DK-50 (1:1)	7	**6**	5 ½	5	4 ½	7 ½	**6 ½**	6	5 ½	
HC-110 (Dil A)	4 ½†	**3 ¾†**	3 ¼†	3†	2 ½†	4 ¾†	**4 ¼†**	4†	3 ¾†	

* Gives greater sharpness than other developers shown in table.
† Avoid development times of less than 5 minutes if possible, because poor uniformity may result.

Note: Do not use developers containing silver halide solvents.

FIGURE 15.7

Graph That Shows Relationships

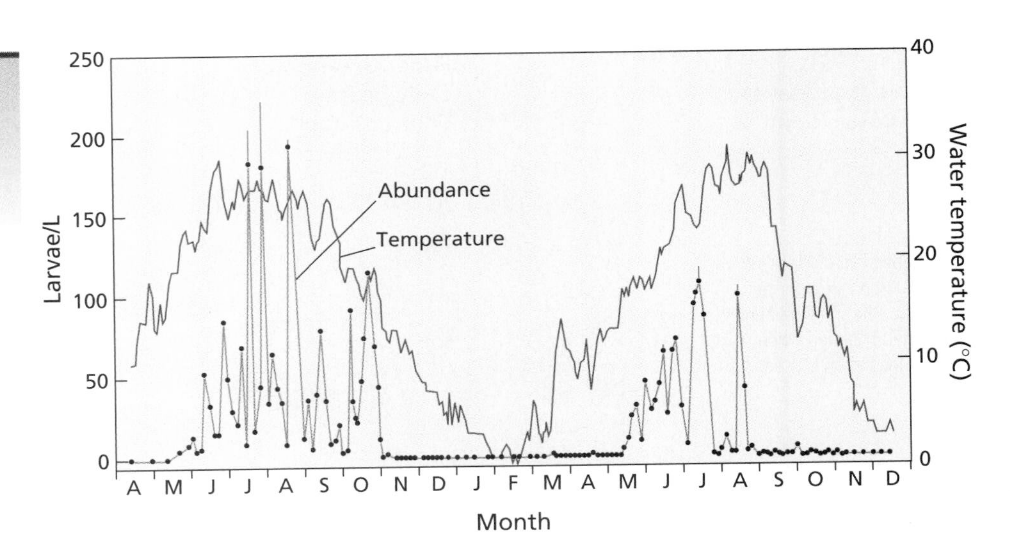

GUIDELINE 2

Look for Places Where Graphics Will Increase Your Communication's Persuasiveness

Graphics can greatly increase a communication's persuasive impact. Often this opportunity arises when a graphic enables you to support your persuasive points by conveying your data in an especially dramatic way (Kostelnick & Roberts, 1998). For example, the manufacturer of a plastic insulating material used the bar graph shown in Figure 15.8 to persuade greenhouse owners that they could greatly reduce their winter heating bills by covering their greenhouses with the company's plastic sheets.

Many other kinds of visual representations can enhance a communication's persuasiveness. A drawing, for instance, can help readers envision the desirable outcomes of projects you propose. A photograph can forcefully portray a problem you want to motivate your readers to address.

GUIDELINE 3

Choose the Types of Graphics That Are Best at Achieving Your Objectives

Once you've decided where to use a graphic, select the type that will be most effective at achieving your objectives. Most information can be presented in more than one type. Numerical data can be presented in a table, bar graph, line graph, or pie chart. The components of an electronic instrument can be represented in a photograph, sketch, block diagram, or schematic. When choosing the type of graphic you will use, consider both the tasks you want to help your readers perform and the way you want to affect their attitudes.

Consider Your Readers' Tasks

Different types of graphics support different reading tasks. Consider, for example, Ben's choices.

FIGURE 15.8

Graphic That Makes a Persuasive Point

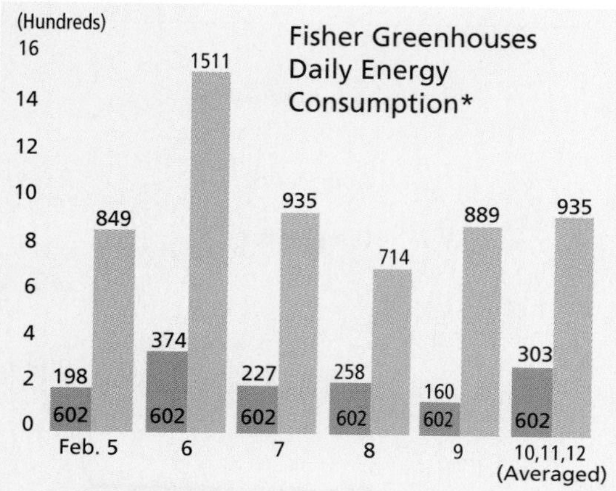

The brightest colour is given to the 602 bars in order to emphasize the small amount of energy consumed when the 602 insulation is used.

Ben must decide how to display salary data.

Ben has surveyed people who graduated over the past three years from three departments in his school. Now he wants to report to the alumni what he has learned about their average starting salaries. He could do this with a table, bar graph, or line graph (see Figure 15.9, page 370). Which type of graphic would be best? To decide, Ben must determine the specific task the alumni will want to perform using the graphic. If they want to learn the average starting salary of people who graduated in their year from their department, the table will help them most. If they want to compare the average starting salaries in their department with the average starting salaries of people who graduated in that same year from the other departments, the bar graph will work best. And if the alumni's task will be to determine how the average starting salary in their department changed over the years and to compare that change with the changes experienced by the other departments, the line graph will be the most useful.

His choice should depend on the way his readers will want to use his data.

Consider Your Readers' Attitudes

In addition to thinking about your readers' tasks, think also about their attitudes when selecting the type of graphic you will use. Pick the type that most quickly and dramatically communicates the evidence that supports your persuasive point. Consider, for instance, Akiko's choices.

Akiko must select the type of graphic with the highest visual impact.

Akiko recommended a change in the design of one of her company's products. To show how the company has benefited from this change, she has tallied the number of phone calls received by the company's toll-free help line during the months immediately before and after implementation of the new design. As Figure 15.10 (page 371) shows, if Akiko presents her data in a table, her readers will have to do a lot of subtracting to appreciate the impact of her recommendation. If she presents her data in a line graph, they will be able to recognize her accomplishment at a glance.

Select graphics your readers know how to interpret.

Of course, your readers will find your graphics useful and persuasive only if they can understand them. Some types are familiar to us all, but others are more specialized, like those in Figure 15.11 (page 372). Although they are very informative to people familiar with the symbols and conventions used, specialized representations will only baffle other people. Either select graphics that your readers will understand, or provide the explanations they need.

GUIDELINE 4 Make Each Graphic Easy to Understand and Use

Having chosen the *type* of graphic you will use, you must design the item itself. When doing so, focus first on usability. Make your graphics, like your text, easy for your readers to understand and use. Here are some suggestions.

Design Your Graphics to Support Your Readers' Tasks

Imagine your readers trying to use your graphic.

First, follow this familiar, reader-centred strategy: Imagine your readers in the act of using your graphic and then design it accordingly. In drawings or photographs for step-by-step instructions, for example, show objects from the same angle that your readers will see them when performing the actions you describe.

Likewise, in a table, arrange the columns and rows in an order that will help your readers find the particular piece of information they are looking for. Maybe that means you should arrange the columns and rows in alphabetical order, according to a logical pattern, or by some other system. Use whatever arrangement your readers will find most efficient.

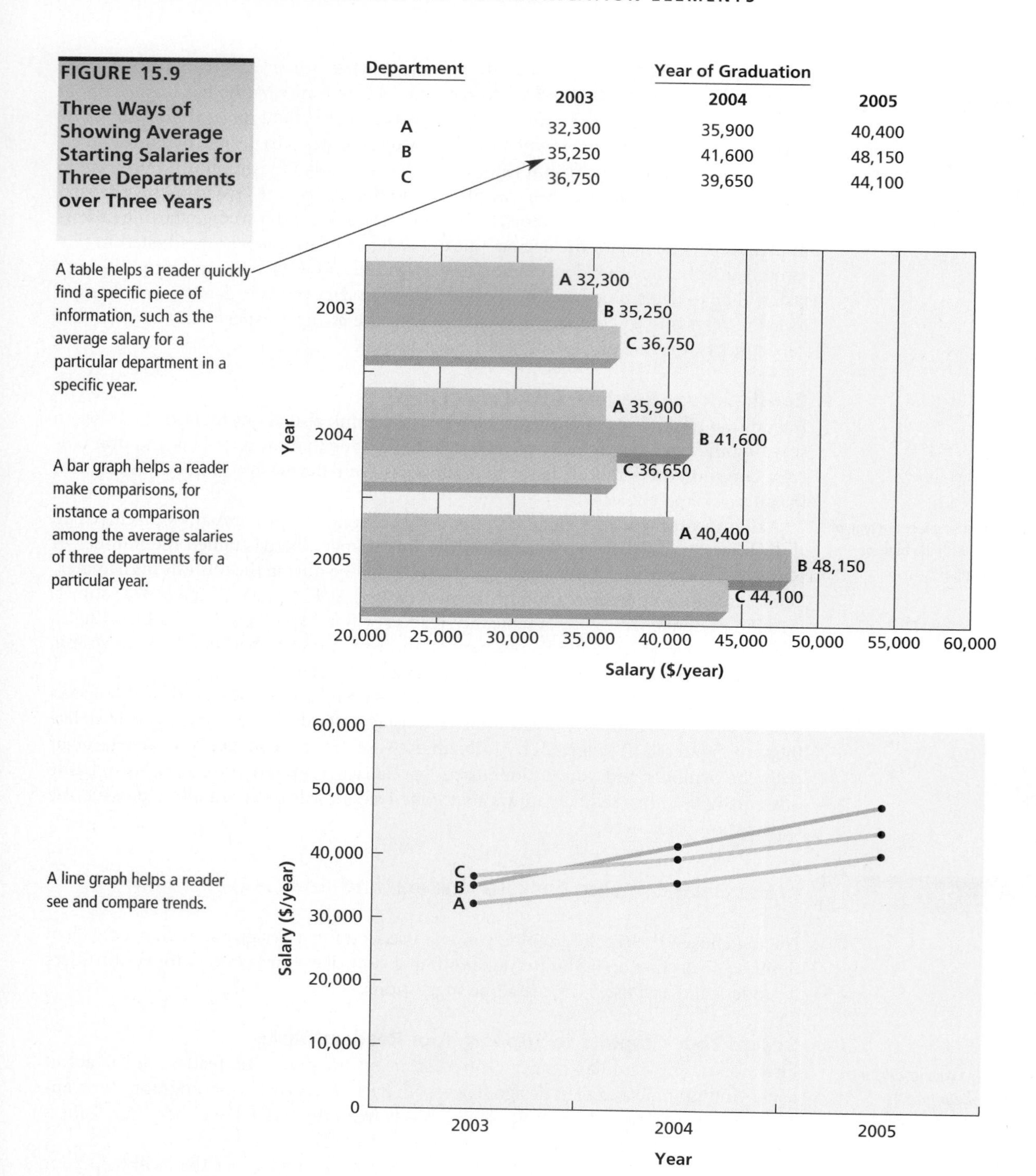

Department	Year of Graduation		
	2003	2004	2005
A	32,300	35,900	40,400
B	35,250	41,600	48,150
C	36,750	39,650	44,100

FIGURE 15.9

Three Ways of Showing Average Starting Salaries for Three Departments over Three Years

A table helps a reader quickly find a specific piece of information, such as the average salary for a particular department in a specific year.

A bar graph helps a reader make comparisons, for instance a comparison among the average salaries of three departments for a particular year.

A line graph helps a reader see and compare trends.

FIGURE 15.10

Comparison of Data Presented in a Table and in a Line Graph

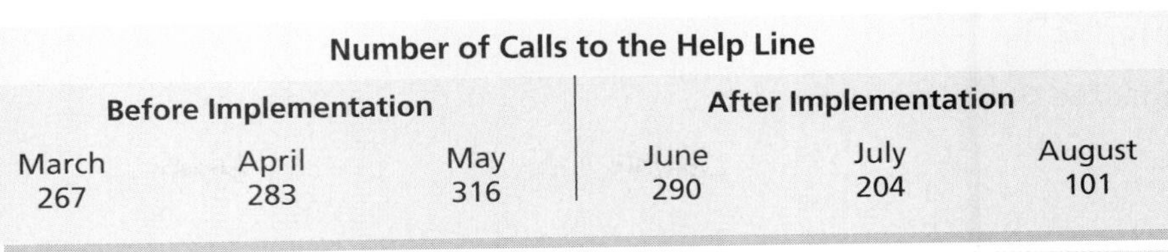

Number of Calls to the Help Line

Before Implementation			After Implementation		
March	April	May	June	July	August
267	283	316	290	204	101

Graphs are much more effective than tables at showing trends.

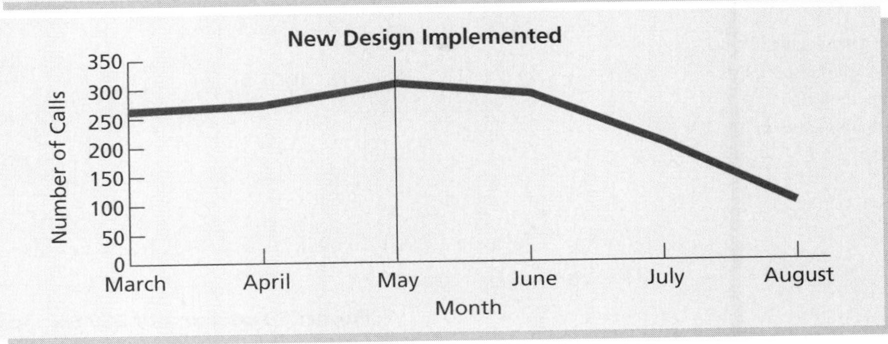

Simplify Your Graphics

A second strategy for making your graphics easy to understand and use is to keep them simple. Simplicity is especially important for graphics that will be read on a computer screen or from a projected image because people have more difficulty reading from these media than from paper. Here are several effective strategies for keeping your graphics simple.

Advice for keeping visuals simple

- **Avoid including too much material.** Sometimes, two or three separate graphics will communicate the same information more effectively than one.
- **Include only the details that are necessary to achieve your purpose.** Like unnecessary words in text, superfluous details in graphics create extra, unproductive work for readers and obscure the really important information. Figure 15.12 (page 373) shows how the elimination of extraneous detail can simplify a graph. As another example, Figure 15.13 (page 374) shows two photographs and a line drawing of the same experimental apparatus used in new-product research.

Label the Important Content Clearly

By providing labels for the parts included in your graphics, you can help your readers find the information they seek and understand what they see. In a table, label every row and column. In a diagram, label every part that your readers will be interested in. But avoid labelling other features. Unnecessary labels clutter a graphic and make it difficult to understand and use.

Word and place labels carefully.

For each label, carefully choose the appropriate word or words and place them where they are easy to see. If necessary, draw a line from the label to the part. Avoid placing a label on top of an important part of your figure. Also avoid labelling parts with letters or numbers that are explained in a separate key. Figure 15.14 (page 375) shows how much more convenient it is for readers to have labels next to the parts they designate.

FIGURE 15.11
Specialized Graphics

Some graphics are very meaningful to specialists, though they are incomprehensible to others.

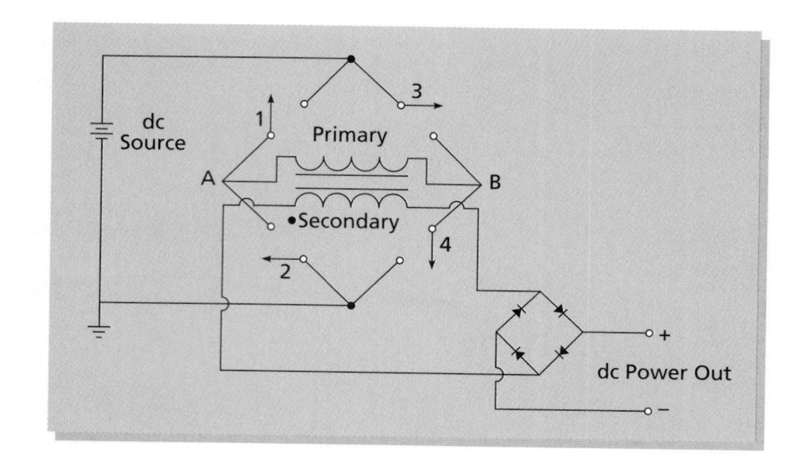

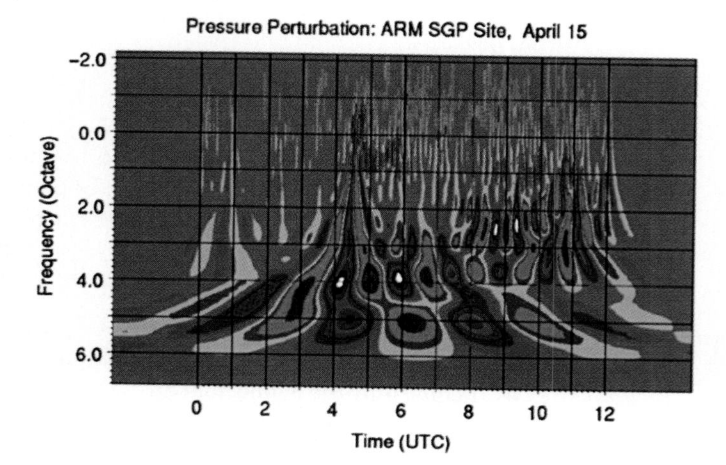

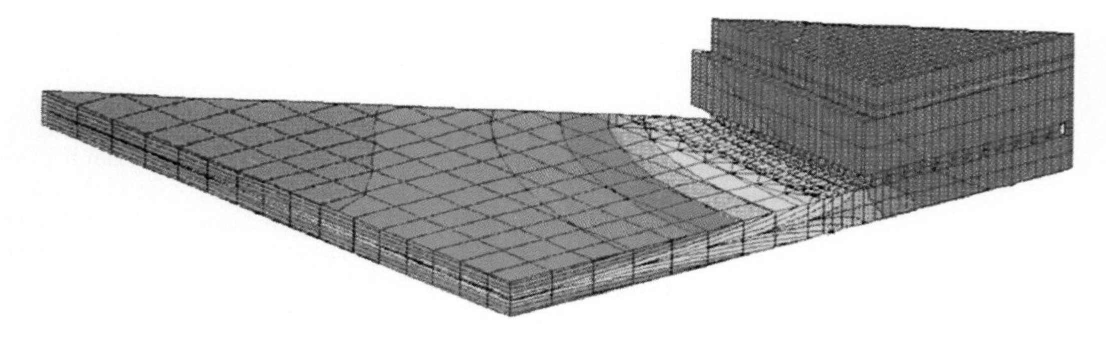

FIGURE 15.12

A Line Graph with and without Grid Lines

Removing the unnecessary grid lines makes the graphed lines easier to follow and the labels easier to read.

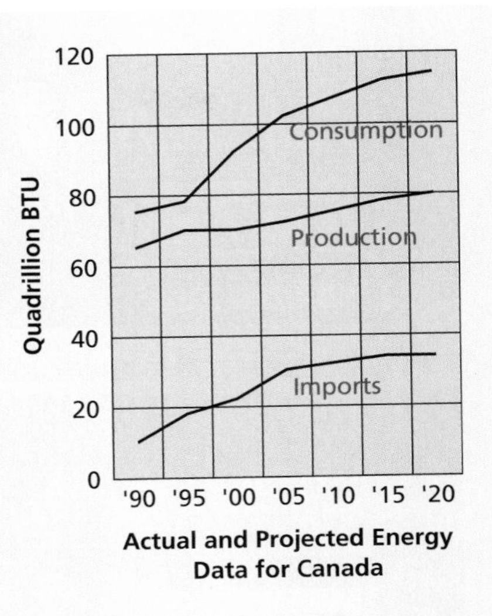

Unnecessary Detail Removed

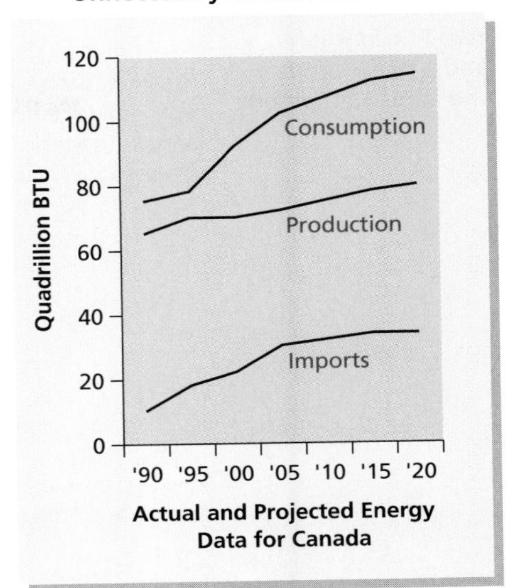

Provide Informative Titles

Titles help your readers find the graphics they are looking for and know what the graphics contain when they locate them. Typically, titles include both a number (for example, "Figure 3" or "Table 6") and a description ("Effects of Temperature on the Strength of M312").

Advice for writing titles

Make your titles as brief—and informative—as possible. Use more words if they are needed to give your reader precise information about your graphic. Don't say, "Information on Computer Programs," but say instead, "Comparison of the Speed and Capabilities of Three Database Programs."

Be consistent in the placement of your titles. Placing titles below your figures is common practice.

For a long communication in which readers might seek a specific figure whose location won't be obvious from the regular contents list, provide a separate list of the figures and the pages where they can be found.

Note, however, that some graphics don't need a title. For instance, when you are including a very short table in your text, you can make its contents perfectly clear with a simple statement in the preceding sentence:

You will be pleased to see how well our top four salespeople did in June:

A table that needs no title

O'Brien	$227,603
Tanaka	195,278
Opara	193,011
Gotceitas	188,590

FIGURE 15.13

Two Photographs and a Drawing of the Same Equipment

LESS EFFECTIVE PHOTOGRAPH

The cluttered background makes it difficult to identify the parts of the testing equipment.

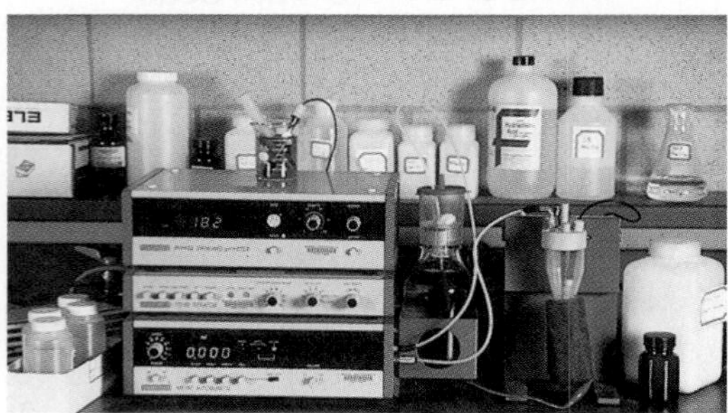

MORE EFFECTIVE PHOTOGRAPH

Removal of the clutter allows the parts to be distinguished.

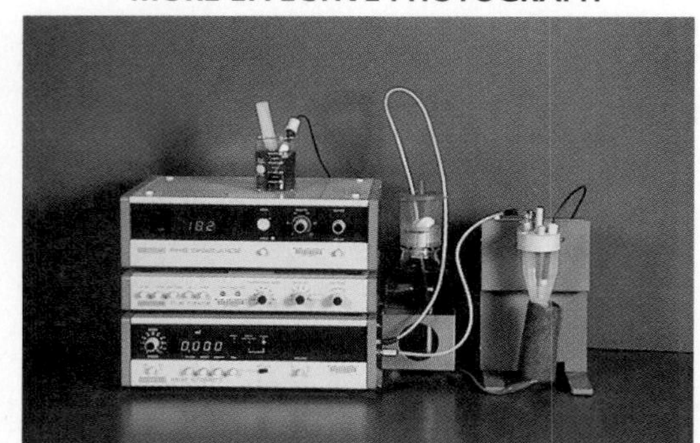

DRAWING

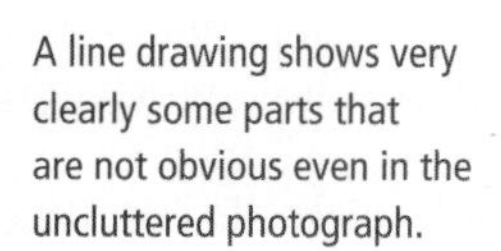

A line drawing shows very clearly some parts that are not obvious even in the uncluttered photograph.

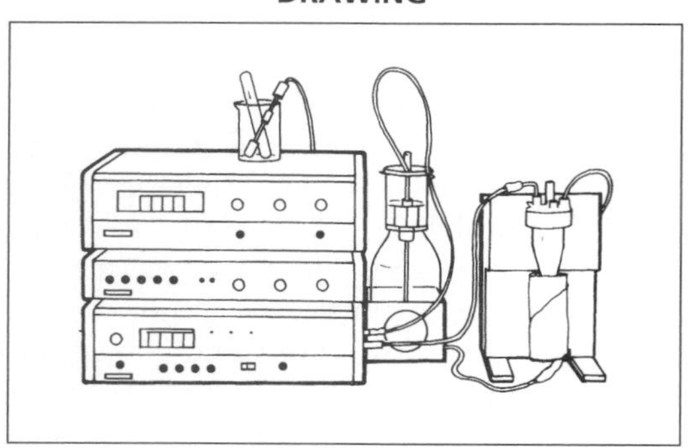

GUIDELINE 5 Use Colour to Support Your Message

In recent years, there's been an explosion of colour in workplace communications. Webpages and other on-screen communications are saturated with colour. Colour enlivens and enhances the many oral presentations made by individuals using software such as PowerPoint. In addition, as colour printing and copying have become less expensive, even routine printed documents now incorporate multicoloured designs in some organizations.

FIGURE 15.14

Labels Provided with and without a Key

When labels are placed next to their parts, readers can easily match each part to its name.

Easy to Read: Labels Placed by Parts

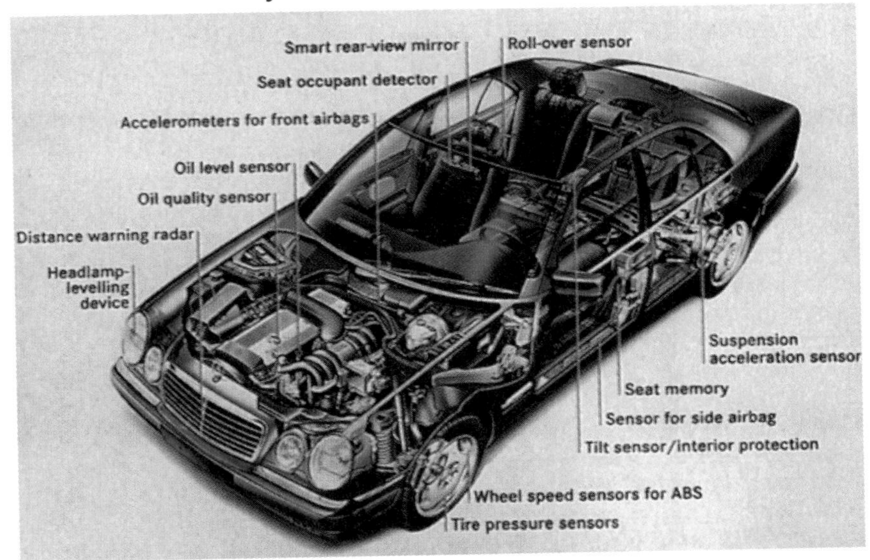

Sensors and Integrated Safety Devices

When labels are placed in a key, readers must repeatedly shift their attention from the diagram to the key in order to learn the names of the parts.

Less Easy to Read: Labels in a Key

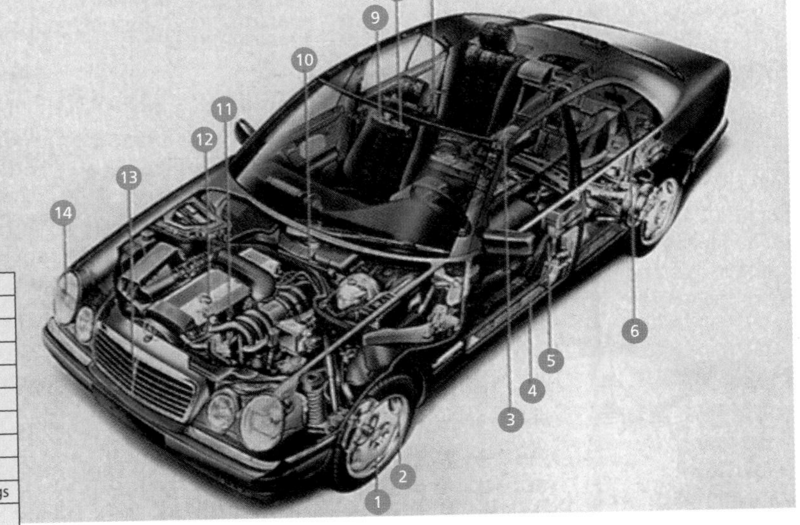

1	Tire pressure sensors
2	Wheel speed sensors for ABS
3	Tilt sensor/interior protection
4	Sensor for side airbag
5	Seat memory
6	Suspension acceleration sensor
7	Roll-over sensor
8	Smart rear-view mirror
9	Seat occupant detector
10	Accelerometers for front airbags
11	Oil level sensor
12	Oil quality sensor
13	Distance warning radar
14	Headlamp-levelling device

Sensors and Integrated Safety Devices

The widespread availability of colour puts a new and powerful aid to communication at your disposal. With colour you can clarify your messages, speed your readers' comprehension, and make your information easy for your readers to use. Among other things, colour can help you do the following:

Some uses of colour

- Highlight a point
- Tell the reader where to look first
- Group related items
- Establish hierarchies of importance
- Provoke an emotional response
- Make your communication look more polished and attractive

Guidelines for Using Colour

The impact of colour on readers is determined partly by physiology and partly by psychology. Based on what researchers know about these responses, the following suggestions will enable you to use colour in a reader-centred way:

Misuse of colour can diminish a communication's effectiveness.

1. **Use colour primarily for clarity and emphasis, not decoration.** When used merely for decoration, colour can create two problems. First, because colour attracts the eye, it can draw a reader's eye to ornamentation rather than to more important content, thereby reducing the reader's ability to grasp your message.

   Second, attractive colours can actually make reading difficult, as communication researcher Colin Wheildon (1995) discovered. He showed people two versions of the same page, one printed in black and one in blue. The people said the blue version was more attractive, but those who actually read it scored substantially lower on a comprehension test than those who read the black version. It is critical that you deploy colours in a way that promotes rather than hinders comprehension.

A colour's appearance can change if the surrounding colours are changed.

2. **Choose colour schemes, not just single colours.** Readers see a colour in terms of its surroundings. To illustrate this point, design expert Jan V. White (1990) uses blocks of colour like those shown in Figure 15.15. The top pair of blocks in Figure 15.15 shows that the same shade of blue appears lighter viewed against a dark colour than against a pale colour. As the pair of blocks on the next page demonstrates, the same pure colour looks much different when it is surrounded by another shade of the same colour than when it is surrounded by a complementary colour. Here are some facts about colour's effects on readers that you should keep in mind in choosing colour schemes:
   - **Bright colours attract the eye more assertively than dull colours.** Use a bright colour for accent and a dull colour for background. Figure 15.16 shows how one writer used bright yellow to focus his readers' eyes to the central component in his illustration of a device for creating silicon-germanium crystals.
   - **Warmer and more intense colours appear closer to the reader than cooler and less intense colours.** When you are trying to communicate depth, use cooler, less intense colours for what appears more distant. Figure 15.17 (page 378) shows how a group of scientists used a dull background and bright colours to make their image of the molecular structure of cytokine appear to advance toward their readers.

- **Contrast creates emphasis.** To make something stand out, put it against a much different colour rather than a very similar colour.

Greater contrast creates more emphasis.

More Emphasis | Less Emphasis

FIGURE 15.15

Some Ways a Colour's Appearance is Affected by the Surrounding Colour

The same blue square appears brighter against a darker colour (such as black) than it does against a lighter colour (such as yellow).

The same orange square appears brighter against a complementary colour (in this case, green) than it does against a different shade of its own colour.

FIGURE 15.16

Bright Colours Used to Focus Attention

The bright colour at the centre of this figure draws the reader's attention there.

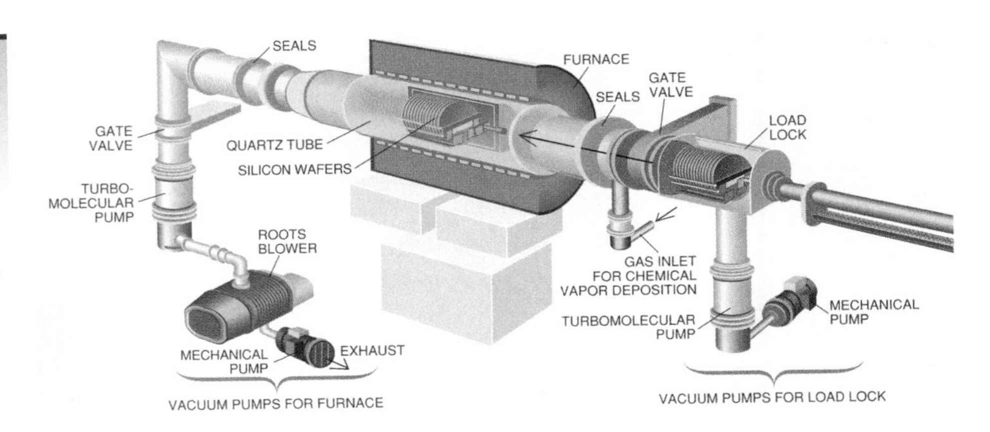

FIGURE 15.17

Bright Colours Used against a Dull Background to Make Important Material Advance toward the Reader

Bright colours appear to advance from a dark background.

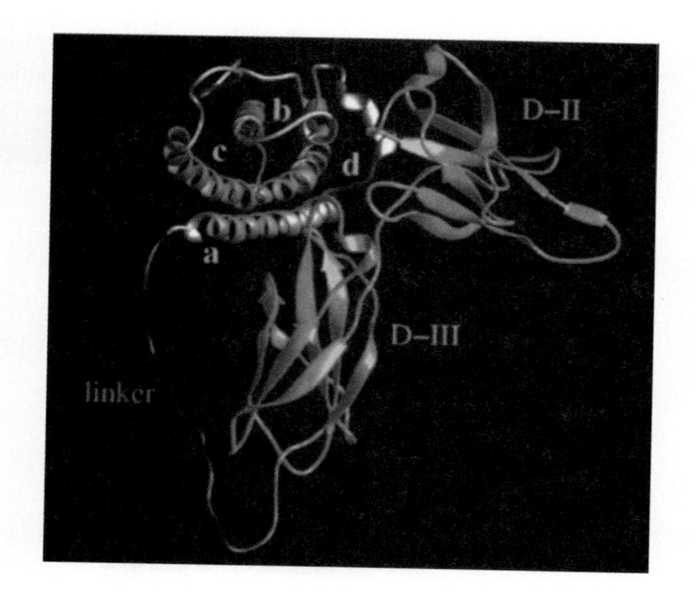

3. **To promote easy reading, use a high contrast between text and background.** To comprehend a written message, readers must first pick out the letters from the background on which they are printed. We traditionally use black print on white paper because this colour combination creates a very high contrast that makes letter identification easy. However, backgrounds of colours other than white are very common in computer-projected graphics for oral presentations. They often appear in printed documents as well. The following chart shows how much more difficult type becomes to read when contrast with the background is reduced.

Reducing contrast reduces readability.

20%	This type is surprinted in black.	This type is dropped out in white.	20%
40%	This type is surprinted in black.	This type is dropped out in white.	40%
60%	This type is surprinted in black.	This type is dropped out in white.	60%
80%	This type is surprinted in black.	This type is dropped out in white.	80%
100%	This type is surprinted in black.	This type is dropped out in white.	100%

The same colour can have different associations in different contexts and different cultures.

4. **Select colours with appropriate associations.** In many contexts, colours have specific associations and even symbolic meaning. Driving through a city, we associate red with "Stop" and green with "Go." As Jan V. White (1990) points out, however, our associations with colour vary from one context to another. In politics, blue is associated with conservatism and in business with stability, but to a doctor it connotes death. In other contexts, blue suggests sky, water, and cold. People's associations with colours are quite powerful, so be alert to them when selecting colours for your graphics. Colour associations also vary from one culture to another. In many Western cultures, black is the colour of death and mourning. In China, white is. If you are addressing an international audience, determine what your readers might associate with the colours you are thinking of using.
5. **Stick to a few colours.** An overabundance of colours can cause confusion. Colours can compete with one another for attention, leaving readers unsure of the visual hierarchy being established among the elements of a graphic. Readers also rely on colour to establish patterns of meaning, as when one colour is employed consistently to represent the raw materials used in a manufacturing process and another colour is employed to represent the products of the process. If too many colours are used, such patterns are obscured—if they even existed in the first place.
6. **Use colour to unify your overall communication.** When choosing colours, think beyond the individual graphics and consider your communication as a whole. By using a limited group of colours, you can create visual unity; by using these colours in a consistent pattern, you can help your readers understand your message.

When writing long documents, you can also employ colour to help your readers use your communications efficiently. Figure 15.18 (page 380) depicts several techniques for doing so for print documents. Consistent, strategic use of colour can also unify the graphics for oral presentations (Figure 15.19, page 381) and online communications (Figure 15.20, page 382).

GUIDELINE 6 Adapt Existing Graphics to Your Readers and Purpose

At work, people often increase their writing efficiency by using existing graphics instead of creating new ones. A major source for reusable graphics is previous communications. When writers create final reports, for instance, they often incorporate tables, diagrams, and similar items from previous proposals and progress reports that they or other employees at their organization previously prepared concerning the same project. Writers occasionally use clip art from the libraries included with software such as Microsoft Word and PowerPoint. They sometimes download images from the Internet or scan them from print publications, although this practice is less common because of the need to obtain permission from the copyright owners.

When you use an existing graphic, be sure to review it from the perspective of your readers and purpose. Determine whether the graphic will be completely usable and persuasive for your readers or whether it needs to be modified to achieve your communication objectives. Perhaps the labels need to be changed or some portion needs to be redesigned. After such examination, writers sometimes discover that a graphic they thought they could reuse won't serve their purpose adequately, so they really do need to create a new one.

FIGURE 15.18

Ways of Unifying a Printed Communication with Colour

There are many ways to use colour to unify a long communication visually.

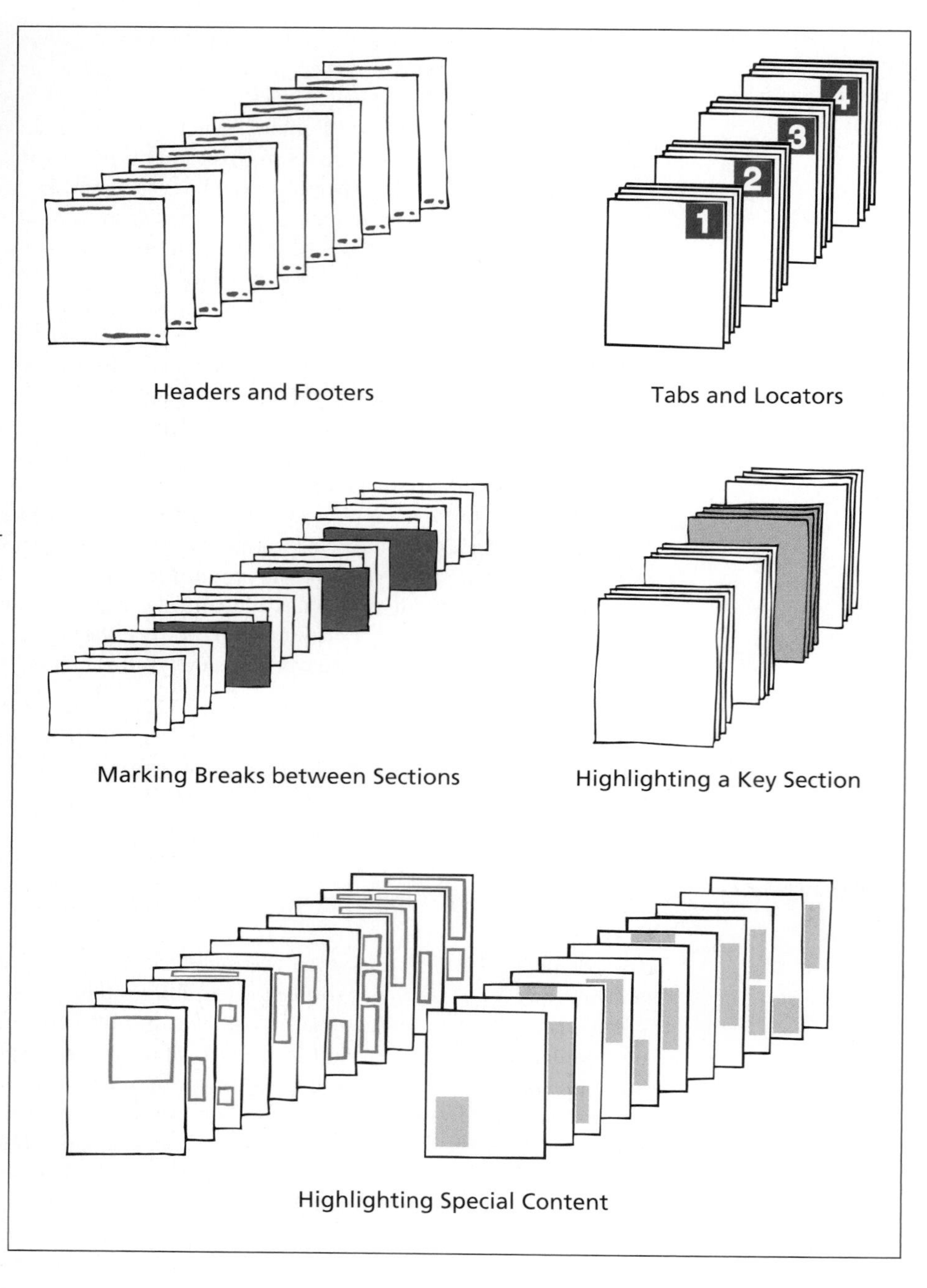

FIGURE 15.19

Ways of Unifying an Oral Presentation with Colour

By using the same colour scheme throughout this presentation, the writer creates visual unity among the slides.

To make reading easy, the writer uses large letters and high contrast between background and text.

Consistent use of design elements such as the red bar unifies these slides visually.

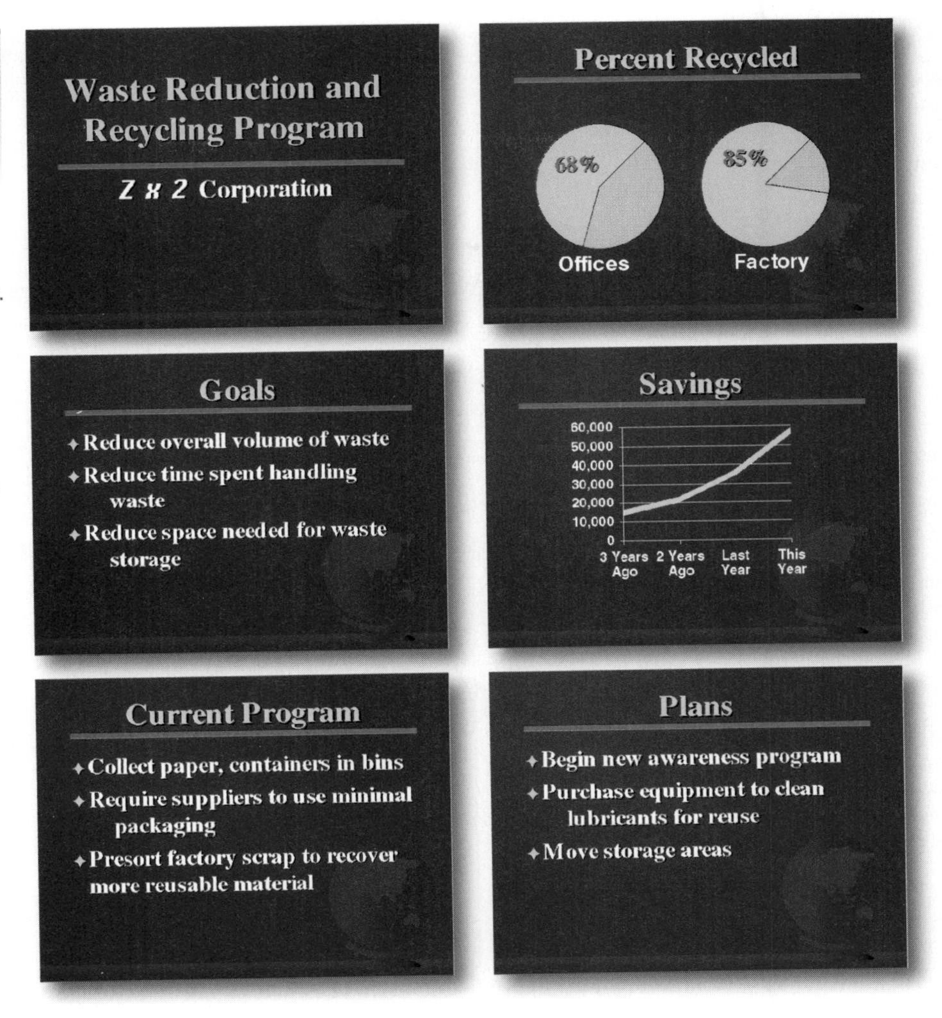

Also, when using existing graphics that aren't owned by your employer, be careful to comply with copyright law. As explained in Chapter 13, you need permission from the copyright owner of each graphic you use unless the owner has placed the graphic in the public domain. Always document the sources of the graphics you use.

GUIDELINE 7 Integrate Your Graphics with Your Text

The first six guidelines in this chapter focus on graphics in isolation from your text. In contrast, this guideline asks you to think about your graphics from the point of view of your readers in the act of reading your overall communication. This guideline suggests that you integrate your graphics with your text so that they work together harmoniously to create a single, unified message. Here are four strategies you can follow:

FIGURE 15.20

Ways of Unifying an Online Communication with Colour

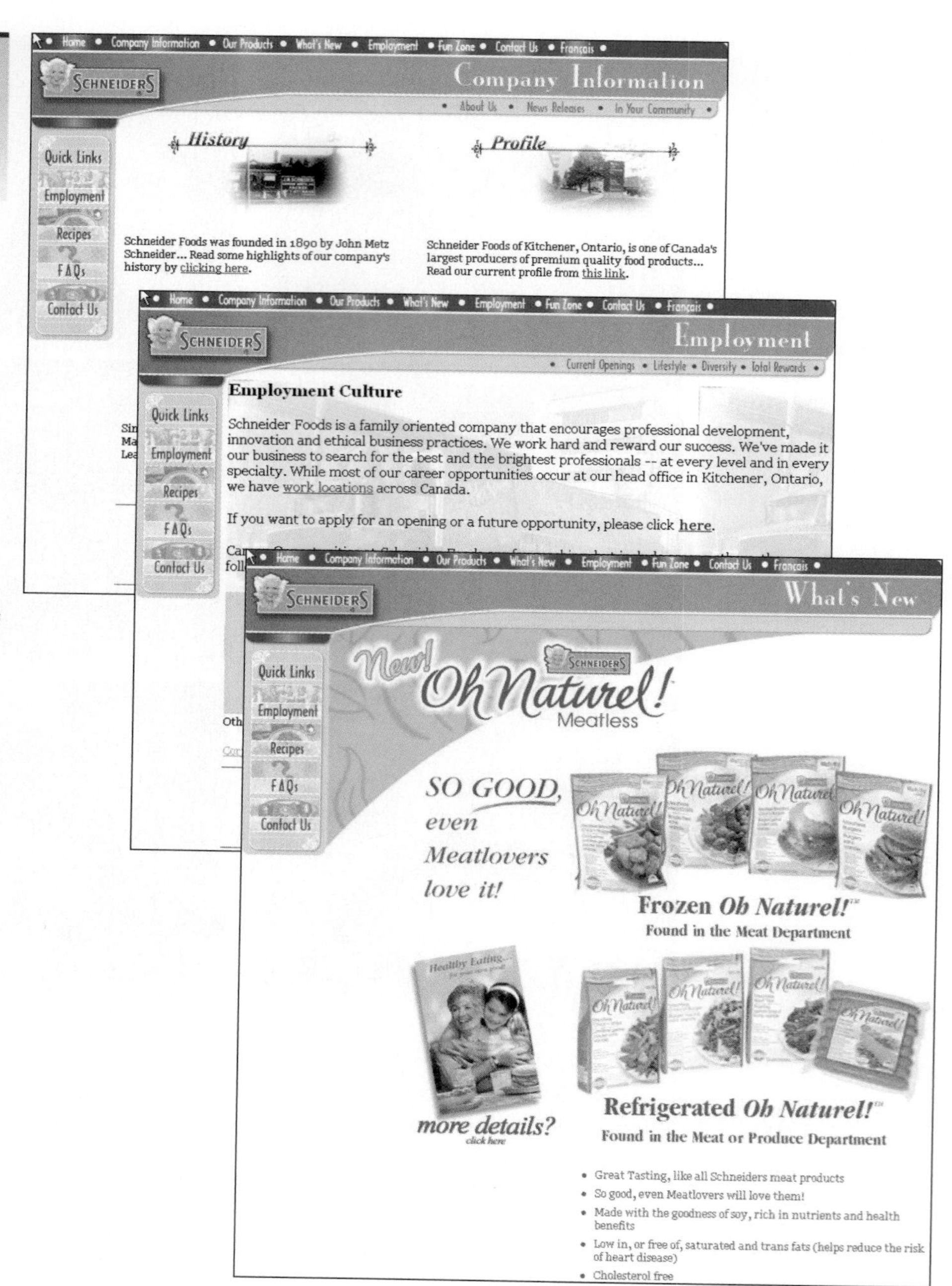

These pages demonstrate the use of colour to unify a website.

Note the various ways that black and the same shades of blue and orange are worked into each of these pages.

The pages are also unified through the repetition of certain design elements and through the use of the same grid (see page 394).

- **Introduce your graphics in your text.** When people read sequentially through a communication, they read one sentence and then the next, one paragraph and then the next, and so on. When you want the next element they read to be a table or chart rather than a sentence or a paragraph, you need to direct their attention from your text to the graphic and also say what they will find in it. There are various ways of doing this:

Two sentences

> In a market test, we found that Radex was much more appealing than Talon, especially among rural consumers. See Figure 3.

One sentence with an introductory phrase

> As Figure 3 shows, Radex was much more appealing than Talon, especially among rural consumers.

One sentence with the reference in parentheses

> Our market test showed that Radex was much more appealing than Talon, especially among rural consumers (Figure 3).

Sometimes your text reference to a graphic will have to include information your readers need in order to understand or use the graphic. For example, here is how the writers of an instruction manual explained how to use one of their tables:

Writer tells reader how to use the graphic

> In order to determine the setpoint for the grinder relay, use Table 1. First, find the grade of steel you will be grinding. Then read down column 2, 3, or 4, depending upon the grinding surface you are using.

Whatever kind of introduction you make, place it at the exact point where you want your readers to focus their attention on the graphic.

- **State the conclusions you want your readers to draw.** Another way to integrate your graphics into your text is to state explicitly the conclusions you want your readers to draw from them. Otherwise, readers may draw conclusions that are quite different from the ones you have in mind.

  For example, one writer included a graph that showed how many orders she thought her company would receive for its rubber hoses over the next six months. The graph showed that a sharp decline would occur in orders from automobile plants, and the writer feared that her readers might focus on that fact and miss the main point. So she referred to the graph in the following way:

Writer tells readers what conclusion to draw from the graphic

> As Figure 7 indicates, our outlook for the next six months is very good. Although we predict fewer orders from automobile plants, we expect the slack to be taken up by increased demand among auto parts outlets.

- **When appropriate, include explanations in your figures.** Sometimes you can help your readers understand your message by incorporating explanatory statements into your figures. Figure 15.21 (page 384) shows how paint powder is applied to cars in an automobile assembly line.
- **Place your graphics near your references to them.** When your readers come to a statement asking them to look at a graphic, they lift their eyes from your text and search for the graphic. You want to make that search as short and simple as possible. Ideally, you should place the graphic on the same page as your reference to it. If there isn't enough room, put the graphic on the page facing or the page that follows. If you place the figure farther away than that (for instance, in an appendix), give the number of the page on which the figure can be found.

FIGURE 15.21

Explanatory Text Incorporated in a Figure

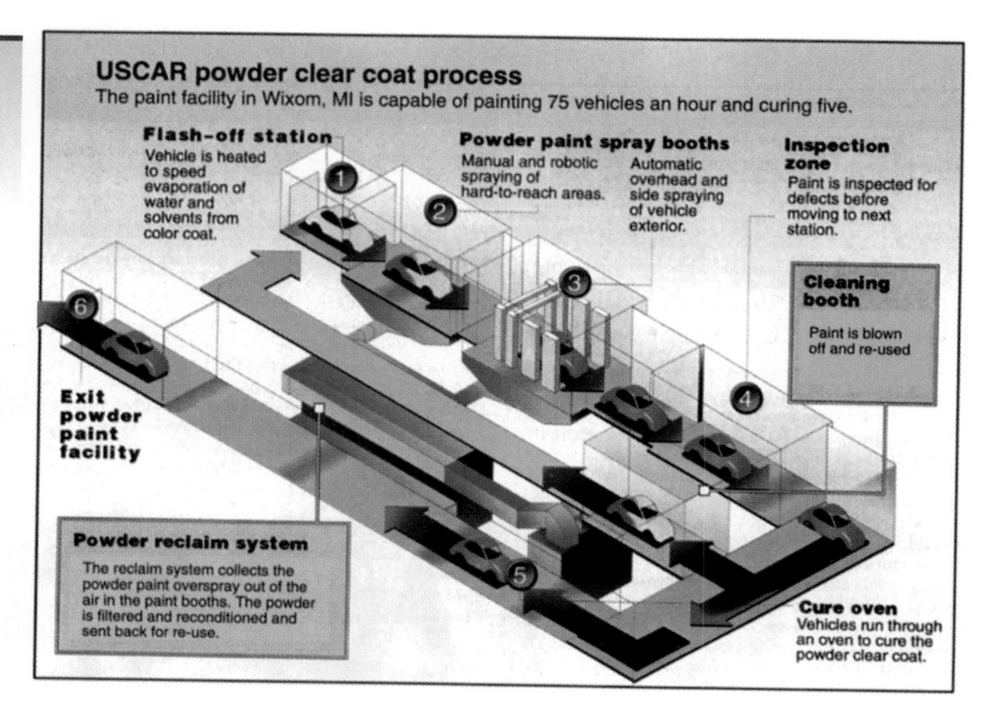

The numbered explanations help readers understand the process by describing it step by step.

The boxed explanations help readers identify and understand the function of two important parts of the paint facility.

GUIDELINE 8 Before Addressing an International Audience, Check Your Graphics with People from Other Cultures

Visual language, like spoken and written language, differs from culture to culture. The design of many ordinary objects differs from country to country. If you use a picture of any object that looks odd to people in your target audience, the effect you are striving for may be lost.

Graphics are interpreted differently in different cultures.

Colours, too, have different connotations in different cultures. While yellow suggests caution in Canada, the use of this colour is considered auspicious and even festive in India. However, in China, yellow is associated with imperialism.

The point is simple: Whenever you are writing for readers in another country, discuss your plans and review your draft graphics with people familiar with that country's culture. If possible, also test your drafts with these individuals.

GUIDELINE 9 Ethics Guideline: Avoid Graphics That Mislead

Graphics can mislead as easily as words can. When representing information visually, you have an ethical obligation to avoid leading your readers to wrong conclusions. This means not only that you should refrain from intentional manipulation of your readers but also that you should guard against accidentally misleading readers. The following paragraphs describe positive steps you can take when creating several common types of graphs. To communicate ethically, follow the spirit of this advice when you create any type of graphic.

Ethical Bar Graphs and Line Graphs

To design bar graphs and line graphs that convey an accurate visual impression, you may need to include zero points on your axes. In Figure 15.22, the left-hand graph makes the difference between the two bars appear misleadingly large because the scale on the vertical axis begins at 85 percent instead of at zero. The centre graph of Figure 15.22 gives a more accurate impression of the data represented because the scale on the vertical axis begins at zero.

If you cannot use the entire scale, indicate that fact to your readers by using hash marks to signal a break in the axis and, if you are creating a bar graph, in the bars themselves. The right-hand graph in Figure 15.22 shows an example.

Note, however, that zero points and hash marks are sometimes unnecessary. For example, in some technical and scientific fields, certain kinds of data are customarily represented without zero points, so readers are not misled by their omission.

Ethical Pictographs

Pictographs can also mislead readers. For example, the graphs in Figure 15.23 (page 386) represent the average percentage of an apple harvest that a grower should expect to be graded Extra Fancy. This left-hand graph makes the percentage of Red Delicious apples seem much larger than the percentage for Golden Delicious, even though the actual difference is only 20 percent. That's because the picture of the Red Delicious apple is larger in height and width, so its area is much greater. The right-hand graph shows how to accurately represent the data: by making columns that differ in height alone.

Ethical Use of Colour

Like any other element of a graph, colour can be used unethically. For example, because bright colours attract the eye, they can be used to distract the reader's focus from the most important point. For example, the left-hand graph in Figure 15.24 (page 386) uses a bright line to de-emphasize the crucial bad news by emphasizing relatively unimportant news. The right-hand graph shows how the problem can be corrected.

FIGURE 15.22

Creating Ethical Bar Graphs

Because the scale on its vertical axis begins above zero, the graph on the left can mislead.

The two graphs on the right illustrate ways to avoid misleading.

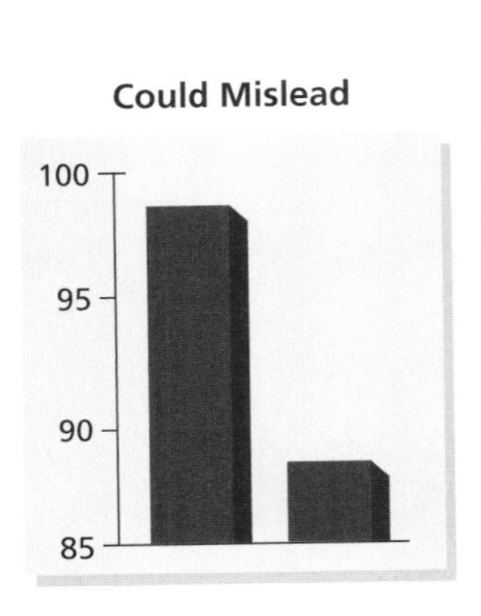

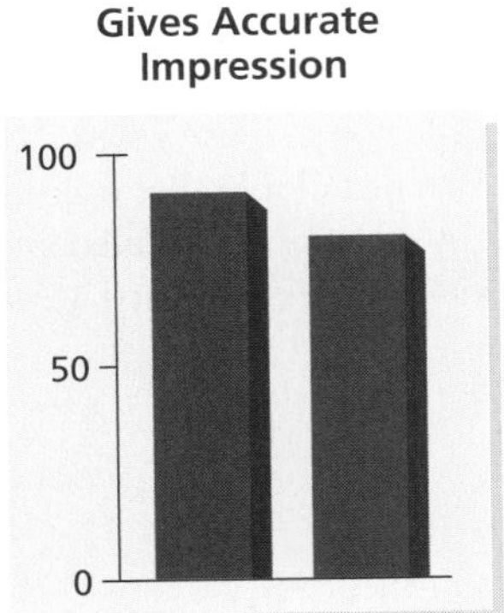

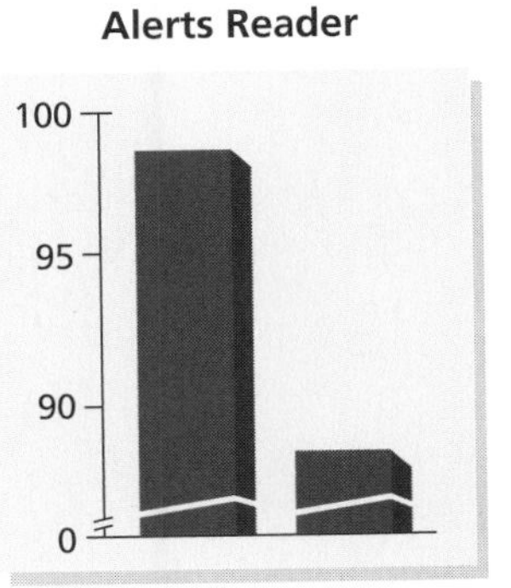

FIGURE 15.23

Creating Ethical Pictographs

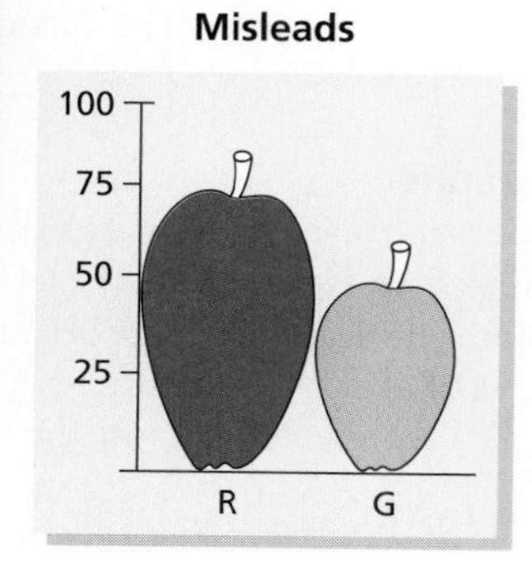

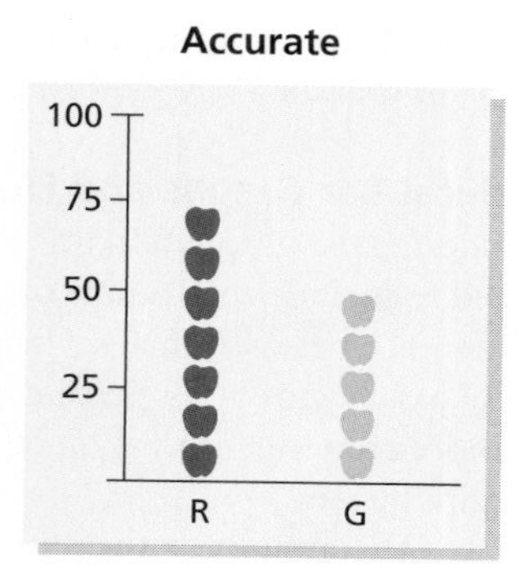

FIGURE 15.24

Using Colour Ethically

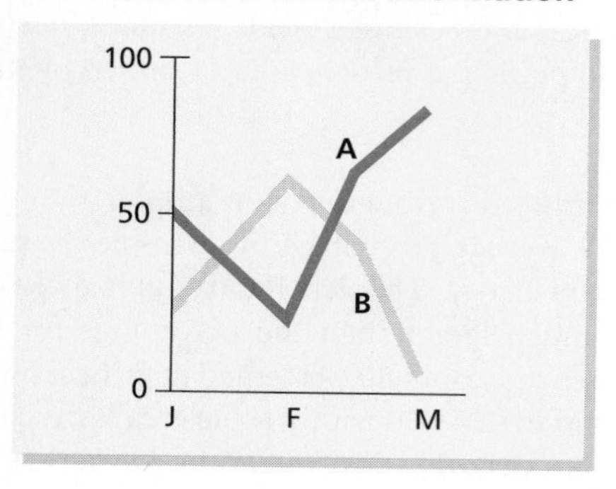

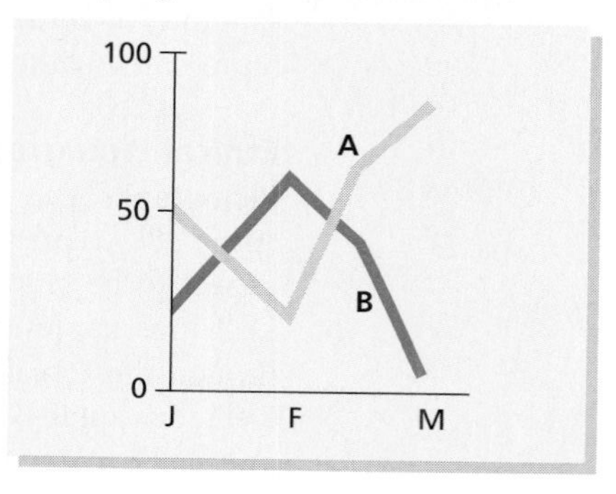

The left-hand graph misleads readers if Trend B represents the crucial information.

The right-hand graph shows how to use colour to avoid misleading.

CONCLUSION

www.techcomm.nelson.com

Graphics can greatly increase the clarity and impact of your written communications. To use graphics well, you need to follow the same reader-centred strategy that you use when writing your text: Think about the tasks your readers will perform while reading and think about the ways you want your communication to shape your readers' attitudes. Doing so will enable you to decide where to use graphics, determine the most effective types of graphics to use, make them easy to understand and use, and integrate them successfully with your text.

A guide you can use when designing your graphics is available on this text's website. The Reference Guide that accompanies this section supplements this chapter's general advice by providing detailed information about how to construct thirteen types of graphics that are often used at work.

EXERCISES

For additional exercises, visit www.techcomm.nelson.com.

Expertise

Create an outline for a communication you will prepare for your course. On the finished outline, list each place where a graphic could make your communication more usable and persuasive in your reader's eyes. For each place, identify the type of graphic that will be most effective in helping you achieve your communication objectives.

Online

1. Using software, make a table that displays the number of hours in a typical week that you spend in each of several major activities, such as attending class, studying, eating, and visiting with friends or family. The total number of hours should equal 168. Be sure to include a brief heading for the columns and rows. Next, convert your table to a graph or chart. Refine your table so that the graph is as easy to read as possible.

2. Using a software program's drawing feature, create a simple illustration of a piece of equipment used in your field of study or in one of your hobbies or activities.

Collaboration

1. Team up with another student. First, each of you is to locate two different types of graphics in journals, textbooks, or other sources in your field (including websites). Working together, evaluate each of the graphics from the perspective of the guidelines in this chapter. Which features of the designs are effective? How could each graphic be improved?

2. Working with another student, exchange drafts of a project in which you are both using one or more graphics. Using the Guide for Creating Graphics, review the graphics in the drafts. Which features are most effective in achieving the writer's usability and persuasive goals? How could the graphics be made more effective from the target readers' perspective?

Ethics

Graphs and various types of charts appear frequently in newspapers and popular magazines. Occasionally, they are criticized for presenting data in a misleading fashion. Find a graphic that you feel may mislead readers. Explain the reason for your assessment and tell how the graphic could be redesigned. Also, identify people who might be adversely affected or who might unfairly benefit if some action were taken on the basis of the misinterpretation. Present your results in the way your professor requests.

CHAPTER 16

Designing Pages and Screens

GUIDELINES

1. Begin by considering your readers and purpose
2. Use proximity to group related elements
3. Align related visual elements with one another
4. Use contrast to establish hierarchy and focus
5. Use repetition to unify your communication visually
6. Select type that is easy to read
7. Design your overall document for ease of use and attractiveness

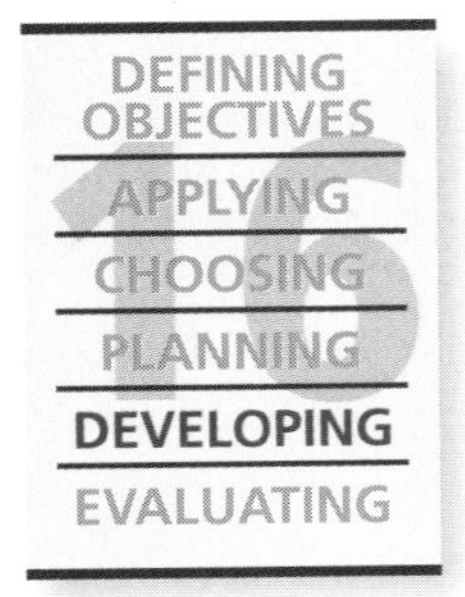

You build your communications out of graphic elements: your words, sentences, and paragraphs; and your drawings, graphs, and tables. Whether you are writing for printed pages or computer screens, your readers see the visual design of these elements before they read and understand your message. And what they see has a powerful effect on the success of your communications.

In fact, visual design has a very substantial impact on the usability and persuasiveness of your communication. Here, for example, are some of the ways that good design enhances usability.

Good design increases usability

- **Good design helps readers understand your information.** For example, you can use visual design to signal the hierarchy of ideas and information in a report. This helps your readers understand what you are saying and what its significance is to them. Similarly, when you write instructions, you can place a direction and a figure next to each other to indicate that the two work together to explain a step.
- **Good page design helps readers locate information.** At work, readers often want to find part of a communication without reading all of it. With headings and other design elements, you can help them do that quickly.
- **Good design helps readers notice highly important content.** With good design, you can emphasize for readers the content that is especially important to them, such as a warning in a set of instructions or a list of actions to take in a recommendation report.

Here are some of the ways good design affects readers' attitudes, thereby increasing a communication's persuasiveness.

Good design increases persuasiveness.

- **Good design encourages readers to feel good about the communication itself.** You've surely seen pages—perhaps in a textbook, a set of instructions, or a website—that struck you as uninviting, even ugly. As a result, you may have been reluctant to read them. And undoubtedly you've seen other printed or online pages that you found attractive, which you approached more eagerly, more receptively. Good design has the same impact on readers of work-related communications. It increases readers' willingness to read reports, proposals, and similar documents carefully, and it extends their willingness to read instructions and websites at all.
- **Good design encourages readers to feel good about the communication's subject matter.** The impact of design on readers' attitudes toward subject matter was dramatically demonstrated by an experiment in which researcher Karen Schriver (1997) asked people to comment on two visual designs for a set of instructions for a microwave oven. When people commented on the design they preferred, they also volunteered that the oven was easy to use—even though they had not seen the oven itself. Regardless of what you are writing about, your communication's design can influence—for better or worse—your readers' attitudes toward your subject.

A READER-CENTRED APPROACH TO DESIGN

Because good design can have such a huge impact on a communication's usability and persuasiveness, you should approach design with the same focus on your readers that you maintain while drafting other elements of your message.

For additional chapter resources, visit Chapter 16 at www.techcomm.nelson.com.

This chapter's seven guidelines will help you do that. Each reminds you to look at your communication from your readers' perspective, think about the way your design decision will affect your readers while they are in the act of reading, and choose the design that, during the readers' dynamic interaction with your communication, is most likely to elicit the response you desire.

Taken together, these guidelines will help you develop expertise at designing printed and online pages and also at integrating these pages into visually unified documents and websites.

DESIGN ELEMENTS OF A COMMUNICATION

Whether you are designing printed pages or webpages, it's helpful to think of yourself as working primarily with six types of elements.

The six design elements of a page

- **Text.** Paragraphs and sentences.
- **Headings and titles.** Labels for sections of your communication.
- **Graphics.** Drawings, tables, photographs, and so on—including their captions.
- **White space.** Blank areas.
- **Headers and footers.** The items, such as page numbers, that occur at the top or bottom of each page in a multipage document.
- **Physical features.** These include paper, which may take many shapes and sizes, and bindings, which come in many forms.

To see the variety of ways in which these elements can be crafted to create effective, attractive, and interesting designs for reading on paper and onscreen, see Figure 16.1.

FOUR BASIC DESIGN PRINCIPLES

This chapter's guidelines describe easy-to-follow, reader-centred strategies for coordinating these six basic elements to design effective communications. Much of this advice involves the application of four simple principles, identified by graphic artist Robin Williams (1994), for creating designs that readers find to be highly usable and persuasive.

Design Principles

- **Proximity.** When two or more items on a page are closely related to one another, you should signal this relationship by placing them in close proximity with one another. (Guideline 2)
- **Alignment.** Everything placed on a page should have a visual connection with something else. You can establish these connections by aligning related elements with one another. (Guideline 3)
- **Contrast.** To establish a hierarchy of importance or to focus a reader's attention, you must make some things look different from others. You can establish difference by using contrast. (Guideline 4)
- **Repetition.** The pages in a document should look like they belong to a cohesive whole. You can provide this visual unity by repeating design elements from page to page. (Guideline 5)

FIGURE 16.1 Sample Designs for Printed Documents and Webpages

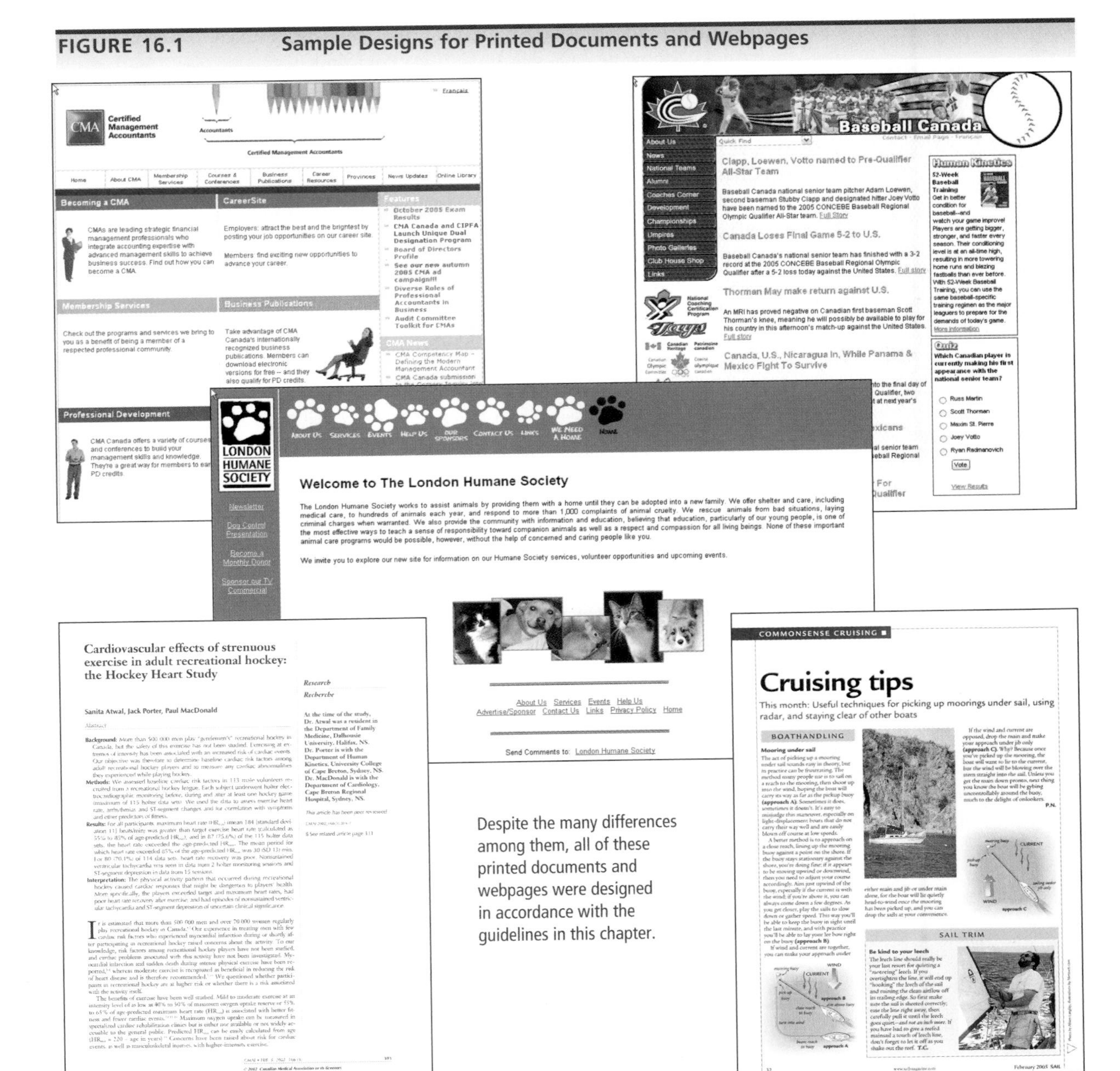

Despite the many differences among them, all of these printed documents and webpages were designed in accordance with the guidelines in this chapter.

GUIDELINE 1 Begin by Considering Your Readers and Purpose

To employ the four design principles effectively, you must have a clear understanding of what you are trying to achieve through your design. Therefore, you should begin your design work in the same way you begin making plans about what information to include and how to organize these contents: by reviewing your communication objectives so that you have the following considerations clearly in mind.

- **Who your readers are.** Review their needs, attitudes, and expectations. All should influence your design decisions.
- **What tasks your communication should enable readers to perform.** You need to understand your readers' tasks if you are going to create a highly usable design that will increase the ease with which they can perform these tasks. For example, if your readers are going to use your communication as a reference work from which they will seek only specific pieces of information, you will need to design differently than if they will simply read it straight through.
- **How you want to influence readers' attitudes.** To design pages that increase your communication's persuasiveness, you need to focus on the ways you want to influence your readers' attitudes and on the goals, values, preferences, and other factors that will shape their responses to your communication.

GUIDELINE 2 Use Proximity to Group Related Elements

One goal of visual design is to increase the usability of your communication by helping your readers see how your information is organized. The first design principle, *proximity,* provides an important strategy for accomplishing this goal: Place related items close to one another. Workplace communicators often talk about proximity in terms of the amount of *white space* (blank space) between items in the documents and on the webpages they create. Less white space places items closer to one another; more white space puts them farther apart.

In the right-hand design, the writer has used white space to establish relationships that are not evident in the left-hand design.

Not Grouped

Jamie Gallant

Shear Logistics Limited

340 Park Place, 777 Burrard St.

Vancouver, BC V6C 2X8

(604) 682-4668

Proximity Establishes Groupings

Jamie Gallant
Shear Logistics Limited

340 Park Place, 777 Burrard St.
Vancouver, BC V6C 2X8
(604) 682-4668

Here are some ways in which you can use white space and the principle of proximity in your on-the-job communications:

- Use less white space below headings than above to place the heading in closer proximity to the text it labels than to the preceding section.
- Use less white space between titles and the figures they label than you use to separate the title and figure from adjacent items.
- In lists that have subgroups, use less white space between items within a subgroup than between one subgroup and another subgroup.

In Figure 16.2, notice how proximity helps to convey the organization of the page.

FIGURE 16.2

Proximity Used to Show the Organization of the Information on a Page

This page illustrates some ways that writers use proximity to help readers see how the information on a page is organized.

By placing these letters close to and left of the screen, the writers enable readers to visually link the captions to the letters.

The writers used less space above these captions than below; this helps readers see that the captions are associated with the screens.

To create a visual link between this heading and the instructions below, the writers put only half as much space below the heading as they put above it.

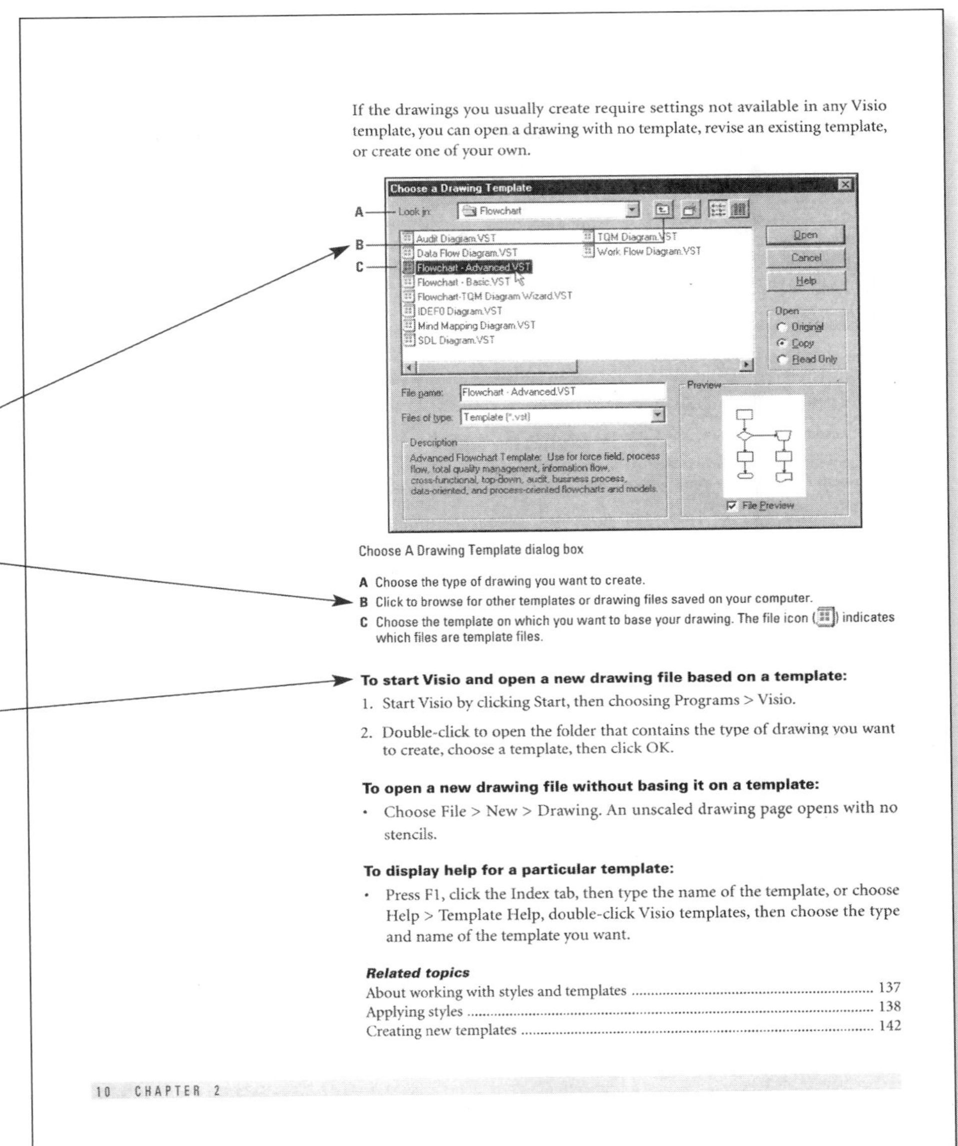

If the drawings you usually create require settings not available in any Visio template, you can open a drawing with no template, revise an existing template, or create one of your own.

Choose A Drawing Template dialog box

A Choose the type of drawing you want to create.

B Click to browse for other templates or drawing files saved on your computer.

C Choose the template on which you want to base your drawing. The file icon ( ) indicates which files are template files.

To start Visio and open a new drawing file based on a template:

1. Start Visio by clicking Start, then choosing Programs > Visio.
2. Double-click to open the folder that contains the type of drawing you want to create, choose a template, then click OK.

To open a new drawing file without basing it on a template:

- Choose File > New > Drawing. An unscaled drawing page opens with no stencils.

To display help for a particular template:

- Press F1, click the Index tab, then type the name of the template, or choose Help > Template Help, double-click Visio templates, then choose the type and name of the template you want.

Related topics

10 CHAPTER 2

GUIDELINE 3 Align Related Visual Elements with One Another

The second principle of visual design—*alignment*—helps you to enhance both the usability and the persuasiveness of your communication. Through alignment, you avoid the appearance that you have placed items arbitrarily on your pages. When the arrangement of the items on a page seems (or is) arbitrary, readers have difficulty discerning meaningful relationships, find it hard to understand and use the information and ideas presented, and may form an unfavourable impression of the communication as well as of the subject matter and the writer.

By comparing the following two designs, you can see how readily alignment establishes connections among related items.

In the right-hand design, the writer has used alignment to establish relationships that are not apparent in the left-hand design.

Arbitrary Placement

Jamie Gallant
Shear Logistics Limited
340 Park Place, 777 Burrard St.
Vancouver, BC V6C 2X8
(604) 682-4668

Alignment Establishes Relationships

Jamie Gallant
Shear Logistics Limited
340 Park Place, 777 Burrard St.
Vancouver, BC V6C 2X8
(604) 682-4668

In this demonstration, the right-hand design connects related items with one another by aligning them along the invisible line of either the left-hand or the right-hand margin. To coordinate the visual elements in the more complex page designs often required on the job, use additional vertical and horizontal lines. The resulting network of lines is called a *grid*. Here are three examples:

Sample grid designs

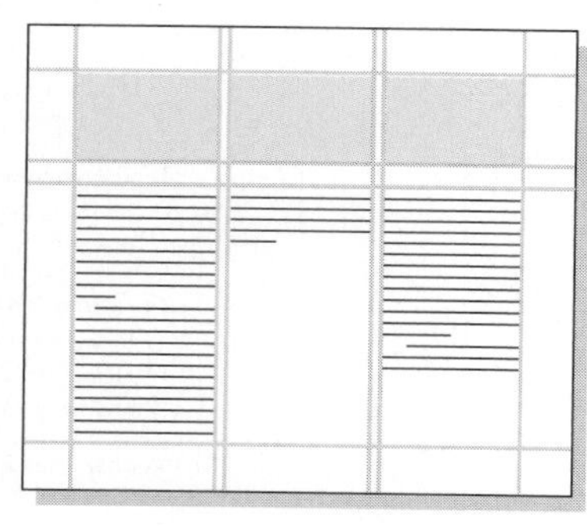

In the left-hand example, a pair of vertical grid lines is used to create a simple, two-column page. The white space between the columns is called a *gutter*.

The centre example represents a page in which the writer presents three figures, each accompanied by an associated block of text. To connect each figure to its text, the writer aligns each text-and-figure pair vertically in a column of its own, leaving varying amounts of white space at the bottoms of the columns.

The right-hand example represents a set of instructions in which the writer places all the directions in the left-hand column and all the illustrations in the right-hand column. To link each direction visually with its corresponding illustration, the writer aligns their tops along the same horizontal grid line. Horizontal gutters separate the direction-and-figure pairs from one another.

Using grid patterns creatively, you can build an unlimited variety of functional and attractive page designs. Figure 16.3 shows a few examples.

FIGURE 16.3

Grid Lines Used to Align the Elements of a Page

Grid lines create the graphic structure of a page.

Graphic elements are aligned with one another by being placed against the grid lines.

When graphic elements extend across vertical grid lines, they usually go all the way to the far side of the adjacent area.

Sometimes writers break out of the basic grid pattern to emphasize one particular graphic element.

Report

3

Chapter Title

Heading XXXXX X XX X X X XX xxx x

Subheading Xxxxx X xxxxx

Subheading Xx XXXXXXXX Xxx

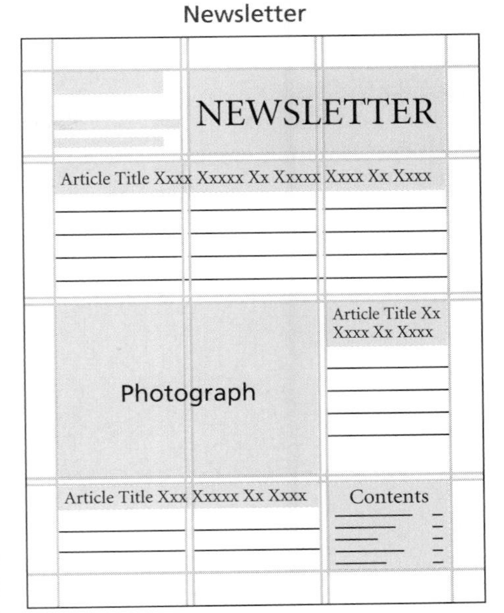

Instructions

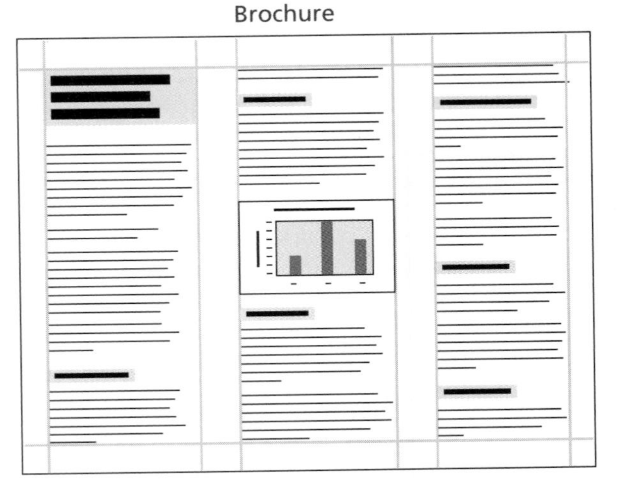

As you look at Figure 16.3, note that you can span two or more columns with a heading, figure, or other graphic item as long as you still align the item within the grid system. Placing items across columns is one way of emphasizing them.

Items aligned against a left grid line are said to be *flush left* and those against a right grid line are *flush right.* A centred item isn't aligned against a left or right grid line, of course, but its midpoint aligns with a line running down the column's centre. Headers and footers align flush right or left in the top and bottom margins.

If your communication includes drawings or other graphics that have irregular outlines, consider enclosing the figures in rectangles that run along the grid lines. With word processing software, you can easily do this by using the borders or frames feature. Figure 16.4 shows how such rectangles can anchor the figures in the framework of the page.

FIGURE 16.4

Comparison of Irregularly Shaped Figures without and with Rectangular Enclosures

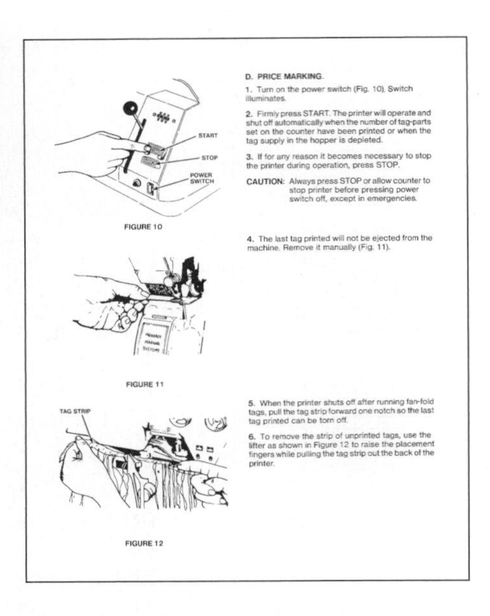

START
STOP
POWER SWITCH
FIGURE 10

D. PRICE MARKING.

1. Turn on the power switch (Fig. 10). Switch illuminates.

2. Firmly press START. The printer will operate and shut off automatically when the number of tag-parts set on the counter have been printed or when the tag supply in the hopper is depleted.

3. If for any reason it becomes necessary to stop the printer during operation, press STOP.

CAUTION: Always press STOP or allow counter to stop printer before pressing power switch off, except in emergencies.

4. The last tag printed will not be ejected from the machine. Remove it manually (Fig. 11).

FIGURE 11

TAG STRIP

5. When the printer shuts off after running fan-fold tags, pull the tag strip forward one notch so the last tag printed can be torn off.

6. To remove the strip of unprinted tags, use the lifter as shown in Figure 12 to raise the placement fingers while pulling the tag strip out the back of the printer.

FIGURE 12

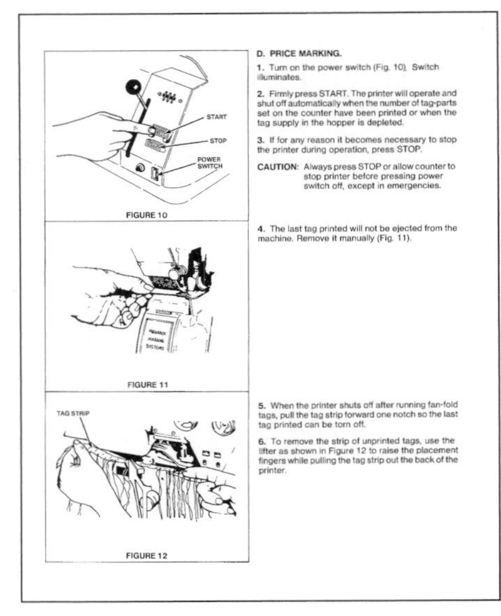

START
STOP
POWER SWITCH
FIGURE 10

D. PRICE MARKING.

1. Turn on the power switch (Fig. 10). Switch illuminates.

2. Firmly press START. The printer will operate and shut off automatically when the number of tag-parts set on the counter have been printed or when the tag supply in the hopper is depleted.

3. If for any reason it becomes necessary to stop the printer during operation, press STOP.

CAUTION: Always press STOP or allow counter to stop printer before pressing power switch off, except in emergencies.

4. The last tag printed will not be ejected from the machine. Remove it manually (Fig. 11).

FIGURE 11

TAG STRIP

5. When the printer shuts off after running fan-fold tags, pull the tag strip forward one notch so the last tag printed can be torn off.

6. To remove the strip of unprinted tags, use the lifter as shown in Figure 12 to raise the placement fingers while pulling the tag strip out the back of the printer.

FIGURE 12

Word processing software makes it easy to create grid lines for aligning the graphic elements of your pages.

To create multicolumn pages, use the menu or button in your software program that lets you choose the number of columns on a page and the width of the gutters that separate the columns. For designs that rely heavily on horizontal grid lines, use the program's feature that lets you create tables. The boundaries of the cells in the table will define your page's vertical and horizontal grid lines.

GUIDELINE 4 Use Contrast to Establish Hierarchy and Focus

Of the various items on a page, you will usually want some to stand out more than others. For instance, you may want to indicate that some items are at a higher level than others within your organizational hierarchy. Or you may want to focus attention on certain items, such as a warning in a set of instructions. However, nothing will stand out if everything on the page looks the same. To make some things stand out, you must use *contrast,* the third design principle.

Colour, type size, and bold are used in the right-hand design to establish a focus and a visual hierarchy.

No Contrast

Jamie Gallant
Shear Logistics Limited

340 Park Place, 777 Burrard St.
Vancouver, BC V6C 2X8
(604) 682-4668

Contrast Establishes Emphasis

Jamie Gallant
Shear Logistics Limited

340 Park Place, 777 Burrard St.
Vancouver, BC V6C 2X8
(604) 682-4668

When deciding how to establish the visual hierarchy of your page, pay special attention to text items: paragraphs, headings, figure titles, and headers and footers. To create distinctions among these items, you can control four variables:

- **Size.** The size of type is measured in *points* and always refers to the distance between the top of the tallest letters and the bottom of descenders (e.g., the "tail" on lowercase *g* or *y*). The following line illustrates a range of type sizes. There are 72 points to an inch.

M	M	M	M	M	M	M	M	M	M
8	10	12	14	18	24	36	48	60	72

A variety of type sizes, measured in points

When trying to distinguish different elements (such as different levels of headings) by using different type sizes, be sure to select sizes that contrast enough for the difference to be immediately noticeable. On the other hand, consider the relationship of all the visual elements on the page to ensure that differences in size aren't so great that the larger items dominate attention. You want to create a hierarchy of meaning and attention, not create one or two elements that either keep pulling the eye to them or appear to be out of proportion with the rest of the page. As a rule of thumb, try keeping each type size no larger than one and one-half times the size of the next smaller size in your hierarchy.
- **Type treatment.** Use plain type (called *roman*) for most type. Create contrast for items that you want to stand out by printing them in bold or italics.
- **Colour.** Emphasize some items by making them a different colour. Black type and grey type aren't as distinguishable as black type and red type.
- **Typeface.** To help readers distinguish one kind of element on your page from another, you can use different typefaces for each. At work, this is often done to help headings stand out from paragraphs.

  Of course, using different typefaces to distinguish different kinds of content will work only if there is enough contrast between the two typefaces. Consider the following six examples.

Times Roman	Helvetica
Baskerville	Arial
Palatino	**Folio**

As you can see, the typefaces in each column look very similar to the others in the same column, so choosing two from the same column would not accomplish your objective of providing contrast. However, all of the typefaces on the left look quite distinct from all of those on the right. Here's why. The left-hand typefaces all have lines drawn across the ends of their strokes. These lines are called *serifs,* and typefaces with these lines are called serif typefaces. In contrast, all of the typefaces in the right-hand column have no serifs. They are called *sans serif* typefaces (in French, *sans* means "without").

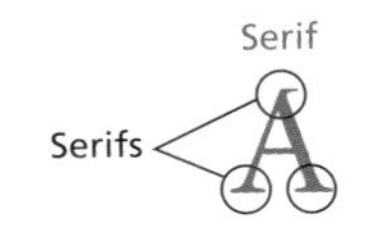

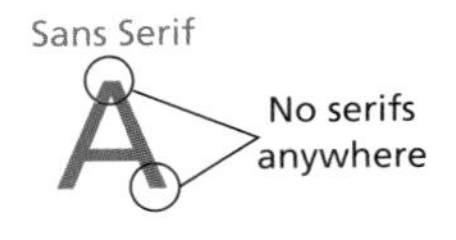

In print documents, writers often use the difference between serif and sans serif typefaces to make their headings stand out more distinctly from their text. Usually, they use a sans serif typeface for the headings and a serif one for the paragraphs. This book provides an example.

To create even greater contrast among the visual elements on the page, you can use the variables of size, type treatment, type category, and colour to reinforce one another. By coordinating these variables with one another, you can create distinctive appearances for a substantial number of visual elements, as the following table illustrates. Figure 16.5 shows a page created using this table.

Element	Type Category	Size	Treatment	Colour
Paragraph	serif (e.g., Times)	12 point	roman	black
Headings				
First level	sans serif (e.g., Arial)	16 point	bold roman	green
Second level	sans serif	14 point	bold roman	green
Figure caption	sans serif	10 point	bold roman	black
Headers	sans serif	14 point	italic	black
Footers	sans serif	10 point	italic	black

You can use the principle of contrast to distinguish other elements as well. For example, in a set of instructions, you can set off warnings in a box that might have a background in colour. In a report, you might set off your tables from your paragraphs by using the same typeface for your tables that you use for your headings.

When working with the page design variables, remember that your reason for creating visually contrasting elements is to simplify the page for your readers so they can readily see the role played by each element on the page. Don't create so many distinctions that the page seems chaotic to readers. For example, in most cases, stick to only two colours for your type and only two typefaces.

Other tools you can use to organize a page visually include rules (lines) that mark the boundaries between sections of text and icons that serve as non-verbal headings to assist readers in locating a certain kind of information. Figure 16.6 (page 400) shows pages that use these tools.

GUIDELINE 5 Use Repetition to Unify Your Communication Visually

In communications that are more than a page long, you should think not only about creating well-designed single pages but also about creating a well-designed set of pages.

When you are creating a set of pages, the chief goal is to make the pages harmonize visually in a way that both supports your readers' use of them and creates an aesthetically pleasing design. To achieve this goal, use the fourth design principle: *repetition.*

For instance, use the same grid pattern throughout. Similarly, use the same type treatment (typeface, size, colour, etc.) for all major headings, for all second-level headings, for all headers and footers, and so on. Such repetition enables your readers to

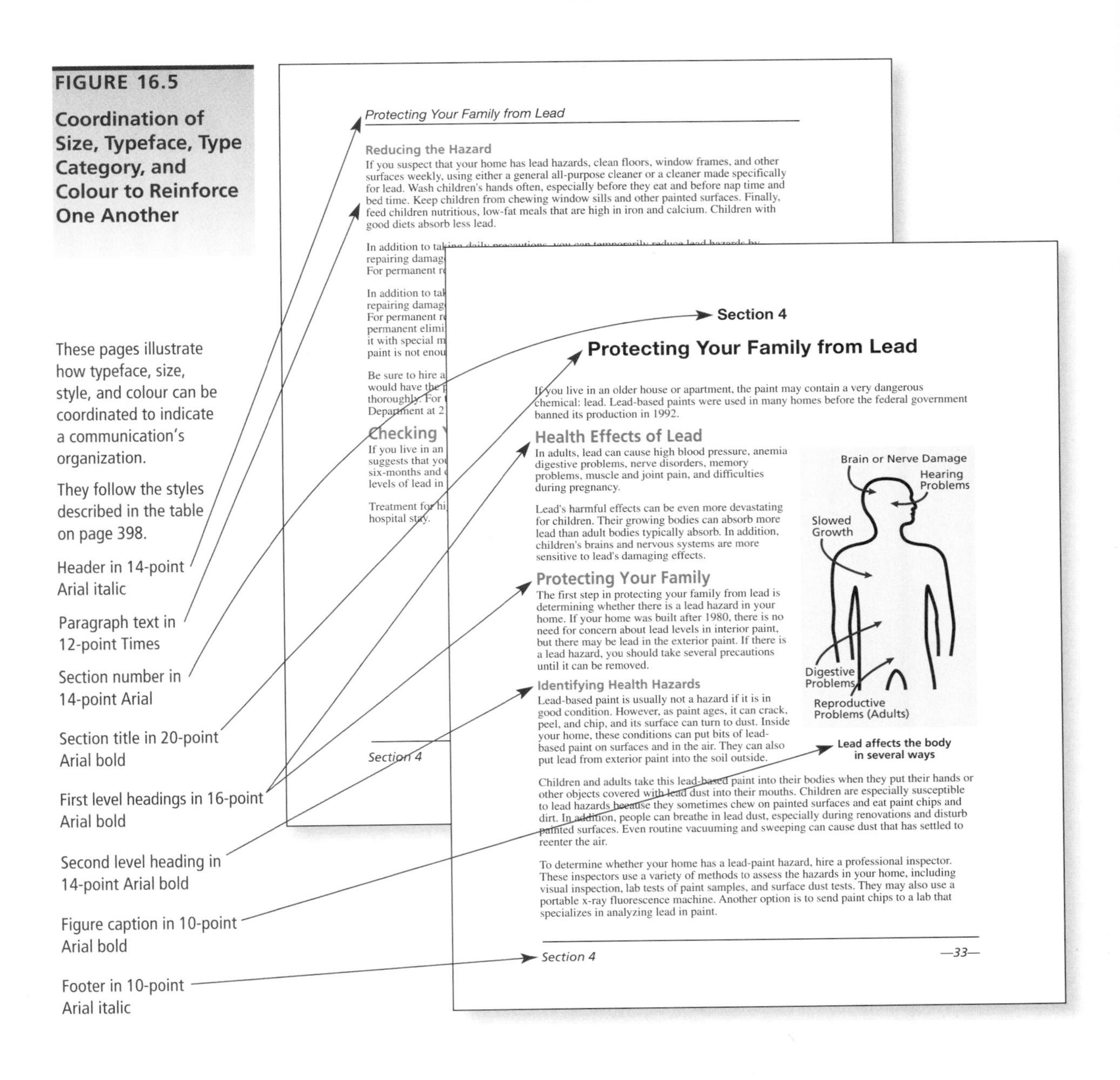

FIGURE 16.5
Coordination of Size, Typeface, Type Category, and Colour to Reinforce One Another

"learn" the structure of your pages so that they immediately understand the organization of each new page as they turn to it. Repetition also creates visual harmony that is more pleasing to the eye than a set of inconsistent designs.

Of course, it's not possible to have exactly the same design for every page in a long communication. Often, such communications will have different types of pages. In long reports, for instance, each chapter may begin with a page that looks different from every other page in the chapter. Instruction manuals often have troubleshooting sections that

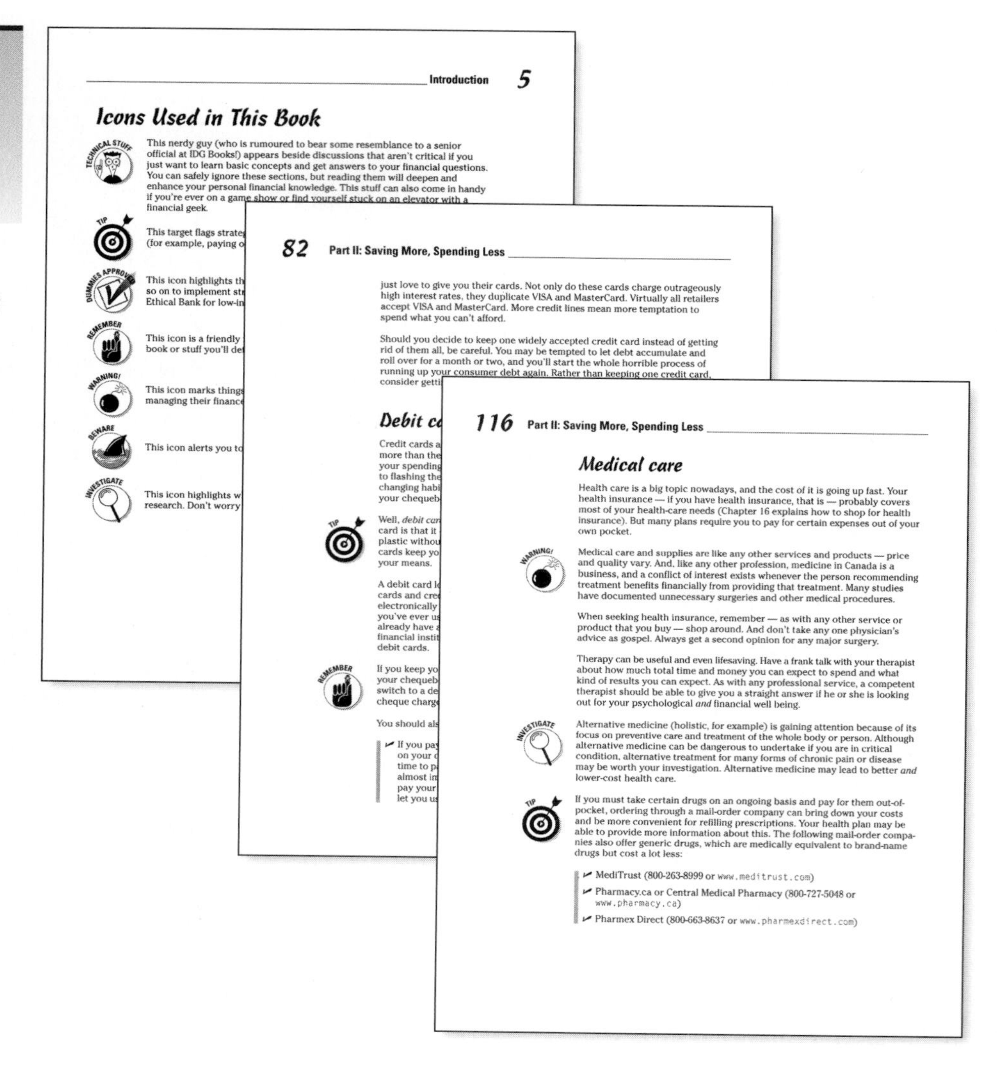

Introduction 5

Icons Used in This Book

TECHNICAL STUFF This nerdy guy (who is rumoured to bear some resemblance to a senior official at IDG Books!) appears beside discussions that aren't critical if you just want to learn basic concepts and get answers to your financial questions. You can safely ignore these sections, but reading them will deepen and enhance your personal financial knowledge. This stuff can also come in handy if you're ever on a game show or find yourself stuck on an elevator with a financial geek.

TIP This target flags strate
(for example, paying o

DUMMIES APPROVED This icon highlights th
so on to implement st
Ethical Bank for low-in

REMEMBER This icon is a friendly
book or stuff you'll de

WARNING! This icon marks things
managing their finance

BEWARE This icon alerts you to

INVESTIGATE This icon highlights w
research. Don't worry

82 Part II: Saving More, Spending Less

just love to give you their cards. Not only do these cards charge outrageously high interest rates, they duplicate VISA and MasterCard. Virtually all retailers accept VISA and MasterCard. More credit lines mean more temptation to spend what you can't afford.

Should you decide to keep one widely accepted credit card instead of getting rid of them all, be careful. You may be tempted to let debt accumulate and roll over for a month or two, and you'll start the whole horrible process of running up your consumer debt again. Rather than keeping one credit card, consider getti

Debit ca

Credit cards a
more than the
your spending
to flashing the
changing habi
your chequeb

TIP Well, *debit car*
card is that it
plastic withou
cards keep yo
your means.

A debit card l
cards and cre
electronically
you've ever u
already have
financial insti
debit cards.

REMEMBER If you keep yo
your chequeb
switch to a de
cheque charg

You should als

- If you pa
on your
time to p
almost in
pay your
let you u

116 Part II: Saving More, Spending Less

Medical care

Health care is a big topic nowadays, and the cost of it is going up fast. Your health insurance — if you have health insurance, that is — probably covers most of your health-care needs (Chapter 16 explains how to shop for health insurance). But many plans require you to pay for certain expenses out of your own pocket.

WARNING! Medical care and supplies are like any other services and products — price and quality vary. And, like any other profession, medicine in Canada is a business, and a conflict of interest exists whenever the person recommending treatment benefits financially from providing that treatment. Many studies have documented unnecessary surgeries and other medical procedures.

When seeking health insurance, remember — as with any other service or product that you buy — shop around. And don't take any one physician's advice as gospel. Always get a second opinion for any major surgery.

Therapy can be useful and even lifesaving. Have a frank talk with your therapist about how much total time and money you can expect to spend and what kind of results you can expect. As with any professional service, a competent therapist should be able to give you a straight answer if he or she is looking out for your psychological *and* financial well being.

INVESTIGATE Alternative medicine (holistic, for example) is gaining attention because of its focus on preventive care and treatment of the whole body or person. Although alternative medicine can be dangerous to undertake if you are in critical condition, alternative treatment for many forms of chronic pain or disease may be worth your investigation. Alternative medicine may lead to better *and* lower-cost health care.

TIP If you must take certain drugs on an ongoing basis and pay for them out-of-pocket, ordering through a mail-order company can bring down your costs and be more convenient for refilling prescriptions. Your health plan may be able to provide more information about this. The following mail-order companies also offer generic drugs, which are medically equivalent to brand-name drugs but cost a lot less:

- MediTrust (800-263-8999 or www.meditrust.com)
- Pharmacy.ca or Central Medical Pharmacy (800-727-5048 or www.pharmacy.ca)
- Pharmex Direct (800-663-8637 or www.pharmexdirect.com)

FIGURE 16.6

Icons Used to Aid Readers in Seeing the Structure of a Page

The icons help readers distinguish special kinds of information, such as Tips, Reminders, and Warnings, from the rest of a page's contents.

present different kinds of information than the manuals' other pages do. For such communications, follow these two strategies:

- Treat in the same way all elements that are shared by different types of pages. For instance, use the same style for all headings, even on different types of pages.
- Use the same design for all pages of the same type (for example, all pages that begin a new chapter).

Figure 16.7 shows pages from a communication that uses these strategies. To see visually unified websites, go to pages 119 and 382.

FIGURE 16.7

Page Design Used to Create Visual Unity in a Communication

By consistently using various design elements, this car owner's manual achieves overall visual unity.

All pages have the same design at the top.

All pages use the same two-column grid, regardless of the type of content.

Wide illustrations exactly fill both columns, thereby keeping the same outside margins.

The same styles are used throughout for the headings.

Colour is applied according to the same pattern throughout.

All right-hand pages have blocks of colour that give the section number.

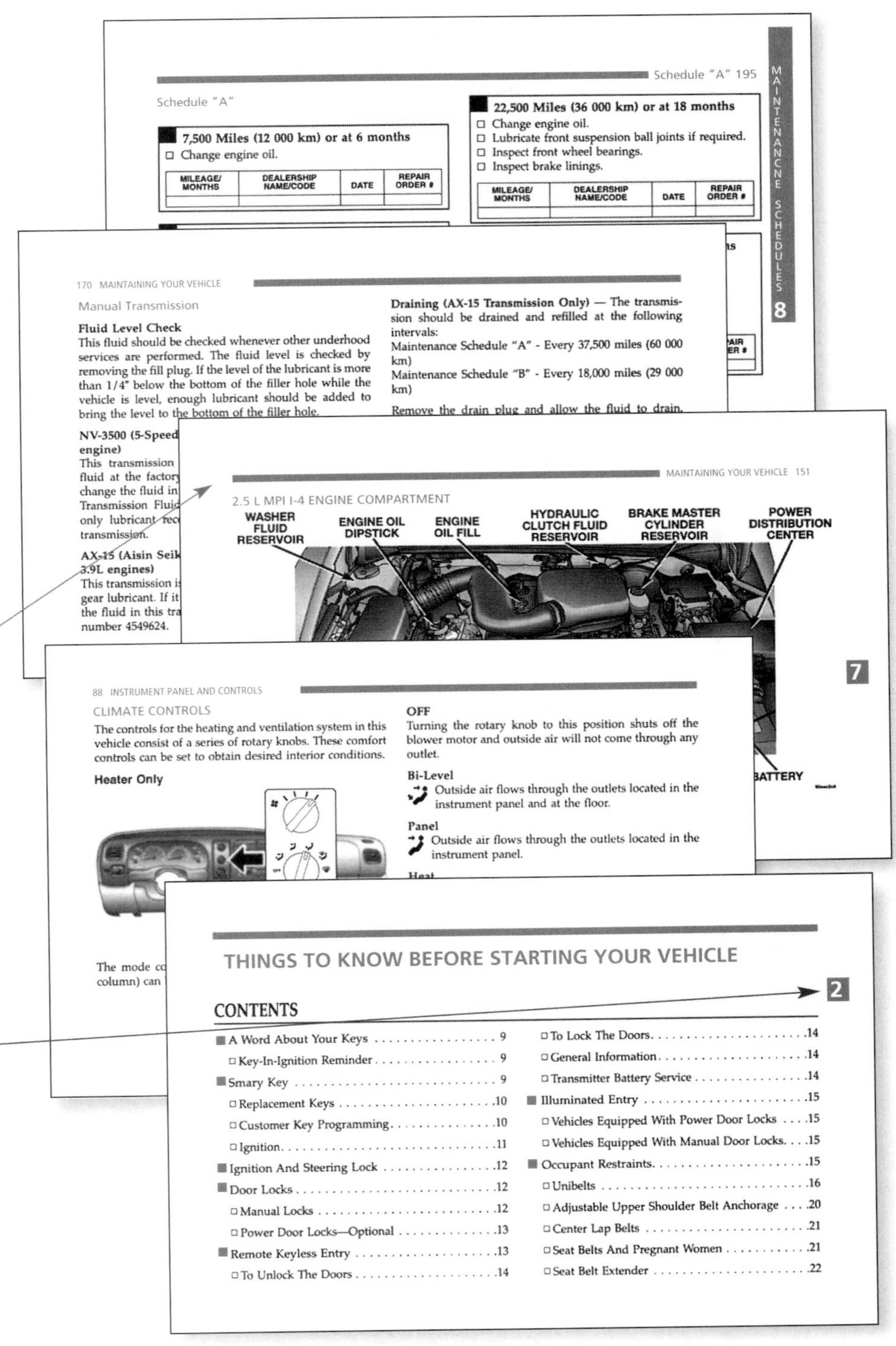

GUIDELINE 6 Select Type That is Easy to Read

We recognize letters by seeing their main lines. These lines are more obvious with some typefaces than with others, as you can see by comparing the two typefaces on the left with the one on the right:

Comparison of three typefaces

Times Roman
All major lines of all the letters are clearly defined.

Ultra
The second leg of the *u* is very thin, as is the cross-stroke of the *t*.

Script
The shape of the *s* is hidden by the ornate lines, and the circle of the *p* is not closed.

When we are reading only a few words, as when looking at a magazine ad, these differences are inconsequential. However, when we are reading an entire page, a full paragraph, or even a few sentences, type with the more distinct main lines is easier to read. Consequently, in choosing typefaces for your text, you should apply the following advice.

Principles of Type Selection

- **Use a typeface with strong, distinct main lines.**
- **Avoid italics for more than a sentence at a time.** Although they are great for emphasis, italics are difficult to read in long passages because they represent a "distorted" or "ornate" version of the alphabet that obscures the main lines.
- **Avoid using all capital letters for more than a few words at a time.** Research has shown that capital letters are difficult to read in long stretches (Tinker, 1969). That's because many capital letters have very similar shapes. For instance, capitals "D" and "P" are the same height and face the same direction, but lower-case "d" and "p" have much different heights and face in different directions. The similarities among capital letters slow down reading because people must quickly identify letters in order to read at their normal pace.

After studying the extensive literature on the readability of different typefaces for printed communications, Karen Schriver (1997) concludes that in extended text serif type may be somewhat easier to read than sans serif, although the two categories can be read equally quickly in small amounts. Schriver notes, however, that several factors may come into play, including cultural preferences and familiarity with the typeface. She concludes that if you choose a typeface with strong, distinct lines, your text will be readable whether you select a serif or sans serif typeface.

Use 10-point or 12-point type for text.

Of course, no type will be legible unless it is large enough. Research has shown that in print communications, it's easiest to read passages printed in type that is between 8 points and 12 points high (Tinker, 1969). In most printed communications written at work, type smaller than 10 points is avoided, and type larger than 12 points is usually reserved for headings, titles, and the like. Type as small as 9 points may work in headers, footers, and the captions for figures.

For onscreen reading, the considerations for type choices are different, as explained in Chapter 5 (see pages 121–24).

PRACTICAL PROCEDURES FOR DESIGNING PAGES

You do not need any special artistic or technical skills to follow the first six guidelines for designing pages. Here are some practical procedures you can follow.

Procedures for Designing Pages

1. **Determine the amount of text and graphics you will include.** Doing this will help you determine how to coordinate your text and graphics. If you are pairing specific blocks of text or graphics across adjacent columns by using horizontal grid lines, identify the longest text passage that goes with a single illustration. Use the length of that passage as the standard for allocating space for all the other text passages.
2. **Draw thumbnail sketches of a variety of grid patterns.** Figure 16.8 (page 404) shows nine thumbnail sketches exploring alternative ways of designing a single page. Sketches like these can help you quickly assess the various grid patterns you might use. Even when making thumbnail sketches, keep your communication's purpose and readers in mind so you will know what material to emphasize and what relationships to make clear. Think, too, about how large you will need to make your illustrations so that all the details are legible. Finally, choose the design most likely to achieve your purpose.
3. **Create a mockup with your word processing software.** Based on your thumbnail sketch, use your word processing to create a full-size model that shows what a typical page or set of pages will look like. Include all the visual elements that will appear in your finished communication. If you haven't written your text and created your graphics yet, type nonsense words and create rectangles where the graphics will go. Experiment with this mock-up until you've refined your design to fully support your communication's usability and persuasive goals.
4. **Use your word processing software's styles feature to simplify creation of your finished communication.** The *styles* feature that is included with word processing software enables you to create consistent, attractive pages very efficiently. For example, you can name your first-level headings "Heading 1" and then use a series of menus and checkboxes to define the typeface, size, colour, margins, indentations, and other features for them. To apply all the details of this style to each appropriate item in your communication, simply highlight the item and select the corresponding style from a menu. The styles feature has the added advantage of letting you adjust styles easily. For example, to change the size and colour of *all* of the first-level headings, you just redefine the style itself. Figure 16.9 (page 405) shows style settings in Microsoft Word.
5. **Test your final pages.** If your design is at all complex, as it might be if you were creating a set of instructions, test it out by asking one or more members of your target audience to try out your communication.

FIGURE 16.8

Nine Thumbnail Page Designs for the Same Material

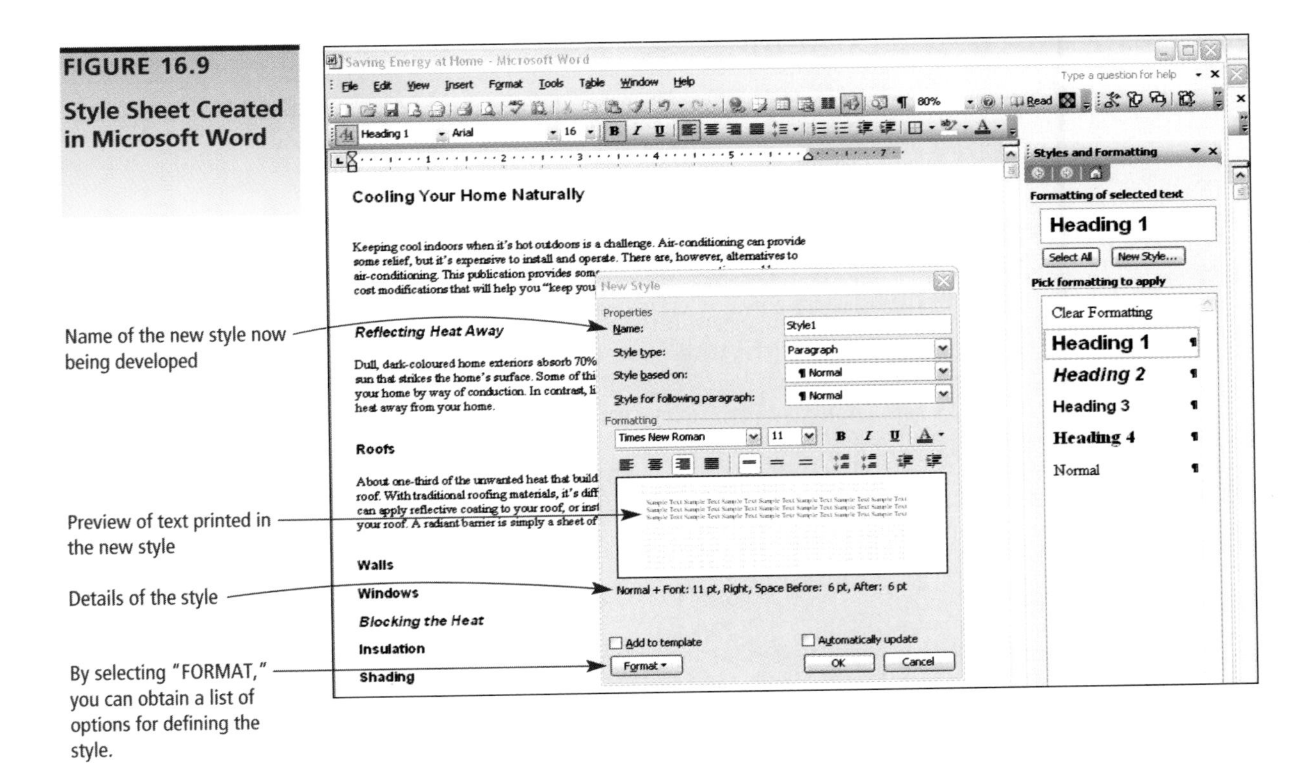

FIGURE 16.9

Style Sheet Created in Microsoft Word

GUIDELINE 7 Design Your Overall Document for Ease of Use and Attractiveness

For some communications, you will need to think about how to design not only the elements on the page but also the overall product that you will provide to your readers. Here, too, consider your communication from your readers' perspective.

- **Size.** Choose a size that your readers will find convenient in the situations in which they will use your communication. For example, if you are creating instructions that readers will use at a desk already crowded by a computer and a keyboard, imitate software manufacturers that print manuals that are smaller than 8½ by 11 inches.
- **Shape.** Choose a shape that will make it convenient for your readers to carry and store. If appropriate, for example, make it a shape that will fit into a shirt pocket or into a special case (such as the case for a notepad computer or a blood glucose monitor).
- **Binding.** Some bindings stay open more easily than others. If your readers will want your communication to lie flat because they need to use both hands for something else, consider a spiral binding or three-ring binder.
- **Paper.** If you are creating a shop manual that might be spilled on or a document that might be used outdoors, consider a coated paper that resists liquids and soil.

GRAPHIC DESIGN OF WEBPAGES

As mentioned at the beginning of this chapter, the guidelines discussed here are just as valid for pages to be read online as for pages to be read on paper. However, there are additional considerations in the graphic design of webpages. For these, review Chapter 5.

CONCLUSION

Every page that you write has some sort of page design. This chapter has explained how you can look at your pages in the way that graphic designers do, thinking about each graphic element in terms of how it affects your readers. By following this chapter's advice, you will create page designs that help your readers read efficiently, emphasize the important contents of your communication, and create a favourable impression.

EXERCISES

For additional exercises, visit www.techcomm.nelson.com.

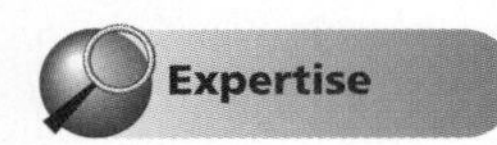

1. Study the design of the sample pages shown in Figure 16.1 (page 391). Identify the number of columns; the placement of text, headings, and graphics; and any other important features of the design.
2. Figure 16.10 shows four package inserts for a prescription medication. All contain the same information, but each was prepared by a different graphic designer. Describe the page designs used in each. What purpose do you think each designer had in mind? Which design do you think works best? Worst? Why?
3. Find three different page designs. (Choose no more than one from a popular magazine; look at instructions, insurance policies, leases, company brochures, technical reports, and the like.) If your professor asks, find all the samples in documents related to your field of study. Photocopy one page illustrating each design. Then, for each page, describe what you think the purpose of the document is and discuss the specific features of the page design that help or hinder the document from achieving that purpose.
4. Do the "Informational Page" project in Appendix C.

Experiment with the page design of a draft or finished communication of yours that has headings and at least one list. Using your software's styles feature, define styles for the communication's design elements. Instead of using the program's default styles, create your own. After you have applied your styles throughout your communication, change the size, colour, and typeface for the headings. Also try different designs for the other elements.

Collaboration

Work with another student on this exercise, which will provide you with practice at evaluating and improving a page design.

a. List the ways in which the design of the page shown in Figure 16.11 (page 411) could be improved.
b. Redesign the page by following steps 1 through 4 of the Procedures for Designing Pages (page 403). Specifically, do the following:
   - Create three thumbnail sketches for the page.
   - Create a full-size mock-up of the best design.
c. Display your mock-up on the wall of your classroom along with the mock-ups prepared by the other students in your class. Decide which mock-ups work best. Discuss the reasons.

This exercise invites you to consider the role that page and screen design can play in your ability to communicate effectively on issues that are important to you for ethical reasons. Visit three or more websites whose creators are advocating a position on an ethical basis. Sites maintained by groups taking various positions on environmental policy are examples. Evaluate the ways that the screen design of each site promotes and hinders the group's ability to achieve the usability and persuasive goals it has for the site.

FIGURE 16.10

Package Inserts for a Prescription Medication

HYGROTON®

chlorthalidone usp

50 mg. Tablets, 100 mg. Tablets

Oral Antihypertensive-Diuretic

DESCRIPTION
HYGROTON (chlorthalidone) is a monosulfamyl diuretic which differs chemically from thiazide diuretics in that a double-ring system is incorporated in its structure. It is 2-Chlor-5-(1-hydroxy-3-oxo-1-isoindolinyl) benzenesulfonamide, with the following structural formula:

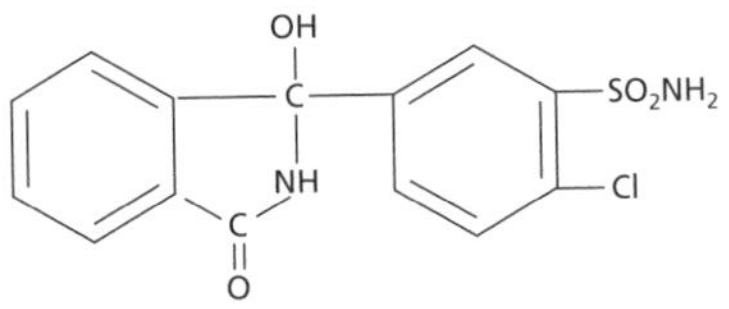

OH C NH C O SO_2NH_2 Cl

ACTIONS
HYGROTON is an oral diuretic with prolonged action (48-72 hours) and low toxicity. The diuretic effect of the drug occurs within two hours of an oral dose and continues for up to 72 hours.

INDICATIONS
Diuretics such as HYGROTON are indicated in the management of hypertension either as the sole therapeutic agent or to enhance the effect of other antihypertensive drugs in the more severe forms of hypertension and in the control of hypertension of pregnancy.

CONTRAINDICATIONS
Anuria. Hypertensitivity to chlorthalidone. The routine use of diuretics in an otherwise healthy pregnant woman with or without mild edema is contraindicated and possibly hazardous.

WARNINGS
Should be used with caution in severe renal disease. In patients with renal disease, chlorathalidone or related drugs may precipitate azotemia. Cumulative effects of the drug may develop in patients with impaired renal function.

USAGE IN PREGNANCY: Reproduction studies in various animal species at multiples of the human dose showed no significant level of teratogenicity; no fetal or congenital abnormalities were observed.

NURSING MOTHERS: Thiazides cross the placental barrier and appear in cord blood and breast milk.

PRECAUTIONS
Periodic determination of serum electrolytes to detect possible electrolyte imbalance should be performed at appropriate intervals.

Chlorthalidone and related drugs may decrease serum PBI levels without signs of thyroid disturbance.

ADVERSE REACTIONS
Gastrointestinal System Reactions:
anorexia constipation
gastric irratation jaundice
nausea (intrahepatic
vomiting cholestatic
cramping jaundice)
diarrhea pancreatitis

Central Nervous System Reactions:
dizziness headache
vertigo xanthopsia
paresthesias

Hematologic Reactions:
leukopenia thrombocytopenia
agranulocytosis aplastic anemia

Other Adverse Reactions:
hyperglycemia muscle spasm
glycosuria weakness
hyperuricemia restlessness
impotence

Whenever adverse reactions are moderate or severe, chlorthalidone dosage should be reduced or therapy withdrawn.

DOSAGE AND ADMINISTRATION
Therappy should be individualized according to patient response. This therapy should be titrated to gain maximal therapeutic response as well as the minimal dose possible to maintain that therapeutic response.

Initiation: Preferably, therapy should be initiated with 50 mg. or 100 mg. daily. Due to the long action of the drug, therapy may also be initiated in most cases with a dose of 100 mg. on alternate days or three times weekly (Monday, Wednesday, Friday). Some patients may require 150 or 200 mg. at these intervals.

Maintenance: Maintenance doses may often be lower than initial doses and should be adjusted according to the individual patient. Effectiveness is well sustained during continued use.

OVERDOSAGE
Symptoms of overdosage include nausea, weakness, dizziness and disturbances of electrolyte balance.

HOW SUPPLIED
HYGROTON (chlorthalidone). White, single-scored tablets of 100 mg. and aqua tablets of 50 mg. in bottles of 100 and 1000; single-dose blister packs, boxes of 500; Paks of 28 tablets, boxes of 6.

CAUTION: Federal law prohibits dispensing without prescription.

ANIMAL PHARMACOLOGY
Biochemical studies in animals have suggested reasons for the prolonged effect of chlorthalidone. Absorption from the gastrointestinal tract is slow, due to its low solubility. After passage to the liver, some of the drug enters the general circulation, while some is excreted in the bile, to be reabsorbed later.

USV PHARMACEUTICAL MFG. CORP.
Manati, P.R. 00701

A

FIGURE 16.10
(continued)

hygroton* chlorthalidone usp

50 mg. tablets
100 mg. tablets

oral antihypertensive-diuretic

Description: *HYGROTON* (chlorthalidone) is a monosulfamyl diuretic which differs chemically from thiazide diuretics in that a double-ring system is incorporated in its structure. It is 2-Chlor-5-(1-hydroxy-3-oxo-1-isoindolinyl) benzenesulfonamide, with the following structural formula:

Actions: *HYGROTON* is an oral diuretic with prolonged action (48-72 hours) and low toxicity. The diuretic effect of the drug occurs within two hours of an oral dose and continues for up to 72 hours.

Indications: Diuretics such as HYGROTON are indicated in the management of hypertension either as the sole therapeutic agent or to enhance the effect of other antihypertensive drugs in the more severe forms of hypertension and in the control of hypertension of pregnancy.

Contraindications: Anuria. Hypersensitivity to chlorthalidone.

The routine use of diuretics in an otherwise healthy pregnant woman with or without mild edema is contraindicated and possibly hazardous.

Warnings: Should be used with caution in severe renal disease. In patients with renal disease, chlorathalidone or related drugs may precipitate azotemia. Cumulative effects of the drug may develop in patients with impaired renal function.

Usage in pregnancy: Reproduction studies in various animal species at multiples of the human dose showed no significant level of teratogenicity; no fetal or congenital abnormalities were observed.

Nursing mothers: Thiazides cross the placental barrier and appear in cord blood and breast milk.

Precautions: Periodic determination of serum electrolytes to detect possible electrolyte imbalance should be performed at appropriate intervals.

Chlorthalidone and related drugs may decrease serum PBI levels without signs of thyroid disturbance.

Adverse reactions:
Gastrointestinal System Reactions:
anorexia constipation
gastric iratation jaundice
nausea (intrahepatic
vomiting cholestatic
cramping jaundice)
diarrhea pancreatitis

Central Nervous System Reactions:
dizziness headache
vertigo xanthopsia
paresthesias

Hematologic Reactions:
leukopenia thrombocytopenia
agranulocytosis aplastic anemia

Other Adverse Reactions:
hyperglycemia muscle spasm
glycosuria weakness
hyperuricemia restlessness
impotence

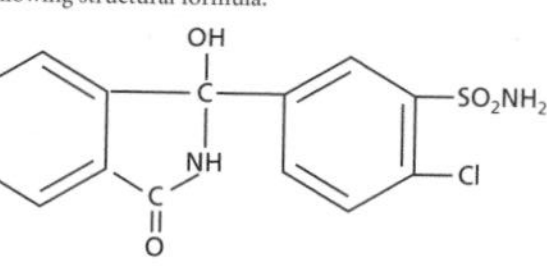

Whenever adverse reaction are moderate or severe, chlorthalidone dosage should be reduced or therapy withdrawn.

Dosage and administration: Therapy should be individualized according to patient response. This therapy should be titrated to gain maximal therapeutic response as well as the minimal dose possible to maintain that therapeutic response.

Initiation: Preferably, therapy should be initiated with 50 mg. or 100 mg. daily. Due to the long action of the drug, therapy may also be initiated in most cases with a dose of 100 mg. on alternate days or three times weekly (Monday, Wednesday, Friday). Some patients may require 150 or 200 mg. at these intervals.

Maintenance: Maintenance doses may often be lower than initial doses and should be adjusted according to the individual patient. Effectiveness is well sustained during continued use.

Overdosage: Symptoms of overdosage include nausea, weakness, dizziness and disturbances of electrolyte balance.

How supplied: HYGROTON (chlorthalidone). White, single-scored tablets of 100 mg. and aqua tablets of 50 mg. in bottles of 100 and 1000; single-dose blister packs, boxes of 500; Paks of 28 tablets, boxes of 6.

Caution: Federal law prohibits dispensing without prescription.

Animal pharmacology: *Biochemical studies* in animals have suggested reasons for the prolonged effect of chlorthalidone. Absorption from the gastrointestinal tract is slow, due to its low solubility. After passage to the liver, some of the drug enters the general circulation, while some is excreted in the bile, to be reabsorbed later.

B

FIGURE 16.10 (continued)

Hygroton®

chlorthalidone USP

50 mg. Tablets 100 mg. Tablets	**Oral** **Antihypertensive-Diuretic**	
Description	**Hygroton** (chlorthalidone) is a monosulfamyl diuretic which differs chemically from thiazide diuretics in that a double-ring system is incorporated in its structure. It is 2-Chlor-5-(1-hydroxy-3-oxo-1-isoindolinyl) benzenesulfonamide, with the following structural formula	OH, C, NH, C, O, SO_2NH_2, Cl
Actions	**Hygroton** is an oral diuretic with prolonged action (48-72 hours) and low toxicity. The diuretic effect of the drug occurs within two hours of an oral dose and continues for up to 72 hours	
Indications	Diuretics such as **Hygroton** are indicated in the management of hypertension either as the sole therapeutic agent or to enhance the effect of other	antihypertensive drugs in the more severe forms of hypertension and in the control of hypertension of pregnancy
Contraindications	Anuria. Hypertensitivity to chlorthalidone	The routine use of diuretics in an otherwise healthy pregnant woman with or without mild edema is contraindicated and possibly hazardous
Warnings	Should be used with caution in severe renal disease. In patients with renal disease, chlorathalidone or related drugs may precipitate azotemia	Cumulative effects of the drug may develop in patients with impaired renal function
Usage in Pregnancy:	Reproduction studies in various animal species at multiples of the human dose showed no significant level of teratogenicity; no fetal or congential abnormalities were observed	
Nursing Mothers:	Thiazides cross the placental barrier and appear in cord blood and breast milk	
Precautions	Periodic determination of serum selectrolytes to detect possible electrolyte imbalance should be performed at appropriate intervals	Chlorthalidone and related drugs may decrease serum PBI levels without signs of thyroid disturbance
Adverse reactions: Gastrointestinal System Reactions:	anorexia vomiting gastric irritation cramping nausea diarrhea	constipation pancreatitis jaundice (intrahepatic cholestatic jaundice)
Central Nervous System Reactions:	dizziness headache vertigo xanthopsia paresthesias	
Hematologic Reactions:	leukopenia thrombocytopenia agranulocytosis aplastic anemia	
Other Adverse Reactions:	hyperglycemia weakness glycosuria restlessness hyperuricemia impotence muscle spasm	Whenever adverse reaction are moderate or severe, chlorthalidone dosage should be reduced or therapy withdrawn
Dosage and Administration	Therapy should be individualized according to patient response. This therapy should be titrated to gain maximal therapeutic response as well as	the minimal dose possible to maintain that therapeutic response
Initiation:	Preferably, therapy should be initiated with 50 mg. or 100 mg. daily. Due to the long action of the drug, therapy may also be initiated in most cases with a dose of 100 mg. on alternate days or	three times weekly (Monday, Wednesday, Friday). Some patients may require 150 or 200 mg. at these intervals
Maintenance:	Maintenance doses may often be lower than initial doses and should be adjusted according to	the individual patient. Effectiveness is well sustained during continued use
Overdosage	Symptoms of overdosage include nausea, weakness, dizziness and disturbances of electrolyte balance	
How Supplied	**Hygroton** (chlorthalidone). White, single-scored tablets of 100 mg. and aqua tablets of 50 mg in	bottles of 100 and 1000; single-dose blister packs, boxes of 500; Paks of 28 tablets, boxes of 6
Caution:	Federal law prohibits dispensing without prescription	

C

FIGURE 16.10 (*continued*)

Hygroton®

Chlorthalidone USP

50 mg. Tablets
100 mg. Tablets
Oral Antihypertensive-Diuretic

DESCRIPTION
HYGROTON (chlorthalidone) is a monosulfamyl diuretic which differs chemically from thiazide diuretics in that a double-ring system is incorporated in its structure. It is 2-Chlor-5-(1-hydroxy-3-oxo-1-isoindolinyl) benzenesulfonamide, with the following structural formula:

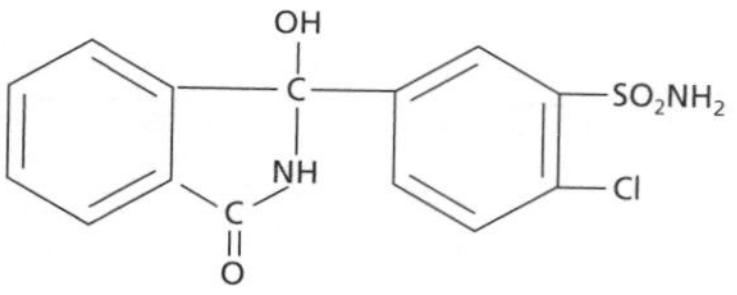

OH
C
NH
C
O
SO_2NH_2
Cl

ACTIONS
HYGROTON is an oral diuretic with prolonged action (48-72 hours) and low toxicity. The diuretic effect of the drug occurs within two hours of an oral dose and continues for up to 72 hours.

INDICATIONS
Diuretics such as HYGROTON are indicated in the management of hypertension either as the sole therapeutic agent or to enhance the effect of other antihypertensive drugs in the more severe forms of hypertension and in the control of hypertension of pregnancy.

CONTRAINDICATIONS
Anuria.
Hypertensitivity to chlorthalidone. The routine use of diuretics in an otherwise healthy pregnant woman with or without mild edema is contraindicated and possibly hazardous.

WARNINGS
Should be used with caution in severe renal disease. In patients with renal disease, chlorathalidone or related drugs may precipitate azotemia. Cumulative effects of the drug may develop in patients with impaired renal function.
Usage in Pregnancy: Reproduction studies in various animal species at multiples of the human dose showed no dignificant level of teratogenicity; no fetal or congenital abnormalities were observed.
Nursing Mothers: Thiazides cross the placental barrier and appear in cord blood and breast milk.

PRECAUTIONS
Periodic determination of serum selectrolytes to detect possible electrolyte imbalance should be performed at appropriate intervals.
Chlorthalidone and related drugs may decrease serum PBI levels without signs of thyroid disturbance.

ADVERSE REACTIONS
Gastrointestinal Systems Reactions:
anorexia
gastric irritation
nausea
vomiting
cramping
diarrhea
constipation
jaundice (intrahepatic cholestatic jaundice)
pancreatitis
Central Nervous System Reactions:
dizziness
vertigo
paresthesias
headache
xanthopsia
Hematologic Reactions: leukopenia
agranulocytosis
thrombocytopenia
aplastic anemia

Other Adverse Reactions:
hyperglycemia
glycosuria
hyperuricemia
impotence
muscle spasm
weakness
restlessness
Whenever adverse reaction are moderate or severe, chlorthalidone dosage should be reduced or therapy withdrawn.

DOSAGE AND ADMINISTRATION
Therapy should be individualized according to patient response. This therapy should be titrated to gain maximal therapeutic response as well as the minimal dose possible to maintain that therapeutic response.
Initiation: Preferabley, therapy should be initiated with 50 mg. or 100 mg. daily. Due to the long action of the drug, therapy may also be initiated in most cases with a dose of 100 mg. on alternate days or three times weekly (Monday, Wednesday, Friday). Some patients may require 150 or 200 mg. at these intervals.
Maintenance: Maintenance doses may often be lower than initial doses and should be adjusted according to the individual patient. Effectiveness is well sustained during continued use.

OVERDOSAGE
Symptoms of overdosage include nausea, weakness, dizziness and disturbances of electrolyte balance.

HOW SUPPLIED
HYGROTON (chlorthalidone). White, single-scored tablets of 100 mg. and aqua tablets of 50 mg. in bottles of 100 and 1000; single-dose blister packs, boxes of 500; Paks of 28 tablets, boxes of

CAUTION: Federal law prohibits dispensing without prescription.

ANIMAL PHARMACOLOGY
Biochemical studies in animals have suggested reasons for the prolonged effect of chlorthalidone. Absorption from the gastrointestinal tract is slow, due to its low solubility. After passage to the liver, some of the drug enters the general circulation, while some is excreted in the bile, to be reabsorbed later.

**USV PHARMACEUTICAL
MFG. CORP.
Manati, P.R. 00701**

D

FIGURE 16.11

Sample Page for Use with Collaboration Exercise

4) Flatten the clay by pounding it into the table. Remember, work the clay thoroughly!!

4)

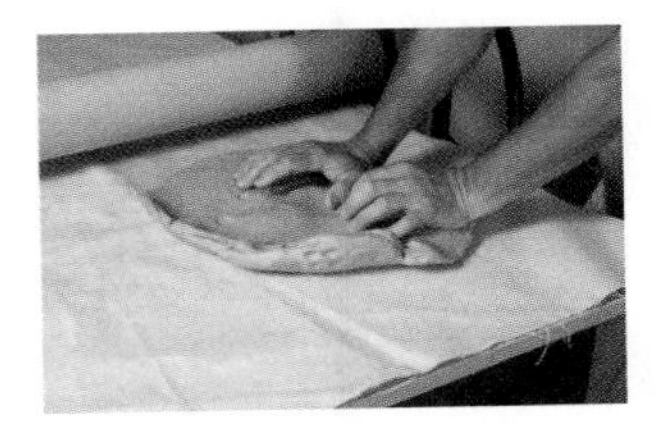

Slab Roller

The slab roller is a simple but very efficient mechanical device. A uniform thickness of clay is guaranteed. The ease of its operation saves countless hours of labor over hand rolling techniques.

1) The thickness of the clay is determined by the number of masonite® boards used. There are three 1/4" boards and one 1/8" board. They can be used in any combination to reduce or increase the thickness of the clay. If no boards are used, the clay will be 1" thick. If all the boards are used, the clay will be 1/8" thick. So, for a 1/2" thickness of clay, use two 1/4" boards.

1)

2) The canvas cloths must be arranged correctly. They should form a sort of envelope around the wet clay. This prevents any clay from getting on the rollers and masonite® boards.

2)

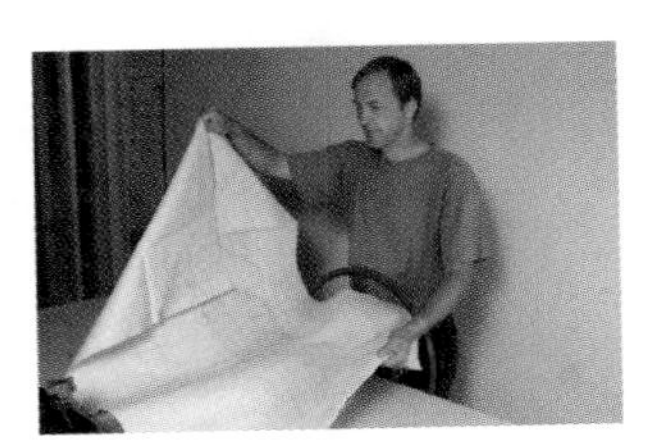

3) Place the clay flat on the canvas near the rollers. It may be necessary to trim some of the clay since it will spread out as it passes through the rollers.

3)

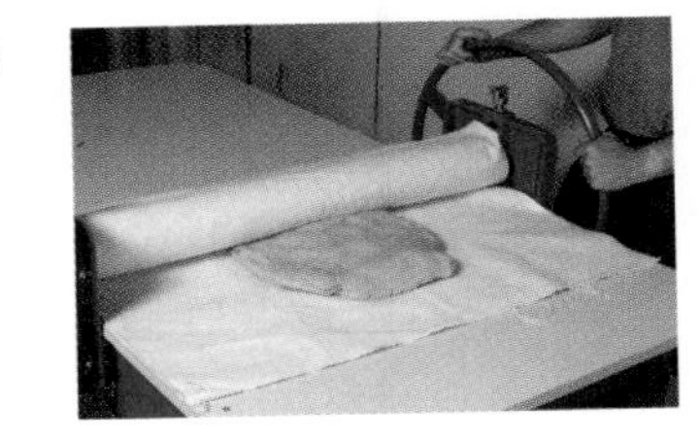

4.

CHAPTER 17 Creating and Delivering Oral Presentations

GUIDELINES

1. Define your presentation's objectives
2. Select the form of oral delivery best suited to your purpose and audience
3. Focus on a few main points
4. Use a simple structure—and help your listeners follow it
5. Use a conversational style
6. Look at your audience
7. Prepare for interruptions and questions—and respond courteously
8. Fully integrate graphics into your presentation
9. Rehearse
10. Accept your nervousness—and work with it

CHAPTER 17

DEFINING OBJECTIVES
APPLYING
CHOOSING
PLANNING
DEVELOPING
EVALUATING

For additional chapter resources, visit Chapter 17 at www.techcomm.nelson.com.

In your career, you will almost certainly make many oral presentations. In some, you may briefly and informally address individuals with whom you work closely every day. At other times, you may speak more formally to upper-level managers, customers, professional organizations, or the general public. Usually you will convey your entire message orally, but on other occasions your talk may supplement a report or proposal you have prepared on the same subject.

Compared to written communications, oral presentations are more personal and more interactive. You'll be able to see your audience's reactions even as you are speaking. Your listeners will usually have an opportunity to make comments and ask questions when you have finished, perhaps even before.

This chapter focuses on a listener-centred approach to oral presentations.

This chapter presents advice that will help you plan and deliver oral presentations in any on-the-job situation. Its ten guidelines teach a listener-centred approach to speaking that parallels the reader-centred approach to writing.

GUIDELINE 1 Define Your Presentation's Objectives

Begin work on an oral presentation in the same way you begin work on a written communication—by figuring out exactly what you want to accomplish and studying the situation in which you will communicate:

Consider these key factors.

1. **Think about your listeners and goals.** Think about who your listeners are and how you want to affect them. To review the procedures for defining objectives, turn to Chapter 1.

Know and stay within your time limit.

2. **Think about what your listeners expect.** In addition to considering their expectations about such things as the topics you will cover and the way you will treat them, focus on your listeners' expectations about how long you will speak. Even when you aren't given a time limit, your listeners are likely to have firm expectations about your presentation's length. Don't ruin an otherwise successful presentation by talking too long.
3. **Assess the scene of your talk.** Three aspects of the scene are especially important:
   - *Size of audience.* The smaller your audience, the smaller your graphics can be, the more likely your audience will expect you to be informal, and the more likely your listeners will interrupt you with questions.
   - *Location of your talk.* If the room has fixed seats, you will have to plan to display your graphics in a place where they are visible to everyone. If you can move the seats, you will be able to choose the seating arrangement best suited to your presentation.
   - *Equipment available.* The kinds of equipment that are available determine the types of graphics you can use. You can't show overheads or PowerPoint slides unless the necessary projector is available.

GUIDELINE 2 Select the Form of Oral Delivery Best Suited to Your Purpose and Audience

At work, people generally use one of three forms of oral delivery: the scripted talk, the outlined talk, or the impromptu talk. Sometimes the situation and your listeners' expectations will dictate the form of talk you will give, but other times you will be free to choose. The following paragraphs describe each type of delivery and provide information that will help you select among them.

Scripted Talk

A scripted talk is a speech that you write out word for word in advance and deliver by reading the script or by reciting it from memory.

Scripted talks offer security but can be rigid.

Because the scripted talk lets you work out your exact phrasing ahead of time, it is ideal for presenting complex information and speaking when a small slip in phrasing could be embarrassing or damaging. Also, with a scripted talk you can be sure you will make all your major points and, if you timed your rehearsals, keep within your time limit.

The scripted talk is also a good choice when you expect to be extremely nervous, as might happen when you are addressing a large or unfamiliar audience. Even if you become too unsettled to think straight, you will have all your words spelled out in front of you.

On the other hand, scripted talks take a long time to prepare, and you cannot alter them in response to audience reactions. Even more importantly, they are difficult to compose and deliver in a natural speaking style that maintains the interest of your audience. Rehearsals are a must.

Outlined Talk

To prepare an outlined talk, you do what the name implies: Prepare an outline, perhaps very detailed, of what you plan to say.

Outlined talks offer flexibility within a general framework.

At work, outlined talks are more common than scripted talks. You can create them more quickly, and you can deliver them in a "speaking voice," which helps increase listener interest and appeal. Furthermore, outlined talks are very flexible. In response to your listeners' reactions, you can speed up, slow down, eliminate material, or add something that you discover is needed. Outlined talks are ideal for situations in which you speak on familiar topics to small groups, as in a meeting with other people in your department.

The chief weakness of the outlined talk is that it is so flexible that unskilled speakers can easily run over their time limit, leave out crucial information, or have difficulty finding the phrasing that will make their meaning clear. For these reasons, you may want to avoid giving outlined talks on unfamiliar material or in situations where you might be so nervous that you become tongue-tied or forget your message.

Impromptu Talk

An impromptu talk is one you give on the spur of the moment with little or no preparation. At most, you might jot down a few notes beforehand about the points you want to cover.

Impromptu talks are ideal for topics you know well.

The impromptu talk is well suited to situations in which you are speaking for a short time on a subject so familiar that you can express yourself clearly and forcefully with little or no forethought.

The chief disadvantage of the impromptu talk is that you prepare so little that you risk treating your subject in a disorganized, unclear, or incomplete manner. You may even miss the mark entirely by failing to address your audience's concerns at all. For these reasons, the impromptu talks given at work are usually short, and they are usually used in informal meetings where listeners can interrupt to ask for additional information and clarification.

GUIDELINE 3 Focus on a Few Main Points

For a variety of reasons, it is often more difficult for people to understand what they hear than what they read:

Difficulties listeners face

- Listeners have more difficulty than readers in concentrating for extended periods. People can read for hours at a time, but many listeners have trouble concentrating for more than 20 minutes.
- A talk proceeds at a steady pace, so listeners have no chance to pause to figure out a difficult point.
- If listeners fail to understand a point or let their attention wander, they cannot flip the pages back and reread the passage. The talk goes forward regardless.

Pick points relevant to your listeners' interests and needs.

For these reasons, you must take special care to make your oral presentations easy to understand. One important strategy is to concentrate on only a few major points. Many experienced speakers limit themselves to three or, at most, four. Of course, major points may have subpoints, but their number should also be limited. And it is always important for you to choose points that are directly relevant to your listeners' interests and needs. For instance, when making a recommendation for overcoming a problem in your employer's organization, your first point might be that the problem has two major features, your second might be that there are three principal causes of the problem, and your third might be that you have three recommendations for overcoming it.

GUIDELINE 4 Use a Simple Structure—and Help Your Listeners Follow It

If your listeners are to understand and remember your presentation fully, they must be able to organize your points in their own minds. They stand the best chance of doing this if they can follow the structure of your talk. However, as listeners try to discern your presentation's structure, they must proceed without the help of paragraph breaks, headings, indentations, and similar aids that readers have.

Consequently, you can increase the understandability and memorability of your presentations by employing a simple structure that's easy to follow. A widely used structure consists of three major parts:

Simple structure for oral presentations

Introduction Introduces your topic and discusses its background briefly.
Body Organized around your three or four main points, which you elaborate and explain.
Conclusion Sums up your main points.

Even when you employ a very simple organizational pattern, you must still help your readers discern it. This means forecasting the structure in the introduction, clearly signaling transitions throughout, and summarizing the structure (and the main points) in the conclusion. Figure 17.1 describes and illustrates techniques for providing these forms of guidance for your listeners.

Note that although the advice you have just read will serve you very well when you are speaking to audiences in Canada, it may not work in some other countries, where traditions about oral presentations may be much different. For example, a talk in China has a longer introduction, containing more background and rationale than a talk in Canada has. The speaker spends more time substantiating each point, defending its value from different angles and emphasizing its importance to the main message of the presentation.

In contrast, a presentation in India is likely to be more circular in nature. Although the presentation will contain an introduction, a body, and a conclusion, these elements may not appear in the linear order expected by a Canadian audience. The talk may begin with cultural allusions and references (to mythology or literature, for example), or the conclusion may preface the body of the presentation. To create an effective talk for an international audience, you will need to research the traditions and expectations of your listeners' culture.

FIGURE 17.1

Signaling the Structure of Oral Presentations

General Strategies	Techniques
Forecast the structure	■ In the introduction, tell what the structure will be: "In the rest of my talk, I will take up the following three topics . . . " ■ Show a graphic that outlines the major parts of your talk.
Signal transitions	■ Announce transitions explicitly, "Now I would like to turn to my second topic, which is . . . " ■ Show a graphic that announces the next topic and, perhaps, lists its subtopics. ■ Highlight your main points. Your discussion of each main point is a major section of your talk. "I'm going to shift now to the second cause of our problem, a cause that is particularly important to understand." ■ Pause before beginning the next topic. This pause will signal to your listeners that you have completed one part of your presentation. ■ Slow your pace and speak more emphatically when announcing your major points, just as you would when shifting to a new topic in conversation. ■ Move about. If you are speaking in a setting where you can move around, signal a shift from one topic to another by moving from one spot to another.
Review	■ In your conclusion, remind listeners explicitly about what you've covered: "In conclusion, I've done three main things during this talk: . . . "

GUIDELINE 5 Use a Conversational Style

For most talks, a conversational style works best. It helps you express yourself clearly and directly, and it helps you build rapport with your listeners. In contrast, a more abstract, impersonal style can lead you to make convoluted statements that are difficult to understand and can cause you to speak in ways that leave your listeners feeling that you are talking at them rather than with them.

Although people often associate a conversational style with informality, the two are not synonymous. The two keys to a conversational style are (1) to speak directly to your listeners, and (2) to express yourself in the same simple, natural, direct way you do in conversation. The following suggestions will help you master this style.

Strategies for Developing a Conversational Style

- **Create your talk with your audience "present."** While preparing your talk, imagine that members of your audience are right there listening to the words you are planning to speak.
- **Use the word *you* or *your* in the first sentence.** Thank your listeners for coming to hear you, praise something you know about them (preferably something related to the subject of your talk), state why they asked you to address them, or talk about the particular goals of theirs that you want to help them achieve.
- **Continue using personal pronouns throughout.** Let the use of *you* or *your* in the first sentence establish a pattern of using personal pronouns (*I, we, our, you, your*) throughout.
- **Use shorter, simpler sentences than you might use when writing.**
- **Choose words your listeners will understand immediately.** This will be especially important when your listeners include persons whose first language is not English. Check with a member of the audience or with someone who is very familiar with your listeners when you are not sure whether your listeners will understand your terminology.

Use your voice as you would in conversation.

As you strive to create a conversational style, remember that the way you use your voice can be as important as the words and phrasing you select. Listen to yourself and your friends converse. Your voices are lively and animated. To emphasize points, you change the pace, the rhythm, and the volume of your speech. You draw out words. You pause at key points. Your voice rises and falls in the cadence of natural speech.

Using your voice in the same way during your oral presentations can help you keep your listeners' interest, clarify the connections between ideas, identify the transitions and shifts that reveal the structure of your talk, and distinguish major points from minor ones. It can even enhance your listeners' estimate of your abilities. Researcher George B. Ray (1986) found that listeners are more likely to believe that speakers are competent in their subject matter if the speakers vary the volume of their voices than if the speakers talk in a monotone.

Show your feelings.

Exhibit enthusiasm for your subject. One characteristic of conversation is that we let people know how we feel as well as what we think. Do the same in your oral presentations, especially when you are advocating ideas, making recommendations, or promoting your employer's products and services. If you express enthusiasm about your topic, you increase the chances that your listeners will share your feelings.

Use gestures.

One last way to achieve a conversational speaking style is to use gestures. In conversation, you naturally make many movements—pointing to an object, holding out your arms to show the size of something, and so on. Similarly, when making oral presentations, avoid standing stiffly and unnaturally. Use natural gestures to help hold your listeners' attention and to make your meaning and feelings clear.

GUIDELINE 6 Look at Your Audience

One of the most effective ways of building rapport with your listeners is to look at them while you speak. There are several reasons why this eye contact is so important:

Benefits of maintaining eye contact with your listeners

- **Eye contact enables you to create a personal connection with your listeners.** You show that you are interested in them as individuals, both personally and professionally.
- **Eye contact helps you make a favourable impression on your listeners.** In one study, researcher S. A. Beebe (1974) asked two groups of speakers to deliver the same seven-minute talk to various audiences. One group was instructed to look often at their listeners, the other group was told to look rarely. Beebe found that the speakers who looked more often at their listeners were judged to be better informed, more experienced, more honest, and friendlier than those who used less eye contact.
- **Eye contact enables you to judge how things are going.** You can see the eyes fastened on you with interest, the nods of approval, the smiles of appreciation, or the puzzled looks and the wandering attention. These signals enable you to adjust your talk, if necessary.

Of course, you don't need to look at your audience constantly. If you are using an outline or a script, you will want to refer occasionally to your notes, but not rivet your gaze on them.

If you have difficulty looking at your listeners when you speak, here are some strategies you can use.

Strategies for Looking at Your Listeners

- **Look around at your audience before you start to speak.** This will give you an opportunity to make initial eye contact with your listeners when you aren't also concentrating on what to say.
- **Follow a plan for looking.** For instance, at the beginning of each paragraph of your talk look at a particular section of your audience—to the right for the first paragraph, to the left for the second, and so on.
- **Target a particular feature of your listeners' faces.** You might look at their eyes, but you could use their foreheads or noses instead. Unless they are very close, they won't notice the difference.
- **When rehearsing, practise looking at your audience.** For instance, develop a rhythm of looking down at your notes, then up at your audience—down, then up. Establishing this rhythm in rehearsal will help you avoid keeping your head down throughout your talk.
- **Avoid skimming over the faces in your audience.** To make someone feel that you are paying attention to him or her, you must focus on an individual. Try setting the goal of looking at a person for four or five seconds—long enough for you to state one sentence or idea.

Prepare for Interruptions and Questions—and Respond Courteously

Questions are very common at work.

Audiences at work often ask questions and make comments. In fact, most of the talks you give there will be followed by discussion periods during which members of your audience will ask you for more information, discuss the implications of your talk, and even argue with you about points you have made. This give-and-take helps explain the popularity of oral presentations at work: They permit speaker and audience to engage in a discussion of matters of common interest. Part of preparing to deliver a talk is preparing yourself for questions and discussions. In a sense, you do this when you plan the presentation itself, at least if you follow Chapter 2's advice that you begin planning your communication by thinking about the various questions that your readers will want it to answer. Usually, however, you will not have time in your talk to answer all the questions you expect that your listeners might ask. The questions you can't answer in your talk are ones your listeners may raise in a question period. Prepare for them by planning your responses.

Respond to questions in ways that maintain good relations.

When a member of the audience asks a question—even an antagonistic one—remember that you want to maintain good relations with all your listeners. If you are speaking in a large room, be sure that everyone hears the question. Either ask the questioner to speak loudly or repeat the question yourself. If people hear only your answer, they may have no idea what you are talking about. Respond to all questions courteously. Remember that the questions and comments are important to the people who ask them, even if you don't see why. Give the requested information if you can. If you don't know how much detail the questioner wants, offer some and then ask the questioner if he or she wants more. If you don't know the answer to a question, say so.

Some speakers ask that questions and comments be held until after they have finished. Others begin by inviting their listeners to interrupt when they have a question. By doing so, they are offering to help listeners understand what is being said and relate it to their own concerns and interests. Sometimes listeners will interrupt without being invited to do so. Such interruptions require special care. Speak to the person immediately. If you are planning to address the matter later in your talk, you may want to ask the questioner to wait for your response. If not, you may want to respond right away. After you do so and resume your talk, be sure to remind your listeners of where you broke off: "Well, now I'll return to my discussion of the second of my three recommendations."

GUIDELINE 8 Fully Integrate Graphics into Your Presentation

You can greatly increase the effectiveness of your oral presentations by using graphics to present your message to your listeners' eyes as well as their ears. A study conducted at the Wharton Applied Research Center found that speakers who use graphics are judged by their listeners to be

- Better prepared
- More professional
- More persuasive
- More credible
- More interesting

than speakers who don't use graphics (Andrews & Andrews, 1992; Jewett & Margolis, 1987).

The following sections provide advice for planning, designing, and displaying graphics.

Look for Places Where Graphics Can Contribute

Begin planning your graphics very early.

Graphics are most effective when they are fully integrated into your oral presentation. To achieve this total integration, begin them early. If you begin by sketching out the verbal portion of your talk, for instance, begin looking for places where graphics will help as soon as you know how your talk will be structured. Alternatively, you can begin work on your presentation by deciding which graphics will work best and then building the verbal portion around them.

In either case, look early for places where graphics can contribute to your presentation in each of the following ways:

Functions that graphics can perform

- **Explain your subject matter.** In an oral presentation, graphics can be as useful as in a written communication for helping you convey data, portray the appearance of an object, illustrate the steps in a process, and explain a difficult concept. Look for places where a picture really will be worth a thousand words.
- **Highlight your main points.** By displaying key data, summarizing your conclusions, listing your recommendations, and making other critical points in your graphics, you give them more emphasis than you could with spoken words alone.

- **Describe the structure of your talk.** Some of the graphics in your presentation can serve the same purpose as the headings in your printed communications: They can help your listeners see the structure of your talk and understand how each part fits in. For example, you might begin your talk with a slide that announces the major topics you will be discussing. If your talk is long, you might preface each major part with a drawing or other image that identifies the subtopics you will explore.
- **Hold your listeners' attention.** Listeners are more likely than readers to let their eyes wander—and when they look away from you, their thoughts may drift off as well. Graphics provide them with a second place to gaze that is directly related to your message. If the graphics are interesting (visually, intellectually, or both), so much the better.
- **Help you remember what you want to say.** Graphics can function as your speaking notes. If you forget what you planned to say next, a glance at the graphic you are displaying can bring you back on track.

Avoid Using Too Many or Too Fancy Graphics

Be cautious. You can overdo graphics.

One of the major challenges you face when planning a presentation is deciding how many graphics to use and how fancy they should be. It's possible to go overboard with any medium. However, people using presentation software such as PowerPoint seem especially liable to go overboard. Your listeners can become dulled by too many graphics. At times, it can be best to turn off the projector. Similarly, your listeners can become distracted and even irritated if you overuse the glitzy features of presentation software or otherwise allow your graphics to distract from your message. Remember that graphics should never call attention to themselves for their own sake. Their purpose is solely to help you achieve your communication's objectives.

Use a Storyboard

Devised by people who write movie and television scripts, storyboards are an excellent tool for planning the verbal and visual dimensions of a presentation simultaneously. Storyboards are divided into two parts, one showing the words and other sounds an audience will hear, and the other describing (in words or sketches) what the audience will see as those words are spoken. By reading a storyboard, someone can tell what an audience will hear and view at each moment.

Two methods of creating storyboards

To make a storyboard on paper, create two columns, one for your talk and the other for the corresponding graphics. If you are using PowerPoint or other presentation software, you can create slides and then make note pages that show an image of each slide along with the points you want to make orally while the slide is projected. Figure 17.2 (page 422) illustrates both methods of making a storyboard.

Select the Medium Best Suited to Your Purpose, Audience, and Situation

In the workplace, people use many different media for the graphics that are part of their oral presentations. The four most common are computer projections made with programs such as PowerPoint, overhead transparencies, chalkboards or dryboards, and handouts. Figure 17.3 (page 423) summarizes the chief advantages and disadvantages of each.

FIGURE 17.2

Storyboards for an Oral Presentation

Using storyboards, you can carefully coordinate your words and graphics.

In this storyboard, the writer identifies the graphic that the writer will show while she speaks the corresponding text.

This notes page shows one slide from a PowerPoint presentation and also the point she plans to make while the slide is projected.

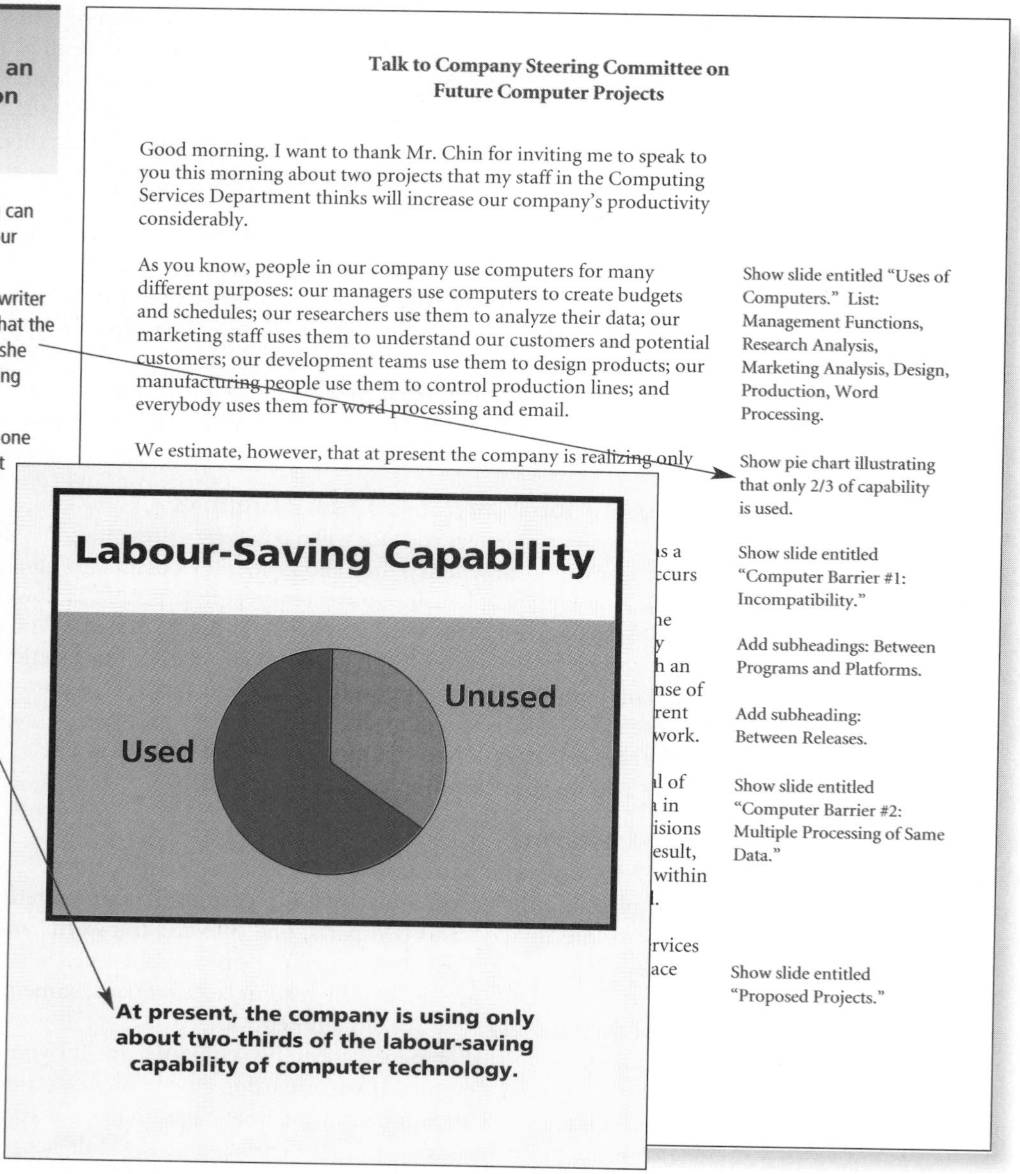

Talk to Company Steering Committee on Future Computer Projects

Good morning. I want to thank Mr. Chin for inviting me to speak to you this morning about two projects that my staff in the Computing Services Department thinks will increase our company's productivity considerably.

As you know, people in our company use computers for many different purposes: our managers use computers to create budgets and schedules; our researchers use them to analyze their data; our marketing staff uses them to understand our customers and potential customers; our development teams use them to design products; our manufacturing people use them to control production lines; and everybody uses them for word processing and email.

Show slide entitled "Uses of Computers." List: Management Functions, Research Analysis, Marketing Analysis, Design, Production, Word Processing.

We estimate, however, that at present the company is realizing only

Show pie chart illustrating that only 2/3 of capability is used.

Show slide entitled "Computer Barrier #1: Incompatibility."

Add subheadings: Between Programs and Platforms.

Add subheading: Between Releases.

Show slide entitled "Computer Barrier #2: Multiple Processing of Same Data."

Show slide entitled "Proposed Projects."

Labour-Saving Capability

Used

Unused

At present, the company is using only about two-thirds of the labour-saving capability of computer technology.

FIGURE 17.3

Advantages and Disadvantages of Several Media

Medium	Advantages and Disadvantages
Computer projections (for example, PowerPoint)	**Advantages** Enable you to create very polished-looking slides without the aid of a graphic artist Enable you to provide visual harmony to your presentation by using a single colourful design for all your slides Allow you to prepare slides quickly by typing words and inserting other content directly into them Enable you to expand your media by incorporating tables, graphs, photos, sound, animation, and movies Allow you to use special aids such as a remote control for changing slides and a laser pointer for highlighting elements on the screen **Disadvantages** Special equipment is required for projecting the slides for your audience Slides cannot be altered or reordered during a presentation to accommodate audience response Preparation can be time consuming Ability to create special effects leads some speakers into making overly elaborate presentations that detract from their content
Overhead transparencies	**Advantages** Can be made simply by copying a word-processed page onto an acetate sheet with an ordinary copy machine Projectors are available in almost every organization Can reorder the slides and draw on them with markers even as you are giving your talk **Disadvantages** Require some preparation Look plain, especially compared to computerized presentations Can't include motion or sound
Chalkboard and dryboard	**Advantages** Require no preparation Very flexible Can be used to record contributions from audience Work well for small meetings and discussion sessions **Disadvantages** Can be used only for words and line drawings Can delay presentation while you do your writing Can leave you speaking to the board rather than your audience
Handouts	**Advantages** Give readers something to take away that provides key information from your talk Aid listeners with notetaking **Disadvantages** Require preparation May tempt audience to read ahead rather than listen to what you are saying

When choosing the medium that is most suitable to your presentation, consider the following criteria:

Criteria for selecting media for graphic presentations

- **Your purpose and audience.** A computerized presentation is appropriate when you need to impress, as with a formal presentation or in a sales meeting. At a small meeting in your own department, overhead transparencies or a handout may be much more suitable.
- **Your listeners' expectations.** If you provide glitz when your listeners expect simplicity, your credibility will diminish, just as it will if you keep pausing to draw diagrams on the board for listeners who expect you to have prepared your graphics in advance.
- **Your resources.** The amount of time you have, the size of your budget, and the unavailability of equipment may limit your choices.

Remember that you can use more than one medium in a presentation. For instance, when speaking to a group of managers, a chemical engineer used slides to show photographs of the crystals grown in an experiment, overhead transparencies to explain the process used, and a handout containing detailed data that her audience could study later.

Make Your Graphics Easy to Read and Understand

For additional advice about creating graphics, see Chapter 15 and Reference Guide 5.

It is much more difficult to comprehend the graphics in an oral presentation than it is to comprehend those in a printed communication such as a proposal or report. Listeners must study the graphics from a distance, which may make your writing and symbols difficult for them to see. Also, listeners must look at the graphics while you are talking, so they have to concentrate on both at once. In addition, listeners have only a brief time to see the graphics—and that time is determined by your pace in speaking, not their pace in comprehending. Consequently, you should design your tables, drawings, and other graphics for immediate understanding.

Following are some ways to make it easy for your listeners to read and understand your graphics. Figure 17.4 shows slides from a PowerPoint presentation that illustrate their application.

Making Your Graphics Understandable and Readable

- Use type large enough to be read throughout the room.
- Use a light background and dark letters for high contrast.
- Use colour to highlight key points and focus attention.
- Avoid overcrowding.
  - Break larger topics into several graphics.
  - Use key words and phrases rather than whole sentences.
  - Use lists.
  - Leave plenty of white space (blank area) between items and around margins.
  - Keep tables, graphs, and drawings simple.
- Provide a brief title for every graphic.
- Use a consistent design for all your graphics.

FIGURE 17.4

Presentation Slides Made with PowerPoint

With PowerPoint and similar programs, you can change the format of your slides with a few clicks.

These formats were chosen from a library provided by PowerPoint; you can also make your own quite easily.

The figure is a piece of clip art that comes with the program and can be easily inserted.

It's also possible to insert tables, graphs, pictures, movies, and sounds.

These slides are for projection from a computer; the program also makes overhead transparencies and 35 mm slides.

Test Your Graphics

Like any other element of a communication, graphics should be tested beforehand to make sure that they will do what you want them to do. Show them to other people, preferably members of your target audience. Such testing is particularly important if you are speaking to an international audience. Especially in graphics that show people or workspaces, it is easy to accidentally violate a taboo or otherwise offend or irritate listeners from another culture. Even if you blunder in a way that only draws chuckles, you will have reduced the effectiveness of your presentation.

Prepare Your Stage

When you deliver a talk at work, you are a stage performer, with some words to deliver and some props (your graphics) to manage. Any difficulties with your graphics can interfere with your listeners' ability to concentrate on and understand your message. Therefore, arrive at the site of your presentation with plenty of time to set up and test any equipment you will use and to arrange your materials so you can work with them smoothly during your presentation.

Present Your Graphics Effectively

During your talk, use your graphics in ways that support and reinforce your presentation rather than detract from it:

- **Display a graphic only when you are talking about it.** If you talk about one thing but display a graphic about something else, each person in your audience must choose whether to listen to your words or read your graphic. Either way, your audience may miss part of your message.
- **Leave each graphic up long enough for your listeners to digest its contents.** Sometimes this pacing will require you to stop speaking while your listeners study your graphic.
- **Explain the key points in your graphic.** If you want your readers to notice a particular trend, compare certain figures, or focus on a particular feature in a drawing, say so explicitly.
- **Do not read from your graphics.** Your overhead transparencies and PowerPoint slides should provide key words and concepts on which you elaborate. Your listeners will become very restless if you merely read your graphics.
- **Use the same words and phrases in your graphics and in your spoken comments.** When speaking, let your spoken words reinforce the words in your graphics. If you don't, your listeners will become distracted as they try to figure out how what they are reading corresponds to what you are saying.

Look at Your Audience Most of the Time

Some speakers break their connection with their listeners when they are displaying graphics. Usually this happens because the speakers lose eye contact with their listeners by turning to look at a chalkboard or projection screen. For this reason, some speakers feel that they must face forward absolutely all the time.

However, it is sometimes natural and appropriate to turn and look at your graphics—just as your listeners are looking at them. For instance, you may want to locate the answer to a question or point to an important piece of information. When you turn away from your listeners, however, do so only briefly and speak loudly enough that you can still be heard by the people who are now behind you.

Give Your Listeners Something to Take Away

No matter how skillfully you design the oral and visual dimensions of your presentation, your listeners are likely to forget some of your message after they leave the room where you are speaking. You can overcome this problem by passing out a summary of your key points—or even your entire script—for your listeners to file for future reference.

However, handouts can also be a distraction. Your listeners may start to read the handout instead of listening to you. For that reason, some communication specialists urge speakers to distribute their handouts only after they have finished speaking. However, if you give your listeners an outline of your talk before you begin, you help them understand your talk's structure and scope and assist them in taking notes. To help you prepare an outline for distribution during your talk, PowerPoint and similar programs automatically generate pages with miniature versions of your slides and blank spaces for notes.

Handouts are especially useful when you are addressing an international audience whose members may not be able to follow everything you say. In this case, if you are speaking from a complete script, the script itself is the best handout, along with copies of your key graphics. If you are not speaking from a script, create a more detailed outline of what you plan to say than you otherwise would. Consider using full sentences—even in an outline—rather than words or brief phrases.

GUIDELINE 9 Rehearse

All of your other good preparations can go for naught if you are unable to deliver your message in a clear and convincing manner. Whether you are delivering a scripted or outlined talk, consider the following advice.

Guidelines for Rehearsing

- **Rehearse in front of other people.** They can help you identify weak spots and make suggestions for improvement.
- **Pay special attention to your delivery of the key points.** These are the points where stumbling can cause the greatest problems.
- **Rehearse with your graphics.** You need to practise coordinating your graphics with your talk.
- **Time your rehearsal.** But be sure that you speak at the same pace you will use in your actual presentation so you have an accurate sense of how long your talk will require.

GUIDELINE 10 Accept Your Nervousness—and Work with It

This guideline is difficult for many novice speakers to follow. But it is very important. Not only is nervousness unpleasant to experience, it can lead to distracting and unproductive behaviours that greatly impair the effectiveness of your talk. These include the following:

- Looking away from your listeners instead of looking into their eyes
- Speaking in an unnatural or forced manner
- Exhibiting a tense or blank facial expression
- Fidgeting, rocking, and pacing

How should you deal with nervousness? First, accept it. It's natural. Even practised speakers with decades of experience sometimes feel nervous when they face an audience. If you fret about being nervous, you merely heighten the emotional tension. Also keep in mind that your nervousness is not nearly so obvious to your listeners as it is to you. Even if they do notice that you are nervous, they are more likely to be sympathetic than displeased. Furthermore, a certain amount of nervousness can help you. The adrenaline it pumps into your system will make you more alert and more energetic as you speak.

Here are some strategies for reducing your nervousness and controlling the pacing, fidgeting, and other undesirable mannerisms it fosters.

Strategies for Controlling Nervousness

- **Arrive early.** Avoid rushing from a previous activity to your talk. Give yourself plenty of time to set up and look around before you begin.
- **Devote a few minutes before your talk to relaxing.** Take a walk or spend a quiet moment alone.
- **Speak with audience members before your presentation begins.** Doing this enables you to make a personal connection with at least some of your listeners before you begin.
- **Remind yourself that your listeners are there to learn from you, not to judge you.**
- **When it's time to begin, pause before you start your talk.** Look at your audience, say "Hello," and adjust your outline or notes as you accustom yourself to standing before your listeners.

MAKING TEAM PRESENTATIONS

The following sections provide advice for creating and delivering effective team presentations.

Plan Thoroughly

When developing a team presentation, plan as carefully as you would if preparing a team-written document. Devoting a team meeting to making plans can be very helpful. Decide which topic each team member is to discuss, which points are to be made, and how each part fits with the others. Set a time limit for each part so that the total presentation doesn't run too long. Also, decide whether the team is going to ask the audience to hold questions until the end or invite the audience to interrupt the speakers with questions. In either case, provide time for the interchange between the team and its listeners.

Decide on Graphic Use

If the team will be using presentation software, plan for a consistent appearance of slides throughout the presentation. In the same way that the use of a style guide leads to a consistent team-written document, decide on a slide template and develop a slide master for all team members to use. If the team will not be using presentation software, ensure that graphics or visuals used in the presentation will be consistent.

Allow for Individual Differences

In a team-written project, the goal is usually to produce a document with a single voice for the entire document. In team presentations, however, each speaker can speak in his or her own style and voice, provided that the general tone of the presentation and the appearance of the graphics are relatively consistent.

Make a Cohesive Presentation

A team presentation consists of several team members speaking with the same goal and audience in mind. To that end, introduce an overall theme, thread, or common set of objectives at the beginning of the team presentation. You may wish to use an agenda for this purpose. It may also be helpful to introduce team members in the order they will be speaking, paired with their part of the presentation. This overall introduction will help the audience discern the overall structure of the presentation. In the same way, conclude the presentation by summarizing the overall findings or topics that were discussed. You may wish to link team members to the topics they covered as you conclude the presentation.

Make Effective Transitions between Speakers

By carefully planning the transitions from one speaker to the next, a team can substantially increase the effectiveness of its presentation. Switch speakers where you are making a major shift in topic. Such a change of speaker will further the audience's understanding of the overall structure of the presentation. Ensure that each speaker's presentation has its own introduction, body, and conclusion. Also, have both speakers explain how the two parts of the presentation fit together. For example, the speaker who is finishing might say, "Now, Clara will explain how we propose to solve the downtime problem I have just identified." The next speaker might then say, "In the next few minutes, I'll outline our three recommendations for dealing with the downtime problem Simon has described." To further the transitions between speakers, you may wish to reiterate the team presentation's agenda, highlighting the placement of the speaker and topic in that agenda.

Show Respect for One Another

Your team can increase the effectiveness of your presentation if each member shows respect for the others. If team members seem confident of one another's contributions and capabilities, your audience is more likely to adopt the same attitude. Ensure that team members appear attentive and enthusiastic while other members are presenting.

Rehearse Together

Rehearsals are crucial to the success of team presentations. Team members can help one another polish their individual contributions and ensure that all the parts are coordinated in a way that the audience can easily understand. Moreover, they can see to it that the entire presentation can be completed in the time allotted. Running overtime is a common and serious problem for groups that have not worked together previously.

CONCLUSION

WWW To download a planning guide for oral presentations, visit Chapter 17 at www.techcomm.nelson.com.

Making oral presentations can be among your most challenging—and rewarding—experiences at work. By taking a listener-centred approach that is analogous to the reader-centred approach described elsewhere in this book, and by striving to communicate simply and directly with your audience, you will prepare talks that your listeners will find helpful, informative, interesting, and enjoyable. A planning guide is available on this text's website.

EXERCISES

For additional exercises, visit www.techcomm.nelson.com.

Expertise

1. Outline a talk to accompany a written communication that you have prepared or are preparing in one of your classes. The audience for your talk will be the same as for your written communication. The time limit for the talk will be ten minutes. Be sure that your outline indicates the following:

   - The way you will open your talk
   - The overall structure of your talk
   - The main points from your written communication that you will emphasize
   - The graphics you will use

   Be ready to explain your outline in class.

2. Imagine that you must prepare a five- to ten-minute talk on some equipment, process, or procedure. Identify your purpose and listeners. Then write a script or an outline for your talk (whichever your professor assigns). Be sure to plan which graphics you will use and when you will display each of them.

3. Do one of the "Oral Briefing" projects in Appendix C.

Online

1. Using PowerPoint or a similar program, prepare a brief set of slides for the talk you will give in one of the preceding exercises. Follow the advice given in Guideline 8.

2. Using a search engine, find an online source for advice about oral presentations. Compare the advice given there with the advice you have found in this chapter.

Collaboration

At work, you will sometimes be asked to contribute to discussions about ways to make improvements. For this exercise, you and a classmate are to deliver a five-minute impromptu talk describing some improvement that might be made in some organization you are familiar with. Topics you might choose include ways of improving efficiency at a company that employed you for a summer job, ways of improving the operation of some club you belong to, and ways that a campus office can provide better service to students.

Plan your talk together, but divide your content so that you both talk. In your talk, clearly explain the problem and your solution to it. Your professor will tell you which of the following audiences you should address in your talk:

- Your classmates in their role as students. You must try to persuade them of the need for, and the reasonableness of, your suggested action.
- Your classmates, playing the role of the people who actually have the authority to take the action you are suggesting. You should take one additional minute at the beginning of your talk to describe these people to your classmates.

Ethics

Prepare a brief, three- to five-minute presentation concerning an ethical issue that you or someone you know has confronted on the job. Describe the situation, identify the stakeholders, tell how the stakeholders would be affected, and tell what was done by you or someone else. If nothing was done, tell why. Be sure to keep within your time limit. According to your professor's assignment, give your report to the entire class or to two or three other students.

Reference Guide 5: Thirteen Types of Graphics

CONTENTS

This Reference Guide presents detailed advice for constructing thirteen types of graphics that are widely used on the job. The thirteen types are organized according to the functions they most often perform in workplace writing.

For additional resources, visit Reference Guide 5 at www.techcomm.nelson.com.

Types of Graphics

Photographs and drawings appear in the list twice because, like many other forms of graphics, they can be used for a variety of purposes.

GRAPHICS FOR DISPLAYING DATA

When you need to display data, consider using one of the following types of graphics: tables, bar graphs, pictographs, line graphs, and pie charts. Figure VA.1 identifies applications for which each type is typically used.

TABLES

Tables are everywhere: on the side of your cereal box; in the owner's manual for your sound system, television, car, or computer; and in the reports, proposals, memos, letters, and instructions you will read (and write) at work.

FIGURE VA.1

Advice for Choosing among Various Ways of Displaying Numerical Data

Type	Typical Application	Comments
Table	Helps readers find particular facts in large sets of information	Not good for emphasizing trends or making comparisons unless only a few points are involved
Bar Graph	Helps readers compare quantities at a glance	Can also show trends if readers can easily draw lines mentally between the ends of the relevant bars
Pictograph	A type of bar graph that makes data more vivid by calling to mind the people and things discussed	In some situations, reader may think that a pictograph is unprofessional or otherwise inappropriate
Line Graph	Highlights trends and relationships among variables	Can display complex relationships more clearly than other types of visual aids described here
Pie Chart	Shows how a whole has been divided into parts and indicates the relative sizes of the parts	Sometimes difficult for readers to compare the size of wedges accurately

Tables are used so often because they help writers achieve several common communication objectives:

- **Tables present detailed facts in a concise, readable form.** Consider, for instance, the following passage:

Data presented in prose

> In 2002, Canada spent $114.0 billion on health care, which amounted to $3,635 per capita and 9.9 percent of the gross domestic product (GDP). In 2003, total expenditure increased to $123.0 billion, which was $3,885 per capita and 10.1 percent of the GDP. In 2004, the total expenditure was $130.3 billion, equalling $4,078 per capita and 10.1 percent of the GDP.

Placed in a table, this same information is organized and displayed in a manner that is much easier to read.

The same data presented in a table

Canadian Health Care Expenditures

Year	Total Dollars (in billions)	Dollars Per Capita	Percent of GDP
2002	114.0	3,635	9.9
2003	123.0	3,885	10.1
2004	130.3	4,078	10.1

- **Tables also help readers find specific facts quickly.** To demonstrate this point, first use the text version and then the table to locate the total amount of money spent in Canada on health care in 2004. Or try to determine how the percentage of the gross domestic product spent on health care services changed from 2002 to 2004.

Tables are often used in passages that group facts or present comparisons. See pages 41–45 and 50–54.

Tables can be just as effective at presenting information in words as in numbers. They are often used to help readers understand options and choose among them (Figure VA.2). They are also used in the troubleshooting sections of instruction manuals (Figure VA.3).

FIGURE VA.2

A Table that Helps Readers Understand Options

This interactive web-based table compares digital camera models made by one manufacturer.

The table enables readers to quickly compare models by sorting on key features. In the Web-based version of the table, the reader can click on the column heading to sort the models by the feature of interest.

Click on a column heading to sort					
Cameras	MP	Max. Print Size	Zoom	Memory	Docks
P880	8.0 MP	75 × 100 cm (30" × 40")	11.6X (5.8X optical 2X digital)	32MB built-in (SD/MMC card slot)	Optional: Camera Dock Series 3 Printer Dock Plus
DX7630	6.1 MP	75 × 100 cm (30" × 40")	12X (3X optical 4X digital)	32MB built-in (SD/MMC card slot)	Optional: Camera Dock 6000 Printer Dock 6000
P850	5.1 MP	50 × 75 cm (20" × 30")	39.6X (12X optical 3.3X digital)	32MB built-in (SD/MMC card slot)	Optional: Camera Dock Series 3 Printer Dock Plus
CX7525	5.0 MP	50 × 76 cm (20" × 30")	15X (3X optical 5X digital)	16MB built-in (SD/MMC card slot)	Optional: Camera Dock 6000 Printer Dock
C340	5.0 MP	50 × 75 cm (20" × 30")	15X (3X optical 5X digital)	16MB built-in (SD/MMC card slot)	Optional: Camera Dock Series 3 Printer Dock Series 3
C360	5.0 MP	50 × 75 cm (20" × 30")	15X (3X optical 5X digital)	32MB built-in (SD/MMC card slot)	Optional: Camera Dock Series 3 Printer Dock Series 3
CX7530	5.0 MP	50 × 75 cm (20" × 30")	15X (3X optical 5X digital)	32MB built-in (SD/MMC card slot)	Optional: Camera Dock 6000 Printer Dock
DX4530	5.0 MP	50 × 75 cm (20" × 30")	10X (3X optical 3.3X digital)	32MB built-in (SD/MMC card slot)	Optional: Camera Dock II Printer Dock 4000
DX7590	5.0 MP	50 × 75 cm (20" × 30")	30X (10X optical 3X digital)	32MB built-in (SD/MMC card slot)	Optional: Camera Dock 6000 Printer Dock

How to Construct a Table

To construct a table, you systematically arrange information in rows and columns, adding headings and other explanatory information as necessary. Figure VA.4 (page 436) shows the basic structure of a table.

Word processing software makes it very easy to create tables within a draft you are preparing or to import tables you've created in a spreadsheet program such as Microsoft Excel. Your challenge is to use the power of this software to design tables that are easy to understand, easy to use, and attractive. The following guidelines will enable you to do that.

FIGURE VA.3

A Table that Provides Troubleshooting Information

The vertical and horizontal lines help readers to quickly find the information they need.

The problems that may be experienced by readers are grouped into categories.

Computer Monitor		
Problem	**Cause**	**Solution**
The image on your monitor is jagged or inappropriately colored.	The monitor is not set correctly.	Go to Start → Settings → Control Panel → Display Settings. Under Color, select a higher color depth (High Color, 16 bit or higher)
EZ200 Software screen takes up too much space on the screen.	Screen resolution is set too low.	Go to Start → Settings → Control Panel → Display Settings. Under Screen Area, slide the indicator toward More.
Camera Communications		
Problem	**Cause**	**Solution**
Computer cannot communicate with the camera.	The USB cable is not properly connected to the camera or computer.	Connect the cable to the camera and computer ports.
	There is a problem with your USB connection.	Visit our web site at www.kodak.com and click Service and Support.
	The software is not installed.	Install the software.
Camera		
Problem	**Cause**	**Solution**
Camera does not turn on.	The batteries are not installed properly.	Remove and reinstall the batteries.
	The batteries are exhausted.	Replace the batteries.

Guidelines for designing tables

- **Order items in a way your readers will find useful.** For example, in a table designed to show the budgets of Canada's fifty largest biotechnology companies, you could arrange the companies alphabetically if your readers want to quickly locate the sales output of a few specific companies. However, if your readers want to compare companies in terms of the amounts they spend on research, you could order the companies from the highest to the lowest in terms of the research and development expenses.
- **Make the key information stand out visually.** Column and row headings are very important to readers. Consider using boldface, larger type, colour, and rules (lines) to make them stand out. Use these same tools if you want to emphasize certain data within your table or to help readers find a particular row or column.
- **Label your columns and rows clearly.** From a reader's perspective, clear and precise labels are essential. If several columns are related, span them with a higher level heading that indicates that they go together. Similarly, use indentation or similar devices to group row labels that go together.

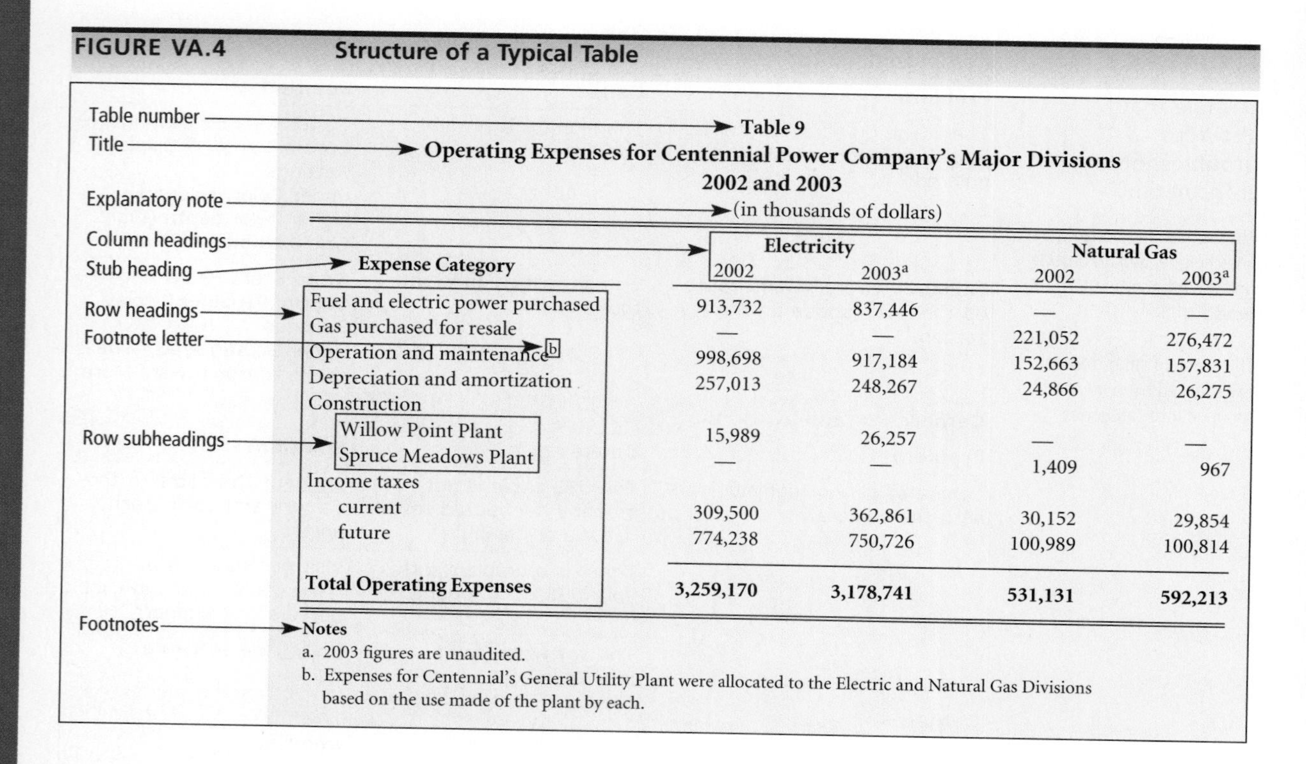

FIGURE VA.4 Structure of a Typical Table

Table 9

Operating Expenses for Centennial Power Company's Major Divisions 2002 and 2003

(in thousands of dollars)

Expense Category	Electricity		Natural Gas	
	2002	2003[a]	2002	2003[a]
Fuel and electric power purchased	913,732	837,446	—	—
Gas purchased for resale	—	—	221,052	276,472
Operation and maintenance[b]	998,698	917,184	152,663	157,831
Depreciation and amortization	257,013	248,267	24,866	26,275
Construction				
Willow Point Plant	15,989	26,257	—	—
Spruce Meadows Plant	—	—	1,409	967
Income taxes				
current	309,500	362,861	30,152	29,854
future	774,238	750,726	100,989	100,814
Total Operating Expenses	**3,259,170**	**3,178,741**	**531,131**	**592,213**

Notes

a. 2003 figures are unaudited.

b. Expenses for Centennial's General Utility Plant were allocated to the Electric and Natural Gas Divisions based on the use made of the plant by each.

Sort row labels for easy reading.

Avoid Unsorted Lists of Row Labels

Fruits
Grains
Legumes
Commercial baked goods
Deep-fried foods

Use Headings and Indentation To Sort Row Labels

Nutritious Foods
  Fruits
  Grains
  Legumes
Non-nutritious Foods
  Commercial baked goods
  Deep-fried foods

- **In longer tables, guide your readers' eyes across rows.** Leave a blank or insert a horizontal rule after every five rows or so. This will help your readers' eyes move across the rows. Alternatively, you can use shading of rows to support this motion. Of course, if the rows are grouped by some logical relationship, place the breaks between the groupings.
- **When presenting data, indicate the units.** Make sure your readers know what each entry in your table represents: dollars, kilograms, percentages, and so on. When all the entries in a long column or row use the same units, it's usually best to identify the units in the heading to avoid cluttering the table. Use abbreviations only when you are certain your readers will understand them.

- **Align entries in a way that supports easy reading.** Align numerical entries either on the units or on the decimal point:

Numbers can be aligned on the units column or on the decimal point

```
23,418      2.79
 5,231    618.0
    17     13.517
```

Align words and phrases on the left-hand margin or centre them in the column.

Words can be aligned flush left or centred in their column

```
Acceptable                          Acceptable
Marginally Acceptable          Marginally Acceptable
Unacceptable                       Unacceptable
```

- **Place explanatory notes where they are accessible but not distracting.** In some tables, you will need to include notes that explain the labels or cite the sources of your information. If your notes are short, you can place them next to the appropriate title or headings. Otherwise, place them in footnotes at the bottom of your table. Use lowercase letters to label the footnotes if your readers might confuse a superscript number with a mathematical exponent. For example, "14^2" could be read as "fourteen squared" rather than the number 14 followed by a superscript. Using a lowercase letter avoids that confusion: "14^b."
- **Where possible, keep your table on one page.** Consider breaking large tables into smaller, separate ones. If your table contains too many columns to fit on the page, turn it sideways.

Informal Tables

Sometimes you can create tables that are much simpler than those just described. Such "informal tables" need no row or column headings, no titles or notes. They are useful where the preceding sentence explains what the table is about and where the interpretation of the table is obvious. Here is an example showing how an informal table can be built into a paragraph.

Informal tables fit into the flow of your prose without figure numbers or titles

> Even more important, the sales figures demonstrate how our investment in technical research has helped several of our divisions become more competitive. The following figures show the increase in the market share enjoyed since the same quarter last year by three major divisions that have extensive research programs:
>
> Strausland Microchips 7%
> Minsk Machine Tools 5%
> PTI Technical Services 4%

Note that such tables work only when they contain a single column of facts and when readers will readily understand them. If you think that such an informal table might confuse your readers, even momentarily, use a formal table instead.

BAR GRAPHS

Bar graphs are often used in passages that present comparisons. See pages 50–54.

Like a table, a bar graph can represent numerical quantities. Instead of numbers, however, a bar graph uses rectangles called *bars*. The greater the quantity, the longer the bar.

FIGURE VA.5
Bar Graph Showing Comparison

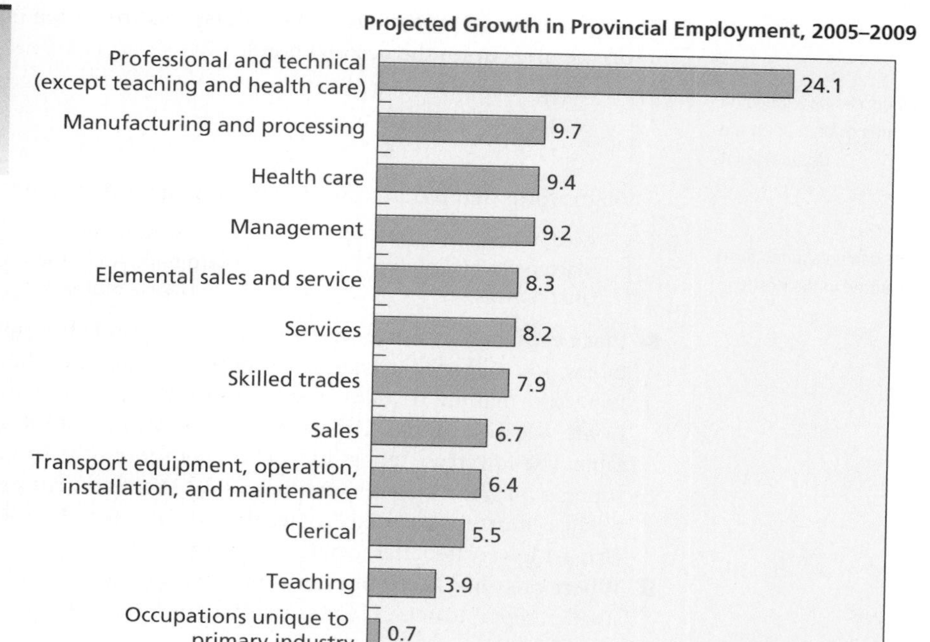

Here are some uses of bar graphs:

Uses of bar graphs

- **To compare quantities at a glance.** Because the bars are drawn to scale, bar graphs can help readers tell immediately not only which quantities are the larger or smaller, but also how great the differences are (see Figure VA.5).

A bar graph enables readers to compare quantities instantly.

- **To show trends.** If a series of bars is used to represent a quantity over time (such as world airline fleet growth), readers will be able to detect overall trends (see Figure VA.6).

Because the heights of the bars change steadily, the trend becomes obvious to readers.

- **To indicate the composition of a whole.** For instance, a bar that represents diabetes patients' hospital visits might be subdivided to show treatment and non-treatment visits. A series of such bars can show the changing (or unchanging) composition of visits according to age group (see Figure VA.7).

How to Construct a Bar Graph

Word processing, spreadsheet, and database programs enable you to create bar graphs quickly. The following guidelines and Guideline 9 in Chapter 15 will help you use these computer tools to design bar graphs that represent your data clearly and accurately:

Guidelines for creating bar graphs

- **Draw the horizontal and vertical axes so that your graph will be roughly square.**
- **Decide whether to extend the bars vertically or horizontally.** Vertical bars are often used for height and depth, whereas horizontal bars are often used for distance, length, and time.
- **Use tick marks to indicate quantities on the axis parallel to the bars.** Label them ($5 million, $10 million, etc.; 50 psi, 100 psi, etc.). Plan the tick marks so that the longest bar will extend nearly to the end of its parallel axis.

FIGURE VA.6

Bar Graph Showing a Trend

Because the heights of the bars change steadily, the trend becomes obvious to readers.

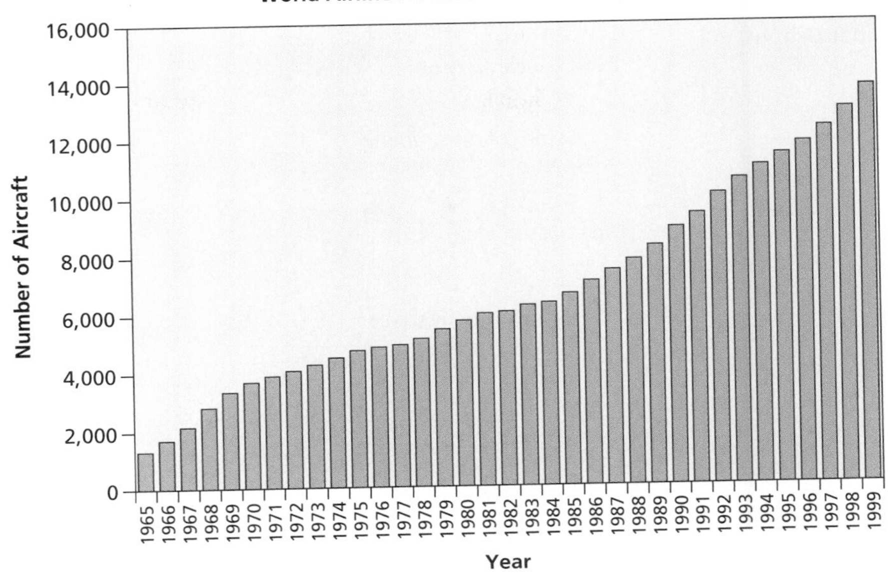

FIGURE VA.7

Bar Graph with Subdivided Bars

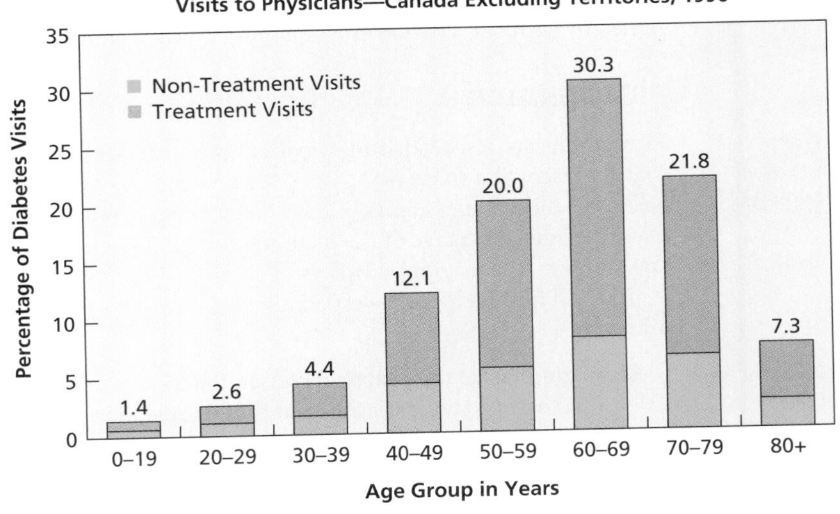

*Diabetes medication prescribed

- **Order the bars to suit your communication purpose.** For example, if you want your readers to discern rank orders, arrange the bars in order of length. If you want your readers to compare subgroups of quantities, group the appropriate bars together.

FIGURE VA.8
Multibar Graph

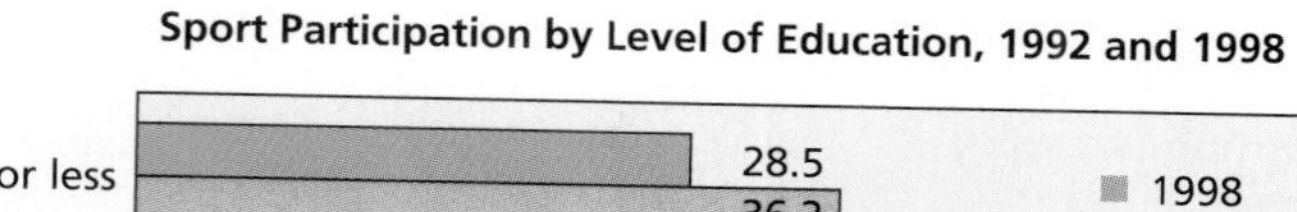

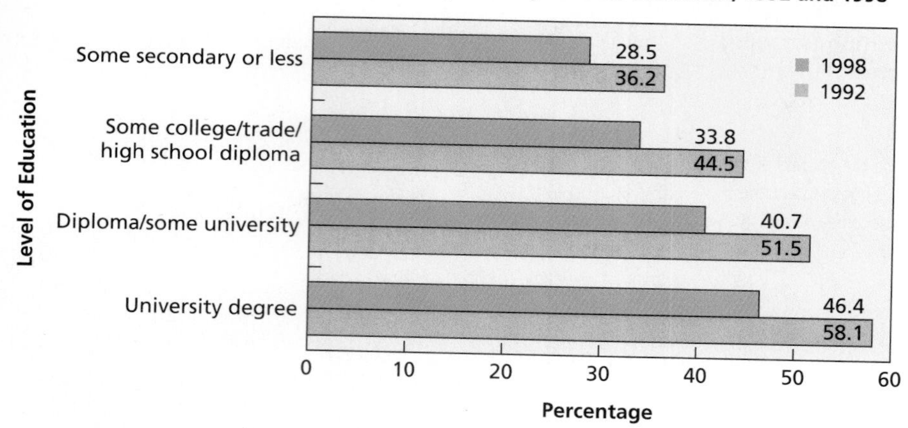

- **Place labels next to the bars, when possible.** Generally, readers find such labels easier to use than a separate key. An exception occurs when you use the same groups of bars repeatedly, as in Figure VA.8. In that case, use distinctive shading, colour, or cross-hatching to indicate each category within the group of bars and provide a key to the categories.

Usually, it is unnecessary to include numbers that indicate the exact quantity represented by each bar. Provide exact quantities only if your readers are likely to want them.

PICTOGRAPHS

Pictographs are often used in passages that present comparisons. See pages 50–54.

Pictographs are a special kind of bar graph in which the bars are replaced by drawings that represent the thing being described. In Figure VA.9, for example, the number of barrels of oil used per capita in the United States is represented by drawings of oil barrels. The chief advantage of the pictograph is that the drawings symbolize concretely the quantities that your graph displays.

You will find pictographs especially useful where you want to do one or both of the following:

Pictographs can suggest the significance of data

- **To emphasize the practical consequences of the data represented.** For example, a pictograph that uses silhouettes of people to represent the workers who will be employed in a new plant emphasizes a benefit that the plant will bring to the community. You can select from a multitude of clip art symbols in word processing and spreadsheet software, as well as other sources such as the Internet.
- **To make your data visually interesting and memorable.** Visual interest is especially important when you are addressing the general public. In some situations at work, however, readers expect a more abstract representation of information and would consider pictographs inappropriate.

How to Construct a Pictograph

To create a pictograph, follow the procedure for creating a bar graph but substitute drawings for the bars.

FIGURE VA.9

Pictograph (combined with line graph)

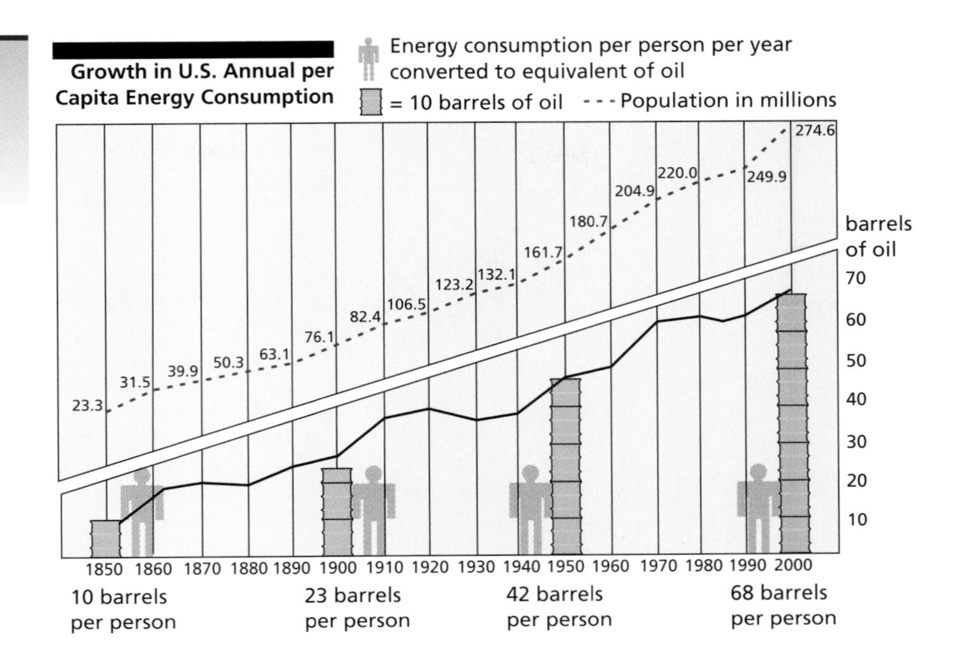

LINE GRAPHS

A line graph shows how one quantity changes as a function of changes in another quantity. You can use line graphs for many purposes, including the following:

Line graphs are often used in passages that explain cause and effect. See pages 54–56.

- **To show trends and cycles.** When you want to show a pattern of change, line graphs can be very helpful.
- **To compare trends.** Line graphs are also very useful for showing readers how two or more trends relate to one another. For instance, using the graph shown in Figure VA.10 (page 442), economists were able to demonstrate quite dramatically how the number of people employed as bank tellers has plummeted as the number of transactions at automated teller machines has risen.
- **To show how two or more variables interact.** Figure VA.11 (page 442) shows a line graph used by a group of physicists studying cosmic rays.

How to Construct a Line Graph

In line graphs, you usually want to show how variations in one factor (the *dependent variable*) are affected by variations in another factor (the *independent variable*).

Guidelines for creating line graphs

- **Draw the horizontal and vertical axes so that your graph will be roughly square.**
- **Indicate the quantities along each axis with tick marks placed at regular intervals.** Label the tick marks. Usually your graph will be less cluttered if you make the tick marks short (but still clearly visible). In some situations, however, your readers may find it easier to read a line graph if you extend the tick marks all the way across the graph to form a grid. If you do this, draw the grid with thinner lines than your plotted lines so that the plotted lines stand out.

FIGURE VA.10

Line Graph Comparing Trends

The relationship between the two trend lines suggests a strong relationship between the trends they represent.

The hash marks on the vertical axis alert readers that it doesn't begin at zero.

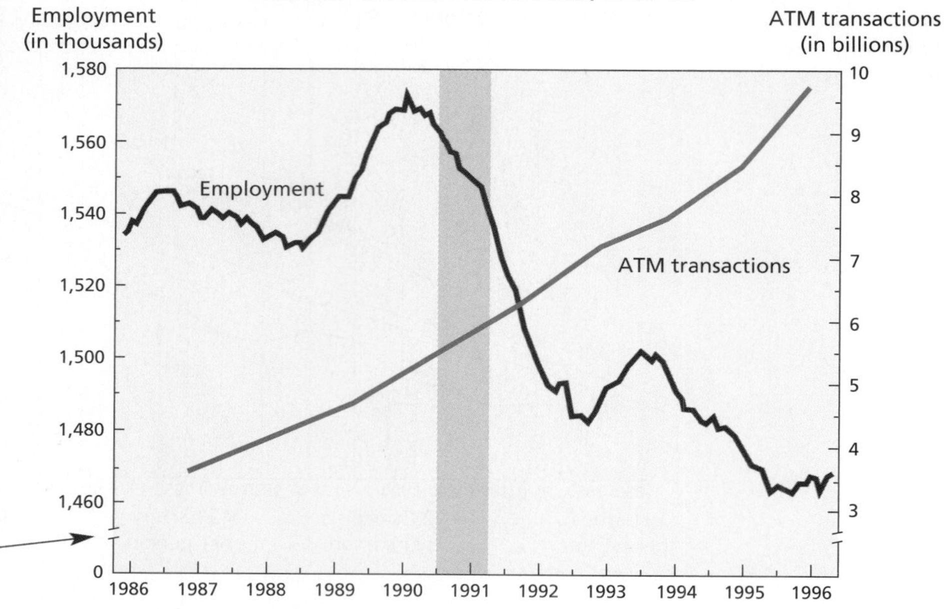

NOTE: The shaded area denotes a recession, as identified by the National Bureau of Economic Research.
SOURCES: Unpublished data from the Bureau of Labor Statistics Current Employment Statistics program; and *Bank Network News*.

FIGURE VA.11

Line Graph Showing the Interaction among Variables

Careful placement of the labels and the use of colours make it easy to identify the lines in this complex graph.

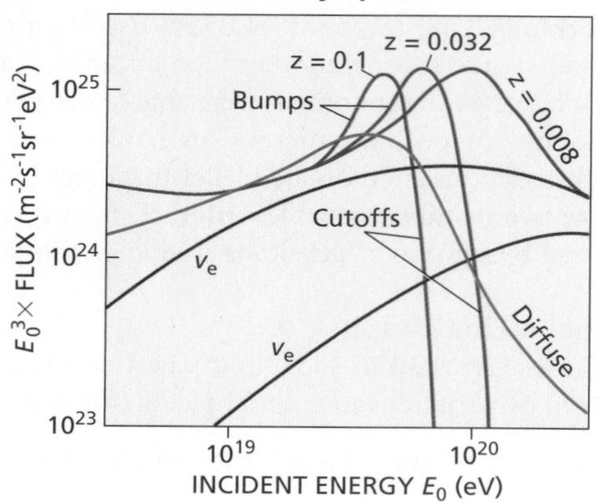

- **Start the axes at zero, where possible.** Otherwise, readers may be misled about the proportions of changes shown (see Guideline 9 in Chapter 15). If the vertical axis does not begin at zero, alert your readers to that fact by using hash marks to indicate that the scale is not continuous from zero. Where you need to indicate

FIGURE VA.12

Pie Chart

The slices are arranged in descending size to make comparisons easy.

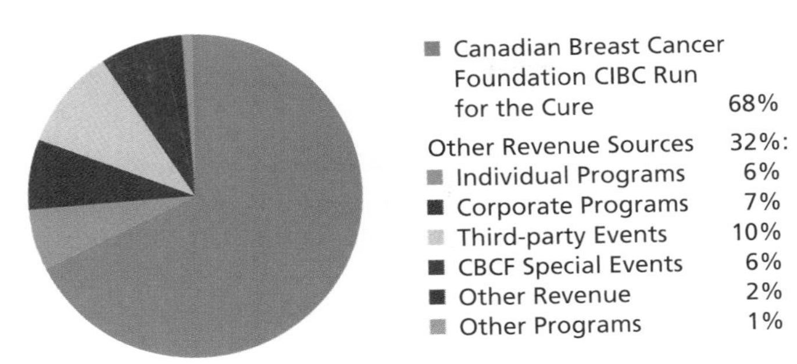

some negative quantities, such as losses or degrees below zero, use both positive and negative values on whichever of the axes needs them.

- **Put labels next to the plotted lines and data points.** Generally, readers find such labels easier to use than a separate key.

PIE CHARTS

Pie charts are often used in passages that present comparisons. See pages 50–54.

Pie charts are a good choice when you want to show the composition of a whole—for example, the sources of revenue for the Canadian Breast Cancer Foundation (see Figure VA.12).

How to Construct a Pie Chart

A pie chart is simply a circle divided into wedges proportional to the amount of the total that each quantity represents:

- **Arrange the wedges in a way that will help your readers perceive the rank order of the wedges and compare their relative sizes.** Often that means starting with the largest wedge near the top and proceeding clockwise in descending order of size.
- **Label each wedge and give its percentage of the whole.** Depending on the size of a wedge, place its label inside or outside the circle.
- **Limit the number of wedges to about eight.**
- **Create an "Other" wedge if you have a collection of small quantities that would be difficult to distinguish.**

GRAPHICS FOR SHOWING HOW SOMETHING LOOKS OR IS CONSTRUCTED

When you want to tell your readers what something looks like or how it is constructed, you may be able to communicate most effectively with a photograph or drawing. For communications involving software, screenshots can be very useful to readers.

PHOTOGRAPHS

Photographs have become a common element in workplace writing with the incorporation of digitized images into email messages, documents, and presentations. By showing *exactly* what something looks like, photographs can help you achieve many goals, including the following:

Photographs and drawings are often used in passages that describe an object. See pages 45–47.

Uses of photographs

- **To indicate the appearance of something the audience has never seen.** Perhaps it's the inside of a human heart, the surface of one of the moons of Saturn, a new building purchased in another city, or a new product your company has just begun manufacturing.
- **To show how something is done.** For example, the photograph shown in Figure VA.13 shows readers how one of the steps in genetic engineering is performed.
- **To show the condition of something.** Photographs can help portray the condition of an object when that condition is indicated by the object's appearance. Maybe you want to show the result of the treatment of a skin ailment with a new drug your company is marketing, the damage caused by improper handling of a shipment of products, the progress being made on the construction of a new factory, or the difference between corn that has received sufficient water and corn that has not (see Figure VA.14).
- **To help readers recognize something.** For instance, in a field guide for biologists, photographs like those displayed in Figure VA.15 help readers identify various species of frogs.

FIGURE VA.13

Photograph Used to Show How Something Is Done

The photographer chose to record this action from an angle that shows all the important details about how the researcher performs this step.

FIGURE VA.14

Photographs Used to Show the Condition of Something

Watering regimen: 1 inch of irrigation water per week. Leaves green and pliable to the edges. Sheen on leaves indicates a healthy cuticle that is able to resist insects and disease.

Watering regimen: 3 weeks without water. Necrosis on leaf tips and margins, where water stress typically shows itself first. No sheen. Inner necrotic spots caused by disease that was able to penetrate the weakened cuticle.

Captions point out key features of the photographs and interpret their meaning for readers.

FIGURE VA.15

Photographs Used to Help Readers Recognize Something

Poison arrow frog, ***Dendrobates leucomelas,*** found in the rainforest of Guyana.

Red-eye leaf frog, ***Agalychnis callidryas***, found on the Caribbean side of Costa Rica.

- **To help readers locate something.** In an instruction manual, for instance, you might employ a photograph of a cell phone to help users find the components used to operate the unit (see Figure VA.16, page 446).

FIGURE VA.16

Photograph Used to Help Readers Locate Something

The labels are printed on a white background to make them stand out.

They are carefully placed where they will not cover an important part.

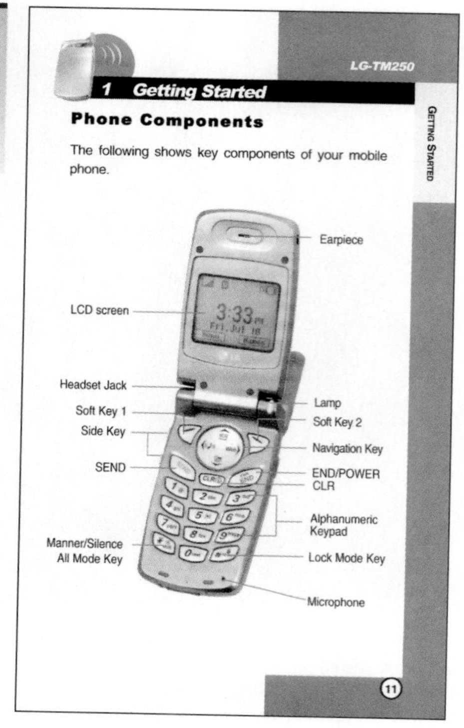

How to Create a Photograph

Here are some suggestions that will help you create effective photographs:

Guidelines for creating photographs

- **Choose an appropriate angle of view.** If you want to help your readers recognize or find something, photograph it from the angle that your readers would take when looking for it.
- **Eliminate unnecessary or distracting details.** Before taking your photograph, remove irrelevant objects from the area to be photographed. If such objects show up in a digital image, use photo editing software to eliminate the irrelevant objects from the image to focus your readers' attention on what you want them to see.
- **Draw attention to key parts with colour.** Use shading, contrast, and highlighting to emphasize key components.
- **Ensure that all relevant parts are clearly visible.** Don't let important parts get hidden (or half hidden) behind other parts. Ensure that the important parts are in focus.
- **Choose an appropriate distance from your subject.** If you want to help your readers focus their attention on what you want them to see, fill the frame of the photograph with the object of interest. If necessary, use photo editing software to crop, reorient, or enlarge the image.
- **Provide whatever labels your readers will need.** Be sure that labels stand out from the background of the photograph. With some word processing, drawing, or photo editing software, you can add these labels electronically.

DRAWINGS

Drawings are often more effective than photographs.

In some situations, drawings can be even more effective than photographs at showing how something looks or is put together. Often, you can prepare drawings more quickly, and they allow you to readily emphasize important details and omit distracting ones. Moreover, drawings enable you to show things that photographs cannot, such as the inner parts of an object. Figure VA.17 (page 448) shows four drawings—one that takes an external view and three that take internal views: cutaway, cross-section, and exploded.

How to Create a Drawing

Tailor your drawing to the use your readers will make of it:

Guidelines for creating a drawing

- **Choose the angle of view that your readers will find most helpful.** In many cases, this will be a "three-cornered" view, which shows, for example, two sides and the top of an object drawn in perspective. Such a view helps people see at a glance what the object looks like and how its major parts fit together. If you are preparing a drawing for a set of instructions, show the object from the same point of view that your readers will have when working with it.
- **Select your details.** Remember that your purpose is not to produce a perfectly realistic image of the object, but to highlight its significant parts or features. If you want to make a relatively realistic drawing, you might manipulate a digital image with drawing or photo editing software. Alternatively, you might scan an object to produce a three-dimensional image, which you can then edit electronically.
- **Draw attention to key parts.** To emphasize significant details, you might draw them slightly larger than they actually are, draw them with a heavier line than you use for other parts, or point them out with labels. You might also use shading or highlighting to emphasize particular components.

Drawings allow you to include only the parts that are important to the reader.

Figure VA.18 (page 450) shows a drawing that uses several of these techniques.

SCREENSHOTS

Screenshots are images captured from a computer screen. They are especially useful in instructions for using computer software. After you've imported the image into your word processing or presentation software, you can enlarge it, shrink it, or enhance it by cropping away unneeded elements or by making helpful additions. See Figure VA.19 (page 450).

How to Use Screenshots

When using screenshots, follow this advice, which is illustrated in Figure VA.19:

Guidelines for using screenshots

- **Show only what's important to readers.** Don't show the whole screen if only one window is important, and don't show a whole window if readers need to see only a part of it.
- **Add arrows, colour, or other devices to guide your readers' attention to the key elements.**
- **Fill in fields with information that will help your readers.** For example, if readers are supposed to fill in a particular field in a certain way, let your screenshot illustrate that field filled in properly.

FIGURE VA.17

Four Types of Drawings

EXTERNAL VIEW

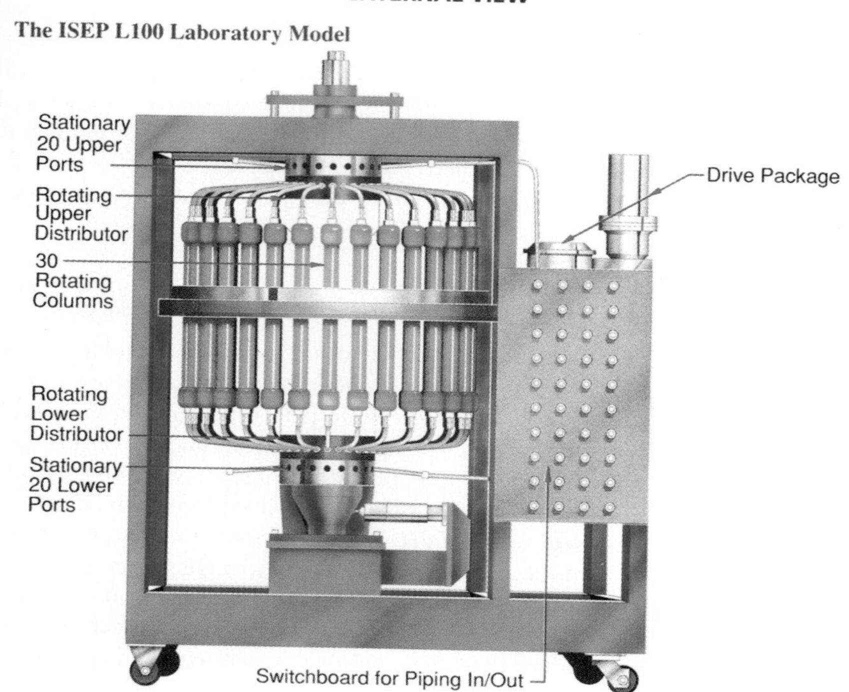

CUTAWAY

FIGURE VA.17
(continued)

CROSS-SECTION

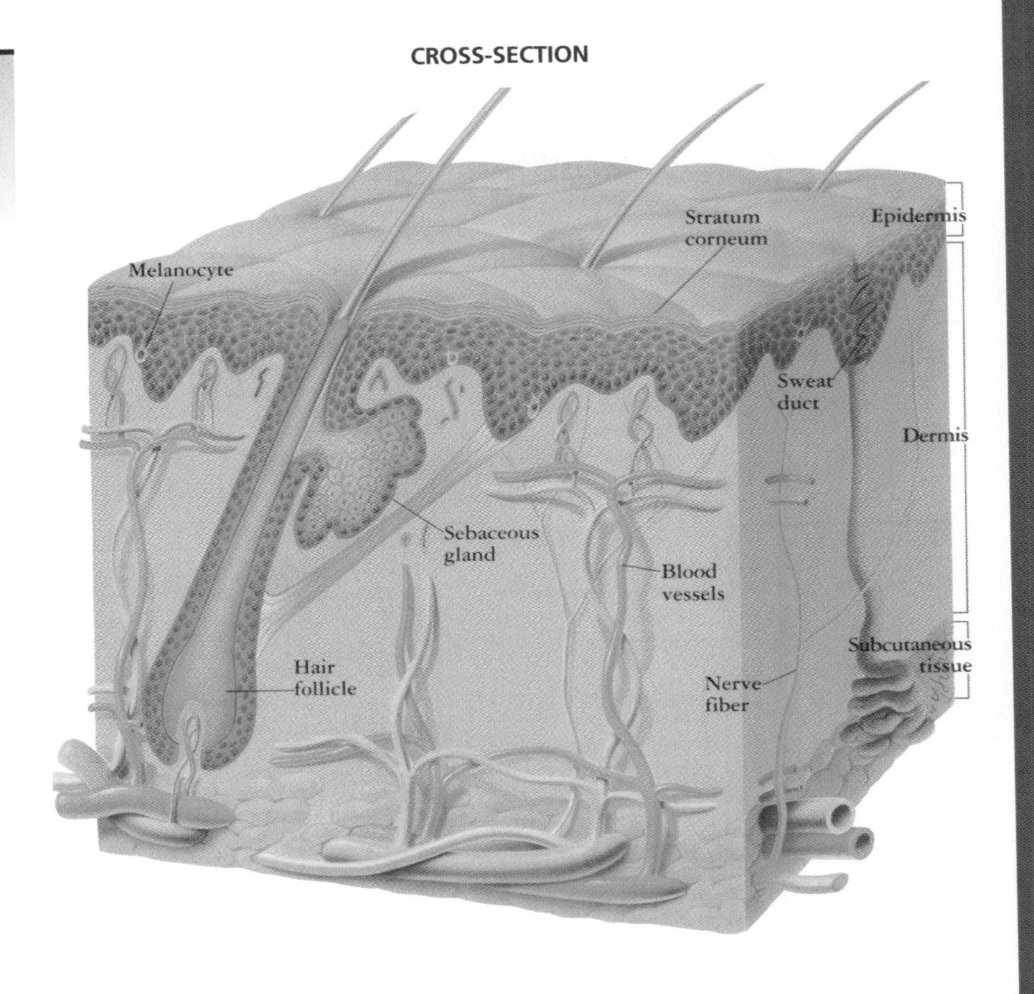

EXPLODED VIEW

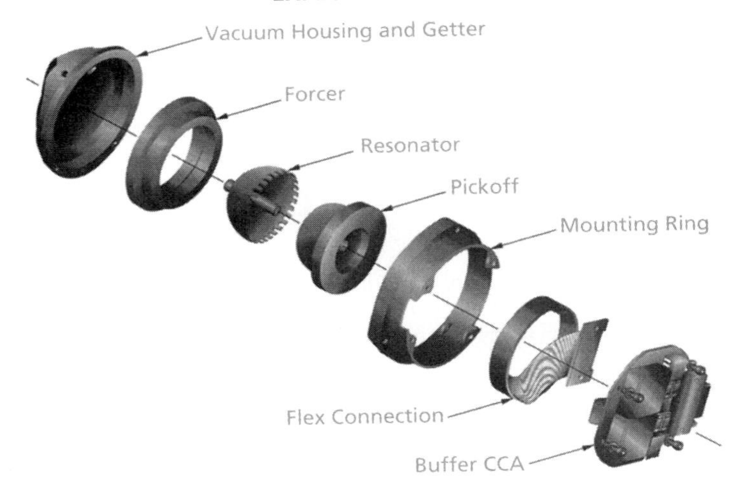

FIGURE VA.18

Drawing That Emphasizes Key Parts

This drawing helps car owners locate the reserve tank for radiator fluid.

Other parts are shaded to focus the readers' attention on the reserve tank.

The surrounding parts are sketched in vaguely to orient readers without confusing them with too much detail.

The drawing's angle of view approximates the one that owners would have when performing this task.

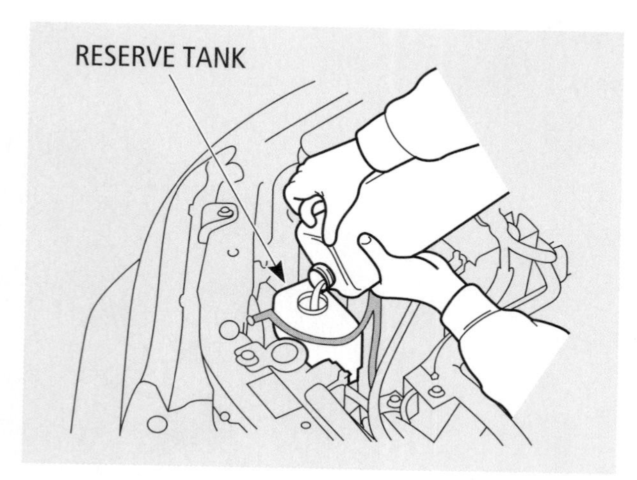

FIGURE VA.19

Screenshots

This screenshot shows what users will see when they open a blank worksheet in a new workbook using Microsoft Excel.

The first workbook is called Book1 in the title bar at the top of the window until it is saved with its own title.

The first cell into which data will be entered is outlined, as it appears in the application.

The entire spreadsheet is not shown.

The relevant sheet tab at the bottom of the workbook window is bolded, as it appears in the application.

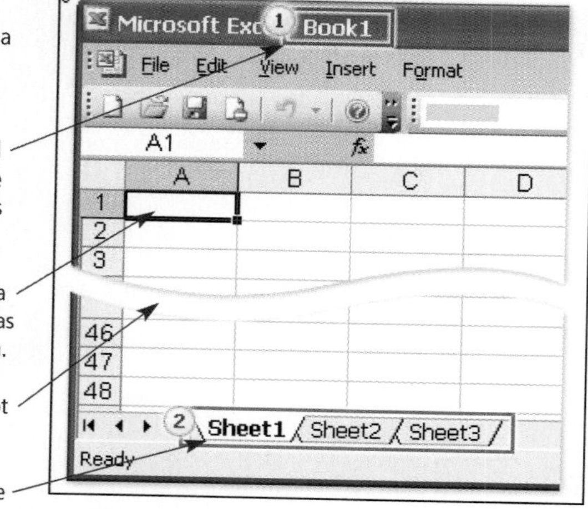

Figure 1 A blank worksheet in a new workbook.

1 The first workbook you open is called Book1 in the title bar at the top of the window until you save it with your own title.

2 Sheet tabs at the bottom of the workbook window.

GRAPHICS FOR SHOWING HOW TO DO SOMETHING

When writing instructions, you will often need to show your readers how to perform some task. Photographs and drawings provide excellent means of doing this.

PHOTOGRAPHS AND DRAWINGS

Photographs and drawings are often used with passages that explain a process. See pages 47–50.

For general advice about preparing these forms of graphics, see the preceding discussions of them (pages 444–50). Figure VA.20 presents a drawing that shows how to do something. It might have been replaced by a photograph with essentially the same design.

FIGURE VA.20

Drawing That Shows How to Do Something

This drawing and its caption work closely together to help the reader.

The drawing shows exactly how to hold the switch by using the same angle of view that readers would have when performing the task.

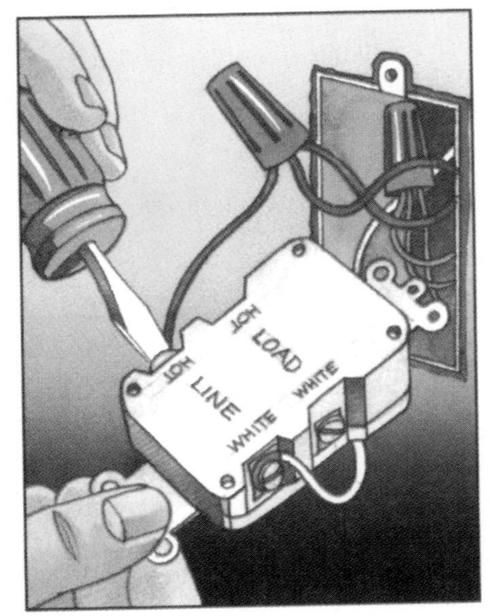

The caption provides additional detail. Note that the drawing shows the result of pigtailing all the black wires.

2 Pigtail all the black hot wires together and connect them to the terminal marked HOT LINE on the GFCI.

When preparing photographs and drawings for instructions, be especially careful to use the same angle of view that your readers will have and to include all details necessary to enable your readers to understand the action they are going to perform.

GRAPHICS FOR EXPLAINING A PROCESS

When you need to explain the succession of events in a process or procedure, consider using a flowchart or diagram. Flowcharts can be especially useful for describing processes that have a fixed beginning and a fixed conclusion. Diagrams work well for *ongoing* processes, such as the operation of the food chain.

FLOWCHARTS

Flowcharts are often used with passages that describe a process. See pages 47–50.

The simplest flowcharts use rectangles, circles, diamonds, or other geometric shapes to represent events and arrows to indicate the progress from one event to another. Sketches suggesting the appearance of objects also can be used (see Figure VA.21, page 452).

Some technical fields, such as computer science and engineering, have developed special techniques for their flowcharts (see Figure VA.22, page 453). They use agreed-upon sets of symbols to represent specific kinds of events and outcomes, and they follow agreed-upon rules for arranging these symbols on the page. If you are in a field that uses such specialized flowcharts, include them when addressing your fellow specialists but not when addressing people outside your specialty.

FIGURE VA.21
General Flowchart

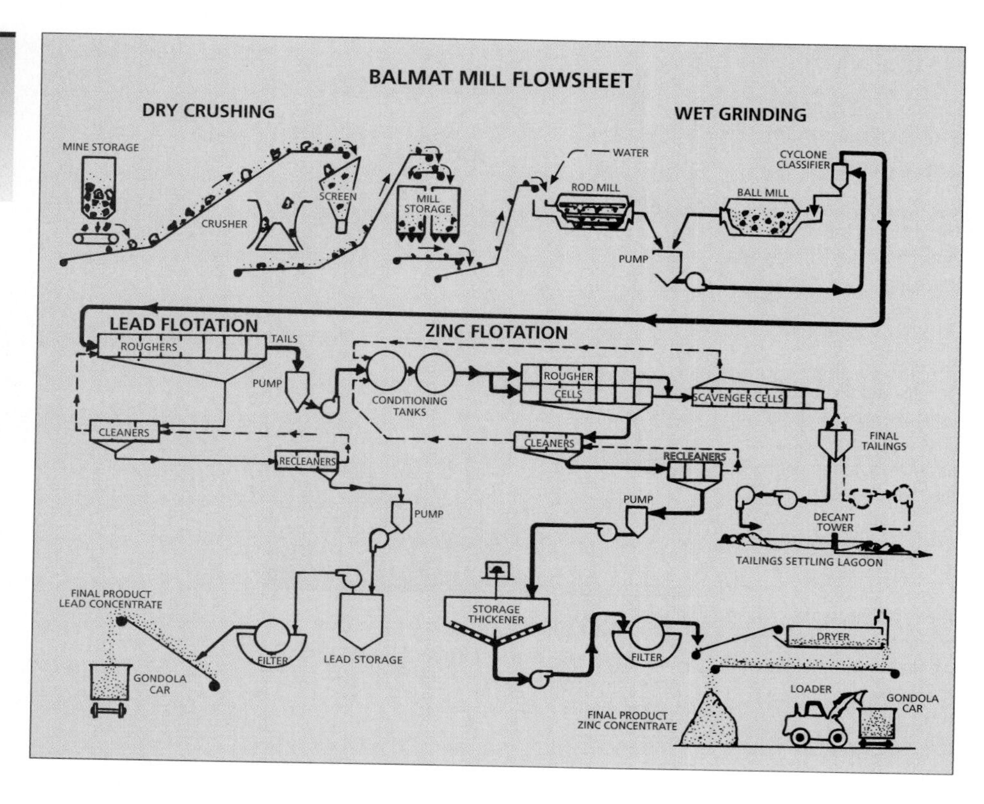

Because the key parts are clearly labelled, readers can understand this figure even if they don't understand some of the special symbols.

How to Create a Flowchart

A few conventions govern the design of flowcharts:

Guidelines for creating flowcharts

- **Place labels that identify the activities *inside* the boxes that represent those activities.**
- **Arrange the boxes so that activity flows from left to right, or from top to bottom.** If your flowchart continues for more than one line, begin the second and subsequent lines at the top of the chart or at the left-hand margin (as in Figure VA.21).

When you create a flowchart, be patient. It may take you several drafts to get the boxes the right size, to place the labels neatly inside them, and to arrange the boxes and arrows in an attractive and readily understandable way.

DIAGRAMS

Diagrams are often used with passages that describe a process. See pages 47–50.

Like flowcharts, diagrams can be used to explain a process. Some diagrams are very pictorial, almost like drawings, and others are quite abstract. For example, Figure VA.23 (page 454), which illustrates the carbon cycle, includes sketches of trees and an oil well, though it does not represent an actual scene.

How to Create a Diagram

The following suggestions will help you create diagrams that clearly explain processes to your readers:

FIGURE VA.22 Flowchart Showing Special Techniques Used in Computer Science and Engineering

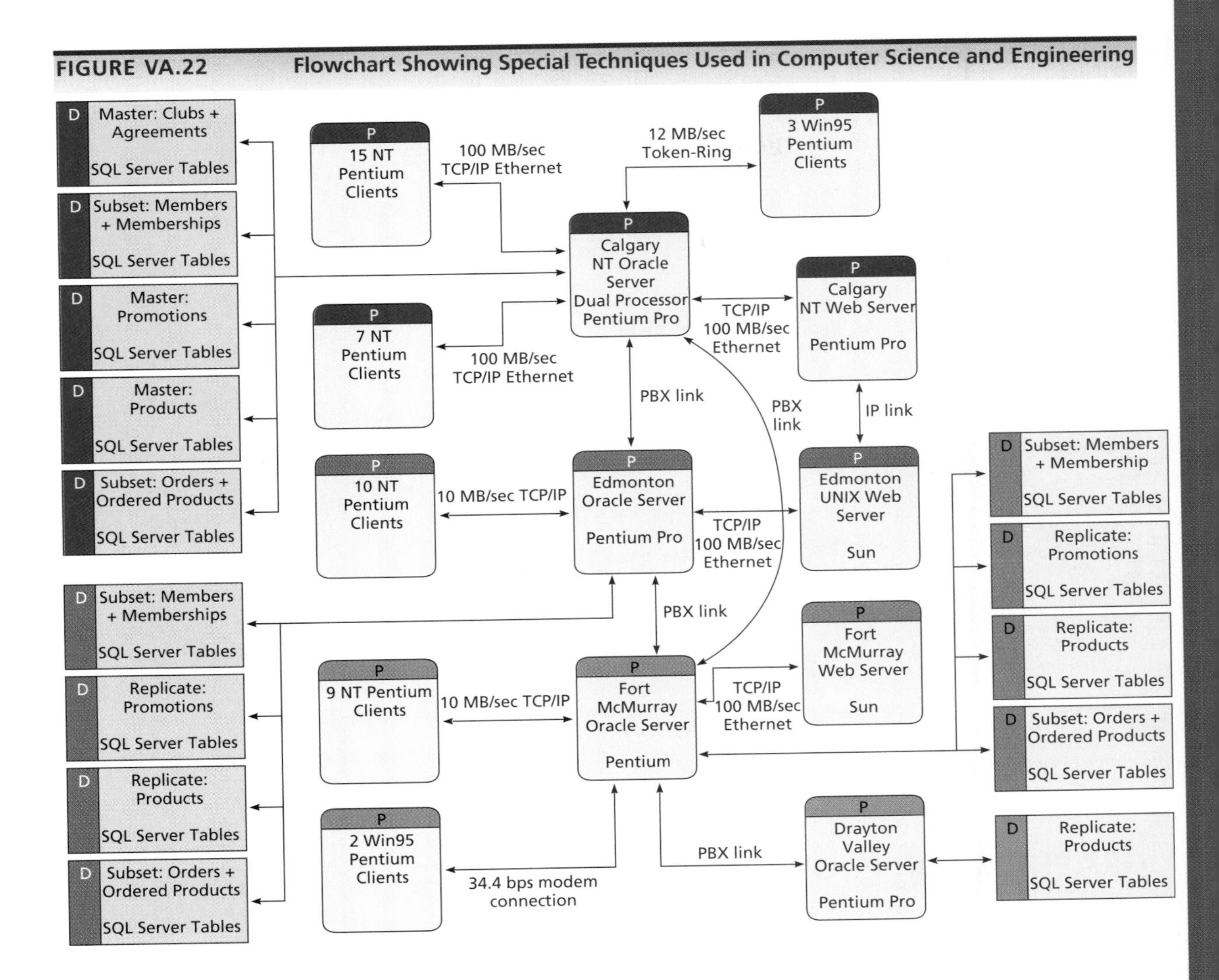

Much creativity is required to design effective diagrams.

- **Decide exactly what you want to show.** What are the events and objects that you want to show, and what relationships among them do you want to convey?
- **Create an appropriate means to represent your subject.** You can represent objects and events with geometric shapes or with sketches that suggest their appearance. You can show relationships among objects and events by the way you arrange the shapes and by drawing arrows between them. Figure VA.23 illustrates these techniques. When designing a diagram, you may find it helpful to examine communications similar to yours that were written by other people.
- **Provide the explanations readers will need in order to understand your diagram.** You may provide necessary explanations in the diagram itself, in a separate key, in the title, or in the accompanying text.

FIGURE VA.23 Diagram That Does Not Use Specialized Symbols

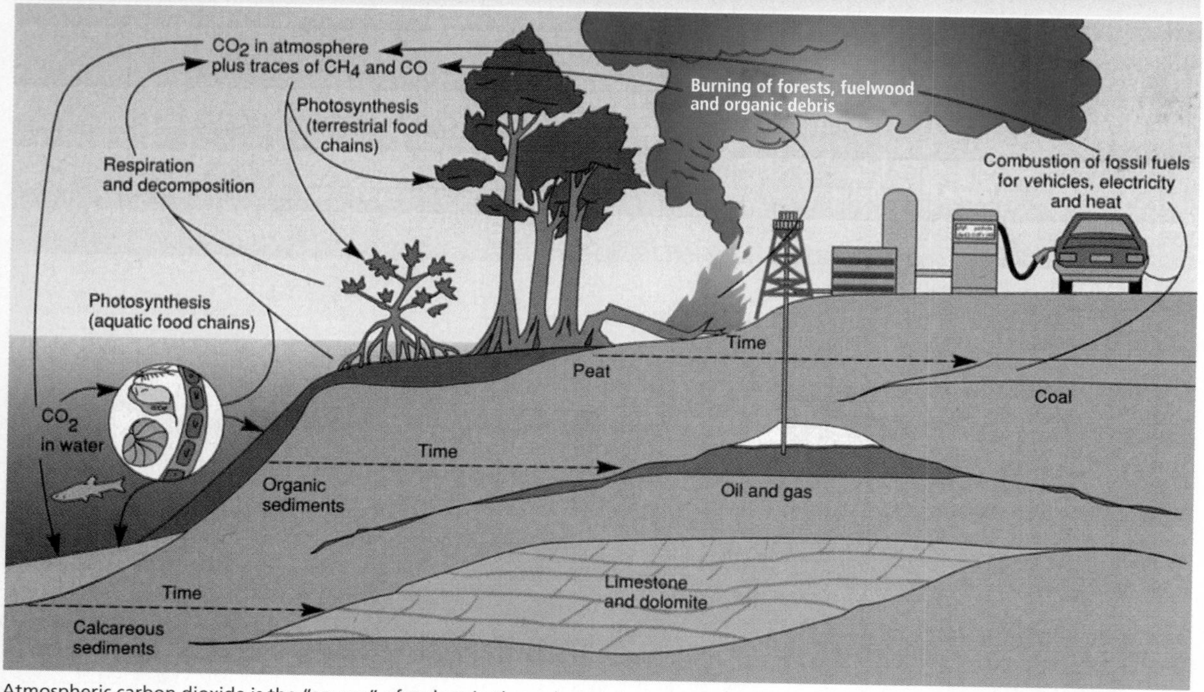

Atmospheric carbon dioxide is the "source" of carbon in the carbon cycle. It passes into ecosystems through photosynthesis and is captured in the bodies and products of living organisms. It is released to the atmosphere by weathering, respiration, and combustion. Carbon may be locked up for long periods in both organic (coal, oil, gas) and inorganic (limestone, dolomite) geological formations, which are, therefore, referred to as carbon "sinks."

GRAPHICS FOR PROVIDING MANAGEMENT INFORMATION

Writers often need to explain the managerial aspects of the projects they are proposing or reporting about. The graphics most often used for this purpose are organizational charts, schedule charts, and budget statements.

ORGANIZATIONAL CHARTS

An organizational chart uses rectangles and lines to tell how the people and departments in an organization are arranged into a managerial hierarchy. It shows, for example, how smaller units (such as departments) are combined into larger ones (such as divisions), and it indicates who reports to whom and who gives direction to whom (see Figure VA.24). Among the uses of organizational charts are these:

Uses of organizational charts

- To show the scope and arrangement of an organization
- To show the formal lines of authority and responsibility in an organization
- To provide a map of an organization so that readers can readily locate the people they want to contact

FIGURE VA.24

Organizational Chart

An organizational chart like this can be created very quickly with software.

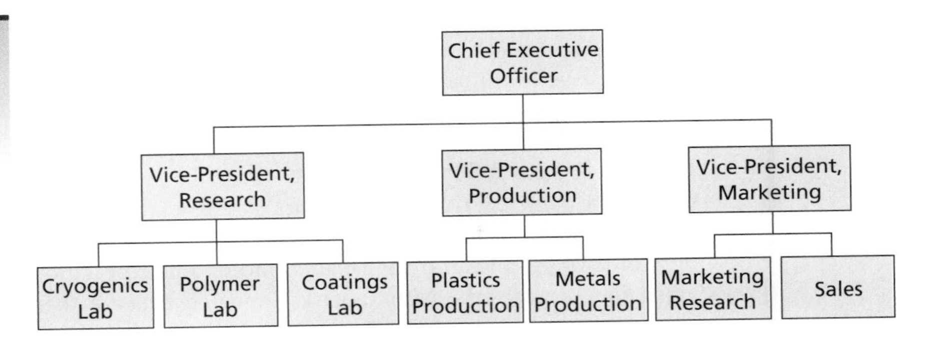

How to Create an Organizational Chart

Because of the hierarchical nature of most business organizations, organizational charts usually are pyramidal. You do not need to show every part of the organization, only those parts that are of interest to your readers. Sometimes, you may need to represent more than one kind of relationship. You can, for example, use solid lines to indicate relationships of direct authority and dashed lines to indicate close consultation or cooperation.

SCHEDULE CHARTS

A schedule chart identifies the major steps in a project and tells when they will be performed. As Figure VA.25 illustrates, a schedule chart enables readers to see what will be done, when each activity will start and end, and when more than one activity will be in progress simultaneously. (This type of schedule chart is called a *Gantt chart,* named for its inventor.)

FIGURE VA.25

Schedule Chart

Indenting is used to distinguish major tasks and subtasks.

Bars overlap because some tasks are performed simultaneously.

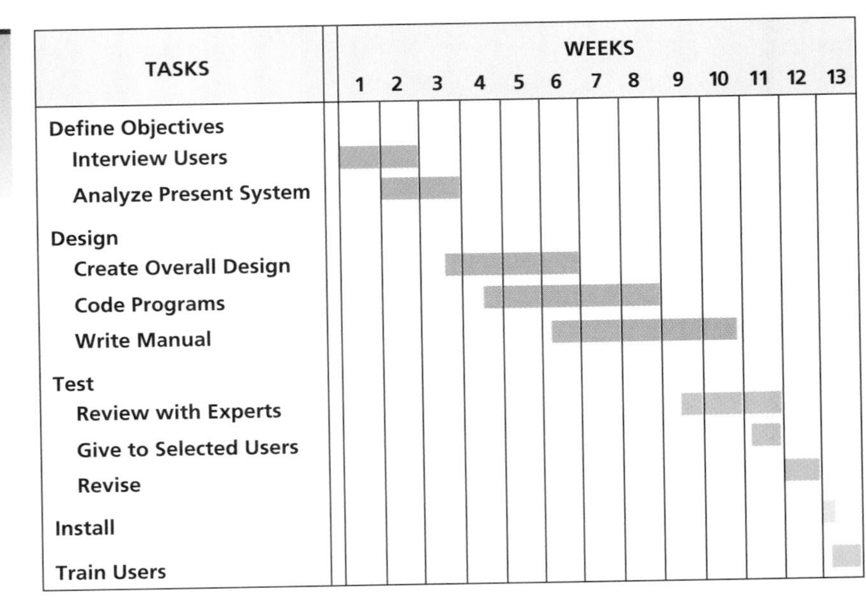

Schedule charts are often used in proposals to show the projected plan of work. You can also use them in progress reports to show what you have accomplished and what you still have to do, and in final reports to describe the process you followed while working on a completed project.

How to Create a Schedule Chart

One of the principal considerations in creating a schedule chart is deciding how much detail to include—something that you can determine only on a case-by-case basis in light of your purpose and your readers' needs and expectations. Tasks are always listed along the vertical axis, with subtasks indented beneath them. The intervals (weeks, months) are usually marked off with vertical lines to help readers see exactly when tasks begin and end.

BUDGET STATEMENTS

A budget statement is a table that shows how money will be acquired or spent. It may be either very simple or very elaborate.

On the job, you may use budget statements in the following situations:

Uses of budget statements

- **To explain the expenses involved in a project or purchase.** You might prepare a budget statement when requesting funds or when reporting on the financial feasibility of a particular course of action.
- **To summarize the savings to be realized by following a recommendation you are making.** Because profits are a major goal of most organizations, budget statements are often prepared for this purpose.
- **To report the costs that have been incurred in a project for which you are responsible.** Figure VA.26 shows an example of a budget statement prepared for this purpose.

FIGURE VA.26
Detailed Budget Statement

PROJECT COSTS	
Equipment	
4 KRN 3781 Robots ($37,000 apiece)	$148,000
1 Super server	6,500
4 Power supplies ($5,100 apiece)	20,400
Constructing (includes labour and supplies)	
Rewiring	7,100
Construction pads for mounting robots	3,000
Initial Programming (ten days at $125 per day)	1,250
Travel for Installation Personnel	
Airfare (two round trips)	400
Car rental	200
Living expenses (two people for five days at $150 per day)	1,500
TOTAL	$188,350

- **To explain the sources of revenue associated with some project or activity.** For instance, you may use a budget statement in reporting your department's sales of products and services.

How to Create a Budget Statement

The following steps will enable you to prepare simple, informative budget statements (when accountants and auditors prepare budget statements, they sometimes use specialized conventions that are not discussed here):

Guidelines for creating budget statements

1. Divide your page into two vertical columns.
2. In the left-hand column, list the major categories of *expense.* If your readers will want additional detail, list the principal expenses that make up each major category. Make the major categories most prominent visually.
3. In the right-hand column, write the *amount* of each expense. Align the figures on the decimal point.
4. Indicate the total.
5. To show *income* also, repeat steps 2 through 4.

EXERCISES

For additional exercises, visit www.techcomm.nelson.com.

Expertise

1. Figure VA.27 (page 459) shows two tables containing information about educational attainment and enrollments in Canadian universities. A writer might use that information in a variety of ways. For each of the uses listed here, decide whether the writer should present the information in a bar graph, pictograph, line graph, or pie chart.
   a. To show the educational attainment of women aged 15 and over in 2001.
   b. To enable readers to make a rough comparison of enrollments of women at each university level with enrollments of men at each university level in the 2000–01 year.
   c. To enable educational planners to see the proportion of women and men who are projected to be enrolled in the 2010–11 year at each level of study. To make these projections, assume that enrollment levels were 10% less in the 1980–81 year, and 5% less in the 1990–91 year, for both women and men than in 2000–01.

2. Figure VA.28 (page 460) shows world gas production in the years 1992 to 2002. Using that information, make the following graphics. Be sure to provide all appropriate labels and a title for each.
   a. A bar graph showing how much gas was produced in 2002 by the top five producers that year.
   b. A pictograph showing how much gas was produced by Canada in each of the five years from 1998 to 2002.
   c. A line graph showing trends in gas production in the following countries: Canada, the USA, the UK, Algeria, and the Netherlands.
   d. A pie chart showing the proportions of the world's gas produced in 2002 by the following countries: the former Soviet Union, the USA, Canada, the UK, Algeria, the Netherlands, and all other countries.

3. In textbooks, journals, or other publications related to your field of study, find three photographs. For each photograph, answer the following questions:
   a. What does the photograph show, who is the intended audience, and how is the reader supposed to use or be affected by the photograph?
   b. What angle of view has the photographer chosen, and why?
   c. What, if anything, has the writer done to eliminate unnecessary detail?
   d. Are all the relevant parts of the subject visible? If not, what is missing?
   e. Has the writer supplied helpful labels? Would any additional labels be helpful? Are any of the labels unnecessary?
   f. Could the writer have achieved his or her purpose more effectively by using a drawing or a diagram instead of a photograph? Why or why not?

4. Create a drawing to include in a set of instructions for operating one of the following pieces of equipment. Explain who your readers are and, if it isn't obvious, how they will use the equipment. Be sure to label the significant parts and include a figure title.

- An instrument or piece of equipment used in your field of study
- A mulching lawnmower
- The subject of a set of instructions you are preparing for your writing class
- Some other piece of equipment that has at least a half-dozen parts that should be shown in a set of instructions

5. Draw a diagram that you might use in a report, proposal, instruction manual, or other communication written on the job. The diagram might be an abstract representation of some object, design, process, or other subject. Use special symbols if you wish. Explain what use you might make of the diagram; explain also how your readers might use it. Be sure to provide all appropriate labels and a figure title.

6. Create a flowchart for one of the following processes and procedures. State when you might use the flowchart and why. Be sure to provide appropriate labels and a figure title. Show no more than sixteen steps. If the process calls for more, show the major steps and omit the substeps.

   - Applying for admission to your school
   - Changing a component in a car or computer
   - Making paper in a paper mill (begin with the trees in the forest)
   - Preparing to make an oral presentation in class or on the job. Your presentation should include graphics. Start at the point where you decide that you are going to speak or are given the assignment to do so.
   - Explaining a process or a procedure that you will describe in a communication you are preparing for your writing class
   - Explaining some process or procedure that is commonly used in your field

7. Create an organizational chart for some organization that has at least three levels—perhaps a club you belong to or a company that employs you. Or visit an office or a store and ask someone there to provide you with information about its organizational hierarchy.

8. Make a schedule chart for an assignment you are preparing for your writing class. Cover the period from the date you received the assignment to the date you will turn it in. Be sure to include all major activities, such as planning, gathering information, bringing a draft to class for review, and so on.

9. Create a budget statement using the following data on the monthly costs that an electronics company would incur if it opened a new service centre in a new city. Remember to group related expenses and to provide a total.

   Salary for centre manager $5,450. Rent $1,900. Business tax (prorated) $1,250. Electricity (year-round average) $875. Water $245. Administrative assistant's salary $2,100. Office supplies $200. Car lease $500. Salary for technician $3,000. Salary for technician's trainee $1,800. Supplies for technician $300. Travel for monthly trip to main office by centre manager $300. Car driving expenses $250. Internet and phone $330. Depreciation on equipment $1,700.

FIGURE VA.27 Tables for Use with Exercise 1

Educational Attainment of the Population, aged 15 and Over, 2001

	Women %	Men %	Total %
Less than Grade 9	10.2	9.4	9.8
Some high school	20.9	22.0	21.4
High school graduation	15.1	13.1	14.1
Trades certificate or diploma	11.0	10.7	10.8
Some postsecondary	7.8	14.1	10.9
Non-university certificate or diploma	20.2	14.6	17.4
University degree	14.9	16.0	15.4
Total	100.0	100.0	100.0
Total population (000s)	12,274.6	11,626.8	23,901.4

Full-time University Enrollment, by Level, 2000–01

	Women	Men	Total	Women as % of total enrollment
Bachelor's and first professional degree	280,821	208,638	489,459	57.4
Master's	24,702	23,031	47,733	51.8
Doctorate	11,114	13,301	24,415	45.5
Total enrollment	316,637	244,970	561,607	56.4

Source: Statistics Canada, Centre for Educational Statistics.

FIGURE VA.28 **Table for Use with Exercise 2**

World gas production
(bn cubic metres)

	2002$	2001$	2000	1999	1998	1997	1996	1995	1994	1993	1992
Former Soviet Union*		—	—	—	685.31	669.68	714.13	705.21	719.35	159.60	772.80
Former Soviet Union+	741.34	725.76	723.24	696.82	642.50	625.87	669.00	659.80	671.10	710.20	728.60
Of which: Russia*	595.30	581.60	584.20	589.48	590.68	571.10	600.30	594.90	606.80	617.60	640.10
Turkmenistan*	53.50	51.30	47.00	22.73	13.25	17.30	35.18	32.30	35.80	60.91	56.06
Uzbekistan*	57.00	57.00	56.40	50.27	49.55	50.24	48.20	48.00	47.20	41.98	39.93
Ukraine*	18.80	18.30	18.04	18.07	17.97	18.10	18.40	18.20	18.30	23.10	20.90
Azerbaijan*	5.15	5.52	5.66	5.98	5.59	5.96	6.36	6.64	6.38	6.34	7.37
Kazakhstan*	11.15	11.60	11.50	9.97	7.95	6.09	4.24	4.80	4.50	6.25	7.56
USA	538.75	548.61	537.62	531.05	532.64	535.24	538.18	526.66	530.85	521.20	505.10
Canada	183.50	186.81	183.19	176.80	173.26	165.58	164.08	158.66	149.12	139.00	128.00
Netherlands	71.24	72.26	67.75	75.00	80.44	81.84	89.65	78.35	78.41	84.00	83.00
UK	102.50	106.30	108.33	104.96	95.50	85.84	89.40	75.48	69.83	65.50	56.00
Indonesia	70.56	66.30	66.80	67.68	65.11	67.15	66.66	63.36	62.48	55.90	54.20
Algeria	84.05	83.75	83.24	83.19	76.36	68.90	62.13	58.10	50.30	53.90	55.80
Saudi Arabia	56.70	53.69	49.81	49.01	49.65	45.34	41.34	38.04	37.70	35.90	34.00
Iran	64.50	63.30	60.24	54.82	52.21	47.00	40.21	35.30	31.80	27.10	25.00
Norway	65.40	53.90	52.88	50.99	47.60	45.94	40.91	30.47	29.46	27.40	28.30
Malaysia	48.50	46.94	48.35	39.80	37.67	38.40	36.68	28.88	26.13	24.90	22.60
Australia	34.65	33.80	32.30	31.24	31.09	30.43	28.35	27.29	25.55	22.00	22.00
Mexico	35.31	35.31	35.82	36.44	35.85	31.70	31.00	26.61	25.85	25.40	26.20
Venezuela	29.80	31.71	30.47	30.70	31.03	28.14	27.20	25.20	24.81	23.10	21.60
Argentina	36.11	37.14	37.41	38.98	33.35	27.38	28.93	25.01	22.27	21.50	15.40
UAE	46.14	45.14	27.50	37.33	36.35	25.32	25.73	21.60	16.74	14.40	14.10
Italy	14.58	15.52	15.70	17.49	19.01	19.46	20.22	20.38	20.64	19.50	18.20
Germany	21.38	21.44	21.24	23.32	21.88	21.52	21.89	20.21	19.59	18.70	18.80
Romania	11.00	13.00	13.60	13.62	14.00	15.70	17.96	19.35	19.59	21.10	22.10
India	22.50	22.75	22.50	20.98	21.18	20.30	19.72	17.77	16.81	15.10	15.30
Pakistan	22.70	21.93	22.80	19.91	18.19	19.78	19.72	18.29	17.76	17.30	15.50
China	32.63	30.30	27.10	25.38	23.28	21.20	18.99	17.03	16.67	15.80	15.20
Qatar	29.30	27.90	29.10	23.79	19.43	17.40	13.70	13.50	13.50	13.50	12.60
Egypt	26.67	24.60	18.30	15.68	14.07	13.50	13.40	12.43	12.00	11.30	9.80
Thailand	19.10	18.37	18.62	19.46	17.60	15.23	12.13	10.41	9.68	8.80	7.80
Brunei	10.41	10.35	9.94	9.59	9.28	9.18	9.37	9.35	8.42	8.30	8.10
World Total	2581.60	2551.06	2492.31	2420.71	2364.25	2295.75	2305.47	2198.30	2167.70	2158.60	2105.70

* Russia and the CIS countries measure gas volumes at 20°C rather than the 15°C used by most other countries. At 20°C gas volumes are around 7% greater than at 15°C for 1992–1998.

+ Figures adjusted to 15°C to facilitate direct comparison.

CASE

INCREASING ORGAN AND TISSUE DONATIONS

You've been working for the past six months at Trillium Gift of Life Network (TGLN), a provincial agency responsible for organ and tissue donations in Ontario. Since working at TGLN, you have learned that more than 2,000 people have their lives enhanced or saved by tissue transplants annually in this province. Moreover, each donor may provide up to six organs and tissue for up to seventy-five recipients.

However, you have also learned that at any given moment, another 2,000 people in Ontario are on a waiting list for organ donations and transplants. To discover ways of increasing the number of organ and tissue donors in the province, Jinny Chu, executive director of TGLN, asked staff member Aaron Weinstein to report on a recent Tactic Ltd. survey of attitudes toward organ and tissue donation. However, Aaron suddenly left TGLN to take another job. Jinny has asked you to prepare the report using the information Aaron had collected prior to his departure. The information includes the following data and notes.

YOUR ASSIGNMENT

According to your professor's directions, do one of the following:

A. For one of the survey questions, or for a group of the questions identified by your professor, state the important conclusion(s) that you draw(s), explain the evidence that supports your conclusion, and make a recommendation.

B. Using all of the questions or a group identified by your professor, write a full report to Jinny. Include
   - An introduction
   - A brief introduction of the survey method used
   - The key conclusions that you draw, and the specific results that support them
   - Your recommendation.

Notes Aaron Left

- Largest provincial survey conducted on attitudes toward organ and tissue donation
- Survey conducted by Tactic Ltd.
- Procedures used make results representative of the provincial adult population
- Telephone survey of 800 Ontarians during one week in March 2006
- Survey conducted using random digit dialing technique
- Everyone is a potential organ and tissue donor, regardless of their age
- If you are 16 years of age or over, you can register your intent with a centralized provincial data bank
- People can also sign an organ donor card to indicate their desire to donate organs
- Next of kin's final consent is always needed even if a person has indicated a desire to donate organs
- Studies show that donating the organs and tissues of a loved one who has died can provide immediate comfort and long-lasting consolation to family members in their grieving
- Recovery of organs and tissue does not interfere with funeral practices, and no one will know about the donation unless the family communicates the information
- Living organ donation of a kidney or part of a liver allows a donor to live an active life

Survey Questions

1. Do you support or oppose the donation of organs and tissue for transplants?

Support	85%
Oppose	6%
Don't know	9%

2. Are you willing or unwilling to donate your own organs and tissue for transplant purposes?

Willing	77%
Unwilling	13%
Don't know	10%

3. If you are unwilling, is there a particular reason you are not likely to want to have your organs or tissue donated upon your death? What might that reason be?

Medical reasons	13%
Believe I'm too old	10%
Don't want body cut up	9%
Don't feel right about it	6%
Against religion	5%
No reason/don't know/haven't given it much thought	57%

4. Would you be likely to consent to donating a family member's organs or tissue if a request has been made by that family member?

Very likely	85%
Somewhat likely	8%
Somewhat unlikely	3%
Very unlikely	4%

5. Would you be likely to consent to donating a family member's organs or tissue if you were unsure of the family member's donation wishes?

Very likely	10%
Somewhat likely	66%
Somewhat unlikely	15%
Very unlikely	9%

6. Have you signed an organ and tissue donor card?

Yes	53%
No	47%

7. If you have not signed an organ and tissue donor card, would you be willing to do so?

Yes	45%
No	55%

8. If you are willing to donate your organs or tissue, have you told some member(s) of your family about your wish to donate after your death?

Yes	55%
No	45%

9. Have you participated in a family discussion about organ and tissue donation?

Yes	44%
No	56%

10. If you have participated in a family discussion about organ and tissue donation, have all the participants decided to donate?

Yes	47%
Undecided	51%
No	2%

11. Are you interested in learning more about organ and tissue donation?

Yes	69%
Somewhat	17%
No	14%

12. How do you perceive the need for organ transplants, on a scale of 1 (little or no need) to 5 (extreme need)? Score 4.5

13. In the past year, have you read, seen, or heard any information about organ and tissue donation?

Yes	58%
No	42%

14. If you have read, seen, or heard information about organ and tissue donation, where did you receive that information?

From a health care professional	43%
From a religious or community organization	19%
From a funeral director	12%
From a magazine, newspaper, or print media	10%
From the radio or television	7%
From the Internet	5%
Other	4%

15. Do you believe that you can be a living organ donor and live an active life?

Yes	71%
No	25%
Don't know	4%

PART VI

Evaluating a Communication

CHAPTER 18

Checking, Reviewing, and Testing Drafts

REFERENCE GUIDE 6

Four Steps for Revising

CHAPTER 18 Checking, Reviewing, and Testing Drafts

GUIDELINES

Guidelines for Checking

1. Check from your readers' point of view
2. Check from your employer's point of view
3. Distance yourself from your draft
4. Read your draft more than once, changing your focus each time
5. Use software to find (but not to cure) possible problems

Guidelines for Reviewing

1. Discuss the objectives of the communication and the review
2. Build a positive interpersonal relationship with your reviewers or writer

3. Rank suggested revisions—and distinguish matters of substance from matters of taste
4. Explore fully the reasons for all suggestions
5. Use software for reviewing in a reader-centred way

Guidelines for Testing

1. Establish your test objectives
2. Pick test readers who truly represent your target readers
3. Focus on usability: Ask your test readers to use your draft the same way your target readers will
4. Focus on persuasiveness: Learn how your draft affects your readers' attitudes
5. Interview your test readers after they've read and used your draft
6. Avoid biasing your test results
7. Test early and often
8. Ethics Guideline: Obtain informed consent from your test readers

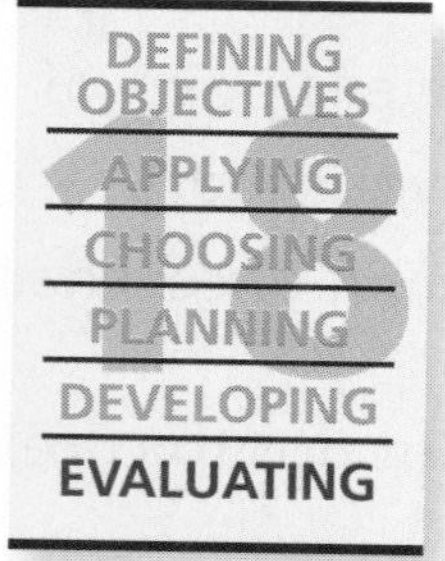

Successful workplace writers know that even though they follow a thoroughly reader-centred approach when drafting, they can still increase the usability and persuasiveness of almost any draft they create. Consequently, before sending their communications to their readers, these writers pause to evaluate what they've produced so far, looking for ways to improve it. Some pause to evaluate each part of their draft as they finish it. Others wait until they've completed an entire draft before evaluating. The most important point for any writer is not *when* to evaluate, but *how* to do so.

This chapter describes reader-centred strategies that will help you develop your expertise at using the three types of evaluation that are most common in the workplace:

- **Checking.** You carefully examine your draft yourself.
- **Reviewing.** You ask for advice, principally about your draft's usability and persuasiveness, from people who are not part of your target audience.
- **Testing.** You ask members of your target audience to use your draft so you can determine how usable and persuasive they find it to be.

WWW For additional chapter resources, visit Chapter 18 at www.techcomm.nelson.com.

Reference Guide 6 describes strategies for revising your drafts in light of what you learn from checking, reviewing, and testing them.

GUIDELINES FOR CHECKING

The first thing to know about checking over your own drafts is this: It is extremely difficult to recognize problems in your own writing. Several obstacles hamper your ability to do so. The following guidelines identify these obstacles and suggest ways of overcoming them.

GUIDELINE 1 Check from Your Readers' Point of View

Writers sometimes create the first obstacle by defining the focus of their checking too narrowly. They concentrate on spelling, punctuation, and grammar, forgetting to consider their communication from the perspective of readers. Consequently, you should always begin your checking by reminding yourself of your communication's objectives. If you wrote them down, pull out your notes. If you recorded them mentally, review them now. What tasks do you want your communication to help your readers perform? How do you want it to alter their attitudes?

Then, with your communication's objectives fresh in mind, read your draft while imagining your readers' moment-by-moment responses. As they proceed, how usable will they find each section, paragraph, and sentence to be? How persuasive will they find each of the writing strategies by which you hope to shape their attitudes? Look especially for places where the readers may not respond in the way you desire. That's where you need to revise.

GUIDELINE 2 Check from Your Employer's Point of View

In many working situations, you must consider your communication not only from your readers' perspective, but also from your employer's. Here are three questions you should ask:

Things to consider when evaluating from your employer's perspective

- **How will my communication impact others in the organization?** By anticipating conflicts and objections that your communication might stir up, you may be able to reduce their severity or avoid them altogether.
- **Does my communication promise something on my employer's behalf?** For example, does it make a commitment to a client or customer? If so, ask yourself whether you have the authority to make the commitment, whether the commitment is in your employer's best interest, and whether the commitment is one your employer wishes to make.
- **Does my communication comply with my employer's policies?** First, check your draft against your employer's policies on writing style and format. Second, determine whether it complies with the employer's regulations concerning what you can say. For example, for legal reasons, your employer may want to restrict when and how you reveal information about a certain aspect of your work. This might happen, for instance, if you are working on a patentable project or a project regulated by a government body or agency.

GUIDELINE 3 Distance Yourself from Your Draft

When we check, we often see what we intended, not what actually appears.

Another obstacle to effective checking is that we are too "close" to what we write. Because we know what we meant to say, when we check for errors we often see what we intended to write rather than what is actually there. A word is misspelled, but we see it spelled correctly. A paragraph is cloudy, but we see clearly the meaning we wanted to convey.

To distance yourself from your draft, try the following measures:

Ways to distance yourself from your draft

- **Let time pass.** As time passes, your memory of your intentions fades. You become more able to see what you actually wrote. Set your draft aside, if for only a few minutes, before checking it.
- **Read your draft aloud, even if there is no one to listen.** Where you stumble when saying your words, your readers are likely to trip as well. Where the "voice" you've created in your draft sounds a little off to you, it will probably sound wrong to your readers, too.

GUIDELINE 4 Read Your Draft More Than Once, Changing Your Focus Each Time

Yet another obstacle to effective checking is summed up in the adage, "You can't do two things at once." To "do" something, according to researchers who study the way humans think, requires "attention." Some activities (like walking) require much less attention than others (like solving calculus problems). We can attend to several more or less automatic activities at once. However, when we are engaged in any activity that requires us to concentrate, we have difficulty doing anything else well at the same time (J. R. Anderson, 1995).

This limitation has important consequences. When you concentrate on one aspect of your draft, such as the spelling or consistency of your headings, you diminish your ability to concentrate on the others, such as the clarity of your prose.

Read separately for substantive and mechanical matters, and for shareholders' perspective.

To overcome this limitation on attention, check your draft at least twice, once for substantive matters (such as clarity and persuasive impact) and once for mechanical ones (such as correct punctuation and grammar). If you have time, check your draft again to ensure that you have kept your stakeholders' concerns in mind.

GUIDELINE 5 Use Software to Find (But Not to Cure) Possible Problems

Word processing software offers a variety of tools that can help you check your drafts:

- **Spell checkers.** Spell checkers identify possible misspellings by looking for words in your draft that aren't in their dictionaries.
- **Grammar checkers.** Grammar checkers identify sentences that may have any of a wide variety of problems with grammar or punctuation.
- **Style checkers.** Style checkers analyze various aspects of your writing that are related to style, such as average paragraph, sentence, and word length. Using special formulas, they also compute scores that attempt to reflect the "readability" of a draft.

All three types of tools can be helpful, if used with caution. They bring problems to your attention and even suggest possible solutions. But all three also have a serious shortcoming: They just don't "know" enough.

After using a spell, grammar, or style checker, read over your draft carefully yourself.

For example, spell checkers ignore words that are misspelled but look like other words, such as *forward* for *foreword* or *take* for *rake*—a shortcoming that can have serious consequences if you don't also perform your own careful proofreading. Similarly, grammar checkers can't tell whether you are using the word *fly* as a noun (the insect) or a verb (the way birds travel), nor can they distinguish a good use of the passive voice from a poor one. And style checkers can't tell a clearly written long sentence from a murky short one. Moreover, they don't even consider such writing problems as poor organization, poor use of topic sentences, or poor use of headings.

WWW To download a checklist for checking your drafts, go to www.techcomm.nelson.com.

Because of these and similar limitations, use software tools for checking with great care. Don't accept any of their suggestions without careful review. Always read over your drafts carefully yourself.

GUIDELINES FOR REVIEWING

Reviews enable employers to ensure the quality of the communications their employees write.

In reviewing, writers give their drafts to someone else to look over. You may be familiar with this process through the peer reviewing often done in classes. At work, writers are often required to have their drafts reviewed to ensure that their communications are well written, conform to the employer's policies, and say things that serve the organization's interests. For routine communications, reviews are usually conducted by the writer's manager. For sensitive communications, the reviewers may also include company executives, lawyers, public relations specialists, and many others. Commonly, the people performing required reviews usually have the authority to demand changes.

Reviews can help you learn to write more effectively.

Even when a review is not required, employees often ask coworkers or managers to look over their drafts. Through these voluntary reviews, employees learn additional ways to increase the usability and persuasiveness of their communications.

You can influence the quality of the reviews of your drafts.

The number of changes you may be required to make and the helpfulness of the suggestions you receive will depend largely on the way you manage your interactions with your reviewers. The five guidelines below will help you develop expertise at eliciting supportive and helpful reviews of your drafts.

You should learn how to give helpful reviews of other people's drafts.

Several of the guidelines also include suggestions for reviewing other people's drafts effectively. When you are promoted to a supervisory or managerial position, the ability to review other people's drafts will be essential. Even when you are not a manager, you will almost surely be asked to "look over" someone else's draft, perhaps as early as your first months at work. Moreover, you will be evaluated on the value of the advice you offer.

The guidelines are as useful for the peer reviewing done in class as for the reviewing conducted in the workplace.

GUIDELINE 1 Discuss the Objectives of the Communication and the Review

Reviewing needs to be as reader-centred as any other writing activity. Don't just hand your draft to reviewers with a request that they "look it over." Share the detailed information you developed while defining your communication's objectives. Describe the tasks you want to help your readers perform, the ways you want to influence your readers' attitudes, and the facts about your readers and their situation that will affect the way they respond. Tell the reviewers who your stakeholders are and how you think these individuals might be affected by your communication. Ask them to help you discover other stakeholders or impacts that you might have missed. Only if they have this information can reviewers suggest effective strategies for increasing your communication's usability and persuasiveness.

Also discuss the scope and focus of the review. As the writer, you may have a good sense of areas that most need attention, whether organization, selection of material, accuracy, tone, page design, or something else. Without discouraging your reviewers from noting other kinds of improvement, guide them to examining what you think is most critical.

Similarly, if you are the reviewer and your writers don't tell you their communications' objectives or their desires for the reviews, ask for this information. You need it in order be able to give the best possible assistance.

Build a Positive Interpersonal Relationship with Your Reviewers or Writer

Emotions are an important element in reviewing.

The relationship between writers and their reviewers is not only intellectual but also interpersonal and emotional. The quality of this human relationship can greatly affect the outcome of the review process. When writers feel criticized and judged rather than helped and supported, they become defensive and closed to suggestions. When reviewers feel their early suggestions have been rebuffed rather than welcomed, they cease giving the advice the writer needs. The following strategies will help you build positive, productive relationships with your reviewers and writers.

When You Are the Writer

Ways to encourage your reviewers to help as much as they can.

As a writer, interact with your reviewers in ways that encourage them to be thoughtful, supportive, and generous. Reviewers are more helpful if they believe you welcome their comments. Throughout your discussions with them, project a positive attitude, treating your reviewers as people who are on your side, not as obstacles to the completion of your work. Begin meetings by thanking them for their efforts. Paraphrase their comments to show that you are listening attentively and express gratitude for their suggestions.

When your reviewers offer comments, stifle any temptation to react defensively—for instance, by explaining why your original is superior to a suggested revision. Even if a particular suggestion is wrongheaded, listen to it without argument. On the other hand, don't avoid dialogue with your reviewers. When they misunderstand what you are trying to accomplish, explain your aims while also indicating that you are still open to suggestions. If your reviewers misunderstood what you were attempting, it's probably a sign that you aren't being clear and would benefit from their advice.

When You Are a Reviewer

When you are the reviewer, use these strategies to build a good relationship with the writer:

Ways to encourage openness to your suggestions

- **Begin with praise.** By doing so, you show the writer that you recognize the good things he or she has done, and you indicate your sensitivity to the writer's feelings. In addition, some writers don't know their own strengths any better than they know their weaknesses. By praising what they have done well, you encourage them to continue doing it and you reduce the chances they will weaken strong parts of their communication by revising them.

- **Focus your suggestions on goals, not shortcomings.** For example, instead of saying, "I think you have a problem in the third paragraph," say, "I have a suggestion about how you can make your third paragraph more understandable or persuasive to your readers."
- **Use positive examples from the writer's own draft to explain suggestions.** For example, if the writer includes a topic sentence in one paragraph but omits the topic sentence in another, cite the first paragraph as an example of a way to improve the second. By doing this, you indicate that you know the writer understands the principle but has slipped this one time in applying it.
- **Project a positive attitude toward the writer.** Even if you have the authority to demand changes, think of the writer as a person you want to help, not judge. If the writer feels you are judgmental, he or she will almost surely resent and resist the revisions you suggest.

GUIDELINE 3 Rank Suggested Revisions—and Distinguish Matters of Substance from Matters of Taste

By ranking your suggestions, you help both the writer and yourself work effectively.

When you review a writer's draft, one of the most helpful things you can do is rank your suggestions. By doing so, you help the writer decide which revisions will bring the greatest improvement in the least amount of time. Also, many writers can face only a limited number of suggestions before feeling overwhelmed and defeated. As a reviewer, look over your suggestions to determine which will make the greatest difference. Convey those to the writer. Keep the rest to yourself, or make it clear that they are less important. Ranking also helps *you* work productively. Often, you will not have time to review an entire draft in detail. Begin by scanning the draft to identify the issues that most need attention, then focus your effort on them.

Distinguish matters of substance from matters of taste.

When ranking suggestions, distinguish those based on substantial principles of writing from those based on your personal taste. If you suggest changes that would merely replace the writer's preferred way of saying something with your preferred way, you will have done nothing to improve the writing—and you will almost certainly spark the writer's resentment. Unfortunately, it can be difficult to determine whether we like a certain way of saying something because it is good in itself or because it matches our own style. When you are faced with this uncertainty, try to formulate a reason for the change. If all you can say is, "My way sounds better," you are probably dealing with a matter of taste. If you can offer a more objective reason—for instance, one based on a guideline in this book—you are dealing with a matter of substance.

One caution, however. Sometimes your sense that something doesn't sound right is a clue that a problem exists. For example, you may stumble over a sentence that doesn't sound right and discover, after closer examination, that it contains a grammatical error. If something sounds wrong, see whether you can find an objective reason for your dissatisfaction. If so, tell the writer. If not, let the matter drop.

So far, the discussion of this guideline has focused on the reviewers' perspective. When you are the writer, you can be helped greatly by your reviewers' rankings of their suggestions. If your reviewers don't volunteer a ranking, request one.

GUIDELINE 4 Explore Fully the Reasons for All Suggestions

Explanations can be more valuable than the suggestions themselves.

The more fully writers and reviewers explore the reasons for the reviewers' suggestions, the more helpful the review will be.

When you are acting as a reviewer, it's in your interest to provide a full explanation of your suggestions. If you don't explain the reasons for your suggestions, the writer may think that you are simply expressing a personal preference. As a result, the writer may dismiss your suggestions, leaving the draft weaker than it might otherwise have been.

Full explanations also help the writer learn to write better. For example, imagine that you suggest rephrasing a sentence so that the old information is at the beginning and the new information is at the end. The writer may agree that your version is better but may not know the reason why unless you explain the *principle* that you applied.

Some reviewers withhold such explanations because they think the reasons are obvious. However, reasons that are obvious to the reviewer are not necessarily obvious to the writer. Otherwise, the writer would have avoided the problem in the first place.

Phrase your suggestions from the intended readers' viewpoint.

Whenever possible, phrase your suggestions from the perspective of the intended readers. Instead of saying, "I think you should say it like this," say, "I think your intended readers will be able to understand your point more clearly if you phrase it like this." Such a strategy will help the writer take a reader-centred approach to revising the draft. It will also help you present yourself not as a judge of the writer's writing, but as a person who wants to help the writer achieve his or her communication objectives.

GUIDELINE 5 Use Software for Reviewing in a Reader-Centred Way

Word processing software includes features designed to support communication among writers and their reviewers.

For example, a feature to track changes allows reviewers to highlight insertions or deletions and to insert comments addressed to the writer.

Furthermore, some software includes an online collaboration feature that allows writers and reviewers to simulateneously communicate on a document during online meetings.

When you use these and similar tools, remember that you must write your review comments with the same attention to usability and persuasiveness that you devote to anything else you write on the job. While the word processing software will make your comments easy for the writer to find, make sure you explain suggestions succinctly but supportively. Long explanations can be difficult for writers to use as they try to revise efficiently. However, don't let brevity make you curt. In your written comments, be as considerate and diplomatic as you would be in a conversation.

GUIDELINES FOR TESTING

Testing lets you see how readers actually respond to your draft.

In testing, you ask one or more people, called *test readers,* to read and use a draft of your communication in the way your intended readers will read and use your finished communication. Then, by various means, you gather information from your test readers that will enable you to predict how other readers will probably respond. If your test is constructed effectively, where things worked well for your test readers, they will probably work well for your target readers. Where the test readers encounter problems, your target audience probably will, too.

The simplest tests occur when you are addressing only one person or a few people. In this situation, you can actually give your draft to your entire target audience and ask for feedback. Testing becomes more complex when you are preparing communications such as instructions, websites, and brochures that are addressed to a larger audience so that you can't give your draft to each person in it. To test such communications, some companies employ testing specialists and build sophisticated testing facilities. Even without such facilities, however, you can use the same basic strategies that testing experts employ. Testing a draft using the following seven guidelines will enable you to identify problem areas and make revisions before you give the final version to your readers. Then a final guideline will help you ensure that you treat your test readers ethically.

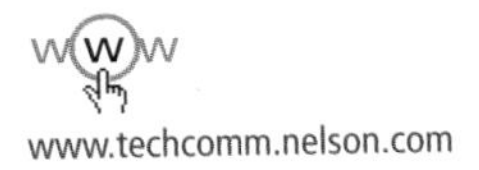

GUIDELINE 1 Establish Your Test Objectives

At a general level, all your tests will aim to answer two questions:

- **How can I improve my draft?** In the workplace, most tests emphasize the diagnostic goal of learning ways the draft can be improved.
- **Is my communication good enough?** It is always helpful to learn how close your draft is to a "final" draft you can deliver to your target audience.

Whatever your overall emphasis, your first step in creating a test is to decide exactly what you want to find out so that you can tailor your test to provide the specific information you need. Here are three steps for defining your test's objectives.

Defining Your Test's Objectives

1. **Review your communication's usability and persuasiveness objectives.** These objectives tell exactly how your communication is intended to affect your readers while they are reading it. Your test should help you determine where your draft succeeds in achieving these objectives and where it needs improvement.
2. **Identify the draft's features you hope will contribute most to its usability and persuasiveness.** For longer communications, you may not be able to test everything. Therefore, you need to focus on what you believe will have the greatest impact on your target readers' responses.
3. **Identify any features about which you feel unsure.** Your test should settle the issue.

You can encapsulate your test's objectives in a set of questions you want your test to answer. How well does my draft support my readers as they perform their first task? Their second one? How effective is it in influencing their attitudes in the way I want? How well do the headings work? Are the technical terms I used in the introduction understandable to them?

If you wish to determine whether your draft is "good enough," you may also want to establish measurable criteria for your test. For example, if you are drafting an instruction manual, you might decide that the manual is good enough if it enables test readers to perform their task in less than two minutes or with only one error that is easily corrected. However, in many cases your subjective judgment will be fully adequate: Your draft is good enough if it enables your test readers to perform their tasks without much difficulty and if they say your draft had a positive impact on their attitudes.

GUIDELINE 2 Pick Test Readers Who Truly Represent Your Target Readers

The selection of test readers is a critical part of test design.

To construct a test that will accurately predict how your target audience will respond to your finished communication, you must pick test readers who truly represent your target readers. If you are writing instructions for plumbers or computer programmers, choose plumbers or computer programmers for your test readers. If you are creating a website for adults who suffer from asthma, choose adult asthma sufferers. You could actually make your communication *less* effective for your intended readers if you made revisions based on responses from people who are not in your target audience.

If it is impossible to pick test readers from your target audience, choose people as similar as possible to that audience. In a class, for instance, you may be writing a report to engineers who design robots. If you can't enlist such engineers as test readers, ask engineering seniors who've had coursework in robotics. Consider especially your target readers' familiarity with your subject. There's no point in using test readers who already know what your communication is intended to tell them. They will be able to move past problematic parts based on previous knowledge, so you will not be alerted to things that need fixing. On the other hand, if your test readers have a significantly lower level of knowledge about your subject than your target readers do, your test results could induce you to include background information and explanations your target readers neither need nor want.

Even a single test reader can provide valuable insights.

How many test readers are enough? Because there is generally little variation in the ways readers respond to step-by-step instructions, two or three test readers may be sufficient for them. In contrast, people vary considerably in the ways they interpret and respond to a report or informational website. Common practice is to use about a dozen test readers for such communications. However, testing with even a single reader is infinitely better than no testing at all.

GUIDELINE 3 Focus on Usability: Ask Your Test Readers to Use Your Draft the Same Way Your Target Readers Will

Whenever you test a draft, one of your primary goals will be to evaluate its usability for its target readers. To learn where you need to make revisions, you must construct a test in which the test readers use your draft to perform the same tasks your target readers will want to perform with the finished communication.

At work, most readers' tasks usually fall into three major categories.

- **Perform a procedure,** as in reading instructions.

- **Locate information,** as in looking for a certain fact in a reference manual or website.
- **Understand and remember content,** as when trying to learn about something through reading.

For many communications, all three types of tasks are important. For example, when people consult the manual for a software package, they usually want to *locate* directions for a particular procedure, *understand* the directions, and then use the directions as a guide while they *perform* the procedure. Although the three types of tasks are often bound together, each can be tested in a different way. Consequently, the following sections discuss separately the design of performance, location, and understandability tests.

Performance Tests

To test a draft's effectiveness at helping its readers perform a procedure, you could simply give the draft to your test readers and watch them use it. To get the best results, however, you must construct several elements in your test so that your test readers' reading situation resembles as closely as possible the situation of your target readers. The following strategies will help you plan the four major elements of *performance tests* designed to achieve this goal.

Strategies for Performance Tests	
Tasks	Ask your test readers to perform the same tasks your target readers will perform.
Location	Conduct the test in the same setting as the target readers would use.
Resources	Provide the test readers with the same tools, equipment, reference materials, and other resources that the target readers would have—but not additional ones.
Information gathering	Gather information in ways that enable you to observe the details of the test readers' efforts without interfering in their use of your draft.

Holly picked test readers from her target audience.

She asked them to work in the same setting her target readers would use.

To see how you might apply these guidelines, consider the way Holly tested instructions that would enable ordinary consumers to install a sound system manufactured by her employer. First, Holly recruited two friends as her test readers. Both owned cars and neither had made a similar installation before, so both represented her target readers. Holly asked Rob to install the equipment in his garage at home and Janice to install hers in the parking lot of her apartment building. These are work areas that would be used by Holly's target readers. Rob already had the necessary tools. Holly supplied Janice with a power drill and a Phillips head screwdriver, but nothing else because these might be the only tools an ordinary consumer might have.

Holly made detailed notes, and she attempted to learn what her test readers were thinking throughout the procedure.

While Rob and Janice worked, Holly took detailed notes throughout, using the form shown in Figure 18.1 (page 475). During the test, she focused only on the left-hand column, writing down everything that seemed to indicate that Rob or Janice was having difficulty. So that she could pay full attention to the details of Rob's and Janice's efforts,

FIGURE 18.1

Test Observation Form

After the test, you can ask the test reader to tell you what caused the problems you observed.

Observation Sheet for Performance Tests

Problem (What difficulty did the reader have?	**Interpretation** (What might have caused the difficulty?)	**Solution** (What might prevent this difficulty?)
After completing a step, Rob sometimes hunted around the page for his place.	1. The steps don't stand out plainly enough.	1. Enlarge the step numbers. 2. Print the steps in bold so they are easier to distinguish from the notes and cautions.
At Step 7, Rob couldn't find the hole for the mounting bracket.	1. Rob couldn't understand the figure.	1. Simplify the figure by eliminating unnecessary details.

she waited to fill in the other columns until after the test had been completed. To obtain a full sense of what Rob and Janice were thinking as they used her instructions, she asked them to "think aloud," verbalizing their thoughts throughout the process.

Holly assisted only when the test would otherwise have had to stop.

As he worked on the installation, Rob asked Holly several questions. Instead of answering, she urged him to do his best without her help. If she had begun to provide oral instructions, she would no longer have been testing her written ones. However, she did assist at a point where Rob was completely stumped. Without her help, he would have had to stop work altogether, and Holly wouldn't have been able to find out how well the rest of her instructions worked.

Reza used a scenario to tell his test readers what he wanted them to do.

For some communications, you may need to create *scenarios,* or stories, that tell your test readers the tasks you want them to perform. For example, Reza has drafted a website for a company that sells outdoor sports equipment on the Internet. In his performance test, Reza asked his test users to imagine that they want to purchase a product for some particular activity, such as day hikes along easy trails or a weeklong ascent of a glacier-wrapped mountain peak. He then requested that they use his draft of the website to determine the appropriate products and order them.

Reza used an alternative technique for learning what his test readers were thinking.

Several other features of the test Reza conducted illustrate some of the variations that are possible when applying the strategies for performance tests. Because Reza's website offered so many different kinds of products, it was not feasible for Reza to ask his test readers to use every part of the site. Instead, he asked them to use only certain carefully selected parts of the site. Also, instead of using a worksheet to record information, Reza placed a video camera behind his test users to record the entire test for later study. To learn what his test users were thinking as they used his website, Reza engaged them in a conversation in which he continuously asked them what they were trying to do and how they were attempting to do it. He did not, however, tell them how to overcome problems they encountered. He chose this technique, developed by researchers M. Ted Boren and Judith Ramey (2000), because it is especially useful for testing online communications, where users often work so quickly that it is difficult to follow their actions.

Location Tests

Location tests closely resemble performance tests: You give your draft to your test readers, asking them to find specific pieces of information or the answer to a particular question as rapidly as possible. This procedure enables you to evaluate the effectiveness of the headings, topic sentences, contents lists, and other guideposts that reveal the organization of print communications. For websites and other online communications, location tests can help you assess the organization of your site and the navigational aids it contains. With location tests, as with performance tests, you must simulate your target readers' reading situation as closely as possible in order to achieve the most helpful results.

Understandability Tests

In an understandability test, you ask your test readers to read your draft and then you ask them questions designed to determine whether they understood it accurately. Understandability tests often are combined with the other kinds of tests. For example, because readers cannot use information unless they understand it, understandability is an element in every performance test, such as those designed by Holly and Reza. Similarly, an understandability test can be combined with a location test by asking test readers to use the information they find.

If you want to test understandability in isolation, ask your test readers to read your draft and then pose questions about what they've read. Several types of questions are illustrated below. Most were used by Shane, an employee at a paralegal firm, to test the following paragraph from a draft version of a traffic ticket defence policy.

A passage Shane tested

> In return for your retainer fee and signed retainer authorization forms, we will obtain your traffic ticket information from you, file your ticket for you at the appropriate court location, request the disclosure of the prosecutor's evidence against you, and attend court on your behalf for an adjournment if we do not receive the disclosure prior to the scheduled court date. We will then review the disclosure with you and inform you of the relevant defences and plea resolutions available to you. We will help you to determine your defence or plea. Finally, we will attend court on your behalf to present your defence or plea.

Ways to test understandability

- **Ask your test readers to recognize a correct paraphrase.**

Sample questions

> **True or false:** We will always request an adjournment on your behalf.
>
> When will we attend court on your behalf? (a) When we believe it's the best thing to do. (b) When we present your plea or defence. (c) To obtain your traffic ticket information.

- **Ask your test readers to create a correct paraphrase.** By asking them to paraphrase, you can find out whether they understood well enough to explain your message in their own words.

Request for a paraphrase

> In your own words, tell when we start work on your traffic ticket.

- **Ask your test readers to apply your information.** To test your readers' ability to apply information, create a fictional situation in which the information must be used. Shane developed the following question to test the understandability of one section of his draft (not the section quoted above).

Situation used by Shane to determine whether test readers understood his draft well enough to apply its information

> You receive a speeding ticket for a clocked speed of 130 km/h in an 80 km/h zone. The ticket outlines that you will receive 4 demerit points, and a fine of $175, plus administrative costs. What are your options?

GUIDELINE 4 Focus on Persuasiveness: Learn How Your Draft Affects Your Readers' Attitudes

To test the persuasive impact of a draft, compare your test readers' attitudes before reading with those after reading. At work, user tests usually focus on the readers' attitudes toward one or both of the following:

- **The *subject matter* of the communication.** Among other things, Holly would look for positive changes in the readers' attitudes toward the sound system and the job of installing it by themselves. Among other things, Reza would want to determine his website's ability to prompt positive changes in his test readers' attitude toward products sold there.
- **The *quality* of the communication.** Holly would be concerned with test readers' attitudes toward the instructions she drafted and Reza toward the website he created.

There are several ways to obtain information about your test readers' attitudes before and after they read your draft. Once they've finished reading, you can ask how their attitudes have changed. This approach may yield flawed results, however, because readers sometimes misremember their earlier attitudes.

To obtain more reliable results, ask the same question before and after their reading. Avoid questions that can be answered with "Yes" or "No." Attitude changes are usually a matter of degree: "I felt a little positive, but now I feel very positive." Here are two effective types of questions:

- **Open-ended questions.** "How do you feel about buying outdoor sports equipment online?" "Describe your level of confidence that you could easily install a sound system in your car using a set of printed instructions."
- **Questions constructed around a scale.** "On a scale of 1 (very appealing) to 5 (very unappealing), how do you feel about buying sports equipment online?"

GUIDELINE 5 Interview Your Test Readers after They've Read and Used Your Draft

By interviewing your test readers after they've used your draft, you can often obtain a wealth of insights you couldn't otherwise have discovered. Here are five especially productive questions.

Things to ask when interviewing test readers

- **What were you thinking when you encountered each problem?** Even when test readers try to speak their thoughts aloud as they read your draft, their account may be incomplete. For example, some test readers become so engrossed in their task that they forget to speak. Or, their thoughts fly faster than they can report them.

- **How did you try to overcome each problem?** Test readers' strategies may suggest ways you can improve the presentation of your information.
- **How do you respond to specific elements of my communication?** Ask about any element about which you'd like feedback. Holly might ask, "Do you find the page design appealing?" Reza could inquire, "Do you like the website's colours?"
- **What do you suggest?** Readers often have excellent ideas about ways a communication can better meet their needs.
- **How do you feel about the communication overall?** When responding to such an open-ended question about their feelings, test readers may reveal some very helpful information about your communication's usability or persuasiveness that your other questions didn't give them an opportunity to express.

GUIDELINE 6 Avoid Biasing Your Test Results

The value of your test depends on how closely your test readers' responses to your draft match those of your target audience. Anything that diminishes the closeness of this match can cause you to miss needed changes or to revise in counterproductive ways.

Here are some strategies for avoiding bias.

Avoiding Bias

- **Remain unobtrusive.** Let the test readers focus on their tasks. If they are overly conscious of being observed, they may have difficulty concentrating.
- **Refrain from intervening unless absolutely necessary.** Even if your test readers ask for help, request that they rely only on the draft to solve problems. Intervene only if you must do so to enable them to continue with other parts of the test or if they are about to injure themselves or others.
- **Phrase questions in an unbiased way.** When interviewing test readers, avoid phrasing questions in ways that seem to steer them toward a certain answer. Your test readers may comply, thereby depriving you of accurate information needed as the basis for solid, reader-centred revisions.
- **Ask someone else to conduct your test—if you can watch.** Research indicates that test readers are reluctant to criticize a draft to the writer. By asking someone else to conduct your test, you increase your chances of obtaining unbiased responses. Note, however, that someone else's written or oral summary of a test cannot provide you with as many detailed, helpful insights as actually watching the test. Have someone else conduct your test only if you can watch, whether by sitting nearby, viewing the test through a one-way mirror, or watching a videotape of it.
- **De-emphasize your relationship to the draft.** If you conduct the test yourself, avoid emphasizing that you are the writer. When interviewing your test readers, welcome criticisms. If you seem disappointed or unreceptive to criticisms, test readers will stop making them. Assure the readers that all comments are helpful to you.

Test Early and Often

If necessary, use more than one test and more than one type of test.

Often, you can write much more efficiently and effectively if you test the parts of your draft as they are completed rather than waiting until you have a full draft that you can test all at once.

For instance, as soon as Reza tentatively decided on the navigational features and visual design for his website, he drafted the home page and several other main pages. He then asked test readers what they thought would happen if they clicked on each of the navigational buttons that would link to other pages. He also asked what they thought of the images he'd chosen. Through these early tests, Reza learned things he could use not only when revising the pages he tested but also while creating other parts of his website.

GUIDELINE 8

Ethics Guideline: Obtain Informed Consent from Your Test Readers

An ethical principle developed for research in medicine, psychology, and many other fields also applies to the testing of the drafts you write at work or in class. This principle states that the people involved should be volunteers who have agreed to participate after being fully informed about what you are asking them to do.

For certain kinds of research, this is not only an ethical requirement but also a legal one. Researchers must tell potential volunteers about the study in writing and obtain in writing the volunteers' consent to participate.

Although such legal requirements probably will not apply to the testing you do of draft communications you prepare at work, the ethical principle still does. When asking others to try out something you have drafted, let them know the following:

- The test's purpose
- The things you will ask them to do during the test
- The time required for the test
- Where the test will take place
- Any potential risks to them (some lab procedures, for instance, do involve risk if not done properly)
- Their right to decline to participate
- Their right, if they do volunteer, to stop participating at any time

A sample informed consent form can be found on this book's website. Also on the website is a planning guide that will help you learn how your readers will respond to a communication you are planning.

CONCLUSION

WWW To download a sample consent fom and planning guide for testing drafts, visit Chapter 18 at www.techcomm.nelson.com.

This chapter has provided guidelines for three of the most common evaluation procedures used in the workplace: performing your own quality check, reviewing, and testing for usability and persuasiveness. The primary strategies underlying all of the guidelines are the same ones you should employ when you are planning and drafting any communication on the job: Consider the communication from the perspective of its target readers, the context in which you are writing, and the communication's ethical impact on its stakeholders.

EXERCISES

For additional exercises, visit www.techcomm.nelson.com.

Expertise

1. Following the advice given in this chapter, carefully check a draft you are preparing for this class. Then, give your draft to one or more of your classmates to review. Make a list of the problems they find that you overlooked. For three of these problems, explain why you missed them. If possible, pick problems that have different explanations.

2. Explain how you would test each of the following communications:
   a. A display in a provincial park that is intended to explain to the public how the park's extensive limestone caves were formed.
   b. Instructions that tell homeowners how to design and construct a patio. Assume that you must test the instructions without having your test readers actually build a patio.
   c. Instructions that tell homeowners how to build and operate a gas barbecue.

Online

1. Exchange electronic copies of the drafts for your current assignment with a classmate. Using the features of the word processing software, insert comments in your partner's assignment and return it to him or her in electronic form. Evaluate the comments your partner provided for their clarity and ease of use, while he or she does the same for your comments. Share praise for comments that were presented effectively, and share suggestions concerning comments that could have been presented more effectively.

2. Design a user test for your school's website. Imagine that the target audience is high school students who are deciding which post-secondary institution to attend. Whom would you select as test readers? How would you define the objectives of the test? What would you ask the testers to do, and how would you gather information from them concerning the site's usability and persuasiveness?

Collaboration

1. This exercise will strengthen your constructive reviewing ability.
   a. Exchange drafts with a classmate, together with information about the purpose, audience, and stakeholders for the drafts.
   b. Carefully read your partner's draft, playing the role of a reviewer. Ask your partner to read your draft in the same way.
   c. Offer your comments to your partner.
   d. Evaluate your success in delivering your comments in a way that makes the writer feel comfortable while still providing substantive, understandable advice. Do this by writing down three specific points that you think you handled well and three ways that you think you can improve. At the same time, your partner should be making a similar list of observations about your delivery. Both of you should focus on such matters as how you opened the discussion, how you phrased your comments, and how you explained them.
   e. Talk over with your partner the observations that each of you made.
   f. Repeat steps c through e, but have your partner give you his or her comments on your draft.

2. Interview a classmate about a project he or she is drafting. Then design a test that will help your partner identify ways that he or she could revise the draft to make it more usable and persuasive for its target audience.

Ethics

Find a print or online communication that you consider unethical or that you believe might be unethical. Imagine that you are seeing it not as a finished communication, but in draft form, because its creator has asked you to review it. Plan your approach to raising and discussing the ethical issue with the person who drafted the communication. Because this person is free to accept or ignore your suggestions, you will need to convey your recommendations in a way that this person finds persuasive. Present your plan in the manner your professor requests.

Reference Guide 6: Four Steps for Revising

CONTENTS

WWW For resources related to Reference Guide 6, visit www.techcomm.nelson.com.

This Reference Guide provides reader-centred advice about the final activity of writing: revising. This advice consists of the following four steps:

1. Read and understand the reviewers' comments
2. Adjust your efforts to fit the situation
3. Make the most significant revisions first
4. Retain copies of your communication as revisions are made

When you revise, you improve a draft by changing it in ways that have been suggested by the checking, reviewing, and testing that you did while following the guidelines in Chapter 18. If you have not already done so, you should review those guidelines on which this advice relies.

READ AND UNDERSTAND THE REVIEWERS' COMMENTS

Many writers fail to read and understand the reviewers' comments before they rush into making the revisions. The writer begins by looking at the first paragraph, makes revisions there, and then proceeds through the rest of the communication in the same hasty and sequential way. This procedure works well if the most serious shortcomings or inconsistencies and organizational problems occur in only the opening paragraphs. However, if the problems require simultaneous attention to passages scattered throughout the communication, a sequential approach to revising may ignore them altogether. In the same way that you would read over a test before beginning to answer its questions, reading over reviewers' comments to get a sense of the level and applicability of their suggestions allows you to make consistent and appropriate revisions.

Reading and understanding the reviewers' suggestions may require you to explore the reasons for the suggestions. If you do not understand the reasons for a suggestion, stifle any temptation to act defensively. Rather, discuss the suggestions with the reviewer. Understanding the reasons for the suggestions can help you recognize and avoid similar problems in the future. Even when a suggestion seems useless, knowing the reviewers' reasons for making it can help you identify a problem you previously overlooked so you can devise your own way of solving it.

Revising needs to be as reader-centred as any other writing activity. Don't just accept and incorporate all the reviewers' comments—particularly if they seem to give contradictory advice. Consider the reviewers' comments in light of the tasks you want to help your reader perform, the ways you want to influence your readers' attitudes, the facts about your readers and their situation that will affect the way they respond, and the effect on the stakeholders. Retain your focus on usability and persuasiveness as you approach your revisions to ensure that the effectiveness of your communication is enhanced, not diminished, by the suggested changes.

ADJUST YOUR EFFORTS TO FIT THE SITUATION

As you begin revising any communication, you must determine how much time and energy you will invest in making those revisions, rather than in performing your other duties. While all your communications should present your central points clearly and achieve an appropriate tone, different communications need different levels of additional polish in the workplace. In a survey of workers in a diverse mix of professions, researchers Barbara Couture and Jone Rymer (1993) found that three times as many employees "often" or "very often" make major revisions in communications they consider to be "special" as make major revisions in communications they consider to be "routine."

To determine how good a communication you are writing needs to be, do the following:

- **Think about your communication's usability and persuasive objectives.** How carefully does your communication need to be crafted to help your readers perform their tasks? To affect your readers' attitudes in the way you want?
- **Consider general expectations about quality.** Usually, you need to polish your communications more thoroughly when addressing people at a higher level than at your own level in your organization, when addressing people outside your organization rather than inside, and when trying to gain something rather than give something (for example, when writing proposals to prospective clients, rather than reports to existing clients).
- **Look at similar communications.** Often, the level of quality needed is determined largely by what's customary. By looking at communications similar to the one you're preparing, you can see what level of quality has succeeded in the past.
- **Ask someone.** Your manager or coworkers can be excellent sources of information about the level of quality needed.

That said, there are good reasons for wanting your communication to go beyond the minimum level needed to make your communication "good enough."

Reasons for making your communications better than "good enough."

First, your writing is a major factor on which your job performance will be evaluated. In fact, people who are not involved with your work from day to day may base their opinion of you solely on your writing. There's a real advantage to having readers say not only, "Your report provided us with lots of useful information," but also, "And you wrote it so well!"

Second, communications prepared at work often have a wider audience than the writer assumes. Notes to coworkers and to your immediate manager may be passed along to upper management or to members of other departments for whom you would want to prepare more polished communications. When determining what level of quality to strive for, consider who all your possible readers may be.

MAKE THE MOST SIGNIFICANT REVISIONS FIRST

After you've read and understood your reviewers' comments and decided how much effort to invest in revising, identify the revisions that will bring the greatest improvement in the shortest time. Unfortunately, you may not have enough time to make all the revisions that checking, reviewing, or testing suggest. As a result, you will need to use the time you have to make revisions that result in large improvements rather than small ones. Ranking the revisions will help you invest your revising time wisely.

The relative importance of a particular revision depends on several factors.

How to Rank Revisions

No single ranking applies to all communications. The importance of most revisions depends on the situation. For one thing, the relative importance of a certain revision is determined by the other revisions that are needed or desirable. Consider, for instance, the place of "supplying missing topic sentences" in the ranked list (Figure RV.1) that Paolo created after surveying the comments made by four reviewers of a proposal he had drafted. In comparison with the other items in the list, supplying topic sentences is relatively unimportant. However, in a draft with fewer problems in more critical areas, providing topic sentences might bring more improvement than any other revision Paolo could make.

Another factor affecting the relative importance of a revision is its location. Improvements in an opening summary would increase the usability and persuasiveness of a long report more than a similar revision to an appendix. Revisions in passages that convey key points have greater impact than similar revisions to passages that provide background information and explanations.

A third factor can come into play when you are ranking revisions requested by reviewers: the importance of the person making the request. For obvious and practical reasons, you may want to give first attention to revisions requested by people in positions of authority, even if you would not otherwise have given high priority to these same revisions.

When you have limited time for revising, some of the revisions may never be made. By creating a ranked list, you ensure that the revisions you do complete are the ones that contribute most to your communication's effectiveness.

Be Sure to Correct Mechanical Problems

No matter how you rank other possible improvements, place the correction of mechanical problems—those that involve a clear right and wrong—at the top of your list, unless you are writing in a situation where you are sure that errors won't matter. Mechanical problems diminish your communication's persuasiveness by undermining your

FIGURE RV.1

Amount of Improvement That Various Revisions Would Make in Paolo's Proposal

MAKE GREAT IMPROVEMENT

- ________ Correcting errors in key statements
- ________ Adding essential information that was overlooked
- ________ Correcting misspellings
- ________ Repairing obvious errors in grammar
- ________ Fixing major organizational difficulties
- ________ Supplying missing topic sentences
- ________ Revising sentences that are tangled but still understandable
- ________ Correcting less obvious problems in grammar

MAKE SMALL IMPROVEMENT

credibility as a person capable of careful work. They can also reduce usability by distracting readers from your main points, slowing comprehension, and even causing misunderstanding.

Here are the major mechanical problems you should focus on:

Mechanical issues to focus on

- **Correctness.** Fix all errors in spelling, grammar, and punctuation as well as such things as the numbering of figures and the accuracy of cross-references.
- **Consistency.** Where two or more items should have the same form, be sure they do. For instance, revise parallel tables and parallel headings that should look the same but don't.
- **Conformity.** Correct all deviation from your employer's policies on such matters as the width of margins, the use of abbreviations, and the contents of title pages. Even if your employer hasn't developed written policies on these things, remember that your communication must still conform to your employer's informal policies and expectations.
- **Attractiveness.** Be sure your communication looks neat and professional.

Check mechanical matters last.

Often, it's best to postpone revising mechanical matters until the end of your work on a communication. There is no point in investing time in correcting passages that may later be deleted or changed. On the other hand, it's usually best to clear up mechanical problems in the drafts you give to a reviewer because they can distract him or her from more substantive issues.

RETAIN COPIES OF YOUR COMMUNICATION AS REVISIONS ARE MADE

As you begin incorporating the revisions that your reviewers have suggested, ensure that you retain a version of the original communication. You may wish to electronically save several versions of your communication while you are in the process of revising it. However, it is easy to grow confused when you have more than one copy of a document, and to inadvertently make changes to an old copy rather than to the most recent copy (Hackos, 1994).

To avoid this confusion when revising documents, you need to develop a system for naming your electronic copies. To implement this type of naming convention, you may wish to include the date of revision in the file name (for example, **Roseglen Proposal 15 February 06**), or a version number, as used in software releases (such as **Roseglen Proposal V 2.0**). Adopting a consistent system for naming and subsequently saving your in-progress electronic files will ensure that you incorporate your reviewers' comments in a manner that can be tracked and reviewed, if necessary.

CASE

FIXING UP THE PROPOSAL

The time has just flown by since you received the draft proposal back from your boss after lunch today. There has been a constant flow of staff in and out of the office all afternoon at Duggan Foundations, a small start-up civil engineering firm that hired you after graduation. Your boss's boss, Glory, needs to sign the letter of transmittal for the proposal you wrote in response to the borough's RFP. The deadline for the proposal is noon the day after tomorrow. Everyone on staff is hoping your firm will land this contract with the municipality, because your firm could use the public-sector experience and contacts.

You need to make all the changes from all the reviewers as soon as you can, so that Glory will have time to review the proposal before she signs the letter of transmittal. At least everyone had a chance to look at it: your boss, Erik, your coworker, Julie, and your finance guy, Pierre. But it seems like every one of your reviewers has different things that they want changed. And Erik's comments and Julie's suggestions seem to contradict each other. Now, where to start?

YOUR ASSIGNMENT

Decide how you will approach making the revisions to the proposal. How will you figure out what changes need to be made? How will you make all the changes in time for Glory to read the proposal? How will you reconcile the conflicting suggestions of Erik and Julie? Assume that Glory will also want you to make some final revisions.

Appendixes

APPENDIX A

Formats for Email, Memos, Letters, and Reports

APPENDIX B

Documenting Your Sources

APPENDIX C

Projects

APPENDIX A

Formats for Email, Memos, Letters, and Reports

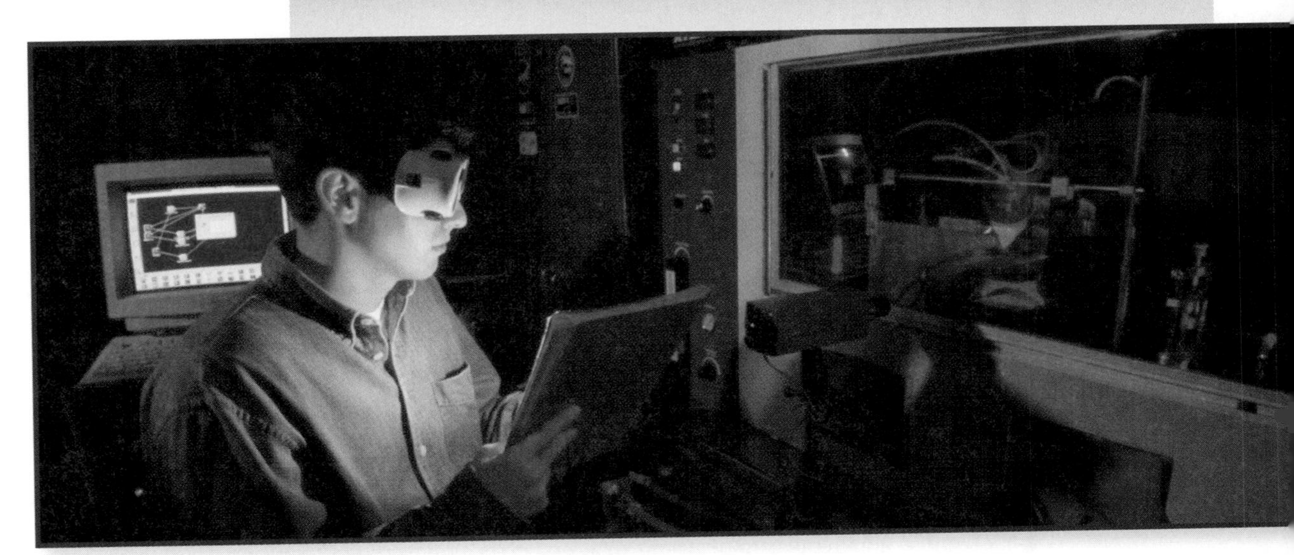

CONTENTS

Formats are conventional packages for presenting messages. This appendix is about the four formats most often employed at work:

Formats	
Email message	Used primarily for short communications inside and outside the organization
Memo	Used primarily for short internal communications
Letter	Used primarily for short communications sent outside the organization
Report	Used for communications addressed inside or outside the organization that are long enough to have a title page (i.e., longer than five pages)

By following the detailed conventions for each of these formats, you can create communications that look like the documents your readers expect to see. This is one of the easiest ways to build your credibility as a workplace writer.

Many employers distribute instructions that tell employees how to prepare each of these formats. These instructions, sometimes called *style guides*, describe everything from the width of margins to the placement of headings. Of course, the instructions vary somewhat from company to company. For instance, one employer may want the titles of all reports to be placed in a certain spot on the cover, and another employer may want titles in another spot. Despite these variations, most letters, memos, and formal reports written on the job look very much alike. The formats described in this chapter reflect common practices in the workplace. You will have little trouble adapting them to other, slightly different versions as the need arises.

EMAIL FORMAT

At work, you will spend a great deal of time writing email messages to individuals both within and outside your own organization. Chapter 5 offers several guidelines to help you work effectively with this vital communication format. Figure A.1 illustrates a business-appropriate format for an email message. However, email conventions vary from organization to organization, and furthermore, email messages sent within an organization may be formatted differently from email messages sent outside that organization. It is important to observe the email conventions of your workplace so that you meet your readers' expectations while using the format.

MEMO FORMAT

On the job, you will use memos when writing to coworkers, managers, and other people inside your employer's organization. The distinguishing feature of the memo is the heading, which looks like a form that has slots for your name, your reader's name, the date, and (perhaps) the memo's subject. Figure A.2 (page 494) shows an example. You don't include addresses in a memo because the message is being sent internally. You may

FIGURE A.1

Business Format for Email

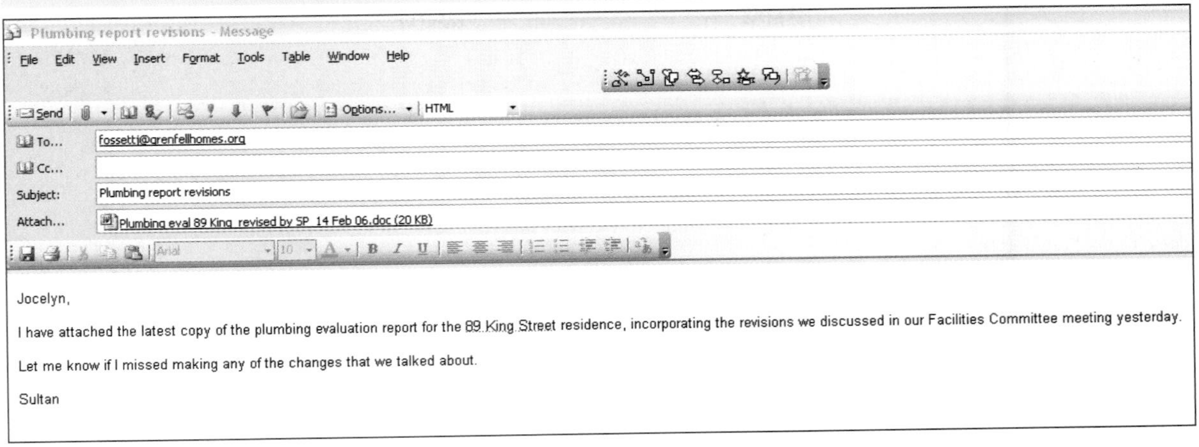

Plumbing report revisions - Message

File Edit View Insert Format Tools Table Window Help

Send | Options... | HTML

To... fossettj@grenfellhomes.org

Cc...

Subject: Plumbing report revisions

Attach... Plumbing eval 89 King revised by SP 14 Feb 06.doc (20 KB)

Jocelyn,

I have attached the latest copy of the plumbing evaluation report for the 89 King Street residence, incorporating the revisions we discussed in our Facilities Committee meeting yesterday.

Let me know if I missed making any of the changes that we talked about.

Sultan

write your memos on paper that has the heading already printed on it; more likely, however, you may store the heading in your word processing software so that you can summon it each time you want to draft a new memo.

Many memos are quite short, only a few sentences. Begin the body of your memos two or three lines below the heading, regardless of the amount of blank space this placement might leave on the rest of the page. When a memo runs to more than one page, provide the additional pages with the same sort of header that you use in letters. See page 497.

Memo format may also be used for short reports written to coworkers, managers, and other people inside your employer's organization. If the content of the report is brief (five pages or less), the report may not require a cover or a contents list. Instead, use the heading of a memo to address your readers and to orient them to the subject matter, and proceed directly to write the body of the report on the first page of the memo. If necessary, use the continuation header for subsequent pages of the report.

To decide how to sign your memo, see what others in your organization are doing. In some, writers sign memos with their initials next to their names in the heading. In others, they sign their full name or initials at the bottom.

LETTER FORMAT

At work, you will usually use letters when writing relatively short messages to customers, clients, and other readers outside your employer's organization. There are two major variations of the letter format: the *block* format (Figure A.3, page 495) and the *modified block* format (Figure A.4, page 496). Both figures show letters prepared on paper with preprinted letterheads.

FIGURE A.2

Memo on Preprinted Stationery

Heading preprinted on the stationery

If your employer's memo paper doesn't include a subject line, you can help your reader by adding one when you prepare your message

ALGONQUIN COLLEGE

Memorandum

To: Robert McBride, Professor, School of Technology and Building Trades

From: Marni Squire, Student Success Committee, B315b
squirem@algonquincollege.com

Date: 19 February 2006

Subject: Midterm Audit of 05/06 Approved QAF Retention Initiatives

The Student Success Committee is conducting a midterm audit of the retention initiatives funded by the Quality Assurance Funds for the 05/06 year. This year, as usual, we are conducting a quantitative audit to ensure all projects are on track and within budget. To help us accomplish this, could you please answer the questions below concerning your particular project(s) by 7 March 2006.

Study to Identify and Weight College-wide Predictors of Class Attendance Funding approved: $2,960

1. Name and contact information of person primarily responsible for the implementation, if not yourself.
2. Has there been any change in the target group (i.e., program, course or special population) for which this initiative was designed?
3. Is the work of the initiative proceeding as described in your original proposal? If not, please briefly describe any needed changes that have been made.
4. Has there been any change to your original budget projection as approved?

A final audit will be performed at the end of term. In the meantime, good luck with your project and thanks for your cooperation.
If you have any questions, don't hesitate to contact me at extension 5902.

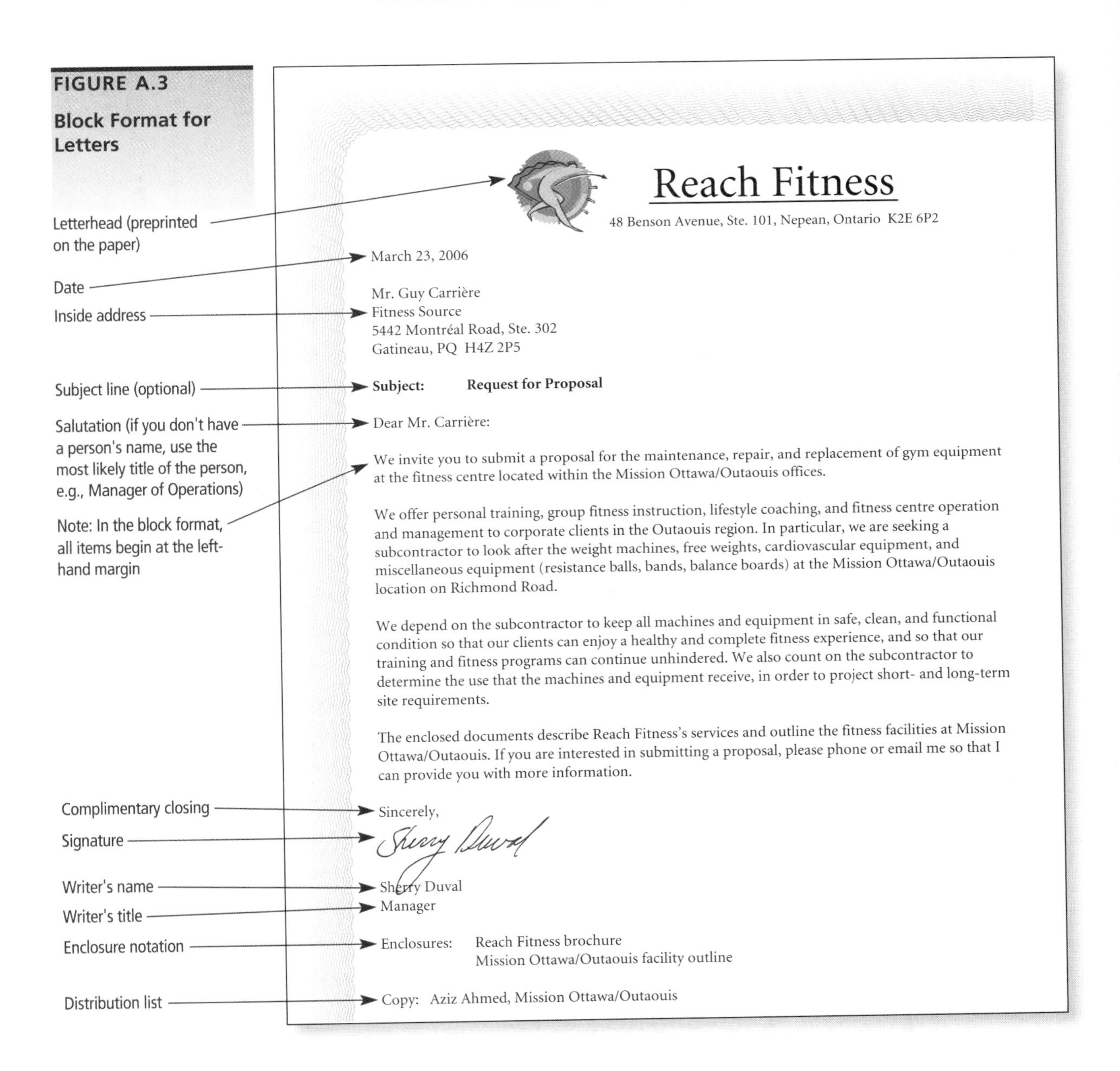
FIGURE A.3

Block Format for Letters

Reach Fitness

48 Benson Avenue, Ste. 101, Nepean, Ontario K2E 6P2

March 23, 2006

Mr. Guy Carrière
Fitness Source
5442 Montréal Road, Ste. 302
Gatineau, PQ H4Z 2P5

Subject: Request for Proposal

Dear Mr. Carrière:

We invite you to submit a proposal for the maintenance, repair, and replacement of gym equipment at the fitness centre located within the Mission Ottawa/Outaouis offices.

We offer personal training, group fitness instruction, lifestyle coaching, and fitness centre operation and management to corporate clients in the Outaouis region. In particular, we are seeking a subcontractor to look after the weight machines, free weights, cardiovascular equipment, and miscellaneous equipment (resistance balls, bands, balance boards) at the Mission Ottawa/Outaouis location on Richmond Road.

We depend on the subcontractor to keep all machines and equipment in safe, clean, and functional condition so that our clients can enjoy a healthy and complete fitness experience, and so that our training and fitness programs can continue unhindered. We also count on the subcontractor to determine the use that the machines and equipment receive, in order to project short- and long-term site requirements.

The enclosed documents describe Reach Fitness's services and outline the fitness facilities at Mission Ottawa/Outaouis. If you are interested in submitting a proposal, please phone or email me so that I can provide you with more information.

Sincerely,

Sherry Duval
Manager

Enclosures: Reach Fitness brochure
Mission Ottawa/Outaouis facility outline

Copy: Aziz Ahmed, Mission Ottawa/Outaouis

FIGURE A.4
Modified Block Format for Letters

Reach Fitness
48 Benson Avenue, Ste. 101, Nepean, Ontario K2E 6P2

March 23, 2006

Mr. Guy Carrière
Fitness Source
5442 Montréal Road, Ste. 302
Gatineau, PQ
H4Z 2P5

Subject: Request for Proposal

Dear Mr. Carrière:

We invite you to submit a proposal for the maintenance, repair, and replacement of gym equipment at the fitness centre located within the Mission Ottawa/Outaouis offices.

We offer personal training, group fitness instruction, lifestyle coaching, and fitness centre operation and management to corporate clients in the Outaouis region. In particular, we are seeking a subcontractor to look after the weight machines, free weights, cardiovascular equipment, and miscellaneous equipment (resistance balls, bands, balance boards) at the Mission Ottawa/Outaouis location on Richmond Road.

We depend on the subcontractor to keep all machines and equipment in safe, clean, and functional condition so that our clients can enjoy a healthy and complete fitness experience, and so that our training and fitness programs can continue unhindered. We also count on the subcontractor to determine the use that the machines and equipment receive, in order to project short- and long-term site requirements.

The enclosed documents describe Reach Fitness's services and outline the fitness facilities at Mission Ottawa/Outaouis. If you are interested in submitting a proposal, please phone or email me so that I can provide you with more information.

Sincerely,

Sherry Duval

Sherry Duval
Manager

Enclosures: Reach Fitness brochure
Mission Ottawa/Outaouis facility outline

Copy: Aziz Ahmed, Mission Ottawa/Outaouis

In the modified block format, the date may be on the right-hand side or centred.

In the modified block format, the first line of each paragraph may begin at the margin, or it may be indented 1.25 cm (1/2 inch).

In the modified block format, the signature is usually aligned slightly to the left of the middle of the page.

Margins for letters

The customary margins for letters are 2.5 cm on the top and sides. Typically, the bottom margin is 1 cm deeper than the top margin. With letters shorter than a full page, you might position the middle line a few lines above the middle of the page.

Second and subsequent pages

If your letter is longer than one page, use plain paper rather than letterhead stationery for the second and subsequent pages. At the top of the second and subsequent pages, provide the name of the addressee, the page number, and the date. Here is one commonly used arrangement:

Header for top of second and subsequent pages

Hasim K. Lederer 2 November 16, 2007

On second and subsequent pages, begin the text a line or two below the header, even if that leaves a lot of blank space at the bottom of the page. If the complimentary close and signature block would be the only items to appear on the final page, insert blank lines at the top of the first page so that more material is pushed to the final page.

REPORT FORMAT

Communications in the report format may have many of the features of published books: bindings, protective and informative covers, a contents page, and separate sections for each major block of information. Reports and proposals prepared in this format are often called *formal reports* or *formal proposals.*

At work, you may use the report format for communications ranging from six to hundreds of pages. Usually, you will use 8½ × 11-inch paper, usually printing on only one side of the sheet. Your binding may be a set of staples, a plastic spine, or some other device.

When writing in the report format, think of your communication as having three parts: front matter, body, and back matter. You may also need to write a letter of transmittal to accompany your communication when you present it to your readers.

FRONT MATTER

The front matter of the communication you write in the report format may consist of some or all of the following items, depending on your readers' needs and your employers' expectations.

- **Cover.** At a minimum, include a specific, informative title that helps readers know exactly what the communication will discuss. Of course, if the cover is a plastic sheet that reveals the title page underneath, no information will appear on the cover.
- **Title page.** On your title page, repeat the information on the cover. At a minimum, include your name and employer, the readers for whom the communication is intended, and the date of preparation. You may also include additional details about the report, such as its contract or project number.
- **Executive summary or abstract.** You can often help your readers by providing an executive summary or abstract, of one page or less, that conveys the essence of your communication. The summary will not contain any information that is not contained in the report; it will, however, contain your communication's topic, its purpose, its methodology, and its conclusions and recommendations.
- **Contents list.** By providing a list of contents, you help readers who want to find a specific part of your communication without reading all of it. Your contents list also assists readers who want an overview of the communication's scope and contents before they begin reading it in its entirety. You can use your word processing software to automatically generate a contents list, provided that you indicate the heading levels you wish to include in it.
- **List of illustrations.** By providing a list of illustrations and the pages on which they can be found, you can help readers who are searching for specific figures and

tables. You can use your word processing software to automatically generate a list of figures and a list of tables, provided that you use the software to sequentially label the figures and tables in your report.

- **Glossary.** When you are writing at work, you will sometimes use special terms that are not familiar to some of your readers, or that you are using in a particular way in the context of your communication. If you are using each of these terms only once or in a small segment of your communication, you can include your explanation in the text. However, if you are using a special term repeatedly, you can help readers understand your communication without encumbering their reading by locating the definition where they can locate it. Glossaries are also useful in communications in which some of your readers need definitions of terms and some don't. Those who know the terms can skip the definitions; those who don't know the terms can find out what they mean.

  Glossary terms are usually printed in boldface so that readers can find them easily.
- **List of symbols.** By providing a list of abbreviations and the terms they denote, you can help readers understand the use of these abbreviations in the context of your communication. As in the case of your glossary, if you are using these symbols only once or in a small segment of your communication, you can include your explanation in the text.

BODY

In the report format, the body is made up of the *content* or *discussion* sections. You divide your communication into these sections in order to help your readers find information and understand the communication's structure. The divisions among your sections should reflect this logic even if one section is only a single page, while others are many times that size. Despite such variation in length between sections, begin each section of your report on a new page.

In many reports, the first section is an introduction to the rest of the report. An introduction helps your readers by orienting them to the organization and contents of the report through an elaboration on its purpose, background, and scope. You may also wish to state any limitations or disclaimers that apply to your communication.

Similarly, in many reports, the final section consists of conclusions, and perhaps recommendations. Your conclusions state the major points that can be drawn from your communication; any recommendations you provide outline what should be done as a result of the conclusions you have found.

BACK MATTER

When writing communications in the report format, you may also create some elements that appear after the body. These elements are collectively called the *back matter.*

- **List of references.** You can place your references, endnotes, or bibliography immediately after the body of your communication, or after the appendixes. To learn what sources to mention and how to construct reference lists, endnotes, and bibliographies, see Appendix B.

- **Appendixes.** Use appendixes to present information you want to make available to your readers even though you know it won't interest all of them. For instance, in a research report you might create an appendix for a two-page account of the calculations you used for data analysis. By placing this account in an appendix, you help readers who will want to use the same calculations in another experiment, but you save readers who aren't interested from the necessity of wading through the calculations while reading the body of your report.

  When you create appendixes, list them on your contents page and give each an informative title that indicates clearly what it contains. Also, mention each appendix in the body of your report at the point where your readers might want to refer to it: "Printouts from the electrocardiogram appear in Appendix II."

  Begin each appendix on its own page. Arrange and label the appendixes in the same order in which they are mentioned in the body of the communication. If you have only one appendix, label it simply "Appendix." If you have more than one, you may use Roman numerals, Arabic numerals, or capital letters to label them.
- **Index.** In long communications, you can use an index to provide your readers with a quick route to specific pieces of information. Some types of word processing software enable you to automatically generate an index, provided you mark the entries in your communication that you want included.

LETTER (OR MEMO) OF TRANSMITTAL

When you prepare communications in the report format, you will often send them to readers outside your organization. In such cases, you will want to write a letter of transmittal to accompany them. If your readers are within your own organization, you will want to attach a memo of transmittal to your report. The letter (or memo) of transmittal is not bound to your communication; rather, it is loosely attached to the cover of your report, or placed just inside the cover.

Although the precise contents of a letter (or memo) of transmittal depend on your purpose and situation, a typical one contains some of the following elements:

- **Introduction.** Here, mention the accompanying communication and perhaps explain (or remind your readers about) its topic and rationale.
- **Body.** In the body, describe the purpose, contents, or special features of the communication. For instance, the writers of a research report who want their readers to note the most important consequences of their findings might briefly explain their findings and list their recommendations. It is common for there to be some repetition of content between the letter (or memo) of transmittal and the summary of the report.
- **Closing.** Transmittal letters (or memos) commonly end with a short paragraph (or sentence) that states the writer's willingness to work further with the reader or to answer any questions that the reader may have. In the closing, you may also wish to acknowledge or thank those individuals who helped you research and write the communication.

APPENDIX B

Documenting Your Sources

CONTENTS

In many of the communications you will write at work, you will want to tell your readers about other sources of information concerning your subject. You may have any of the following reasons for wanting to do so:

- **To acknowledge the people and sources that have provided you with ideas and information.** For a discussion of this reason for citing your sources, see Chapter 13's Guidelines 7 and 8.
- **To help your readers find additional information about something you have discussed.**
- **To persuade your readers to pay serious attention to a particular idea.** By showing that an idea was expressed by a respected person or in a respected publication, you are arguing that the idea merits acceptance.
- **To explain how your research contributes to the development of new knowledge in your field.** In research proposals and in research reports published in professional journals, writers often include a literature survey to demonstrate how their own research advances knowledge. For more information on using references in this way, see the discussion of literature surveys in the discussion of empirical research reports, pages 129–99.

For additional information, visit www.techcomm.nelson.com.

CHOOSING A FORMAT FOR DOCUMENTATION

There are numerous formats for documentation, some very distinct from one another and some differing only in small details. No one format is the best for all situations. To document correctly, you must use the format required or most appropriate for the particular circumstances in which you are writing. Most organizations specify the documentation style they want their employees to use. Also, when you are writing to people in another organization, you may find that they have their own preferences or requirements, which you should follow.

To describe the documentation format they prefer, many organizations issue style guides that detail their rules and provide sample citations. Other organizations ask writers to follow a style guide published by a professional organization such as the American Psychological Association (APA), the Modern Language Association (MLA), the University of Chicago Press (*Chicago Manual of Style*, CMS), or the Institute of Electrical and Electronics Engineers (IEEE). The rest of this appendix explains the APA, MLA, CMS, and IEEE styles. Most other documentation styles resemble one of these four.

DECIDING WHERE TO PLACE IN-TEXT CITATIONS

In the MLA style, you cite a source by putting the name of the author at the appropriate place in the body of your communication. In the APA and CMS styles, you also include the year of publication. In the IEEE style, you cite a source by using the number assigned to it, sequentially in the order it is scited in your communication. This citation refers your readers to the full bibliographic information that you provide in a list of sources at the end of your communication. Among the four styles there are many differences in the way you write the in-text citations and the entries in the reference list.

Your primary objective when placing citations in your communications is to make clear to your readers what part of your text each citation is referring to. This is easy with citations that pertain to a single fact, sentence, or quotation. You simply place your citation immediately after the appropriate material.

If your citation refers to material that appears in several sentences, place the citations in a topic sentence that introduces the material. Your readers will then understand that the citation covers all the material that relates to that topic sentence. As a further aid to your readers, in the APA, MLA, and CMS styles, you may use the author's name (or a pronoun) in successive sentences:

> A much different account of the origin of oil in the earth's crust has been advanced by Thomas Gold (1983). He argues that. . . . To critics of his views, Gold responds. . . .

USING THE APA DOCUMENTATION STYLE

www.techcomm.nelson.com

The following sections explain how to write in-text citations and the entries in a reference list using the APA style.

WRITING APA IN-TEXT CITATIONS

To write an APA in-text citation, enclose the author's last name and the year of publication in parentheses inside your normal sentence punctuation. Place a comma between the author's name and the date. Use p. if you are citing a specific page, and use pp. if citing more than one page.

> Damage to gums and bones from smoking, stress, and poor nutrition may result in poorly fitting dentures (Parsa-Stay, 1996, p. 80).

If you incorporate the author's name in the sentence itself, give only the year and pages (if any) in parentheses:

> According to Parsa-Stay, damage to gums and bones from smoking, stress, and poor nutrition may result in poorly fitting dentures (1996, p. 80).

Here are some other types of in-text citations in the APA style:

Two authors.

> (Doundoulakis & Strugatch, 2004)

First citation for three, four, or five authors.

> (Mittelman, Mittelman, & Barilla, 2001)

Second and subsequent citations for three to five authors. Omit year in subsequent citations in same paragraph.

> (Mittelman et al., 2001)

First and subsequent citation for six or more authors ("et al." is an abbreviation of the Latin "et alii," which means "and others").

> (Norton et al., 1994)

Government or corporate author.

(Standing Committee on Health, 2003)

No author listed (use the first few words of the title). In the example, the words from the title are in quotation marks because the citation is to an article; if a book is cited, the words would be in italics with no quotation marks.

("Denture cleansers," 2006)

Two or more sources cited together (arrange them in alphabetical order).

(Bugel, 1994; Disipio, 1999)

In some communications, you might cite two or more sources by the same author. If they were published in *different* years, your readers will have no trouble telling which work you are referring to. If they were published in the same year, you can distinguish between them by placing lowercase letters after the publication dates in your citations and in your reference list:

(Burkehardt, 1998a)
(Burkehardt, 1998b)

WRITING AN APA REFERENCE LIST

Illustrated and explained here are APA reference list entries for the most common types of print, electronic, and other sources. To create entries that are not listed here, follow the logic of these examples or else consult the *Publication Manual of the American Psychological Association*. Figure B.1 shows how to arrange entries in your reference list.

Print Sources

1. Book, one author—APA

Parsa-Stay, F. (1996). *The complete book of dental remedies: A guide to nutritional and conventional dental care.* Garden City Park, NY: Avery Publishing.

- Give the author's last name followed by a comma and initials (not full first or middle names).
- Place the copyright date in parentheses, followed by a period.
- Italicize the title, and capitalize only the first word of the title, the first word of the subtitle (if any), and proper nouns.
- Follow the city of publication with a comma and, if it is not well known, the state, province, or country.
- Indent the second and subsequent lines.

2. Book, two or more authors—APA

Doundoulakis, J., & Strugatch, W. (2004). *The perfect smile: The complete guide to cosmetic dentistry from tooth whitening and bleaching to veneers and implants.* New York: Healthy Living Books.

3. Anthology or essay collection—APA

Lutz, P.L., & Musick, J.A. (Eds.). (1997). *The biology of sea turtles.* Boca Raton, FL: CRC Press.

4. Second or subsequent edition—APA

Rodman, L. (2005). *Technical communication* (3rd ed.). Toronto: Thomson Nelson.

5. Government report—APA

Standing Committee on Health (2003). *First Nations and Inuit dental health: Report of the Standing Committee on Health.* (Government of Canada Investigation Report 03–4272). Ottawa: House of Commons.

- If the report doesn't list an author, use the name of the committee, ministry, or agency that published it as the author.
- If the report has an identifying number, place it immediately after the title.

6. Corporate report—APA

Daimler-Benz AG. (1997). *Environmental report 1997.* Stuttgart, Germany: Author.

- List the names of the individual authors rather than the corporation if the names are given on the title page.
- If the names of the individual authors aren't given on the title page, list the corporation as the author. (In the example, "Daimler-Benz AG" is the name of a company.)
- When the author and publisher are the same, use the word *Author* as the name of the publisher.

7. Essay in a book—APA

Sullivan, P. (1996). *Ethics in the computer age.* In J. M. Kizza (Ed.), *Social and ethical effects of the computer revolution* (pp. 288–297). Jefferson, NC: McFarland.

8. Paper in a proceedings—APA

Youra, S. (1996). *Placing writing in engineering education.* In R. Pose & A. Jawary (Eds.), *Proceedings of the Australian Communication Conference* (pp. 59–76). Melbourne, Australia: Monash University.

9. Encyclopedia article—APA

Rich, E. (1996). *Artificial intelligence. In Encyclopedia Americana* (Vol. 2, pp. 407–412). Danbury, CT: Grolier.

10. Pamphlet or brochure—APA

The Wool Bureau of Canada Limited. (1998). *Pull the wool over.* [Pamphlet]. Toronto: Author.

- When the author and publisher are the same, use the word *Author* as the name of the publisher.

11. Article in journal that numbers its pages continuously through each volume—APA

McLaurin, J., & Chakrabartty, A. (1997). *Characterization of the interactions of Alzheimer ß-amyloid peptides with phospholid membrances.* European Journal of Biochemistry, *245,* 355–363.

- After the journal's name, add a comma and the volume number (in italics).
- The word *Alzheimer* is capitalized because it is a proper name.

12. Article in journal that numbers its pages separately for each issue

Bradley, J., & Soulodre, G. (1997). *The acoustics of concert halls.* Physics World, *10* (5), 33–37.

- After the journal title, include the volume number, followed by the issue number (in parentheses).
- Put the volume number—but not the issue number—in italics.

13. Article in a popular magazine—APA

Hessler, P. (2006, February 13 & 20). Hutong Karma. *New Yorker,* 82–89.

- Give the full date of the issue, placing the year first.
- Provide the volume number (in the example: 129) but not the issue number.

14. Newspaper article—APA

Zimonjic, P. (2006, February 18). Animal lovers gone wild. *Ottawa Citizen,* p. B7.

- In front of the page number for newspapers, write *p.* (for one page) or *pp.* (for more than one page).

15. Article with no author listed—APA

Rethinking traditional design. (1997). *Manufacturing Engineering, 118* (2), 50.

- Begin with the article's title. This example gives the issue number in parentheses because the journal numbers its pages separately for each issue (see Example 12).

Electronic Sources

For information about the APA documentation style for other types of electronic sources, go to Appendix B at www.techcomm.nelson.com.

Because placing a period at the end of an Internet address can cause confusion, final periods are omitted in APA entries for online sources.

Because material that is available on the World Wide Web changes continuously, APA entries include the date you retrieved the item.

16. Report available only at a World Wide Web site —APA

International Business Machines. (n.d.). What's user-centered design? Retrieved January 25, 2003, from http://www-3.ibm.com/ibm/easy/eou_ext.nsf/Publish/2

- Because the site doesn't give the date this page was published on the Web, the abbreviation "n.d." is used to mean "no date."

17. Report downloaded from an FTP site—APA

Modjeska, D., & Marsh, A. (1997). Structure and memorability of Web sites. University of Toronto Computer Systems Research Institute. Retrieved June 15, 2003, from ftp://ftp.cs.toronto.edu/csri-technical-repots/364

18. Online journal article that is not available in print—APA

Toll, D. (1996). Artificial intelligence applications in geotechnical engineering. *Electronic Journal of Geotechnical Engineering.* Retrieved July 12, 2002, from http://geotech.civen.okstate.edu/ejge/Abs9608.htm

19. Online journal article that is also available in print—APA

Bernstein, A., Want, Y., Gratta, G., & West, T. (2002). Nuclear reactor safeguards and monitoring with antineutrino detectors [Electronic version]. *Journal of Applied Physics, 91*, 4672–4676. Retrieved April 22, 2003, from http://ojps.aip.org/journals/doc/JAPIAU-ft/vol_91/iss_7/4672_1-div0.html

20. Encyclopedia article from an online service—APA

Haggerty, M. (2002). Asbestosis. In Jacqueline L. Longe (Ed.), *Gale Encyclopedia of Medicine, 1,* 369–371. Retrieved February 19, 2006, from http://find.galegroup.com/gvrl/infomark.do?&contentSet=EBKS&type=retrieve&tabID=T001&prodId=GVRL&docId=CX3405600174&source=gale&userGroupName=otta35732&version=1.0

21. CD-ROM—APA

"Rainforest." (1997). Encarta '97 [CD-ROM]. Redmond, WA: Microsoft.

22. Email—APA

The APA style includes references to emails only in parentheses in the text, not in the reference list. The parenthetical citation in the text includes the author's initials as well as his or her last name and an exact date:

(D. Galway, personal communication, December 4, 2005).

Other Sources

23. Letter—APA

The APA style treats letters the same way it treats emails.

(C. Gray, personal communication, January 21, 2006).

24. Interview—APA

The APA style treats interviews the same way it treats emails.

USING THE MLA DOCUMENTATION STYLE

WWW For additional information and details about the MLA documentation style, including links to other documentation systems, visit Appendix B at www.techcomm.nelson.com.

The following sections explain how to write in-text citations and the entries in a reference list using the MLA style.

WRITING MLA IN-TEXT CITATIONS

A basic MLA citation contains two kinds of information: the author's name, and the specific page or pages on which the cited information is to be found. Enclose these in parentheses—with no punctuation between them—and place them inside your normal sentence punctuation. If you are citing the entire work, omit the page numbers. The following citation refers the reader to page 80 of a work by Parsa-Stay:

> Damage to gums and bones from smoking, stress, and poor nutrition may result in poorly fitting dentures (Parsa-Stay 80).

If you incorporate the author's name in the sentence itself, give only the page number(s) in parentheses:

> According to Parsa-Stay, damage to gums and bones from smoking, stress, and poor nutrition may result in poorly fitting dentures (80).

Here are some other types of in-text citations in the MLA style:

Two authors.

(Doundoulakis and Strugatch 53)

Three authors.

(Mittelman, Mittleman, and Barilla 42)

Four or more authors ("et al." is an abbreviation for the Latin phrase "et alii," which means "and others").

(Norton et al. 776)

Government or corporate author.

(Standing Committee on Health 11)

No author listed (use the first few words of the title). In the example, the words from the title are in quotation marks because the citation is to an article; if a book is cited, the words would be in italics with no quotation marks.

("Denture Cleansers," 2)

Two or more sources cited together (arrange them in alphabetical order).

(Bugel 1; Disipio 1)

In some of your communications, you might cite two or more sources by the same author. You can distinguish between them by placing a comma after the author's name, followed by a few words from the title:

Whole work cited | (Burkehardt, "Gambling Addiction," 99)
Specific pages cited | (Burkehardt, "Obsessive Behaviors," 81–97)

WRITING AN MLA WORKS CITED LIST

The following sections describe how to write entries for the most common types of print, electronic, and other sources. To create entries that are not listed here, follow the logic of these examples or else consult the *MLA Style Manual.* Figure B.2 (page 519) shows how to arrange entries in your list of works cited.

Print Sources

1. Book, one author—MLA

Parsa-Stay, Flora. *The Complete Book of Dental Remedies: A Guide to Nutritional and Conventional Dental Care.* Garden City Park: Avery Publishing, 1996.

- Give the author's last name followed by a comma and then the first and middle names or initials—exactly as they appear on the title page.
- Italicize the title, and capitalize all major words.
- Follow the city of publication with a colon, the publisher's name, a comma, and the publication date.
- Indent all lines after the first one.

2. Book, two or three authors—MLA

Doundoulakis, James, and Warren Strugatch. *The Perfect Smile: The Complete Guide to Cosmetic Dentistry from Tooth Whitening and Bleaching to Veneers and Implants.* New York: Healthy Living Books, 2004.

- Give the first author's name in reverse order (last name first).
- Give the names of additional authors in normal order (first name first).

3. Anthology or essay collection—MLA

Lutz, Peter L, and John A. Musick, eds. *The Biology of Sea Turtles.* Boca Raton: CRC Press, 1997.

- Use the abbreviation "ed." for a single editor and "eds." for multiple editors.

4. Second or subsequent edition—MLA

Rodman, Lilita. *Technical Communication.* 3rd edition. Toronto: Thomson Nelson, 2005.

5. Government report—MLA

Standing Committee on Health. *First Nations and Inuit Dental Health: Report of the Standing Committee on Health.* Government of Canada Investigation Report 03–4272. Ottawa: House of Commons, 2003.

- If the report doesn't list an author, begin the entry with the name of the government, followed by a period and the name of the committee, ministry, or agency that issued the document. Example: "United States. Geological Survey."
- If the report has an identifying number, place it immediately after the title.

6. Corporate report—MLA

Daimler-Benz AG. *Environmental Report 1997.* Stuttgart: Daimler-Benz AG, 1997.

- List the names of the individual authors rather than the corporation if the names are given on the title page.
- If the names of the individual authors aren't given on the title page, list the corporation as the author. (In the example, "Daimler-Benz AG" is the name of a company.)

7. Essay in a book—MLA

Sullivan, Patricia. "Ethics in the Computer Age." *Social and Ethical Effects of the Computer Revolution.* Ed. Joseph Migga Kizza. Jefferson: McFarland, 1996. 228–297.

- Use the abbreviation "Ed." for a single editor and for multiple editors.

8. Paper in a proceedings—MLA

Youra, Steven. "Placing Writing in Engineering Education." *Proceedings of the Australian Communication Conference.* Ed. Ronald Pose and Anita Jawary. Melbourne: Monash University, 1996. 59–76.

9. Encyclopedia article—MLA

Rich, Elaine. "Artificial Intelligence." *Encyclopedia Americana.* 1996 ed.

- If no author is listed, begin with the article's title.
- If entries in the work are arranged alphabetically, do not give volume or page number.
- When citing familiar reference works, give the edition number (if provided) and year of publication, but not the publisher or city of publication.

10. Pamphlet or brochure—MLA

The Wool Bureau of Canada Limited. *Pull the Wool Over.* Toronto: The Wool Bureau of Canada, 1998.

- If the pamphlet or brochure doesn't list an author, begin the entry with the name of the government or other organization that published it, followed by a period and the name of the agency that issued the document.
- If the pamphlet or brochure lists no author and no publisher, begin with the document's title.

11. Article in journal that numbers its pages continuously through each volume—MLA

McLaurin, Joanne, and Avijit Chakrabartty. "Characterization of the Interactions of Alzheimer ß-amyloid Peptides with Phospholid Membranes." *European Journal of Biochemistry.* 245 (1997): 355–63.

- Place the article's title in quotation marks and capitalize all major words.
- After the journal's name, give the volume number, followed by the year (in parentheses).
- For page numbers larger than 99, give only the last two digits unless more are needed for clarity. Examples: "355–63" and "394–405."

12. Article in journal that numbers its pages separately for each issue —MLA

Bradley, John, and Gilbert Soulodre. "The Acoustics of Concert Halls." *Physics World,* 10.5 (1997): 33–37.

- After the volume number, add a period and the issue number. (In the example, "10.5" signifies volume 10, issue 5.)

13. Article in a popular magazine—MLA

Hessler, Peter. "Hutong Karma." *New Yorker,* 13 and 20 February 2006: 82–87.

- Give the full date of the issue, beginning with the day and abbreviating the month.
- Do not give the issue or volume number.

14. Newspaper article—MLA

Zimonjic, Peter. "Animal Lovers Gone Wild." *Ottawa Citizen* 18 February 2006: B7.

- If the newspaper lists an edition (for example, "late edition") in the masthead, place a comma after the date and add the edition's name, using abbreviations where reasonable.

15. Article with no author listed—MLA

"Rethinking Traditional Design." *Manufacturing Engineering* 118.2 (1997): 50.

- Begin with the article's title.
- This example gives the issue number ("2") because the journal numbers its pages separately for each issue (see Example 12).

Electronic Sources

Because material that is available on the World Wide Web changes continually, MLA entries include the date you most recently viewed the item online.

16. Report available only at a World Wide Web site—MLA

International Business Machines. *What's user-centered design?* 25 January 2003 http://www-3.ibm.com/ibm/easy/eou_ext.nsf/Publish/2

17. Report downloaded from an FTP site—MLA

Modjeska, David, and Anna Marsh. "Structure and memorability of web sites," 1997. University of Toronto Computer Systems Research Institute. 15 June 2003 ftp://ftp.cs.toronto.edu/csri-technical-repots/364

18. Online journal article that is not available in print—MLA

Toll, David. "Artificial intelligence applications in geotechnical engineering." *Electronic Journal of Geotechnical Engineering.* 1 1996. 12 July 2002 http://geotech.civen.okstate.edu/ejge/Abs9608.htm

- Give the date the report was posted (if known) after the title.
- After the publisher's name and before the URL, give the date you accessed the source.

19. Online journal article that is also available in print—MLA

Bernstein, A., Y. Want, G. Gratta, and T. West. "Nuclear reactor safeguards and monitoring with antineutrino detectors." *Journal of Applied Physics,* 91 (2002), 4672–4676. 22 April 2003 http://ojps.aip.org/journals/doc/JAPIAU-ft/vol_91/iss_7/4672_1-div0.html

- First, give complete print publication information just as you would if you were describing the print publication only. Follow it with the date you accessed the online source (22 April 2003 in the example), then the URL.

20. Encyclopedia article from an online service—MLA

Haggerty, M. (2002). "Asbestosis." In Jacqueline L. Longe (Ed.), *Gale Encyclopedia of Medicine,* 1, 369–371. 19 February 2006 http://find.galegroup.com/gvrl/infomark.do?&contentSet=EBKS&type=retrieve&tabID=T001&prodId=GVRL&docId=CX3405600174&source=gale&userGroupName=otta35732&version=1.0

21. CD-ROM—MLA

"Rainforest." *Encarta '97.* CD-ROM. Redmond: Microsoft, 1997.

- For a CD-ROM that you accessed through a network (for example, at your library), add the date you last accessed it to the end of the entry.

22. Email—MLA

Galway, Devon. "Proof 1 requirements." Email to Kerry Surman. 4 December 2005.

- Place the title (taken from the subject line) in quotation marks.

Other Sources

23. Letter—MLA

Gray, Cheryl. Letter to the author. 20 August 2005.

24. Interview—MLA

Gray, Cheryl. Telephone interview. 20 August 2005.

USING THE CMS DOCUMENTATION STYLE

The following sections explain how to write the in-text citations and the entries in a reference list using the CMS style. Although the *Chicago Manual of Style* presents two documentation systems, the more concise author-date system in the *Manual* is generally preferred.

WRITING CMS AUTHOR-DATE IN-TEXT CITATIONS

A basic CMS citation contains two kinds of information: the author's name and the date of publication. Enclose these in parentheses—with no punctuation between them—and place them inside your normal sentence punctuation. The following citation refers to page 80 of a work by Parsa-Stay:

> Damage to gums and bones from smoking, stress, and poor nutrition may result in poorly fitting dentures (Parsa-Stay 1996).

If you incorporate the author's name in the sentence itself, give only the year in parentheses:

> According to Parsa-Stay, damage to gums and bones from smoking, stress, and poor nutrition may result in poorly fitting dentures (1996).

Here are some other types of in-text citations in the CMS author-date style:

Two authors.

> (Doundoulakis and Strugatch 2004)

Three authors.

> (Mittelman, Mittleman, and Barilla 2001)

Four or more authors ("et al." is an abbreviation for the Latin phrase "et alii," which means "and others").

> (Norton et al. 1994)

Government or corporate author.

> (Standing Committee on Health 2003)

No author listed (use the first few words of the title). In the example, the words from the title are in quotation marks because the citation is to an article; if a book is cited, the words would be it italics with no quotation marks.

> ("Denture cleansers" 2005)

Two or more sources cited together (arrange them in alphabetical order).

> (Bugel 1994, Disipio 1999)

In some of your communications, you may need to cite two or more sources by the same author written in the same year. You can distinguish between them by placing lowercase letters after the publication dates:

(Burkehardt 1998a)
(Burkehardt 1998b)

WRITING A CMS REFERENCE LIST

Illustrated here are CMS reference list entries for the most common types of print, electronic, and other sources. To create entries that are not listed here, follow the logic of these examples or consult the *Chicago Manual of Style*. Figure B.3 (page 520) shows how to arrange entries in your reference list.

Print Sources

1. Book, one author—CMS

Parsa-Stay, Flora. 1996. *The complete book of dental remedies: a guide to nutritional and conventional dental care.* Garden City Park, NY: Avery Publishing.

2. Book, two or three authors—CMS

Doundoulakis, James, and Warren Strugatch. 2004. *The perfect smile: the complete guide to cosmetic dentistry from tooth whitening and bleaching to veneers and implants.* New York: Healthy Living Books.

3. Anthology or essay collection—CMS

Lutz, Peter L, and John A. Musick, eds. 1997. *The biology of sea turtles.* Boca Raton: CRC Press.

4. Second or subsequent edition—CMS

Rodman, Lilita. 2005. *Technical communication. 3rd edition.* Toronto: Thomson Nelson.

5. Government report—CMS

Standing Committee on Health. 2003. *First Nations and Inuit dental health: report of the Standing Committee on Health.* Government of Canada Investigation Report 03–4272. Ottawa: House of Commons.

6. Corporate report—CMS

Daimler-Benz AG. 1997. *Environmental report* 1997. Stuttgart: Daimler-Benz AG.

7. Essay in a book—CMS

Sullivan, Patricia. 1996. Ethics in the computer age. *Social and ethical effects of the computer revolution.* Edited by Joseph Migga Kizza. Jefferson: McFarland.

8. Paper in a proceedings—CMS

Youra, Steven. 1996. Placing writing in engineering education. *Proceedings of the Australian Communication Conference.* Edited by Ronald Pose and Anita Jawary. Melbourne: Monash University.

9. Encyclopedia article—CMS

Rich, Elaine. 1996. Artificial intelligence. *Encyclopedia Americana.* Danbury: Grolier.

10. Pamphlet or brochure—CMS

The Wool Bureau of Canada Limited. 1998. *Pull the wool over.* Toronto: The Wool Bureau of Canada.

11. Article in journal that numbers its pages continuously through each volume—CMS

McLaurin, Joanne, and Avijit Chakrabartty. 1997. Characterization of the interactions of Alzheimer ß-amyloid peptides with phospholid membranes. *European Journal of Biochemistry* 245: 355–63.

12. Article in journal that numbers its pages separately for each issue—CMS

Bradley, John, and Gilbert Soulodre. 1997. The acoustics of concert halls. *Physics World* 10.5: 33–37.

13. Article in a popular magazine—CMS

Hessler, Peter. 2006. Hutong karma. *New Yorker,* February 12 and 20: 82–87.

14. Newspaper article—CMS

Zimonjic, Peter. 2006. Animal lovers gone wild. *Ottawa Citizen* February 18: B7.

15. Article with no author listed—CMS

"Rethinking traditional design." 1997. *Manufacturing Engineering* 118.2: 50.

Electronic Sources

16. Report available only at a World Wide Web site—CMS

International Business Machines. 2003. "What's user-centered design?" January 25. http://www-3.ibm.com/ibm/easy/eou_ext.nsf/Publish/2

17. Report downloaded from an FTP site—CMS

Modjeska, David, and Anna Marsh. 1997. Structure and memorability of web sites. University of Toronto Computer Systems Research Institute. ftp://ftp.cs.toronto.edu/csri-technical-repots/364

18. Online journal article that is not available in print—CMS

Toll, David. 1996.Artificial intelligence applications in geotechnical engineering." *Electronic Journal of Geotechnical Engineering:* 1. http://geotech.civen.okstate.edu/ejge/Abs9608.htm

19. Online journal article that is also available in print—CMS

Bernstein, A., Y. Want, G. Gratta, and T. West. 2002. Nuclear reactor safeguards and monitoring with antineutrino detectors. *Journal of Applied Physics,* 91: 4672–4676. http://ojps.aip.org/journals/doc/JAPIAU-ft/vol_91/iss_7/4672_1-div0.html

20. Encyclopedia article from an online service—CMS

Haggerty, M. 2002. Asbestosis. Edited by Jacqueline L. Longe .Gale Encyclopedia of Medicine, 1: 369–371. http://find.galegroup.com/gvrl/infomark.do?&contentSet=EBKS&type=retrieve&tabID=T001&prodId=GVRL&docId=CX3405600174&source=gale&userGroupName=otta35732&version=1.0

21. CD-ROM—CMS

Rainforest. 1997. *Encarta '97.* CD-ROM. Redmond: Microsoft.

22. Email—CMS

- Email messages are not usually provided in a reference list using this method.

Other Sources

23. Letter—CMS

- Letters are not usually provided in a reference list using this method.

24. Interview—CMS

- Interviews are not usually provided in a reference list using this method.

USING THE IEEE DOCUMENTATION STYLE

The following sections explain how to write in-text citations and the entries in a reference list using the IEEE style.

WRITING IEEE IN-TEXT CITATIONS

An IEEE citation consists of a citation number in square brackets. The method of citation works in the following way: Assign a citation number to a source the first time that you refer to it in your communication, and refer to this source by its assigned number every time you cite it throughout your communication.

As you assign these citation numbers sequentially, the first source you cite will be known as [1], the second source you cite will be known as [2], and so on throughout your communication.

The following citation refers a work by Parsa-Stay, which is the first source cited in the writer's communication:

> Damage to gums and bones from smoking, stress, and poor nutrition may result in poorly fitting dentures [1].

Later, in the same communication, the writer refers to another passage from the same work by Parsa-Stay. Note that the IEEE style encourages substituting the reference number for the name of the author:

> According to [1], eucalyptus oil helps heal gums that may be sore as a result of wearing ill-fitting dentures.

WRITING AN IEEE REFERENCE LIST

Illustrated here are IEEE reference list entries for the most common types of print, electronic, and other sources. To create entries that are not listed here, follow the logic of these examples or else consult the *Information for Authors* compiled by the IEEE Periodicals (Transactions/Journals) Department, available on the IEEE website. Figure B.4 (page 521) shows how to arrange entries in your reference list.

The most important characteristic of a reference list compiled using the IEEE method is that the order of the entries is determined by their first mentions in your communication. Arrange the entries in numerical sequence by their citation order in your document, not by using alphabetical order.

Print Sources

1. Book, one author—IEEE

[1] F. Parsa-Stay, *The Complete Book of Dental Remedies: A Guide to Nutritional and Conventional Dental Care.* Garden City Park, NY: Avery Publishing, 1996, pp. 4–6.

2. Book, two or more authors—IEEE

[2] J. Doundoulakis, and W. Strugatch, *The Perfect Smile: The Complete Guide to Cosmetic Dentistry from Tooth Whitening and Bleaching to Veneers and Implants.* New York: Healthy Living Books, 2004, pp. 122–124.

3. Anthology or essay collection—IEEE

[3] P.L. Lutz, and J.A. Musick, Eds., *The Biology of Sea Turtles.* Boca Raton: CRC Press, 1997, pp. 17–24.

4. Second or subsequent edition—IEEE

[4] L. Rodman, *Technical Communication.* 3rd ed. Toronto: Thomson Nelson, 2005, pp. 474–475.

5. Government report—IEEE

[5] Standing Committee on Health, *First Nations and Inuit Dental Health: Report of the Standing Committee on Health.* Government of Canada Investigation Report 03–4272. Ottawa: House of Commons, 2003, pp. 2–4.

6. Corporate report—IEEE

[6] Daimler-Benz AG, *Environmental Report 1997.* Stuttgart: Daimler-Benz AG, 1997, pp. 9–10.

7. Essay in a book—IEEE

[7] P. Sullivan, "Ethics in the computer age," *Social and Ethical Effects of the Computer Revolution.* J.A. Kizza, Ed. Jefferson: McFarland, 1996, pp. 288–297.

8. Paper in a proceedings—IEEE

[8] S. Youra, "Placing Writing in Engineering Education," *Proceedings of the Australian Communication Conference.* R. Pose, and A. Jawary, Eds. Melbourne: Monash University, 1996, pp. 59–76.

9. Encyclopedia article—IEEE

[9] E. Rich, "Artificial intelligence," *Encyclopedia Americana.* Danbury: Grolier, 1996, pp. 407–412.

10. Pamphlet or brochure—IEEE

[10] The Wool Bureau of Canada Limited, *Pull the Wool Over.* Toronto: The Wool Bureau of Canada, 1998.

11. Article in journal that numbers its pages continuously through each volume—IEEE

[11] J. McLaurin, and A. Chakrabartty, "Characterization of the interactions of Alzheimer ß-amyloid peptides with phospholid membranes," *European Journal of Biochemistry* 245, pp. 355–63, 1997.

12. Article in journal that numbers its pages separately for each issue—IEEE

[12] J. Bradley, and G. Soulodre, "The acoustics of concert halls," *Physics World* 10.5: pp. 33–37, 1997.

13. Article in a popular magazine—IEEE

[13] P. Hessler, "Hutong karma," *New Yorker,* pp. 82–87, Feb. 13 and 20, 2006.

14. Newspaper article—IEEE

[14] P. Zimonjic, "Animal lovers gone wild," *Ottawa Citizen,* p. B7, Feb. 18, 2006.

15. Article with no author listed—IEEE

[15] Rethinking traditional design. *Manufacturing Engineering,* 118.2:50, 1997.

Electronic Sources

16. Report available only at a World Wide Web site—IEEE

[16] International Business Machines. (2003, January). What's User-Centered Design*?* [Online]. Available: http://wwibm.com/ibm/easy/eou_ext.nsf/Publish/2

17. Report downloaded from an FTP site—IEEE

[17] D. Modjeska and A. Marsh. (1997). Structure and Memorability of Web Sites. University of Toronto Computer Systems Research Institute. [Online]. Available: ftp://ftp.cs.toronto.edu/csri-technical-repots/364

18. Online journal article that is not available in print—IEEE

[18] D. Toll. (1996). Artificial intelligence applications in geotechnical engineering. *Electronic Journal of Geotechnical Engineering*. [Online]. *1*. Available: http://geotech.civen.okstate.edu/ejge/Abs9608.htm

19. Online journal article that is also available in print—IEEE

[19] A. Bernstein, Y. Want, G. Gratta, and T. West. (2002, November). Nuclear reactor safeguards and monitoring with antineutrino detectors. *Journal of Applied Physics.* [Online]. *91*: 4672-4676. Available: http://ojps.aip.org/journals/doc/JAPIAU-ft/vol_91/iss_7/4672_1-div0.html

20. Encyclopedia article from an online service—IEEE

[20] M. Haggerty. (2003). Asbestosis. Jacqueline L. Longe, Ed. Gale *Encyclopedia of Medicine.* [Online]. 1:369-371. Available: http://find.galegroup.com/gvrl/infomark.do?&contentSet=EBKS&type=retrieve&tabID=T001&prodId=GVRL&docId=CX3405600174&source=gale&userGroupName=otta35732&version=1.0

21. CD-ROM—IEEE

[21] Microsoft Corporation. (1997). Rainforest. *Encarta '97.* CD-ROM. Redmond: Microsoft.

22. Email—IEEE

- Email messages are not usually provided in a reference list using the IEEE referencing method. However, your professor may wish you to use the following form of entry:

[22] M. Squire. (2005, December 1). ENL4004 Foundations. [Email message].

Other Sources

23. Letter—IEEE

- Letters and interviews are not usually provided in a reference list using the IEEE method. However, your professor may wish you to use the following form of entry:

[23] D. Galway. (2006, September 19). IEEE Standards Style Versus Article Submission Style. [Personal interview]. Ottawa: Algonquin College.

24. Interview—IEEE

- Interviews are not usually provided in a reference list using this method. However, your professor may wish you to use a format similar to that above.

FIGURE B.1

APA Reference List

Second and subsequent lines are indented.

When the list includes two or more items by the same author, the oldest appears first.

For the second and subsequent items by the same person, the author's name is repeated.

Items by corporate and government groups are alphabetized by the groups' names (spelled out).

Items without authors are alphabetized by the title.

References

Abernethy, B., Kippers, V., Mackinnon, L. T., Neal, R. J., & Hanrahan, S. (1997). *The biophysical foundations of human movement.* Champaign, IL: Human Kinetics.

Gould, S. J. (1989). *Wonderful life: The burgess shale and the nature of history.* New York: Norton.

Gould, S. J. (1995). *Dinosaur in a haystack: Reflections in natural history.* New York: Harmony.

International Business Machines. (n.d.) What's user-centered design? Retrieved January 25, 2003, from http://www-3.ibm.com/ibm/easy/eou_ext.nsf/Publish/2

McLaurin, J., & Chakrabartty, A. (1997). Characterization of the interactions of Alzheimer ß-amyloid peptides with phospholipid membranes. *European Journal of Biochemistry, 245,* 355–63.

Rethinking traditional design. (1997). *Manufacturing Engineering 118*(2), 50.

NEL

FIGURE B.2

MLA List of Works Cited

Second and subsequent lines are indented.

When the list includes two or more items by the same author, they are alphabetized by title.

For the second and subsequent items by the same person, the author's name is replaced by three hyphens, followed by a period.

Items by corporate and government groups are alphabetized by the groups' names (spelled out).

Items without authors are alphabetized by the title.

Works Cited

Abernethy, Bruce, Vaughn Kippers, Laurel Traeger Mackinnon, Robert J. Neal, and Stephanie Hanrahan. *The Biophysical Foundations of Human Movement.* Champaign: Human Kinetics, 1997.

Gould, Stephen Jay. *Dinosaur in a Haystack: Reflections in Natural History.* New York: Harmony, 1995.

---. *Wonderful Life: The Burgess Shale and the Nature of History.* New York: Norton, 1989.

International Business Machines. *What's User-Centered Design?* 25 Jan. 2003 <http://www-3.ibm.com/ibm/easy/eou_ext.nsf/Publish/2>

McLaurin, JoAnne, and Avijit Chakrabartty. "Characterization of the Interactions of Alzheimer ß-amyloid Peptides with Phospholipid Membranes." *European Journal of Biochemistry,* 245 (1997): 355–63.

"Rethinking Traditional Design." *Manufacturing Engineering 118.2* (1997): 50.

FIGURE B.3

CMS List of References

CMS List of References

Bradley, John, and Gilbert Soulodre. 1997. The acoustics of concert halls. *Physics World* 10.5: 33-37.

Daimler-Benz AG. 1997. *Environmental report 1997.* Stuttgart: Daimler-Benz AG.

Doundoulakis, James, and Warren Strugatch. 2004. *The perfect smile: the complete guide to cosmetic dentistry from tooth whitening and bleaching to veneers and implants.* New York: Healthy Living Books.

Hessler, Peter. 2006. Hutong karma. *New Yorker*, February 12 and 20: 82-87.

Lutz, Peter L., and John A. Musick, eds. 1997. *The biology of sea turtles.* Boca Raton: CRC Press.

McLaurin, Joanne, and Avijit Chakrabartty. 1997. Characterization of the interactions of Alzheimer ß-amyloid peptides with phospholid membranes. *European Journal of Biochemistry* 245: 355-63.

Parsa-Stay, Flora. 1996. *The complete book of dental remedies: a guide to nutritional and conventional dental care.* Garden City Park, NY: Avery Publishing.

Rethinking traditional design. 1997. *Manufacturing Engineering* 118.2:50.

Rich, Elaine. 1996. Artificial intelligence. *Encyclopedia Americana.* Danbury: Grolier.

Rodman, Lilita. 2005. *Technical communication.* 3rd edition. Toronto: Thomson Nelson.

---. 2002. You attitude: a linguistic perspective. *Technostyle* 12.2.

Standing Committee on Health. 2003. *First Nations and Inuit dental health: report of the Standing Committee on Health.* Government of Canada Investigation Report 03-4272. Ottawa: House of Commons.

Sullivan, Patricia. 1996. Ethics in the computer age. *Social and ethical effects of the computer revolution.* Edited by Joseph Migga Kizza. Jefferson: McFarland.

The Wool Bureau of Canada Limited. 1998. *Pull the wool over.* Toronto: The Wool Bureau of Canada.

Youra, Steven. 1996. Placing writing in engineering education. *Proceedings of the Australian Communication Conference.* Edited by Ronald Pose and Anita Jawary. Melbourne: Monash University.

Zimonjic, Peter. 2006. Animal lovers gone wild. *Ottawa Citizen*, February 18: B7.

Second and subsequent lines are indented.

Items without authors are alphabetized by the title.

If the list includes two or more items by the same author, they are alphabetized by title.

For second and subsequent items by the same person, the author's name is replaced by three hyphens, followed by a period.

Items by corporate and government groups are alphabetized by the group's name (spelled out).

FIGURE B.4

IEEE List of References

All citation text is indented and aligned so that only the bracketed number of each entry is aligned to the left.

IEEE List of References

[1] F. Parsa-Stay, *The Complete Book of Dental Remedies: A Guide to Nutritional and Conventional Dental Care.* Garden City Park, NY: Avery Publishing, 1996, pp.4–6.

[2] J. Doundoulakis and A. Strugatch, *The Perfect Smile: The Complete Guide to Cosmetic Dentistry from Tooth Whitening and Bleaching to Veneers and Implants.* New York: Healthy Living Books, 2004, pp. 122–124.

[3] P.L. Lutz and J.A. Musick, Eds., *The Biology of Sea Turtles.* Boca Raton: CRC Press, 1997, pp. 17–24.

[4] L. Rodman, *Technical Communication.* 3rd ed. Toronto: Thomson Nelson, 2005, pp. 474–475.

[5] Standing Committee on Health, *First Nations and Inuit Dental Health: Report of the Standing Committee on Health.* Government of Canada Investigation Report 03-4272. Ottawa: House of Commons, 2003, pp. 2–4.

[6] Daimler-Benz AG, *Environmental Report 1997.* Stuttgart: Daimler-Benz AG, 1997, pp. 9–10.

Projects

PROJECTS

Project 1 Job Application Package

Project 2 Promotional Website

Project 3 Informational Page

Project 4 Unsolicited Recommendation

Project 5 Brochure

Project 6 Instructions

Project 7 User Test and Report

Project 8 Project Proposal

Project 9 Progress Report

Project 10 Formal Report or Proposal

Project 11 Oral Briefing I: Project Plans

Project 12 Oral Briefing II: Project Results

This appendix contains writing and speaking assignments that your professor might ask you to complete. All share this important feature: they ask you to communicate to particular people for specific purposes that closely resemble the purposes you will have for writing in your career.

Some of the assignments contain specifications about such things as length and format. Your professor may change these specifications to tailor the assignments to your writing course.

PROJECT 1 Job Application Package

Write a résumé and a job application letter in response to a real job advertisement. Address them to a real person, in response to a real vacancy, in an organization with which you might actually seek employment. If you will graduate this year, you will probably want to write for a full-time, permanent position. If you aren't about to graduate, you may want to apply for a part-time or co-op position. If you are currently working, imagine that you have decided to change jobs, perhaps to obtain a promotion, secure higher pay, or find more challenging and interesting work.

To complete this project, you will need to do some research. Among other things, you will have to find an organization that is hiring people for the kind of job you want, and you will need to learn something about the organization so you can persuade your readers in the organization that you are knowledgeable about it. If the sources you find don't give the name of some specific person to whom you can address your letter, phone the organization to ask for the name of the human resources manager or the manager of the particular department in which you would like to work. While you work on this assignment, keep this real person in mind—even if you will not actually send your letter to him or her.

Remember that the appearance of your résumé and letter will affect your readers, as will your attention to such details as grammar and spelling.

As part of your package, include the job advertisement for which you are applying. Also include the names, job titles, employers, email addresses, and phone numbers of three references. These may be included within your résumé. If you choose instead to have your résumé say, "References available upon request," enclose a copy of the list of references you would send if the list were requested by an employer. Throughout your work on this project, you should carefully and creatively follow the advice given in Chapter 11 on résumés and letters of application.

PROJECT 2 Promotional Website

For this project, you are to create and post a promotional website for a small business that you are setting up. Begin by identifying a target market and product or service. Then reflect that market and product or service as you identify the audience and purpose for your website. The product or service should be related to your area of study or your current employment. Your website should have the following elements:

- Four or more pages, one of which is a home page (titled "index.html")
- Links among your pages

- One or more links to other websites related to your product or service
- At least one image per page

To count as a "page" in this project, a page must fill one window on a computer screen when the window is at its maximum size. The material in the window must include some text and at least one image. You may wish to use a home publishing software application to help you design your website.

All pages, including the home page, should be informative and persuasive. The text is to be text that you have written, not text you have downloaded or otherwise copied from someone else. You may obtain your images by downloading them, provided you are sure that they are not copyrighted and that use of them is not otherwise restricted. Avoid movies, large images, and other items that consume large quantities of memory.

Your site may include a link to your personal webpage. However, this page will not count among the four that must be about your subject.

Your website should have the following characteristics:

- All information is presented in a way that is suited to your audience (your target market)
- All information is presented in a way that is suited to your purpose
- The text is accurate, informative, and well written
- The pages are easy to read, use, and navigate
- The site is well organized
- The pages are unified in terms of visual design, writing style, and purpose

Submit a memo or email to your professor that identifies the URL of your website and the website's audience (target market), its purpose, and the product or service your business provides.

PROJECT 3 Informational Page

A transportation schedule, a chart comparing the features of competing gaming systems, a timeline that uses diagrams and words to explain the events in a geological formation's history—all of these are "informational pages." They convey a complex set of information or ideas mainly through visual design, instead of relying solely on sentences and paragraphs.

For this assignment, create an informational page addressed to a specific set of readers. Present your information visually to achieve your communication objectives. You may wish to use tables or charts, to mix diagrams with words, or to use only diagrams. Use only one side of an 8½" × 11" sheet of paper.

When planning your informational page, think carefully about how your readers will use your information and about how you can use graphics to make the facts you present as accessible, understandable, and useful as possible. Here are some pages you might create:

- A description of a process that would be important for potential or new customers of your employer to know
- An explanation of a concept from your workplace or area of study that is important for members of the public to understand as they make practical decisions

- A reference card for new clients or volunteers at a social service or parks and recreation program in your community
- A study guide that might help a fellow student learn a topic that will be on an exam

PROJECT 4 Unsolicited Recommendation

This assignment is your chance to improve the world—or at least one small corner of it. You are to write a letter or memo of no more than two pages in which you make a recommendation for improving the operation of some organization with which you have personal contact: your work group or department, another department at your workplace, the company for which you worked last summer or work term, or a club or team to which you belong.

There are four important restrictions on the recommendations you make:

1. Your recommendation must concern a real situation in which your letter or memo can really bring about change. Focus on situations that can be improved by modest measures, for which you will argue effectively in your letter or memo. It is not necessary, however, that your communication aim to bring about a complete solution. Instead, you might aim to persuade one of the key people in the organization that your recommendation, if accepted, would serve the organization's best interests.
2. Your recommendation must be unsolicited; that is, it must be addressed to someone who has not asked for your advice.
3. Your recommendation must concern the way an organization operates, not just the way one or more individuals think or behave.
4. Your recommendation must not involve a problem that would be decided in an essentially political manner by elected officials. Do not address a problem that would be raised in a political campaign.

Of course, you will have to write to an actual person, someone who has the power to help make the change you are recommending. You may have to investigate to learn who that person is. Try to learn also how that person feels about the situation you hope to improve. Bear in mind that most people are inclined to reject advice that they haven't asked for—that's part of the challenge of this assignment.

You may choose to write on matters such as the following:

- A way that your employer could more efficiently handle customers, merchandise, orders, or inquiries
- A strategy for your employer or program to increase participation, loyalty, or retention of customers, students, or alumni
- A proposal that your employer or Students' Association establish a service for employees or students

Bear in mind that a recommendation must compare two alternatives: keeping things the way they are now, and changing them to the way you think they should be. You will have to make the change seem the better alternative from your reader's perspective. To do this, you will find it helpful to understand why the organization does things in the present way. By understanding the goals of the present method, you will probably gain insight into the criteria you reader will apply when comparing the present method with the method you recommend.

PROJECT 5 Brochure

Create a brochure about a product or service offered by your employer, or a product or service provided by a small business you might set up. Alternatively, create a brochure for your program, for a student service on your campus, or for a service or an attraction in your community.

Begin by interviewing people at the organization to learn about their aims for such a brochure. Then, if possible, talk to members of the target audience for the brochure. Remember that to be effective, the brochure must meet the needs of both the organization and the readers.

Use a folded 8½" × 14" sheet of paper so that there are three columns (or panels) on each side of the sheet. When the brochure is folded, the front panel should serve as a cover. Each panel should contain textual and graphical elements. You may wish to use a home publishing software application to help you design your brochure. Along with your brochure, turn in copies of any existing materials you used while working on this project.

Your success in this project will depend largely on your ability to predict the questions your readers will have about your subject—and on your ability to answer those questions clearly, concisely, and usefully. Think very carefully as well about how you want your brochure to alter your readers' attitudes toward your subject. Remember that along with the text, the neatness and graphic design of your brochure will have a strong impact on your readers' attitudes.

PROJECT 6 Instructions

Write a set of instructions that will enable your readers to operate some device or perform some process used in your area of study. The procedure must involve at least twenty-four steps.

With the permission of your professor, you may also choose from topics that are not related to your major. Such topics might include the following:

- Upgrading a computer
- Using some special feature of a software package
- Operating a piece of equipment used in your major area of study
- Changing a bicycle or car tire
- Rigging a sailboat
- Installing a component or making adjustments to an item of sports equipment
- Starting an aquarium
- Adopting a pet
- Some other procedure of interest to you that includes at least twenty-four steps

Your instructions should guide your readers through some specific process that your classmates or professor could actually perform. Do not write generic instructions for performing a general procedure. For instance, do not write instructions for "Operating a Microscope" but rather for "Operating the Thompson Model 200 Microscope."

Be sure to divide the overall procedure into groups of steps rather than presenting all the steps in a single list. Use headings to label the groups of steps.

When preparing your instructions, pay careful attention to the graphic design of your finished communication. You must include at least one illustration, and you may rely heavily on figures if they are the most effective way for you to achieve your objectives. In fact, your instructions need not contain a single sentence.

Finally, your professor may require you to use a page design that has two or more columns (rather than having a single column of type that runs all the way from the left-hand margin to the right-hand margin). In a two-column design, you might put all of your steps in the left-hand column and all of your accompanying illustrations in the right-hand column. Alternatively, you might mix text and figures in both columns. Large figures and the title for the instructions can span both columns.

You may use the format for your instructions that you believe will work best—whether it is a single sheet of 8½ × 11-inch paper, a booklet printed on smaller paper, or some other design.

Don't forget that your instructions must be accurate.

PROJECT 7 User Test and Report

As you learned in Part VI, one excellent way to evaluate the usability and persuasiveness of a draft is to conduct a user test in which you give your draft to members of your target audience, asking them to use it in the same way that your target readers will use the final draft. For this assignment, you are to conduct a user test of a nearly finished draft of a project you are preparing in this course (or another course, if your professor permits), then report the results in a memo to your professor. Instructions make an excellent subject for a user test, but other communications can be evaluated in this way also.

For your test, use a draft that is as close as possible to what you envision for your final draft. Ask two people to serve as your test readers, and arrange for them to work independently. If you cannot recruit test readers who are from your target audience, choose people who resemble the target readers as closely as possible. Similarly, if you cannot arrange for your test readers to read in exactly the same circumstances that your target readers will, simulate those conditions in some reasonable way. Also use simulation if your communication involves a potentially dangerous step (such as jacking up a car or pouring a strong acid from one container to another) that might result in injury to your test readers if they make a mistake.

Write your report in the memo format and use the structure for empirical research reports. To make your report readable and informative for your professor, consider the following advice about the sections of your report:

- **Introduction.** Remind your professor of the topic and target audience of the communication you are testing.
- **Objectives.** Identify the objectives of your communication (see Chapter 18) and those of your test. Remember that even instructions have to be persuasive as well as usable.

- **Method.** Describe your draft (how closely does it resemble your planned final draft?), your test readers (who were they, and why were they good representatives of your target audience?), the location of your test (how closely did it resemble the setting in which your target readers will use your communication?), and your procedure (what did you ask your test readers to do, and how did you gather information from them?).
- **Results and discussion.** Report the results your test produced (where did your test readers have difficulties, and what did they say about your communication?) and tell what these indicate about your communication. Be quite specific in this section.
- **Conclusion.** Tell what you learned overall from your test and indicate the specific revisions you will make as a result of what you learned.

PROJECT 8 Project Proposal

Write a proposal seeking your professor's approval for a report or project you will prepare later this term.

Your work on this proposal serves three important purposes. First, it provides an occasion for you and your professor to agree about what you will do for the later report or project. Second, it gives you experience at writing a proposal, a task that will be very important to you in your career. Third, it gives you a chance to demonstrate your mastery of the material in Part I ("Defining Your Communication's Objectives").

Notice that while working on this assignment, you will have to define the objectives of two different communications: (1) the *proposal* you are writing now, which is addressed to your professor, and (2) the report or *project* you are seeking approval to write, whose purpose and audience you will have to describe to your professor in the proposal.

When writing your proposal, you may think of your professor as a person who looks forward with pleasure to working with you on your project or report and wants to be sure that you choose a report or project from which you can learn a great deal and on which you can do a good job. Until your professor learns from your proposal some details about your proposed project or report, however, his or her attitude toward it will be neutral. While reading your proposal, your professor will seek to answer many questions, including the following:

- What kind of communication do you wish to prepare?
- Who will its readers be?
- What is its purpose?
  What is the final result you want it to bring about?
  What task will it enable its readers to perform?
  How will it alter its readers' attitudes?
- Is this a kind of communication you will have to prepare at work?
- Can you write the communication effectively in the time allocated using resources that are readily available to you?

For additional insights into the questions your professor (like the reader of any proposal) will ask, see Chapter 7.

Your proposal should be between 400 and 800 words long. Write it in a memo format (see Appendix A), using headings. Include a Gantt or another type of scheduling chart showing deliverables and due dates to indicate your proposed schedule. Also, if the proposal is for a report, include a sequentially ordered and hierarchical outline. This outline can be considered a proposed contents page for your report.

PROJECT 9 Progress Report

Write a report of between 400 and 800 words in which you tell your professor how you are progressing on the writing project you are currently preparing. Be sure to give your professor a good sense not only of what you have accomplished but also of what problems you have encountered or anticipate. Use the memo format (see Appendix A).

If possible, include an updated Gantt or another type of scheduling chart to show the differences (if any) between your original proposed dates and the actual dates of delivery. Also, if the progress on which you are reporting is for a report, include an updated contents page to reflect changes in the organization of your findings.

PROJECT 10 Formal Report or Proposal

Write an empirical research report, feasibility report, or proposal. Whichever form of communication you write, it must be designed to help some organization—real or imaginary—solve some problem or achieve some goal, and you must write it in response to a request (again, real or imaginary) from the organization you are addressing.

A real situation is one you have actually encountered. It might involve your employer, your school or department, or a service group to which you belong—to name just a few of the possibilities. Students writing on real situations have prepared projects with such titles as:

- "**Feasibility of Using a Computer Database to Catalogue the Art Department's Slide Library.**" The student wrote this feasibility report at the request of the chair of the Art Department.
- "**Attitudes of Participants in Merit Hotel's R.S.V.P. Club.**" The student wrote this empirical research report at the request of the hotel, which wanted to find ways of improving a marketing program that rewarded administrative assistants who booked their companies' visitors at that hotel rather than at one of the hotel's competitors.
- "**Expanding the Dietetic Services at the Campus Health Centre: A Proposal.**" The student wrote this proposal to the college administration at the request of the part-time dietitian employed by the Health Centre.

An imaginary situation is one that you create to simulate the kinds of situations you will find yourself in once you begin your career. You pretend that you have begun working for an employer who has asked you to use your specialized training to solve some problems or answer some questions that face his or her organization. You may imagine that you are a regular employee or that you are a special consultant. Students writing about imaginary situations have prepared formal reports with titles such as:

- "**Improving the Operations of the Gift Shop at Six Flags of Ohio.**" The student who wrote this proposal had worked at this shop for a summer job; she imagined that she had been hired by the manager to study its operation and recommend improvements.
- "**Performance of Three Lubricants at Very Low Temperatures.**" The student wrote this empirical research report about an experiment he had conducted in a laboratory class. He imagined that he worked for a company that wanted to test the lubricants for use in manufacturing equipment used at temperatures below –50°C.
- "**Upgrading the Monitoring and Communication System in the Psychology Clinic.**" The student who wrote this report imagined that she had been asked by the Psychology Clinic to investigate the possibility of purchasing equipment that would improve its monitoring and communication system. All of her information about the clinic and the equipment were real.

For this project, use the report format (see Appendix A). Remember that your purpose is to help your readers make a practical decision or take a practical action in a real or imaginary organization. The body of your report should be between twelve and twenty pages long (not counting cover, executive summary, title page, contents page, appendixes, and similar parts).

Oral Briefing I: Project Plans

At work, you will sometimes be asked to report in brief talks about projects on which you are working. For this assignment, you are to give an oral briefing to the class about your final project. Here are the things you should cover:

- **What kind of communication are you writing?** Who will your readers be? What role will you be playing? Identify your readers by telling what organization they are in and what positions your key readers hold. Describe your role by saying whether you are imagining that you work for the company as an employee or have been hired as a consultant. State whom you report to.
- **What organizational problem will your communication help your readers solve?** What need or goal will it help them satisfy or reach? Provide full background so that your classmates can understand the situation from your readers' point of view.
- **What are you doing to solve the problem?**
  Your research activities: What kind of information are you gathering and how, or what kind of analysis are you providing and why?
  Your writing activities: How do you plan to organize and present your information? What will your communication look like?
- **What is the gist of your message to your readers?** What are the main points you are planning to make?

As you prepare and deliver your oral briefing, pretend that you are interviewing for a job (or for a new job) and that the prospective employer has asked you to give an oral briefing about a project of yours for which you are now writing a report or proposal. The members of your class can play the role of the people your employer has asked to attend your presentation. Pretend that your classmates have not heard about your project as yet, even though you may have already discussed it in class several times. This means that you will have to provide all the background information that will enable your listeners to understand the organizational situation in which you are writing.

As the name implies, a briefing is a brief presentation. Make yours between four and five minutes long—no longer. Gauge the time by doing timed rehearsals.

PROJECT 12 Oral Briefing II: Project Results

At work, people often present the results of their major projects twice: once in a written communication and a second time in an oral briefing that covers the major points of the written document. In some ways, this briefing is like an executive summary—an overview of all the important things presented in more detail in writing.

For this assignment, you are to give an oral briefing on one of your writing projects. Address the class as if it were the same audience that you address in writing, and imagine that the audience has not yet read your communication.

Limit your briefing to four or five minutes—no longer. Gauge the time by doing timed rehearsals.

REFERENCES

Anderson, J. R. (1995). Attention and sensory information processing. *Cognitive psychology and its implications* (4th ed.; pp. 40–48). New York: W. H. Freeman.

Anderson, P. V. (1985). What survey research tells us about writing at work. In L. Odell & D. Goswami (Eds.), *Writing in nonacademic settings* (pp. 3–85). New York: Guilford P.

Andrews, D. C., & Andrews, W. D. (1992). *Business communication* (2nd ed.). New York: Macmillan.

Anonymous. (1994). Personal interview with corporate executive who requested that the company remain anonymous.

AOL Press: Welcome. America On-Line. April 1998 http://www.aolpress.com/press/2.0/usrguide/preface.htm/

Barnum, C. M. (1993). Working with people. In C. M. Barnum & S. Carliner (Eds.), *Techniques for technical communicators* (pp. 122–125). New York: Macmillan.

Barnum, C. M., & Fischer, R. (Second Quarter 1984). Engineering technologists as writers: Results of a survey. *Technical Communication, 31,* 9–11.

Beaufort, A. (1999). *Writing in the real world: Making the transition from school to work.* New York: Teachers College Press.

Beebe, S. A. (1974). Eye contact: A nonverbal determinant of speaker credibility. *Speech Teacher, 23,* 21–25. Cited in M. F. Vargas, *Louder than words.* Ames, IA: Iowa State UP, 1986.

Beer, D. F., & McMurrey, D. (1997). *A guide to writing as an engineer.* New York: Wiley.

Benne, K. D., & Sheats, P. (2000). Functional roles of group members. In W. L. French, C. H. Bell, Jr., & R. A. Zawacki (Eds.), *Organization development and transformation: Managing effective change.* Boston: McGraw-Hill.

Berghel, H. (April 1997). E-mail—the good, the bad, and the ugly. *Communications of the ACM, 40,* 4.

Bereiter, C., & Scardamalia, M. (1993). *Surpassing ourselves: An inquiry into the nature and implications of expertise.* Chicago, IL: Open Court.

Bevlin, M. E. (1989). *Design through discovery.* Ft. Worth, TX: Holt, Rinehart and Winston.

Boiarsky, C. (1993). *Technical writing.* Boston: Allyn & Bacon.

Boren, M. T., & Ramey, J. (September 2000). Thinking aloud: Reconciling theory and practice. *IEEE Transactions on Professional Communication, 43*(3), 261–278.

Bosley, D. S. (1993). Cross-cultural collaboration: Whose culture is it, anyway? *Technical Communication Quarterly, 2,* 51–62.

Bostrom, R. N. (1981). *Persuasion.* Englewood Cliffs, NJ: Prentice-Hall.

Bransford, J. D., & Johnson, M. K. (1972). Contextual prerequisites for understanding: Some investigations of comprehension and recall. *Journal of Verbal Learning and Verbal Behavior, 11,* 717–726.

Coleman, E. B. (1964). The comprehensibility of several grammatical transformations. *Journal of Applied Psychology, 48,* 186–190.

Couture, B., and Rymer, J. (1993). Situational exigence: Composing processes on the job by writer's role and task value. In R. Spilka (Ed.), *Writing in the workplace: New research perspectives.* Carbondale, IL: Southern Illinois UP.

Covey, S. R. (1989). *The seven habits of highly effective people.* New York: Simon & Schuster.

Cross, G. A. (1993). The interrelation of genre, context, and process in the collaborative writing of two corporate documents. In R. Spilka (Ed.), *Writing in the workplace: New research perspectives* (pp. 141–152). Carbondale, IL: Southern Illinois UP.

Dias, P., & Paré, A. (Eds.) (2000). *Transitions: Writing in academic and workplace settings.* Cresskill, NJ: Hampton Press.

Doheny-Farina, S. (1992). *Rhetoric, innovation, technology: Case studies of technical communication in technology transfers.* Cambridge, MA: MIT.

Filho, S. A., et al. (1986). Stable yeast transformants that secrete functional a-Amylase encoded by cloned mouse pancreatic cDNA. *Biotechnology, 4,* 311–315.

General Electric Company. (1995). *Refrigerator use and care guide.* Louisville, KY: General Electric.

Grosse, R., & Kujawa, D. (1988). *International business: Theory and managerial applications.* Burr Ridge, IL: Richard D. Irwin.

Hackos J. T. (1994). *Managing your documentation projects.* New York: John Wiley & Sons.

Harcourt, J., Krizan, A.C., & Merrier, P. (1991). Teaching résumé content: Hiring officials' preferences versus college recruiters' preferences. *Business Education Forum, 45*(7), 13–17.

Hays, R. B. (1985). A longitudinal study of friendship development. *Journal of Personality and Social Psychology, 48,* 909–924.

Herzberg, F. (1968). *Work and the nature of man.* Cleveland: World.

Ishii, S. (1985). Thought patterns as modes of rhetoric: The United States and Japan. In L. A. Samovar & R. E. Porter (Eds.), *Intercultural communication: A reader* (4th ed.; pp. 97–102). Belmont, CA: Wadsworth.

Jewett, M., & Margolis, R. (1987). A study of the effects of the use of overhead transparencies on business meetings. Reported in B. Y. Auger, *How to run better business meetings: A reference guide for managers.* New York: McGraw-Hill.

Kachru, J. (1988). Writers in Hindi and English. In Alan Purves (Ed.), *Writing across languages and cultures: Issues in contrastive rhetoric* (pp. 109–137). Thousand Oaks, CA: Sage.

Kant, I. (1987). *Fundamental principles of the metaphysics of morals.* Trans. T. K. Abbott. Buffalo, NY: Prometheus.

Kelman, H. C., & Hovland, C. I. (1953). Reinstatement of the communicator in delayed measurement of opinion change. *Journal of Abnormal and Social Psychology, 48,* 327–335.

Kiggins, C. (1999). Companies spending $300,000 on communication skills training. *CareerMag.* Retrieved March 2, 1999 from www.careermag.com

Klare, G. R. (1968). The role of word frequency in readability. *Elementary English, 45,* 12–22.

Klare, G. R. (2nd Quarter 1977). Readable technical writing: Some observations. *Technical Communication, 24*(2), 2.

Kleck, R. E., & Nuessle, W. (1967). Congruence between indicative and communicative functions of eye-contact in interpersonal relations. *British Journal of Social and Clinical Psychology, 6,* 256–266.

Kleinke, C. L., Bustos, A. A., Meeker, F. B., & Staneski, R. S. (1973). Effects of self-attributed gaze on interpersonal evaluations between males and females. *Journal of Experimental Psychology, 9,* 154–163.

Kostelnick, C., & Roberts, D. D. (1998). *Designing visual language: Strategies for professional communicators.* Needham Heights, MA: Allyn & Bacon.

Kozyrskyj, A., et al. (2005). High-cost users of pharmaceuticals: Who are they? Winnipeg: Manitoba Centre for Health Policy.

Kulhavy, R. W., & Schwartz, N. H. (Winter 1981). Tone of communications and climate of perceptions. *Journal of Business Communication, 18,* 17–24.

Lauer, J. (1994). Persuasive writing on public issues. In R. Winterowd & V. Gillespie (Eds.), *Composition in context* (62–72). Carbondale, IL: Southern Illinois UP.

Lay, M. M. (1989). Interpersonal conflict in collaborative writing: What can we learn from gender studies. *Journal of Business and Technical Communication, 3*(2), 5–28.

Layton, P., & Simpson, A. J. (1975). Deep structure in sentence comprehension. *Journal of Verbal Learning and Verbal Behavior, 14,* 658–664.

Lustig, M. W., & Koester J. (1993). *Intercultural competence: Interpersonal communication across cultures.* New York: Harper Collins.

Mancusi-Ungaro, H. R., Jr., & Rappaport, N. H. (April 1986). Preventing wound infections. *American Family Physician, 33,* 152.

Marsico, T. Personal communication with the author. 14 April 1997.

Maslow, A. H. (1970). *Motivation and personality.* New York: Harper & Row.

Mathes, J. C., & Stevenson, D. W. (1991). *Designing technical reports* (2nd ed.). New York: Macmillan.

Mehrabian, A. (1972). *Nonverbal communication.* Chicago: Aldine.

Miller, C. R. (1984). Genre as social action. *Quarterly Journal of Speech, 70,* 151–167.

Mohler, J. L. (1997). *Teach yourself how to become a webmaster in 14 days.* Indianapolis: Sams.net.

Munter, M. (1987). *Business communication: Strategy and skill.* Englewood Cliffs, NJ: Prentice-Hall.

Murray, R. L. (1982). *Understanding radioactive waste.* Columbus, OH: Battelle.

Nielsen, J. (2000). *Designing Web usability: The practice of simplicity.* Indianapolis, IN: New Riders.

Nishiyama, K. (1983). Intercultural problems in Japanese multinationals. *Communication: The Journal of the Communication Association of the Pacific, 12,* 58.

Northy, M. (1990). The need for writing skill in accounting firms. *Management Communication Quarterly, 3,* 480.

Office of Cancer Communications. (1983). *Taking time: Support for people with cancer and the people who care about them.* Bethesda, MD: National Cancer Institute.

Patel, V. L., & Groen, G. J. (1991). The general and specific nature of medical expertise: A critical look. In K. A. Cricsson & J. Smith (Eds.), *Toward a general theory of expertise: Prospects and limits* (pp. 93–125). Cambridge: Cambridge UP.

Patterson, V. (1996). Résumé talk from recruiters. *Journal.* Miami University: CPPO.

Petty, R. E., & Cacioppo, J. T. (1981). *Attitudes and persuasion: Classic and contemporary approaches.* Dubuque, IA: William C. Brown.

Petty, R. E., & Cacioppo, J. T. (1986). *Communication and persuasion: Central and peripheral routes to attitude change.* New York: Springer-Verlag.

Pinelli, T. E., Glassman, M., Oliu, W. E., & Barclay, R. O. (1989). *Technical communications in aeronautics: Results of an exploratory study (TM-101626).* Washington, DC: National Aeronautics and Space Administration.

Ramey, J. (1997). Fact, context, communication: The value added to data by information design. *IEEE International Professional Communication Conference (IPCC), IPCC '97 Conference Record* (pp. 385–392). New York: IEEE.

Ray, G. B. (1986). Vocally cued personality prototypes: An implicit personality theory approach. *Communication Monographs, 53,* 266–276.

Ricks, D. A. (1983). *Big business blunders: Mistakes in multinational marketing.* Burr Ridge, IL: Dow Jones-Irwin.

Rogers, C. R. (1952). Communication: Its blocking and its facilitation. *Harvard Business Review, 30,* 46–50.

Roth, L. (1993). Education makes a difference: Results of a survey of writing on the job. *Technical Communication Quarterly, 2,* 177–184.

Roth, R. F. (1982). *International marketing communications.* Chicago: Crain.

Sauer, B. A. (1994). The dynamics of disaster: A three-dimensional view of documentation in a tightly regulated industry. *Technical Communication Quarterly, 3,* 393–419.

Schriver, K. (1997). *Dynamics of document design: Creating text for readers.* New York: Wiley.

Selber, S. Email to the author. 31 March 1998.

Sides, C. H. (1984). *How to write papers and reports about computer technology.* Philadelphia, PA: ISI.

Smart, G. (1993). Genre as community invention: A central bank's response to its executives' expectations as readers. In R. Spilka (Ed.), *Writing in the workplace: New research perspectives* (pp. 124–140). Carbondale, IL: Southern Illinois UP.

Smith, F. (1993). Word identification. In F. Smith, *Understanding reading* (5th ed.; pp. 119–131). Hillsdale, NJ: Erlbaum.

Smith, L. B. (28 June 1993). A user's guide to PCC (politically correct communiqué). *PCWeek, 10*(25), 204.

Souther, J. W. (1985). What to report. *IEEE Transactions on Professional Communication PC-28.*

Spilka, R. (Ed.) (1993). *Writing in the workplace: New research perspectives.* Carbondale: Southern Illinois UP.

Sternthal, B., Dholakia, R., & Leavitt, C. (1978). The persuasive effect of source credibility: Tests of cognitive response. *Journal of Consumer Research, 4,* 252–260.

Suchan, J., & Colucci, R. (1989). An analysis of communication efficiency between high-impact and bureaucratic written communication. *Management Communication Quarterly, 2,* 464–473.

Thorell, L. G., & Smith, W. J. (1990). *Using computer color effectively: An illustrated reference.* Englewood Cliffs, NJ: Prentice Hall.

Tinker, M. A. (1969). *Legibility of print.* Ames: U of Iowa Press.

Toulmin, S., Rieke, R., & Janik, A. (1984). *An introduction to reasoning* (2nd ed.). New York: Macmillan.

van Dijk, T. (1977). Semantic macrostructures and knowledge frames in discourse comprehension. In M. A. Just & P. C. Carpenter (Eds.), *Cognitive processes in comprehension* (pp. 3–32). Hillsdale, NJ: Erlbaum.

Varner, I. I. (1988). A comparison of American and French business correspondence. *Journal of Business Communication, 25*(4), 59.

Velotta, C. (1987). Safety labels: What to put in them, how to write them, and where to place them. *IEEE Transactions on Professional Communications, 30,* 121–126.

Warren, E. K., Roth, L., & Devanna, M. (Spring 1984). Motivating the computer professional. *Faculty R&D.* New York: Columbia Business School, 8.

Watson, J. D. (1968). *The double helix: A personal account of the discovery of the structure of DNA.* New York: Atheneum.

Weisband, S. P., & Reinig, B. A. (1995). Managing user perception of e-mail privacy. *Communications of the ACM, 38*(12).

Wells, B., Spinks, N., & Hargarve, J. (June 1981). A survey of the chief personnel officers in the 500 largest corporations in the United States to determine their preferences in job application letters and personal résumés. *ABCA Bulletin, 14*(2), 3–7.

Westinghouse Corporation. (1981). *Danger, warning, caution: Product safety label handbook.* Author.

Wheildon, C. (1995). *Type & layout: How typography and design can get your message across—or get in the way.* Berkeley, CA: Strathmoor Press.

White, J. V. (1990). *Color for the electronic age.* New York: Watson-Guptil.

Williams, J. M. (1994). *Style: Ten lessons in clarity and grace* (4th ed.). New York: HarperCollins.

Williams, R. (1994). *The non-designer's design book.* Berkeley, CA: Peachpit P.

Williams, R., & Tollett, J. (1998). *The non-designer's Web book.* Berkeley, CA: Peachpit P.

Zimmerman, M. (1985). *How to do business with the Japanese.* New York: Random House.

PHOTO CREDITS

page 3: © RNT Productions/CORBIS; **page 22:** © M Stock/Alamy; **page 40:** Tom Forst/Stock Image/Getty Images; **page 67:** © Royalty-Free/Corbis; **page 84:** © Rosenfeld/zefa/Corbis; **page 108:** Erik Von Weber/Stone/Getty Images; **page 129:** © Lester Lefkowitz/CORBIS; **page 139:** © SuperStock, Inc./SuperStock; **page 164:** Yellow Dog Productions/The Image Bank/Getty Images; **page 179:** © Royalty-Free/Corbis; **page 193:** © Omni Photo Communications Inc./Index Stock Imagery; **page 227:** Jon Riley/Stone/Getty Images; **page 241:** Todd Warnock/Stone/Getty Images; **page 266:** © David Young-Wolff-PhotoEdit; **page 288:** Roger Tully/Stone/Getty Images; **page 302:** © Jeff Greenberg/PhotoEdit; **page 312:** © David Young-Wolff/PhotoEdit; **page 341:** Bob Kramer/Index Stock Imagery; **page 363:** © Jeff Greenberg/PhotoEdit; **page 365 (top right):** NASA Marshall Space Flight Center (NASA-MSFC); **page 388:** © Image Source/SuperStock; **page 412:** © Warren Morgan/CORBIS; **page 431:** © Stewart Cohen/Index Stock Imagery; page 444: © Bill Varie/CORBIS; **page 445 (top left):** © Grant Heilman Photography; **page 445 (top right):** © Grant Heilman Photography; **page 445 (bottom left):** © Michael & Patricia Fogden/CORBIS; **page 445 (bottom right):** © Michael & Patricia Fogden/CORBIS; **page 465:** © Gabe Palmer/CORBIS; **page 482:** © Peter Beck/CORBIS; **page 491:** © Richard T. Nowitz/CORBIS; **page 500:** © Michael Newman/PhotoEdit; **page 522:** © Royalty-Free/Corbis.

ACKNOWLEDGMENTS

Figure 1.1 (page 7):
Copyright © Hewlett Packard. Reproduced with permission.

Figure 1.2 (page 12):
www.dell.ca Copyright © 2006 DELL Computer Corporation.

Figure 5.4 (page 119):
Reprinted by permission of Roche.

Figure 5.5 (page 120):
www.ieee.org Reprinted with permission of IEEE.

Figure 5.6 (page 123):
Reprinted by permission of Nature Conservancy Canada, www.natureconservancy.ca

page 143:
Reprinted with permission from Microsoft Corporation.

Figure 6.1 (page 148):
Used with permission of Eagle-Picher Industries, Inc., Akron Standard.

Figure 6.2 (page 149):
Reproduced with the permission of Sharp Electronics Corporation.

Figure 6.3 (page 151):
Courtesy of Thomson Consumer Electronics.

Figure 6.4:
Courtesy of Johnson & Johnson.

Figure 6.5 (page 153):
Copyright © 1996 by Apple Computer. Reprinted with permission.

Figure 6.6 (page 155):
Used with permission of Eagle-Picher Industries, Inc., Akron Standard.

Figure 9.3 (page 234):
Reprinted by permission of Ontario School Counsellors Association.

Figure 10.1 (page 244):
Reprinted by permission of Bee-O-Sphere Technologies.

Figure 10.2 (page 245):
Reprinted by permission of Ortho Biotech.

Figure RM.5 (page 324):
Reprinted by permission of SirsiDynix.

Figure RM.6 (page 326):
Reprinted by permission of SirsiDynix.

Figure 14.2 (page 349):
Courtesy of DELL Computer Corporation.

Figure 15.2, left (page 365):
Reprinted by permission of Lucent Technologies, from *Bell Labs Technical Journal,* Autumn, 1997, p. 94.

Figure 15.5 (page 366):
Courtesy of Texas Instruments.
www.ti.sc/graphics/wireless/97/enable/rfgraph.gif

Figure 15.6 (page 367):
Reprinted courtesy of Eastman Kodak Company.

Figure 15.7 (page 367):
From J. A. Stroeckel, D. W. Schneider, L. A. Soeken, K. D. Blodgett, and R. E. Sparks, "Larval Dynamics of a Riverine Metapopulation: Implications for Zebra Mussell Recruitment, Dispersal, and Control in a Large-River System," *Journal of the North American Benthological Society,* 16 (1997): 591. Reprinted by permission of the Journal of the North American Benthological Society.

Figure 15.8 (page 368):
Courtesy of the Monsanto Company.

Figure 15.11, top (page 372):
Courtesy of NASA.

Figure 15.11, middle (page 372):
"Thermal Modeling of the Infrared Reflow Process for Solder Ball Connect (SBC)" from *IBM Journal of Research and Development,* 37, 616, 1993. Copyright 1993 International Business Machines Corporation.

Figure 15.11, bottom (page 372):
Courtesy of NASA.

Figure 15.14 (page 375):
Copyright © 1997 Penn Well Publishing Company, reprinted from *Computer Design,* 36, 31, 1997, December, with permission.

Figure 15.15 (page 377):
Based on Jan V. White, *Color for the Electronic Age* (New York: Watson-Guptill, 1990) p. 16. Courtesy of Jan V. White.

Figure 15.16 (page 377):
Copyright © 1994 by Ian Warpole/Scientific American, Inc. All rights reserved.

Figure 15.17 (page 378):
Fig, 1-A, "Molecular model of H-IL-6" from "A bioactive designer cytokine for human hematopoietic progenitor cells expansion" by Martha Fischer, Jutta Goldschmitt et al., *Nature Biotechnology*, Vol. 15, February, 1997.

Figure 15.18 (page 380):
Based on Jan V. White, *Color for the Electronic Age* (New York: Watson-Guptill, 1990) pp. 40–43. Courtesy of Jan V. White.

Figure 15.19 (page 381):
Reprinted by permission from three IBM webpages of IBM Online. Copyright 1998 International Business Machines Corporation.

Figure 15.20 (page 382):
Reprinted with permission of Schneiders.

Figure 15.21 (page 384):
Copyright © 1997 Society of Automotive Engineers International (SAE), Warrendale, PA.

Figure 16.1, top left (page 391):
Reprinted by permission of CMA Canada.

Figure 16.1, top right (page 391):
Reprinted by permission of Baseball Canada.

Figure 16.1, middle (page 391):
Reprinted by permission of London Humane Society.

Figure 16.1, bottom left (page 391):
Reprinted by permission of Canadian Medical Association Journal.

Figure 16.1, bottom right (page 391):
Reprinted by permission of Sail Magazine.

Figure 16.4 (page 396):
Reprinted by permission of Monarch Marketing Systems, a Pitney Bowes Company.

Figure 16.5 (page 399):
Based on United States Consumer Product Safety Commission, Protect Your Family from Lead in Your Home, (Washington, D.C.: United States Environmental Protection Agency, 1995): 33–34.

Figure 16.6 (page 400):
From *Personal Finance for Canadians for Dummies* 3rd edition by Tyson and Martin, pp. 5, 82 and 116. Etobicoke: Wiley, 2001. Reprinted by permission of John Wiley & Sons.

Figure 16.7 (page 401):
Copyright © 1997 Chrysler Corporation.

Figure 16.8 (page 404):
Courtesy of Professor Joseph L. Cox III.

Figure 16.9 (page 405):
Reprinted with permission from Microsoft Corporation.

Figure 16.10 (pages 407–10):
Courtesy of USV Pharmaceutical Manufacturing Corporation.

Figure 17.4 (page 425):
Reprinted with permission from Microsoft Corporation.

Figure VA.1 (page 434):
Reprinted with permission of Eastman Kodak Company.

Figure VA.3 (page 435):
Reprinted by permission of Eastman Kodak Company.

Figure VA.5 (page 438):
Adapted from the Statistics Canada publication, Labour Market Information and Research, Ministry of Training, Colleges and Universities, and Economic Analysis and Information Directorate, Human Resources and Skills Development Canada, Ontario Region.

Figure VA.6 (page 439):
TP 13549—Sharing the Skies—An Aviation Industry Guide to the Management of Wildlife Hazards, Chapter 5, Figure 5.8. Reproduced with the permission of the Minister of Public Works and Government Services Canada, 2006.

Figure VA.7 (page 439):
LCDC 1997—Using IMS Canadian Disease and Therapeutic Index—year ending December, 1996. Reproduced with the permission of the Minister of Public Works and Government Services Canada, 2006.

Figure VA.8 (page 440):
Adapted from the Statistics Canada publication, General Social Survey, 1992 and 1998.

Figure VA.9 (page 441):
From Standard Oil Company, Energy Adventures (Cleveland, Ohio: Standard Oil Company, 1983) and updated with U.S. Bureau of Census data, 1997. Courtesy of Standard Oil Company.

Figure VA.10 (page 442):
From T. L. Morris, "Commercial Banking Transformed to Computer Technology," *Monthly Labor Review* 119 (1996:8), 34, and unpublished data from the Bureau of Labor Statistics Current Employment Statistics program, and Bank Network News.

Figure VA.11 (page 442):
From T. O'Halloran, P. Sokolsky, and S. Yoshida, "The Highest Energy Cosmic Rays," *IEEE Journal of Quantum Electronics*, 34 (1998): 50.

Figure VA.17, top (page 448):
"Carousel Creates Continuous Ion Exchange," from *Chilton's Food Engineering*, Vol. 65, No, 10: 80, June, 1993. Reprinted by permission of Advanced Separation Technologies, Inc.

Figure VA.17, bottom (page 448):
Cutaway of Audi Engine from "Getting More Direct," *Automotive Engineering*, 105, 84 (December, 1997). Artwork courtesy of Audi of America, Inc.

Figure VA.17, top (page 449):
© Tim Peters and Company.

Figure VA.17, bottom (page 449):
From U.S. National Aeronautics and Space Administration (1996). Research and Technology Report: Goddard Space Flight Center, U.S. Government Printing Office, 120.

Figure VA.18 (page 450):
From The Reserve Tank from 1998 Honda Accord Owner's Manual, XXX, Honda, p. 217. Courtesy of Honda Motor Company of Canada.

Figure VA.19 (page 450):
Reprinted with permission from Microsoft Corporation.

Figure VA.20 (page 451):
Courtesy of The Home Depot.

Figure VA.21 (page 452):
Courtesy of Zinc Corporation of America.

Figure VA.22 (page 453):
From Systems Analysis and Design Methods (4th edition), by J. L. Whitten, L. D. Bentley & K. C. Dittman, 1998, p. 384. Copyright © 1998 McGraw-Hill Inc. Reprinted by permission of The McGraw-Hill Companies.

Figure VA.23 (page 454):
Fig. 3.15: The Carbon Cycle from *Environmental Science: A Global Concern* (4th ed.), by William P. Cunningham & Barbara W. Saigo, 1997. Copyright © 1997 McGraw-Hill, Inc. Reprinted by permission of The McGraw-Hill Companies.

Figure VA.27, top (page 459):
Adapted from the Statistics Canada, 2001 Census of Canada, Education and computer use, Figure 19: Educational attainment of the population aged 15 and over, 2001.

Figure VA.27, bottom (page 459):
Adapted from the Statistics Canada, Centre for Education Statistics, Figure 20: Full-time university enrollment, by level, 2000–2001.

pages 511–14:
Based on *Chicago Manual of Style.*

pages 514–17:
From IEEE Standards Style Manual, Copyright © 2005 by the Institute of Electrical and Electronics Engineers, Inc.

INDEX